T0184859

# The Rare Earths in Modern Science and Technology

## Volume 2

# The Rare Earths in Modern Science and Technology

## Volume 2

Edited by

## Gregory J. McCarthy

*North Dakota State University*
*Fargo, North Dakota*
*and*
*Pennsylvania State University*
*University Park, Pennsylvania*

## James J. Rhyne

*National Bureau of Standards*
*Washington, D.C.*

## and

## Herbert B. Silber

*University of Texas at San Antonio*
*San Antonio, Texas*

PLENUM PRESS · NEW YORK AND LONDON

Library of Congress Cataloging in Publication Data

Rare Earth Research Conference.
  The rare earths in modern science and technology.

  Vol. 2 edited by G. J. McCarthy, J. J. Rhyne, and H. B. Silber.
  Includes bibliographical references and indexes.
  1. Earths, Rare—Congresses. I. McCarthy, Gregory J. II. Rhyne, J. J. III. Silber,
Herbert B. IV. Title.
QD172.R2R27  1977                          546'.4                          78-5365
ISBN-13: 978-1-4613-3056-1          e-ISBN-13: 978-1-4613-3054-7
DOI: 10.1007/978-1-4613-3054-7

Proceedings of the 14th Rare Earth Research Conference,
held at North Dakota State University, Fargo, North Dakota,
June 25–28, 1979.

# Foreword

The Fourteenth Rare Earth Research Conference was held June 25–28, 1979, at North Dakota State University in Fargo. The meeting was hosted by the College of Science and Mathematics and the Department of Physics. Since the first conference was held in 1960, subsequent meetings have grown in size and prestige to become one of the leading international forums devoted to disseminating new information relative to rare earth science and technology.

The meeting in Fargo was one of the largest yet held. The Program Committee scheduled over 160 papers representing colleagues from 18 countries in both oral and poster sessions that included Spectroscopy (Luminescence, Fluorescence, Laser, Mössbauer, ESR); Metallurgy and Materials Preparation; Solution, Solvation and Analytical Chemistry; X-ray and Neutron Diffraction; Transport and Thermal Properties; Hydrides; Magnetism; and Rare Earth Technology.

A first and special event which the organizers hope to perpetuate at future meetings was to announce the recipient of the rare earth prize, hereafter called the Frank H. Spedding Award. Governor Arthur A. Link, State of North Dakota, on behalf of the Selection Committee, presented Professor Frank H. Spedding with a special citation. Professor Spedding spoke briefly and introduced the first recipient, Professor W. Edward Wallace from the University of Pittsburgh.

In addition to the Plenary Address given by Professor Wallace, keynote addresses were given by the following distinguished colleagues: Professors F. Pobell, C. K. Jørgensen, K. A. Gschneidner, Jr., G. R. Choppin, E. Parthé, E. F. Westrum, Jr., W. E. Wallace, K. J. Strnat, A. Tauber and D. Givord.

A conference of this size and diversity depends upon the contributions of time and financial support from many participants and institutions. First, my deep appreciation for our outstanding program goes to Professor W. J. James, Chairman of the Program Committee and Chairman of the Selection Committee for the Frank H. Spedding Award. My appreciation extends as well to the following colleagues who served as session chairpersons:

| | | |
|---|---|---|
| B. W. Berringer | J. C. Glass | W. K. Perrizo |
| P. Boudjouk | J. E. Greedan | F. Pobell |
| E. C. Campbell | K. A. Gschneidner, Jr. | F. Rathmann |
| O. N. Carlson | R. G. Haire | R. Reisfeld |
| F. L. Carter | J. M. Haschke | F. Rothwarf |
| R. S. Craig | C. E. Higer | J. M. Sugihara |
| R. K. Datta | I. S. Hirschhorn | S. M. A. Taher |
| T. Donohue | D. A. Hukin | L. Thompson |
| J.-P. Fort | C. K. Jørgensen | W. E. Wallace |
| D. Givord | E. Parthé | E. F. Westrum, Jr. |

Also, I wish to acknowledge those who served on special commit-
tees that helped to make the program possible.

| Selection Committee, The Frank H. Spedding Award | Program Committee | Local Committee |
|---|---|---|
| W. J. James, Chairman | W. J. James, Chairman | J. C. Glass, Chairman |
| J.-P. Fort | J. G. Cannon | F. Bancroft |
| J. E. Greedan | F. L. Carter | M. Busch |
| C. E. Lundin | J.-P. Fort | R. Chenoweth |
| G. J. McCarthy | J. C. Glass | J. B. Gruber |
| J. J. Rhyne | J. E. Greedan | M. Niskanen |
| | K. A. Gschneidner, Jr. | T. Rockne |
| | J. M. Haschke | |
| | R. C. Ropp | |
| | F. Rothwarf | |
| | S. M. A. Taher | |
| | W. E. Wallace | |

Throughout the conference and for many hours since, the Proceedings Co-editors have been hard at work, along with their refer-
ees, preparing this book for its present form. I especially extend
appreciation to these Co-editors:

G. J. McCarthy        J. J. Rhyne              H. B. Silber

The following donors made it possible to hold the conference
here at North Dakota State University:

NDSU Alumni Association, Fargo, North Dakota, U.S.A.
Companhia Industrial Fluminense, Rio de Janeiro, BRAZIL
Davison Specialty Chemical Co., (A Subsidiary of W. R. Grace
  & Co.), Baltimore, Maryland, U.S.A.

Molycorp, Inc., White Plains, New York, U.S.A.

Reactive Metals & Alloys Corp., West Pittsburg, Pennsylvania, U.S.A.

Research Chemicals, (A Division of NUCOR Corp.), Phoenix, Arizona, U.S.A.

Rhone-Poulenc Industries (Chimie Fine Division), Paris Cedex 08, FRANCE

Ronson Metals Corporation, Newark, New Jersey, U.S.A.

TH. Goldschmidt AG, Essen 1, WEST GERMANY

\* Petroleum Research Fund (American Chemical Society), Washington, D.C., U.S.A.

\*\*U.S. Army Research Office, Research Triangle Park, North Carolina, U.S.A.

Finally, I wish to acknowledge conference secretaries, Faye Kalina, Joyce Mortensen and Mary McDonald, for preparation of the Preliminary Program, the Program and Abstracts, and for many hours spent in handling arrangements and registration details. The hospitality, the mini tours, the evening social events, along with the three day program, added up to a most delightful and successful Fourteenth Rare Earth Research Conference.

*John B Gruber*

John B. Gruber
General Conference Chairman
Fargo, North Dakota
July, 1979

---

\* Acknowledgment is made to the donors of The Petroleum Research Fund, administered by the American Chemical Society, for partial support.

\*\*The views, opinions, and/or findings contained in this program are those of the author(s) and should not be construed as an official Department of the Army position, policy, or decision, unless so designated by other documentation.

# Preface

In this second volume of The Rare Earths in Modern Science and Technology one will again find both overviews and in-depth treatments of virtually every aspect of current interest in the rare earths. The 118 refereed papers that follow are of several types. First, there are invited keynote papers by an authority in a particular field. The first of these is the Frank H. Spedding Award Address of Professor W. E. Wallace. The establishment of this award is featured at the beginning of this volume. The other contributions are papers and notes. The latter may report recent results in a continuing series of experiments, summarize a longer paper to be published in full later, or point the reader to the key literature of a particular subject.

Papers are arranged into chapters. The basic chemistry and physics and the preparation and characterization of rare earth-containing material are again strongly represented. There are chapters on sources and applications, hydrides, lasers and intermetallics that highlight the potential or actual utilization of the rare earths in modern science and technology.

J. J. Rhyne, H. B. Silber and I would like to acknowledge the typing and editorial assistance of Sandra McBride at the Pennsylvania State University and Joyce Mortensen at North Dakota State University.

Gregory J. McCarthy  
Co-Editor  
Fargo, North Dakota  
October, 1979

The First Spedding Award is Presented
to W.E. Wallace by Frank H. Spedding,
June 26, 1979

*Establishment of The Frank H. Spedding Award for Outstanding
Contributions to the Science and Technology of the Rare Earths*

FRANK H. SPEDDING

Governor Link, President Loftsgard, Professor
Spedding, Professor Gruber, Professor Glass,
Colleagues, Ladies and Gentlemen:

One of the tasks assigned to me as Program Chairman is to introduce Frank H. Spedding. This, of course, is no task at all as I
deem it an honor and a privilege to say a few words about Professor
Spedding, former director of the Ames Lab at Iowa State University
and now Distinguished Professor Emeritus.

Professor Spedding received his Ph.D. in 1929 under the super-
vision of Professor G. N. Lewis. His subsequent studies on the
spectroscopy of the rare earth elements--in particular, the dependence of the fine structure of the spectra of solids on the crystal
symmetry of the matrices--resulted in his receiving the Langmuir
Award in Pure Chemistry for a young chemist under 31 years of age.

With the advent of World War II, he was called upon to prepare
U metal of high purity for the Manhattan project. He and his co-
workers at Iowa State were successful in preparing some $2 \times 10^6$ pounds
with 4000 pounds being used to power the first atomic pile at the
University of Chicago. He and his colleagues later developed a

method for producing Th necessary as a blanket material for nuclear reactors. The method, still in use today, resulted in the production of over 300 tons.

One of his most notable achievements was the development in the 1950s of a form of ion exchange displacement chromatography which permits the separation of neighboring rare earth elements to the extent that a given element may contain <5 ppm of all other rare earths collectively. Additionally, the Spedding team established the efficacy of EDTA in performing such separations.

From the mid '50s to the early '70s, the results of his studies of the physical and chemical properties of rare earth metals and their aqueous solutions have assisted researchers in better understanding the electronic nature of the 4f elements.

I must also make mention of Professor Spedding's administrative talents displayed in his establishing and administering the Ames Laboratory, a major interdisciplinary center for the preparation, purification, and characterization of high purity metals.

In the educational realm, Professor Spedding has graduated 84 Ph.D.s and was the first recipient of the distinguished professorship of the College of Science and Humanities at Iowa State.

He is one of six persons to be designated Distinguished Citizen of Iowa in recognition of his contributions to research, education, and scientific administration.

The selection committee for the first rare earth prize could not agree more fully with the statement of a colleague to the effect that the chemistry and metallurgy of the lanthanides could not have developed to their present levels without the pioneering efforts of Frank Spedding.

On the basis of the many awards Professor Spedding has already received, the selection committee felt it more fitting that the prize should bear his name so that the pioneering efforts of this dedicated scientist be long remembered as each award is granted over the years.

I would now like Governor Link and Professor Spedding to come to the podium where the governor will present to Professor Spedding a certificate which formally establishes The Frank H. Spedding Award and honors his contributions to the science and technology of the rare earths.

William J. James
Program Chairman
Rolla, Missouri
July, 1979

# Contents

SEMICONDUCTING AND INSULATING COMPOUNDS

SPECTROSCOPY

CONTENTS

LASERS

THE FRANK H. SPEDDING AWARD ADDRESS

STUDIES OF RARE EARTH INTERMETALLIC COMPOUNDS AND RARE EARTH HYDRIDES[*]

W. E. Wallace

Department of Chemistry, University of Pittsburgh

Pittsburgh, PA 15260

## I.  INTRODUCTION

It gives me extraordinary pleasure to accept the Frank H. Spedding Award.  This recognition of our work on rare earth systems is very pleasing to me.  I am doubly pleased to receive an award having associated with it the name of Frank Spedding.  We all know that were it not for his extraordinary accomplishments a few decades back we would not be assembled here today for a Rare Earth Research Conference.

This occasion permits me to touch upon a few highlights of our work at Pittsburgh dealing with rare earth systems and carried out since the middle 1950's.  I have been asked to keep the remarks general and I shall do so despite the risk of appearing superficial.

Our work has been concerned almost exclusively with two kinds of rare earth systems:  (1) rare earth intermetallic compounds and (2) rare earth hydrides.  In my monograph (1) dealing with rare earth intermetallic compounds, which was published in 1973, I described about 1000 compounds involving more than a dozen structural types (see Table 1).  From a structural point of view these materials can be construed as regular inorganic compounds.  They are, in

*The work on rare earth hydrides was supported by the predecessor of the Department of Energy, the Atomic Energy Commission.  The catalysis work was assisted by the National Science Foundation. Much of the work on magnetism was carried out under a series of grants provided by the U.S. Army Research Office.

my opinion, somewhat more interesting than your "run-of-the-mill" inorganic compound.

I acquire the greatest sense of personal satisfaction in engaging in science that is both intellectually challenging and of societal significance - that is, science which offers promise for improving the human condition.  The study of rare earth intermetallics falls nicely in this category.  Tables 2 and 3 indicate some of the applications possibilities - in hand and projected.

## Table 1

### Rare Earth Intermetallic Compounds

| Structural Types | Examples |
|---|---|
| CsCl | GdAg |
| $MgCu_2$ | $PrAl_2$ |
| CrB | CeNi |
| $Fe_3C$ | $Er_3Ni$ |
| $CaCu_5$ | $LaNi_5$ |
| $Th_2Ni_{17}$ | $Ho_2Co_{17}$ |

## Table 2

### Rare Earth Intermetallics - Present Applications

$SmCo_5$, $Sm_2Co_{17}$

| | |
|---|---|
| Improved microwave devices | Communications, National defense |
| Higher efficiency, lighter electric motors | Energy conservation |
| Biomedical materials | Medicine and dental medicine |

$LaNi_5$

| | |
|---|---|
| Utilization of $H_2$ as a fuel | New energy sources |
| Heterogeneous catalysis | Coal gasification and liquefaction |

Table 3

Rare Earth Intermetallics - Projected Applications

Near-term Future

RFe$_x$ Systems

Transducers, Automotive

| | |
|---|---|
| Ignition systems | Fuel economy |
| Hydrogen getters | Improvement in safety of nuclear reactors |

Today I wish to discuss briefly three of our undertakings in
the rare earth field.

1.  Magnetic and electrical properties of rare earth hydrides,

2.  An aspect of heterogeneous catalysis:  Use of rare earth
    intermetallics in the formation of hydrocarbons from CO
    and H$_2$, and

3.  Magnetism of rare earth-cobalt intermetallics.

II.  MAGNETIC AND ELECTRICAL PROPERTIES
OF RARE EARTH HYDRIDES

I will first turn attention to the magnetic and electrical
properties of rare earth hydrides which we and others have studied
for a number of years.  These studies bear upon the fundamental
exchange mechanism operating in the elemental rare earths.  In these
materials, the distance of separation of adjacent ions is large compared to the range of the f-shell.  Accordingly, overlap of the
f-orbitals centered on adjacent atoms and direct exchange is negligible.  The coupling arises from polarization of the conduction
electrons.  This is the well-known RKKY interaction.

It was observed (2-4) in several laboratories about twenty
years ago that hydrogenation of a rare earth element results in a
fall in the electrical conductivity.  Conductivity is observed to
fall by many orders of magnitude during hydrogenation and in the
fully hydrogenated material it approaches the conductivity of an
insulator.  This observation has a simple explanation.  Hydrogen
absorbs electrons from the conduction band during the hydrogenation
process, and in the fully hydrided materials the conduction band is
completely depopulated (Fig. 1).  Accordingly, the electrical conductivity vanishes.  Viewed in this way, the rare earth hydride is
seen to be the non-conducting analog of the elemental rare earth.

Hydriding Depletes the Rare Earth Conduction Band

$$R + \tfrac{3}{2}H_2 \longrightarrow R^{3+}(H^-)_3$$

Fig. 1

Since the hydride lacks conduction electrons, which is essential for the magnetic coupling (RKKY) mechanism, one expects there to be a sharp fall in the magnetic ordering temperature of the hydride compared to the element from which the hydride is formed. A number of years ago in our laboratory we examined the magnetic characteristics of the rare earth hydrides and indeed did observe a sharp reduction in the magnetic ordering temperature for the hydride compared to the parent metal. Representative data are shown in Table 4.

Subsequently Richard Heckman determined the Hall coefficient for hydrogenated cerium (5). He found that the Hall coefficient was positive and became increasingly positive as the hydrogenation approached completion. A positive Hall coefficient is characteristic of a nearly full band, and so these data seem to indicate band filling effects rather than band depopulation effects. In my invited lecture at the 24th Conference on Magnetism and Magnetic Materials about nine months ago, I showed how a proper resolution of the various electrical magnetic properties of the rare earth hydrides could be achieved using the band structure information provided by the APW calculations of Switendick (6). This can be appreciated by reference to Figure 2. There is a new band, hydrogenic in nature, formed during the hydrogenation process, which lies below the conduction band of the elemental rare earths. During hydrogenation electrons are transferred from the conduction band to the hydrogenic band. Accordingly, there is indeed a band filling effect which apparently accounts for the positive Hall coefficient observed by Heckman. The hydrogenic band is formed from the antisymmetrical 1s orbitals of the hydrogen atom. Accordingly, the electron charge tends to be localized around the nucleus. Hence, to a first approximation, the hydrogen in the rare earth hydrides does look anionic in nature in accordance with my conclusions drawn about 15 years ago.

Table 4

Magnetic Ordering Temperatures of the
Elemental Rare Earths

|    | $x^a$ | $T_N(K)$ | $T_c(K)$ |
|----|------|----------|----------|
| Nd | 2.75 | 19       |          |
| Sm | 2.91 | 106      |          |
| Gd | 3.00 |          | 291      |
| Tb | 2.97 | 235      |          |
| Dy | 2.97 | 184      |          |
| Ho | 3.06 | 135      |          |
| Er | 3.02 | 88       |          |
| Tm | 2.95 | 56       |          |

a. x in the formula $RH_x$, where x is the
maximum amount of hydrogen taken up
at 1 atm. pressure. None of these
materials exhibits magnetic ordering
at temperatures down to 4 K. For a
further discussion of this, see W. E.
Wallace, Ber. Bunsen. Gesellschaft
76, 832 (1972) and ibid., 1979, in
press.

Conduction Band Depleted ; New Band Formed

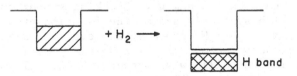

Fig. 2. H band comes from antibonding H 1s orbitals.

The work on the electrical and magnetic properties of the rare earth hydrides constitutes perhaps the best experimental confirmation that we have that the RKKY interaction, which had been postulated by the theoreticians, is indeed the dominant exchange mechanism in the elemental rare earths.

## III.   HETEROGENEOUS CATALYSIS USING RARE EARTH INTERMETALLIC COMPOUNDS

I would now like to leave the studies of rare earth hydrides and turn attention to the use of rare earth intermetallics in the catalytic formation of hydrocarbons from CO and $H_2$. This is one of the important reactions that occurs in coal conversion technology – the production of liquid and gaseous hydrocarbons from coal. Coal is treated with high pressure, high temperature steam and this leads to the formation of a mixture of gaseous CO and $H_2$, among other things. When the carbon monoxide and hydrogen are separated out this gaseous mixture is called synthesis gas. Synthesis gas alone will not react to form hydrocarbons; a catalyst is needed. It is well known that $LaNi_5$ is an excellent solvent for hydrogen (7). When $LaNi_5$ is exposed to hydrogen the hydrogen bond is cleaved and monatomic hydrogen enters the lattice of the intermetallic in large concentration (8). It therefore seems that atomic hydrogen exists at least for a tiny instant of time on the surface of this material. From these considerations it appeared to us a few years ago that $LaNi_5$ might well catalyze the synthesis of hydrocarbons from synthesis gas. We have examined this material and do indeed find that in the presence of $LaNi_5$ and in temperatures in the range 300 to 350°C hydrocarbons are readily formed from synthesis gas (9,10). However it has developed that $LaNi_5$ is not the catalytically active material. During the reaction process it has been transformed into a mixture of Ni and $La_2O_3$. Evidence that this transformation of the intermetallic occurs is obtained from conventional x-ray powder diffraction measurements, by scanning electron microscopy, by EDAX and by Auger spectrometry. It appears that the catalytically active material is not the original intermetallic but is instead the elemental Ni resting on a substrate of $La_2O_3$. We thus have a new way of making supported catalysts – by oxidizing it with synthesis gas or the reaction products of CO and $H_2$, i.e., $CO_2$ and $H_2O$. We can also achieve the oxidation of an intermetallic compound into a transition metal oxide mixture by pretreating the intermetallic with oxygen (11-13). This oxidized material is a very active catalyst. In Table 5 we have an indication of the methane turnover numbers for conventional supported catalysts in the particular case cited, Ni supported on silica, and for the transformed and oxidized intermetallics. The turnover number represents the number of reactions that take place per site per unit of time. It is to be noted that the transformed and oxidized intermetallics have turnover numbers from 5 to 40 times the turnover number of the conventional supported catalyst. The reason for this

Table 5

$CH_4$ Turnover Number x $10^3$, sec$^{-1}$

205°C

A.  Conventional Supported Catalyst

   Ni/SiO$_2$                          0.5 to 1.0

B.  Transformed Intermetallic

   ThNi$_5$      Ni/ThO$_2$            4.7

   LaNi$_5$      Ni/La$_2$O$_3$        2.7

   Ni$_5$Si$_2$  Ni/SiO$_2$           11

C.  Oxidized Intermetallic

   Ni$_5$Si$_2$  Ni/SiO$_2$           18

   ThNi$_5$      Ni/ThO$_2$           10.6

remarkable activity is not clear at the present time. We are begin-
nig to make a serious study of the surface features of new supported
catalysts and hope to clarify their exceptional catalytic activity.

## IV. MAGNETISM OF RARE EARTH-COBALT INTERMETALLICS

I would now like to turn attention to a third topic which I
have selected for discussion - the magnetic characteristics of the
rare earth-cobalt systems.

The rare earths form many compounds with the transition metals.
The rare earth-cobalt compounds are of special interest for many
reasons. An enumeration of the stoichiometries represented by the
rare earth-cobalt compounds is given in Table 6. SmCo$_5$ and Sm$_2$Co$_{17}$,
which are examples of the 1:5 and 2:17 classes of materials, are of
great technical interest. Before turning attention to that, however,
I should point out the coupling mode in the 1:5 RCo systems. This
is illustrated in Fig. 3. It is to be noted that LaCo$_5$ is a ferromagnetic material with all of its magnetism originating with cobalt.
PrCo$_5$ and NdCo$_5$ are ferromagnetic materials with magnetization originating from both the cobalt and the rare earth sublattices. In the
case of SmCo$_5$ it is not entirely clear what the nature of the coupling is. Most likely Sm is coupled parallel to the Co sublattice
(14,15), and this material is hence a ferromagnetic material. However, the Sm moment is quite small and the nature of the coupling
has not been established convincingly. The heavy rare earths, beginning with Gd, coupled antiferromagnetically with the Co

Table 6

Rare Earth–Cobalt Intermetallic Compounds[*]

|  | Structure Type |
|---|---|
| $R_3Co$ | $Fe_3C$(cementite) |
| $R_4Co_3$ | hex. |
| $RCo_2$ | $MgCu_2$(C15) |
| $RCo_3$ | $PuNi_3$ |
| $R_2Co_7$ | hex. |
| $\underline{RCo_5}$ | $CaCu_5$ |
| $\underline{R_2Co_{17}}$ | $Th_2Ni_{17}, Th_2Zn_{17}$ |

---

[*] The technologically significant $SmCo_5$ and
$Sm_2Co_{17}$ compounds are members of the under-
scored $RCo_5$ and $R_2Co_{17}$ Groups

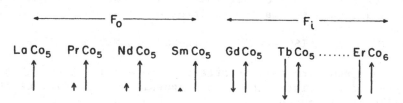

Fig. 3.   Magnetic coupling in RCo5 compounds.

sublattice.[16]  In these materials the magnetization is largely compensated internally and these materials are devoid of technological
interest.  The other materials shown are all of potential technological interest but so far the potential has been realized only
in the case of $SmCo_5$.  The features that make this of exceptional
interest as a high energy permanent magnet material are illustrated
in Table 7.  It is to be noted that its Curie temperature is quite
high, which is of course necessary if this is to be incorporated
in devices operating at elevated temperatures.  It is also to be
noted that it has a powerful anisotropy field.  I shall return to
the significance of the anisotropy field in a moment.

## Table 7

### Characteristics of $SmCo_5$ Which Make It a Premiere High Energy Magnet Material

| | |
|---|---|
| Curie temperature | 1020°K |
| Saturation magnetization | 8.5 $\mu_B$ per formula unit |
| | 1.4 $\mu_B$ per atom |
| Anisotropy field (near 0°K) | 600,000 Oe. |

The significance of $SmCo_5$ does not lie with its saturation magnetization. On a per atom basis the magnetization is only about 60% of that of elemental iron and only about 80% of that of elemental cobalt. The utility of $SmCo_5$ is derived from its powerful anisotropy. For a material to be a good permanent magnet there must be something to clamp the moments in place after the aligning field is removed. That something, in the case of $SmCo_5$, is the crystal field interaction, as I shall indicate in a moment.

At this point it is appropriate to direct attention to some aspects of the anisotropy field of $SmCo_5$. This material is hexagonal. Its atomic moments lie along the hexagonal axis. The anisotropy field is that field required to twist the moment to 90°. This field is about 600 kOe. near absolute zero and about 400 kOe. in the vicinity of room temperature. Work must be done to twist the moment away from the hexagonal axis. The work required to redirect the moments was measured some years back at the Wright-Patterson Air Force Base by Frederick and Garrett (17). In Pittsburgh we have directed attention to the quantum mechanics of the system and calculated that work (17). To evaluate the work requirement we have calculated the energy involved when the moments lie first parallel and second, prependicular to the unique hexagonal axis. The difference between these two energies is the work required to twist the moments through 90° and is often termed the stabilization energy. The operator equivalent method, developed by Stevens at Oxford University some 30 years ago is usually employed to treat the quantum electronics of rare earth systems. However, this calculational procedure cannot be employed for $Sm^{3+}$ because the J multiplets are not well separated. Instead it is necessary to employ the Racah tensorial algebra.

The Hamiltonian employed in the calculation is shown in Table 8. It is diagonalized with the magnetic field applied first parallel and then perpendicular to the c-axis. Ions occupy the crystal field states according to the Boltzmann distribution. The energies of the new systems can be calculated by standard statistical mechanical procedures. With properly chosen values of the crystal field

Table 8

Crystal Field Interaction for $Sm^{3+}$ in $SmCo_5$
Treated by Racah Tensor-Operator Technique

Hamiltonian Used

$$\mathcal{K} = \lambda \, \vec{L} \cdot \vec{S} + \mathcal{K}_{CF} + \mathcal{K}_{ex}$$

1.  $\mathcal{K}_{ex} = 2 \, \mu_B \, S \cdot H_{ex}$

2.  $\mathcal{K}_{CF} = N_2^0 A_2^0 < r^2 > U_0^2 + N_4^0 A_4^0 < r^4 > U_0^4$

    $+ \, N_6^0 A_6^0 < r^6 > U_0^6 + N_6^6 A_6^6 < r^6 > U_6^6$

a.  The $A_q^k$'s are lattice sums which define the crystal
    field intensities.

b.  The $N_q^k$'s are normalization factors.

c.  The $U_q^k$'s are tensorial operators.  They are related
    to spherical harmonics in operator form.

parameters, theory can provide an excellent accounting for experiment (see Fig. 4).  The analysis described in the preceding para-
graphs, the details of which are given in reference 17, brings out
the interesting fact that the powerful magnetism of $SmCo_5$ is a
crystal field effect.

The experimentally achievable energy products of $SmCo_5$ and of
$Sm_2Co_{17-x}Fe_x$ are 25 and 30 MGOe, respectively.  $Sm_2Co_{17}$ alone gives
a poor energy product - so small as to be uninteresting.  This is
a consequence of its small anisotropy field (19) - only about 1/5
of the $SmCo_5$ anisotropy field.  This field can be strengthened by
partial replacement of Co with Fe.

The study of 2:17 systems based on $Sm_2Co_{17}$ is an extremely active field today - in the U.S., in Japan and in Switzerland, particularly Japan.  The 2:17 systems are fascinating systems for study.
There are five magnetic species present - Sm plus four types of Co.
The overall magnetic behavior is a composite of the interactions
between the five sublattices - these include both exchange and
crystal field interactions.  It appears that three of the Co sublattices generate negative anisotropy (20).  Schaller, Craig and
Wallace (21) observed some years ago that anisotropy in $R_2Co_{17}$ systems could be strengthened by partial replacement of Co with Fe or

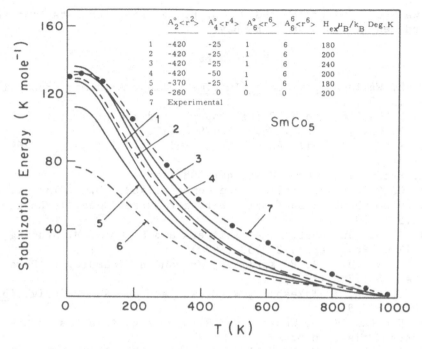

Fig. 4 Calculated and experimental values of the stabilization energy. The several curves are calculated
using a variety of values for the crystal field
intensity parameters appearing in the Hamiltonian
of Table 8.

Mn. This is an example case of the now-recognized general concept
that substitutions of Co by Fe, Mn or Cr can be made use of to selectively suppress deleterious features originating with a particular sublattice (21-23). It is not clear at this time whether this
suppression is a consequence of single ion or band structure effects (22). This is an active field - intellectually challenging
and currently of great technological significance (24).

Before I close I would like to remind you of the obvious -
that work such as I have described is a team effort. We can't
recognize a team and so we choose an individual who symbolizes that
team. Numerous individuals have been members of my team. It is
not feasible to acknowledge all who have contributed, but there are
individuals in the audience who have made very significant contributions to our accomplishments at Pittsburgh and I would like them
to stand and be acknowledged at this time: Professor R. S. Craig
and Drs. A. Elattar and H. Imamura of the University of Pittsburgh,
Professor John Greedan of McMaster University, Dr. T. Takeshita of

the Iowa State University and Dr. S. K. Malik of the Tata Institute
for Fundamental Research in Bombay.

## REFERENCES

1.  W. E. Wallace, Rare Earth Intermetallics, Academic Press, Inc.,
    N.Y. (1973).
2.  W. E. Wallace and T. Peltz, unpublished measurements.
3.  R. Heckman, J. Chem. Phys. 40:2958 (1964).
4.  B. Stalinski, Bull. Acad. Polon. Sci. 5:1001 (1957); 7:269
    (1959).
5.  R. Heckman, J. Chem. Phys. 46:2158 (1967).
6.  A. C. Switendick, Solid State Comm. 8:1463 (1970); Int. J. of
    Quantum Chem. 5:459 (1971); Ber. der Bunsen. Gesellschaft
    76:535 (1972).
7.  J. H. N. Van Vucht, F. A. Kuijpers and H. C. A. M. Bruning,
    Philips Res. Repts. 25:133 (1970).
8.  W. E. Wallace, R. F. Karlicek, Jr. and H. Imamura, J. Phys.
    Chem. 83:1708 (1979).
9.  V. T. Coon, T. Takeshita, W. E. Wallace and R. S. Craig, ibid.,
    80:1878 (1976).
10. A. Elattar, W. E. Wallace and R. S. Craig, Advances in Chem-
    istry Series, in press.
11. H. Imamura and W. E. Wallace, J. Phys. Chem., in press.
12. H. Imamura and W. E. Wallace, J. Phys. Chem., in press.
13. H. Imamura and W. E. Wallace, J. Catal., in press.
14. S. K. Malik, R. Vijayaraghavan and W. E. Wallace, Phys. Rev.
    B19:1671 (1979).
15. S. K. Malik, F. Arlinghaus and W. E. Wallace, ibid., 16:1242
    (1977).
16. Ref. 1, p. 146.
17. S. G. Sankar, V. U. S. Rao, E. Segal, W. E. Wallace, W.
    Frederick and H. Garrett, Phys. Rev. B11:201 (1975).
18. See ref. 1, chapter 3 for a description of the operator
    equivalent method.
19. See, for example, R. L. Bergner, H. A. Leupold, J. T. Breslin,
    F. Rothwarf and A. Tauber, J. Appl. Phys. 50(3):2349 (1979).
20. Private communication from Dr. K. Inomata.
21. H. J. Schaller, R. S. Craig and W. E. Wallace, ibid., 43:3161
    (1972).
22. R. S. Perkins and S. Strässler, Phys. Rev. B15:477 & 490 (1977).
23. M. V. Satyanarayana, W. E. Wallace and R. S. Craig, J. Appl.
    Phys. 50(3):2324 (1979).
24. The technological aspects of rare earth-cobalt systems for
    permanent magnet applications have been dealt with in the four
    Rare Earth-Cobalt Permanent Magnet Workshops organized by Pro-
    fessor Karl Strnat.  Copies of the Proceedings of these work-
    shops are available through the Engineering School of the Uni-
    versity of Dayton, Dayton, Ohio.

# FROM INFINITY TO ZERO AND BACK AGAIN TO INFINITY

## (FROM MIXED RARE EARTHS TO ULTRAPURE METALS TO

## PSEUDO-LANTHANIDES)

K. A. Gschneidner, Jr.
Ames Laboratory-DOE and Department of Materials
Science and Engineering
Iowa State Universtiy, Ames, IA 50011

## ABSTRACT

This paper concerns two topics with which the author is involved. The first is the continuing effort to purify the rare earth metals beyond four nines (99.99 at.%) purity. It appears that we are on the threshold of a major advance — five nines purity on an atomic basis with respect to all the elements in the periodic table.

The second topic is the "contamination" (alloying) of the high purity rare earth metals (99.9 at.%) with each other to form pseudo-lanthanides. The main thrust for this "act of madness" after all the effort to purify them is to use these pseudo-lanthanides to (1) unravel the various contributions to the heat capacity so that we can better understand the electronic and magnetic nature of the lanthanide elements and their compounds; and (2) to prepare new superconducting compounds, which may have interesting and hopefully useful superconducting properties.

## I. INFINITY TO ZERO - ULTRAPURE METALS

The two most common sources of rare earth ores are monazite (20% La, 45% Ce, 6% Pr, 19% Nd, 4% Sm, 2% Gd, 2% Y, and 2% other rare earths) and bastnasite (33% La, 50% Ce, 4% Pr, 12% Nd, 1% other rare earths). These ores and products derived from them, such as mischmetal, rare earth chlorides, etc., should be considered as solid solutions containing anywhere from an infinite amount of solute, i.e. Ce in Tm or Lu (the least abundant rare earths) to concentrated solutions La in Ce. Thus we have the concept of "infinity", especially if one is trying to extract one of the least abundant rare earths from the others in the ore. The methods of separating one rare earth from another were developed in the 1940's and 1950's and today it is possible to prepare a given rare earth element containing less than a total of 30 ppma of all the other rare earths combined, (1) see Table 1. Thus, at least with respect to the other rare earths one can by ion exchange or a combination of liquid-liquid extraction and ion exchange techniques obtain a rare earth element with essentially "zero" other rare earths.

Table 1. Chemical analyses of three cerium stocks. Impurity levels are in ppma. The presence of 82 elements were analyzed for. For these elements not listed the impurity level was 1 ppm or less, except for Nb, Tc, and the rare gases which were not analyzed for.

| Impurity | Ce(1) | Ce(2) | Ce(3) | Impurity | Ce(1) | Ce(2) | Ce(3) |
|---|---|---|---|---|---|---|---|
| H | 139 | 139 | 278 | Ba | < 10 | < 5 | < 0.3 |
| Li | < 20 | ... | < 0.0007 | La | 34 | 4 | 9 |
| C | 152 | 887 | 47 | Pr | 5 | < 4 | < 1 |
| N | 90 | 500 | 100 | Nd | 8 | < 0.4 | 4 |
| O | 639 | 350 | 385 | Sm | < 0.07 | < 0.3 | < 0.4 |
| F | 111 | 103 | 192 | Eu | < 0.05 | < 0.08 | < 0.05 |
| Na | 10 | 2 | 0.04 | Gd | < 1.2 | < 0.9 | 6 |
| Cl | 8 | 6 | 0.7 | Tb | < 0.2 | < 0.5 | < 1 |
| K | 5 | 2 | 0.03 | Dy | < 0.1 | < 0.5 | 1 |
| Ca | 20 | 1 | 0.2 | Ho | < 0.3 | < 0.3 | < 1 |
| Cr | 15 | 0.2 | 3.4 | Er | 1.6 | < 0.5 | 5 |
| Mn | < 1 | < 0.1 | 0.75 | Tm | < 0.05 | < 0.06 | < 0.06 |
| Fe | 7.5 | 7.5 | < 1.3 | Yb | < 0.05 | < 0.2 | < 0.3 |
| Co | < 0.1 | < 0.02 | 0.04 | Lu | 9 | 3 | 0.9 |
| Ni | 3 | 0.5 | 3 | Ta | 7 | 2 | 6.0 |
| Cu | 3 | 0.4 | 2 | Pt | < 0.5 | < 0.5 | 4.0 |
| Y | 5 | < 5 | < 10 | | | | |

|  | Ce(1) | Ce(2) | Ce(3) |
|---|---|---|---|
| Total magnetic rare earths (max.) | 16.6 | 7.7 | 19.8 |
| Total magnetic transition metal (max.) | 26.6 | 8.3 | 8.5 |
| Total non-metallic impurity | 1139 | 1985 | 1003 |
| Total metallic impurity (max.) | 199 | 62 | 96 |
| Total impurity (max.) | 1338 | 2047 | 1099 |
| Cerium purity (at. %) (min.) | 99.87 | 99.80 | 99.89 |

But when one considers the rare earth metals, the presence of these trace amounts of other rare earths no longer has any important effects on the properties of the chosen metal — at least as of today. If they do, their effects are masked by the presence of other impurities, such as the interstitial elements (H, C, N and O) and Fe. Recently the effects of trace amounts of H (2 ppmw or 0.035 at. %) and Fe (20 ppma) on the properties of Lu (2, 3) and Sc, (4, 5) respectively, have been noted. Therefore, it is clear that the rare earth metals need to be purified with respect to the non-rare earth impurities, i.e. to attain "zero", before we can fully appreciate and understand their intrinsic properties and behaviors.

Electrotransport or solid state electrolysis has been shown to be effective in reducing both the interstitial impurity and the 3d transition metal concentrations. (6) The metals La, Nd, Gd, Tb, Lu, Sc and Y have been successfully purified by this method, and this work is being continued at the Ames Laboratory, Iowa State University in the U.S.A. and Centre for Materials Science, University of Birmingham in the United Kingdom (some of it jointly). The major drawbacks of this technique are: the small amount of sample purified ($\sim$ 3 mm diam. x $\sim$ 30 mm long) per run; the long times required to obtain a sample (two to four weeks); and the unfavorable physical properties of some of the rare earth metals (especially their high vapor pressures — Sm, Eu, Dy, Ho, Er, Tm and Yb). Some of these problems can be overcome, but at the expense of another parameter or variable. For example, larger samples can be used, but this requires longer running times to achieve the same level of purification. By using this technique the high purity rare earth metals prepared by Ca reduction of the $RF_3$ and followed by vacuum melting (and sometimes distillation) can be further purified from $\sim$ 99.9 at. % pure (Table 1) with respect to all elements in the periodic table to $\sim$ 99.99 at. %, primarily by reducing the nonmetallic concentration (Table 2). Once the interstitial impurity concentrations have been lowered to this level the metallic impurities, although individually small, when considered in their entirety are comparable to the total amount of nonmetallics present in the metal. Although some of the fast diffuser metallic elements (e.g. Fe, Co, Ni, Cu, Ag and Au) can be removed from the rare earths, this technique is limited in its ability to reduce the total metallic impurity concentration level. In order to achieve five nines purity, it will be necessary to reduce the concentration of the metallic elements, by other methods. One way would be to use higher purity starting materials and better control of the metallurgical processes used to make the metal from the oxide. Although some improvements can be made, it seems unlikely that an order of magnitude reduction can be easily achieved. Experiments at Oxford and Birmingham in England and more recently with cooperative efforts from scientists at the Ames Laboratory indicate that zone refining may be the easiest way to prepare rare

Table 2.  Nonmetallic impurity concentrations in gadolinium before and after electrotransport purification in ppma.  (7)

| Impurity | Before | After |
|----------|--------|-------|
| C | 300 | < 26 |
| N | 314 | < 6 |
| O | 824 | 157 |
| $L_{4.2}$ [a] | 40 | 405 |

[a]Electrical resistance ratio, $R_{300}/R_{4.2}$.

earth metals with a purity approaching 99. 999 at. % at least with
respect to the metallic impurities.

In the last five years Revel and co-workers (8) and Jones et
al (9) showed that in the zone refining of commercial grade rare
earth metals the metallic impurities move with the molten zone
and purification will result if several passes (> 10) are made.
(6, 8, 9)  Furthermore, Jones and co-workers found that the interstitial impurites N and O move, but unfortunately, in the opposite
direction to the molten zone.  The latest results (to be reported on
at this Conference) (10) show that high purity rare earth metals can
also be purified by zoning techniques to levels below 1 ppma for
most of the non-rare earth metallic elements.  These results are
exciting since we are now on the threshold of a significant advance
in the preparation of ultrapure rare earth metals by breaking
through the four nines level and approaching the 99. 999% pure (i. e.
"zero") level.  Such metal can be prepared by zone refining several samples of the highest purity rare earth metal available.
Then by a boot-strap operation, zone refining the highest purity
portions of these samples results in a more pure metal.  The best
portion of this latter sample can then be purified by the electrotransport technique which should yield a metal close to 99. 999 at. %
pure.  After the technique has been developed, physical property
studies will be made to see if indeed these properties have changed
by this multi-purification procedure.

Two advantages of zone refining over the electrotransport technique are larger samples and shorter times.  It is possible that by
using ultra high vacuum systems, interstitial impurity levels may
also be reduced, but probably not to the level obtained by electrotransport.

II.  BACK AGAIN TO INFINITY – PSEUDO-LANTHANIDES

As all rare earthers are aware, there are many similarities in
the physical, chemical and metallurgical properties of the trivalent rare earths, and these properties change in a regular and sys-
temmatic manner across the lanthanide sequence of elements.
Yet, as we know, ~ 10% of the compound series formed by the

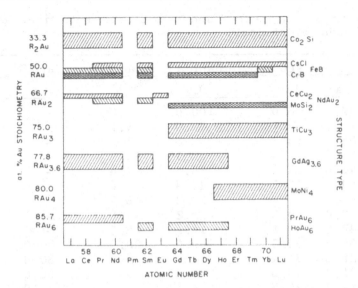

Fig. 1.  Compound formation and crystal structures in the
lanthanide-gold system, after O. D. McMasters,
et al. [reproduced by persmission of the Journal of
the Less-Common Metals 25:139 (1971)].

elements are isostructural from La through Lu.  In the other 90%
we find crystal structure changes across the series and in some
cases no compound exists at a particular stoichiometry for some
of the lanthanides, but for other lanthanides the compound exists.
Examples of all three behaviors are found in the lanthanide-gold
systems, see Fig. 1.  The $R_2Au$ phases all have the $Co_2Si$ type
structure.  Although the RAu and $RAu_2$ phases exist for all trivalent lanthanides three different crystal structures are found in both
compound series, as shown in Fig. 1.  The third behavior – compound stoichiometries for only portions of the lanthanide series –
is exhibited by the phases $RAu_3$, $RAu_{3.6}$, $RAu_4$ and $RAu_6$.  In the
case of the $RAu_6$ compounds a crystal structure change is also
noted.  This diverse behavior of course can be both a blessing and
a curse.  I shall illustrate an important example of each and discuss its impact on two relevant topics of research – low temperature heat capacity and superconductivity.  But before I do I will
need to introduce and discuss the concept of a "pseudo-lanthanide".

## Pseudo-Lanthanides

The term "pseudo-lanthanide" is defined as a material which
contains the proper ratio of two rare earths (except scandium) and
which might be expected to have physical properties similar to the
particular true lanthanide one is trying to duplicate.  The physical
properties which depend directly on the 4f electrons (i. e. optical
properties, crystal field levels, etc.) cannot be generated in the

Table 3.  La to Lu, La to Y and Y to Lu concentrations for preparing pseudo-lanthanides.

| Lantha-nides | Pseudo-Lanthanide Concentrations (at. %) | | | | | |
|---|---|---|---|---|---|---|
|  | La | Lu | La | Y | Y | Lu |
| La | 100.0 | 0.0 | 100.0 | 0.0 | - | - |
| Ce | 92.9 | 7.1 | 88.9 | 11.1 | - | - |
| Pr | 85.7 | 14.3 | 77.8 | 22.2 | - | - |
| Nd | 78.6 | 21.4 | 66.7 | 33.3 | - | - |
| Pm | 71.4 | 28.6 | 55.6 | 44.4 | - | - |
| Sm | 64.3 | 35.7 | 44.4 | 55.6 | - | - |
| Eu | 57.1 | 42.9 | 33.3 | 66.7 | - | - |
| Gd | 50.0 | 50.0 | 22.2 | 77.8 | - | - |
| Tb | 42.9 | 57.1 | 11.1 | 88.9 | - | - |
| Dy≡ Y | 35.7 | 64.3 | 0.0 | 100.0 | 100.0 | 0.0 |
| Ho | 28.6 | 71.4 | - | - | 80.0 | 20.0 |
| Er | 21.4 | 78.6 | - | - | 60.0 | 40.0 |
| Tm | 14.3 | 85.7 | - | - | 40.0 | 60.0 |
| Yb | 7.1 | 92.9 | - | - | 20.0 | 80.0 |
| Lu | 0.0 | 100.0 | - | - | 0.0 | 100.0 |

pseudo-lanthanide, even if both elements comprising the pseudo-lanthanide have unpaired 4f electrons. In general I shall be concerned with pseudo-lanthanides which are made up by alloying La and Lu, or La and Y, or rarely Y and Lu (see Table 3). For scaling purposes Y is assumed to behave like Dy primarily because the alloying behavior of these two elements are quite similar. (11) Thus in order to prepare pseudo-Pm one would make an alloy containing 71.4 at. % La - 28.6 at. % Lu or 55.6 at. % La - 44.4 at. % Y. For the binary intra-rare earth alloys this behavior has been observed for many combinations of rare earth elements properly scaled to give the appropriate pseudo-lanthanide, especially with respect to the observed crystal structures. (12) The magnetic behaviors (13) (i.e. $T_C$ and $T_N$), and the polymorphic phase transformation and melting temperatures of the pseudo-lanthanide also agree reasonably well with the corresponding true lanthanide element. The main exception occurs in the melting and transition temperatures of the pseudo-lanthanide alloys containing Y. This exception can be understood by virtue of the involvement of the 4f electrons in the melting process (14, 15) (the 4f electrons lower the melting point) for the true lanthanides, while yttrium, which has no 4f electron, will not have this same influence.

RM$_x$ compounds containing pseudo-lanthanides, instead of true lanthanides can be prepared, but to date there is a lack of data on compounds. Thus there is some question concerning the validity of this concept and how far one can push it in the case of compounds. Results obtained to date suggest that the pseudo-

lanthanide concept with respect to compounds may be more restricted than that found in the binary intra-rare earth alloys.  An
explanation will be proposed at the end of this paper.

## Low Temperature Heat Capacity

In trying to understand or to analyze the low temperature heat
capacity of lanthanide containing solids, the heat capacity, $C_p$, is
partitioned into the following contributions:

$$C_p = C_L + C_E + C_M + C_N + C_S + C_D \qquad (1)$$

where the subscripts represent L = lattice, E = electronic, M =
magnetic, N = nuclear, S = Schottky, and D = dilation (the difference between $C_v$ and $C_p$).  Depending upon the solid, all terms
except $C_L$ may be zero.  For the nonmagnetic rare earths (Sc, Y,
La and Lu) $C_M = C_N = C_S = 0$, and the $C_L$, $C_E$ and $C_D$ contributions usually can be determined in a straight forward manner from
the experimental data. [a]  For the lanthanide elements containing unpaired 4f electrons $C_M$, $C_N$ and $C_S$ may be zero, but generally are
not and with these additional contributions (along with $C_L$, $C_E$ and
$C_D$) it becomes difficult if not impossible to determine these quan-
ities unequivically from the measured heat capacity of the given
compound.

Generally when the heat capacity of a lanthanide with an unpaired 4f electron is measured $C_M$ or $C_S$ is the quantity of prime
interest.  If the heat capacity of isostructural compounds containing La or Lu is measured, one assumes that the $C_E$ and $C_L$ terms
can be obtained by interpolation. [b]  Since $C_N$ and $C_D$ are usually determined by straight forward calculations, reasonable values for
$C_M$ or $C_S$ can be obtained.  This has been done by Inoue and co-
workers (16) for the RAl₂ compounds which are all isostructural.

But for a series of compounds where there is a crystal structure change, clearly the $C_E$ and $C_L$ values from the compounds
containing the two end-members (La and Lu) cannot be used since
these two heat capacity contributions are different for the two
structures and the interpolation is invalid.  That is, the Fermi
surfaces of the two structures are unlike giving different $C_E$'s.
Likewise the vibrational characteristics of the two structures
would be expected to be dissimilar and thus the $C_L$'s would also
differ.  The situation, of course, becomes more complex if one

---

[a]In the case of Lu containing materials there is a non-zero $C_N$
contribution but it is only important below ~ 1 K.

[b]One cannot use the corresponding Sc or Y containing compound because their atomic masses are too light, relative to the lanthanide
elements and this would lead to an erroneous low value for the $C_L$
term.

or both of the end-members do not form a compound; or if there
is a third crystal structure in the sequence. In some instances
it may be possible to make a reasonable estimate for $C_E$ and $C_L$,
especially if the properties of a compound containing an end-
member have been measured.

Another approach is to prepare the appropriate pseudo-lanthanide compound(s) using La and Lu metals and measure the heat
capacity of the compound(s). For example if the crystal structure
change occurs at Gd, the pseudo-Eu and pseudo-Tb compounds are
prepared. The $C_E$ and $C_L$ contributions for one structure may be
obtained by interpolation between the La and pseudo-Eu compounds,
and for the other by interpolation between the pseudo-Tb and Lu
compounds. In the event three different structures exist along the
lanthanide series three to four pseudo-lanthanide compounds must
be prepared and measured. For the case where the crystal structure exists for only part of the series a pseudo-lanthanide com-
pound(s) which corresponds to both (either) the first and (or) last
lanthanide member of the existing compound sequence needs to be
prepared and measured.

We have attempted to do this for the $\gamma$-phase in the $R_2S_3$ sequence of compounds. This phase is known to exist from La to
Dy, but most of the $R_2S_3$ compounds are polymorphic and as many
as three structures are known for some of the lanthanide compound;
The pseudo-Dy and pseudo-Eu compounds were prepared by alloying the appropriate amounts of La and Lu, however, the resultant
$R_2S_3$ phase did not form the $\gamma$-structure, but one of the other $R_2S_3$
crystal structures. It is possible that under the proper conditions
the $\gamma$-phase for pseudo-Dy and pseudo-Eu can be formed. Then
again, the pseudo-lanthanide concept may not work for the $R_2S_3$
phases. Clearly more work is needed to answer this question.

Superconductivity

The 4f type superconductors represent an interesting and
potentially technologically important class of superconducting
materials. Of the elemental superconductors, La has the highest known superconducting transition temperature, ~ 12 K (at high
pressures). This suggests that suitable La-base alloys might
have superconducting transition temperatures competitive with
the A15 type materials. Since superconductivity in both 4f type
and other materials is many times associated with lattice instabilities, including many materials which undergo a structural
transformation below room temperature, a search for new such
material was begun and is being reported elsewhere at this Conference. (17) As noted earlier many lanthanide compound series
undergo a crystal structure change along the series. The lanthanide compounds at or next to the point at which this change occurs
should exhibit structural instabilities. Another possible location

of structural instabilities in lanthanide compound series is the beginning or the end of a compound stoichiometry.  Thus one would
expect that the lanthanide compounds which are located at or near
these changes might become superconducting if they do not have
any unpaired $4f$ electrons, but since most have unpaired $4f$ electrons, superconductivity has not been found.  However, by choosing the appropriate pseudo-lanthanide one might find the compound
to be a superconductor.  It is recommended that La and Y be used
to make up the pseudo-lanthanide compound, since La is a better
superconductor than Y which in turn is better than Lu.  Furthermore, along this line of reasoning one would expect that the higher
the La content the greater the chances for increasing the superconducting transition temperature, $T_c$.

   For the $RMn_2$ compound sequence Pr and Nd, and Er through
Lu form the hexagonal C14 Laves phase compound, Gd through Dy
form the cubic C15 Laves phase, and Sm and Ho are polymorphic.
Attempts to prepare the $RMn_2$ Laves phases from pseudo-Nd
through pseudo-Sm to pseudo-Dy (100% Y) were a partial success.
The cubic C15 Laves phase was found to exist from ~ pseudo-Sm
to pseudo-Dy, which is what is observed for the pure lanthanide
elements.  With higher La concentrations in the pseudo-lanthanide
$RMn_2$ phase the hexagonal C14 Laves phase was not found, which
is in disagreement with the results reported for the pure lanthanide compounds.  The X-ray and $T_c$ measurements suggest that
these samples consist of R (La, Y solid solution) + $R_6Mn_{23}$.  Although superconductivity was observed when 10 to 40% of Y was
substituted by La it appears to be due to some impurity phase,
such as a La-Y alloy.

## Comments on Pseudo-Lanthanides

   To date the application of the pseudo-lanthanide concept has
been only partly successful.  Perhaps, as our knowledge and skill
of preparing such compounds improves the success rate will be
better — at least I hope so.  It may very well be that there is an
intrinsic limitation associated with this concept.  One of these may
be that in the true lanthanides there is some $4f$ involvement in the
bonding due to some hybridization of $4f$ electrons with the $6s$ and
$5d$ electrons.  Although La might have some $4f$ hybridization, it
is clear that Y (which has no $4f$ electron) and Lu (which has a fully
occupied $4f$ (14) level) will not contribute any $4f$ electrons to the
chemical bond.  This will certainly have some effect, but it will
not completely invalidate this pseudo-lanthanide concept, since the
$4f$ contribution to the bonding is only a small fraction of the total
bonding.

   Another limitation is that the size factor (or as the crystal
chemists call it — the tolerance factor) may be more important in
certain crystal structures than in others.  In the case of solid

solution alloys metals which have size differences greater than
± 15% will not form solid solutions. But since the largest size
difference (excluding Sc) in the rare earth elements is 8.2% (for
La-Lu, considering Lu as the solvent, or 7.6% if La is the solvent)
the size effect for the intra-rare earth alloys should pose no problem for the formation of extensive (> 10 at. %), but not necessarily
complete solid solutions — and this is what is generally observed.
Furthermore, pure metals are rather ductile and can easily tol-
erate different size impurities, but intermetallic and inorganic
(covalent and ionic) compounds are not nearly as ductile and in
general might be expected to be less tolerant of an impurity atom
with a different size. This reasoning is consistent with the observed behaviors we have noted in preparing pseudo-lanthanide
materials in the $R_2S_3$ and $RMn_2$ compound series.

## ACKNOWLEDGMENT

The author wishes to acknowledge the support of the U.S.
Department of Energy, contract No. W-7405-Eng.-82, Office of
Basic Energy Sciences, Division of Materials Sciences (AK-01-02).

## REFERENCES

1.   J. E. Powell, p. 81 in "Handbook on the Physics and Chem-
     istry of Rare Earths", Vol. III, K. A. Gschneidner, Jr. and
     L. Eyring, eds. North-Holland Publishing Co., Amsterdam
     (1979).

2.   D. K. Thome, K. A. Gschneidner, Jr., G. S. Mowry and
     J. F. Smith, Solid State Comm. 25: 297 (1978).

3.   K. A. Gschneidner, Jr. and D. K. Thome, p. 75 in "The
     Rare Earths in Modern Science and Technology (1978)", G.
     J. McCarthy and J. J. Rhyne, eds., Plenum Publishing Corp.
     New York (1978).

4.   T.-W. E. Tsang, K. A. Gschneidner, Jr. and F. A. Schmidt,
     Solid State Comm. 20: 737 (1976).

5.   K. A. Gschneidner, Jr., T.-W. E. Tsang, J. Queen, S.
     Legvold and F. A. Schmidt, p. 23 in "Rare Earths and
     Actinides 1977", W. D. Corner and B. K. Tanner, eds., Inst.
     Phys. Conf. Ser. No. 37, Institute of Physics, Bristol,
     England (1978).

6.   B. J. Beaudry and K. A. Gschneidner, Jr., p. 173 in "Hand-
     book on the Physics and Chemistry of Rare Earths", Vol. I,
     K. A. Gschneidner, Jr. and L. Eyring, eds., North-Holland
     Publishing Co., Amsterdam (1978).

7.   O. N. Carlson, F. A. Schmidt and D. T. Peterson, J. Less-Common Metals 39: 277 (1975).

8.   G. Revel, J.-C. Rouchand and J.-L. Pastol, p. 602 in "Proc. 11th Rare Earth Research Conf. ", J. M. Haschke and H. E. Eick, eds., CONF-741002, Part 2, National Technical Information Service, Springfield, Virginia 22151 (1974).

9.   D. W. Jones, D. Fort and D. A. Hukin, p. 309 in "The Rare Earths in Modern Science and Technology", G. J. McCarthy and J. J. Rhyne, eds., Plenum Publishing Corp., New York (1978).

10.   D. Fort, B. J. Beaudry, D. W. Jones and K. A. Gschneidner, Jr., Zone Refining of High Purity Ce and Gd Metals, paper presented at 14th Rare Earth Research Conf., Fargo, North Dakota (June 25-28, 1979).

11.   E. T. Teatum, K. A. Gschneidner, Jr. and J. T. Waber, U.S. Atomic Energy Commission Rept. LA-4003 (1968).

12.   K. A. Gschneidner, Jr. and R. M. Valletta, Acta Met. 16: 477 (1968).

13.   R. M. Bozorth and R. J. Gambino, Phys. Rev. 147: 487 (1966); and R. M. Bozorth, J. Appl. Phys. 38: 1366 (1967).

14.   B. T. Matthias, W. H. Zachariasen, G. W. Webb, and J. J. Engelhardt, Phys. Rev. Lett. 18: 781 (1967).

15.   K. A. Gschneidner, Jr., J. Less-Common Metals 25: 405 (1971).

16.   T. Inoue, S. G. Sankar, R. S. Craig, W. E. Wallace and K. A. Gschneidner, Jr., J. Phys. Chem. Solids 38: 487 (1977).

17.   R. J. Stierman, O. D. McMasters and K. A. Gschneidner, Jr. Superconductivity in $La_{1-x}Y_xMn_2$, paper presented at 14th Rare Earth Research Conf., Fargo, North Dakota (June 25-28, 1979).

# DETERMINATION OF THE EFFECTIVE SEGREGATION COEFFICIENT

# OF OXYGEN IN TERBIUM DURING ZONE REFINING

D.A. Hukin[+], R.C.C. Ward[+], D.K. Morris[+], K. Davies[*]

[+]Clarendon Laboratory, University of Oxford

[*]Rare Earth Products Ltd., Widnes, England

## INTRODUCTION

Preliminary experiments concerning the application of horizontal cold-boat zone refining to the purification of rare earth metals were reported by Hukin and Jones (1) at the 13th Rare Earth Research Conference.

A more detailed study of the zone refining of terbium has now been undertaken with the determination of the effective segregation coefficient of oxygen being the major objective. Knowledge of $k_{eff}$ facilitates the selection of optimum experimental refining conditions, e.g. the number of zone passes, speed, etc., and establishes the limit of purification possible with this technique.

In order to prepare high quality single crystals of R.E. metals for solid-state research, it is necessary to reduce the high level of oxygen normally present in commercially available starting materials below the room temperature solid solubility limit. Above this limit oxide precipitates are formed which take the form of thin platelets, whose shape, distribution, and crystallographic orientation within the matrix have been studied using optical metallography and scanning electron microscopy. The effect of these precipitates upon the magneto-crystalline anisotropy of gadolinium has been studied by Smith, Tanner, and Corner (2).

Metallic impurities and nitrogen are also present in commercial material at much lower levels than oxygen. These contaminants can also be effectively reduced by cold boat zone refining.

The theory of zone refining is well understood and concentration curves can be predicted for given values of $k_{eff}$, zone length, and successive passes. A computer programme was written for this purpose, based on the treatment of Burris, Stockman, and Dillon (4) following the original zone refining equations of Pfann (3). The effective segregation coefficient can therefore be deduced from a comparison of the theoretical curves with the concentration distribution found experimentally after a zone refining treatment.

Results of the previously reported experiments (1) showed that oxygen is moved in the opposite direction to that of the zone travel and that $k_{eff}$ had a value between 1.1 and 1.5. Since a value of $k_{eff} = 1$ represents zero impurity transport, it is obviously important to optimize the velocity of zone travel and length of zone so that the redistribution can be effected in a reasonable period of time.

For the purpose of the present experiment it was attempted to make $k_{eff}$ as near as practically possible to the equilibrium segregation coefficient $k_0$. The relation between $k_0$ and $k_{eff}$ has been treated by Burton, Prim, and Slichter (5), who show that vigorous stirring of the melt, a high impurity diffusion coefficient in the melt and a slow zoning velocity are necessary for $k_0 \simeq k_{eff}$. R.F. heating produces good electromagnetic stirring within the molten zone, and although there has been no reported measurement of the diffusion coefficient of oxygen in a molten rare earth metal, it is felt that this parameter should not differ greatly from the value of $\simeq 10^{-2} mm^2 sec^{-1}$ shown by several other metals (6). It was estimated that $k_{eff}$ would be within 2% of $k_0$ under the conditions used, with a zone speed of 37.5 mm $hr^{-1}$.

## EXPERIMENTAL

All melting, zone levelling, and zone refining operations were carried out using an HCB 150 watercooled copper boat (7). The samples were outgassed in a vacuum of $10^{-7}$ torr before being processed under static argon at a pressure slightly above atmospheric. The argon was prepurified by passage over titanium sponge held at 800°C. The R.F. Coil was traversed along the stationary boat, and the length of the molten zone was controlled by the level of the R.F. power input.

A full charge of approximately 100 g terbium in the boat produces an ingot 150 mm long, with a cross section of 14 mm major diameter and 10 mm minor diameter.

The starting material was "Sublimed Grade" terbium supplied by Messrs Rare Earth Products Ltd., Widnes, U.K.

Two 100 g bars of terbium, from the same batch of starting material, were subjected to identical zone levelling procedures. One bar was then analysed throughout its length to establish its homogeneity and impurity content with respect to oxygen. The second bar was then checked at the centre point to confirm the starting material impurity levels before being zone refined.

Twenty five passes were made using a zone length of 20 mm and a zone travel speed of 37.5 mm hr$^{-1}$. This produced a bar containing about 7 crystals with the two larger grains being over 30 mm in length. The zone refined bar was sectioned at 10 points along its length and a 2 mm thick slice was removed at each position. Samples (4 mm diameter) were removed from the central region of each slice for analysis.

Gas analyses were carried out using vacuum fusion mass spectrometry. Samples (0.2-0.3g) were dissolved in a Pt/Sn bath (80/20 wt.%) held at 1750°C. under a vacuum of 10$^{-6}$ torr in a carbon crucible. Sample preparation included abrasion followed by electropolishing in Methanol/Perchloric acid at -78°C to minimize surface contamination due to oxidation of cold worked metal.

Sections were prepared for metallographic examination by mechanical polishing, finishing with 1 micron diamond abrasive, followed by a Roman (8) chemical polish. Oriented single crystal samples were also prepared as above for metallographic examination. These were then subjected to a more severe chemical etch in nitric/ acetic acid, which preferentially dissolves the matrix leaving the oxide platelets in relief. The form and orientation of these platelets could then be studied in the scanning electron microscope.

## RESULTS AND DISCUSSION

Table 1 summarizes the impurity levels of oxygen, nitrogen and hydrogen found in the starting material and after zone refining.

### Table 1

Ingot length 147 mm. Condentrations ppm wt.

| Impurity | Starting Material | After Zone Refining (25 passes) | | |
| --- | --- | --- | --- | --- |
| | | Start 6 mm | Middle 72 mm | End 141 mm |
| O | 740 | 1720 | 660 | 210 |
| N | 10 | 45 | 5 | N.D. |
| H | 10 | 10 | 10 | 10 |

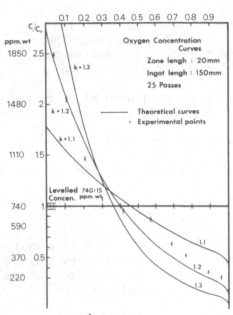

Figure 1.

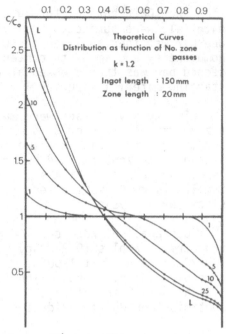

Figure 2.

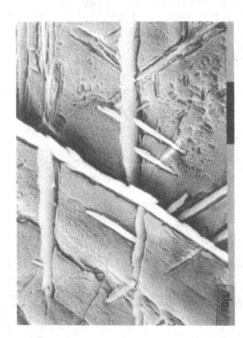

Figure 3.

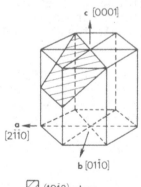

(10$\bar{1}$2) plane

Figure 4.
Tb: Schematic of lattice
showing (10$\bar{1}$2) plane.

The complete plot of the experimental oxygen distribution profile is given in Fig.1, compared with the theoretically derived curves corresponding to $k_{eff}$ values of 1.1, 1.2, and 1.3. This shows that $k_{eff}$ is very close to 1.2 and similar calculations for nitrogen give $N_2$ $k_{eff}$ = 1.3-1.4. Fig.2 shows the theoretical concentration curves for $k_{eff}$ = 1.2 as a function of the number of zone passes, from which it can be seen that the 25 passes used, produced an oxygen segregation very close to that permitted by the theoretical limiting distribution profile.

These figures give a lower limit to the oxygen concentration of $C/C_0$ = 0.25. Since a zone speed of 250 mm hr$^{-1}$ also produced a $k_{eff}$ of 1.2 (1) it is possible to reduce the level of oxygen in the latter part of a bar by a factor of 4 within 24 hrs.

Fig.3 shows an SEM image of basal plane of single crystal terbium containing 1200 ppm wt. of oxygen, and the oxide platelets are clearly visible. Measurements of their intersections with a, b, and c planes show them to lie parallel to the {10$\bar{1}$2} planes, which is primary twin plane of h.c.p. terbium, see Fig.4. The oxide precipitates do not extend across crystal boundaries and there is no evidence of preferential segregation of oxide precipitates to the grain boundaries, in contrast to the gettering action of these for the substitutional impurities. Metallographic examination of the sectioned bar indicates that the room temperature solid solubility of oxygen in terbium is 400-450 ppm.wt. Analysis of the platelets by electron probe microanalysis shows them to consist of terbium and oxygen alone, but their precise structure has yet to be determined.

## REFERENCES

1) D.A. Hukin, D.W. Jones, D. Fort. "Proc.13th Rare Earth Conf." G.J. McCarthy and J.J. Rhyne Eds. Plenum Press. N.Y. (1978).

2) R.L. Smith, B.K. Tanner, W.D. Corner, J.Phys.F: Metal Phys., 7: 8, (1977).

3) W.G. Pfann,"Zone Melting", 2nd. Edn., John Wiley & Sons, N.Y. (1964).

4) L. Burris Jr., C.H. Stockman, I.G. Dillon, Trans AIME, J. Metals, 1017,(Sept.1955).

5) J.A. Burton, R.C. Prim, W.P. Slichter, J.Chem. Phys., 21: 11, 1987 (1953).

6).   P. Protopapas, N.A.D. Parlee, <u>High Temp. Science</u>, 8:
      141, (1976).

7)    Crystalox Ltd., Limborough Road, Wantage, England.

8)    W.A. Roman, <u>J. Less-Common Metals</u>, 10: 150. (1965).

# ELECTROWINNING OF GADOLINIUM METAL BELOW 1,000° C[*]

D. Bratland and K. A. Gschneidner, Jr.

Ames Laboratory-DOE and Department of
Materials Science and Engineering
Iowa State University, Ames, IA 50011

## INTRODUCTION

In the present work solid gadolinium was electrowon directly from $Gd_2O_3$ dissolved in molten $LiF$-$GdF_3$ (65-35 mol %) under inert atmosphere, and then consolidated by arc-melting. In order to minimize the carbon contamination of the metal, inert metal anodes were tried, and the electrolyte was contained in a tantalum crucible. The oxygen contamination was studied as a function of the oxide concentration in the electrolyte, using a graphite container, which also served as the anode. A tantalum cathode was always used.

## ELECTROWINNING WITH INERT ANODES

Three metals, in the form of sheet, were tested as anode material at 820° C: copper, platinum and gold. The electrolyte contained initially 2 wt. % $Gd_2O_3$.

Copper oxidized readily whereas platinum and gold resisted corrosion under certain conditions.

The platinum anode appeared to be passivated below a critical current density (9 A $dm^{-2}$) at which a spontaneous voltage jump took place. The passivated anode surface was black, possibly due to platinization. Above the critical current density the anode corrosion became excessive. The apparent passivation may be ascribed to adsorbed oxygen, which is expelled at a certain anode potential, whereby the platinum surface is exposed to the combined action of nascent oxygen and fluoride ions.

---

[*] A more detailed paper will be published in Electrochimica Acta.

The gold anode behaved essentially the same was as the platinum anode, but even at current densities below the critical value (9 A dm$^{-2}$) the electrowon gadolinium contained about 0.5 wt. % gold, whereas no platinum could be detected in metal electrowon with platinum anode.

The gadolinium metal electrowon contained <200 wt. ppm carbon, but always more than 500 wt. ppm oxygen.

## COMPARTMENTED GRAPHITE CELL WITH LOW OXIDE ELECTROLYTE

The graphite cell used to study the correlation between the oxide concentration in the electrolyte and the oxygen contamination in the metal was operated at constant current at 900° C. For practical reasons the cell was made of graphite, and a perforated graphite sleeve located centrally divided the cell into a cathode compartment and an anode compartment.

In order to keep the oxide concentration constant at constant current, $Gd_2O_3$ was added intermittently in the anode compartment during the electrolysis to make up for consumed oxide. Oxide depletion caused a rapid voltage increase from 2.8V to 3.6V. By monitoring the cell voltage, therefore, it was possible to maintain a proper oxide feed rate. The oxide concentration and the critical current at which this voltage jump took place, were assumed to be roughly proportional.

The oxygen content in the metal was found to increase with increasing cell current above a critical $Gd_2O_3$ concentration (~ 0.1 wt. %) in the electrolyte. By running at the lowest oxide concentration practical (~ 0.2 wt. %), the oxygen contamination was reduced below 200 wt. ppm. However, when electrolysis was carried out with constant current at 3.8 - 4.0V it was reduced to ≤ 100 wt. ppm, and this value remained essentially constant with cell current.

## CONCLUSION

Preparation of gadolinium metal containing less than 200 wt. ppm of each of the two elements oxygen and carbon may be possible by using an anodic platinum cell and an electrolyte with low oxide concentration.

## ACKNOWLEDGMENT

This work was supported by the U.S. Department of Energy, Advanced Nuclear Systems and Projects Division. Thanks are extended to B. J. Beaudry for helpful advice and discussions.

# ZONE REFINING OF HIGH PURITY Ce AND Gd METALS

D. Fort,[a] B. J. Beaudry,[b] D. W. Jones[a]
and K. A. Gschneidner, Jr.[b]

[a]Centre for Materials Sciences
University of Birmingham
Birmingham B15 2TT, United Kingdom

[b]Ames Laboratory-DOE and Department of Materials
Science and Engineering
Iowa State University
Ames, IA 50011, U.S.A.

## ABSTRACT

In work reported at the 13th Rare Earth Conference, zone melting was shown to move impurities in commercial purity rare earth metals. The interstitial impurities O and N moved in the opposite direction from the zone while metallic impurities moved with the zone. We report here the results of further experiments using Ames Laboratory high purity rare earth metals as starting material. Analyses of a zone melted bar of gadolinium showed that the metallic impurities at the starting end of the bar were the lowest ever observed in gadolinium. Over half of the length of the bar had been purified with respect to O and N. The results obtained on cerium metal by this technique are also given.

## INTRODUCTION

The applicability of zone refining in redistributing both metallic and nonmetallic impurities in rare-earth metals has been previously demonstrated (1). However, to date the highest purity rare-earth metals have been prepared using solid state electrotransport (SSE) (2, 3), zoning having been used mainly for crystal growth or the removal of oxide and nitride precipitates from commercially available metals. In the present work Ce and Gd metals, prepared in the Ames Laboratory by methods described in (4) were zone refined in low contamination conditions.

33

## EXPERIMENTAL

The zone refining was done on a cold boat with a 350 kHz R. F. heater. The cold boat was originally designed to fit an ion and sublimation pumped UHV system. All gaskets were metal and the joints in the boat were hard soldered. Except for the boat and the silica tube, the entire system was constructed of stainless steel.

The cold boat was of 'segmented' design, each segment was a 25 cm length of 3.2 x 4.8 mm rectangular copper tubing, cooling water was passed 'up' one tube and back along its neighbor by means of suitable manifolds at each end. Six segments were used in an approximately semi-circular pattern with an internal diameter of $\sim 1$ cm, the gap between segments being $\sim 1$ mm.

Although this equipment was designed for UHV, for these zone refining studies the pumpdown time of this system to base pressure ($\sim 3$ days) was considered excessive. Hence the vacuum system used in the present work was an oil diffusion pump with cryogenic trap, the base pressure of which was $\sim 1.10^{-6}$ torr. However, as pointed out later, the results indicate that an ultra high vacuum is needed to optimize zone melting, or an ultra pure Ar atmosphere.

## CERIUM ZONE REFINING

Ce, prepared at the Ames Laboratory, was loaded into the cold boat and vacuum degassed by gradually increasing the R. F. power in such a way that the vacuum did not deteriorate beyond $5.10^{-6}$ torr. After degassing, the sample was consolidated by melting and zoning was started in a vacuum of $5.10^{-8}$ torr. However, great trouble was experienced in achieving a stable molten zone in vacuum so purified Ar was admitted to a pressure of 1 atmosphere, which effectively stabilized the molten zone and eliminated the rapid freezing-cooling cycles experienced in vacuum. Fifteen zones, all in the same direction, were run at a rate of 11.5 cm/hour, the molten zone width was $\sim 3$ cm. The final length of the bar was 15.9 cm and 3 mm analysis discs were cut at the 0.7, 4.3, 8.0, 11.9, 15.4 cm positions (approximately begin, 1/4, 1/2, 3/4, and end) using a low speed diamond saw. Analyses for O, N and H are given in Table 1.

## GADOLINIUM ZONE REFINING

Two Gd rods (Gd I and Gd II) were zone refined. The general procedure was to degas, melt and consolidate the metals in vacuum which was not allowed to deteriorate beyond $5.10^{-6}$ torr. After these procedures the silica tube surrounding the cold boat was covered with evaporated Gd, so it was removed, cleaned by abrasion with wire wool, quickly replaced, and the system was again evacuated. When base pressure was reached the Gd was quickly zone melted in vacuum at a rate of $\sim 50$ cm/hour. This both degassed

Table 1.   Selected analyses of Ce after zoning (parts per million
           atomic, ppma)

| Position a long bar | Oxygen | Hydrogen | Nitrogen |
|---|---|---|---|
| Begin | 1225 | 1680 | 360 |
| 1/4 | 675 | 980 | 240 |
| 1/2 | 820 | 1120 | 230 |
| 3/4 | 455 | 420 | 130 |
| End | 340 | 140 | 60 |
| Start Material | 395 | 560 | 260 |

the metal in vacuum and evaporated a thin, semi-transparent layer
of Gd onto the silica tube which acted as a getter for any impurities
in the argon, admitted next to a pressure of 1 atmosphere.

For Gd I, 15 zones were run at 11.5 cm/hour followed by one
zone at 2.5 cm/hour, the objective of the last (slow) zone being to
encourage grain growth. The zone width was ~1.5 cm and the
final bar length was 16.9 cm. Analysis discs were cut at the 0.6,
4.8, 8.3, 11.8, 16.0 cm positions (approximately begin, 1/4, 1/2,
3/4 and end). The results are given in Table 2.

Gd II which had a high content of W, was zoned 20 times at
22.5 cm/hour and finally once at 2.5 cm/hour. The final length
was 16 cm. The analyses of the starting material and the zoned
refined metal for some of the major impurities are shown in Table
3. The resistance ratios also given in Table 3, show the interior
sections of the bar have been purified with respect to total impurity content.

## DISCUSSION

In both Ce and Gd the interstitial impurities O, N and H moved
against the direction of the zones, this behavior being typical of

Table 2.   Selected Analyses of Gd I after zoning (ppma)

| Position along bar | Oxygen | Hydrogen | Nitrogen |
|---|---|---|---|
| Begin | 2950 | 2041 | 485 |
| 1/4 | 1965 | 1256 | 330 |
| 1/2 | 1370 | 1413 | 190 |
| 3/4 | 755 | 314 | 90 |
| End | 360 | 314 | 45 |
| Start Material | 595 | 785 | 120 |

Table 3.　Selected Analyses of Gd II after zoning (ppma)

| Impurity | Original Material | Position along bar | | | | |
|---|---|---|---|---|---|---|
| | | Begin | 1/4 | 1/2 | 3/4 | End |
| $O^a$ | 864 | 1770 | 894 | 648 | 432 | 304 |
| $H^a$ | 1089 | 778 | 778 | 778 | 467 | 311 |
| $N^a$ | 34 | 190 | 101 | 101 | 45 | 56 |
| $C^b$ | 367 | 262 | 367 | 524 | 616 | 1087 |
| $W^c$ | 190 | < 1 | < 1 | 55 | 33 | 1433 |
| Ta | 11 | < 0.3 | 1 | 24 | 23 | 10 |
| Fe | 38 | 3 | 10 | 20 | 5 | 160 |
| Al | 10 | < 0.04 | 1 | 4 | 8 | 100 |
| Si | 5 | < 0.3 | 10 | 10 | 2 | 20 |
| Co | 2 | < 0.03 | < 0.05 | < 0.1 | 0.2 | 5 |
| Ni | 3 | < 0.3 | < 0.3 | 3 | 0.8 | 10 |
| Ce | 0.5 | 0.4 | 2 | 3 | 1 | 0.4 |
| Total Impurities | 2613 | 3003 | 2275 | 2170 | 1633 | 3496 |
| Resistivity Ratio | 39 | 40 | 54 | 57 | 58 | 39 |

[a] Determined by vacuum fusion method

[b] Combusion - chromatographic method

[c] Quantitative chemical analysis

that found with other rare earth metals (1).

The lowest levels for O and N are about a factor of ten lower than had previously been obtained for Gd using commercial starting metals. Although impurities were moved, the overall purity of the Ce and Gd I have decreased indicating contamination from the system occurred. Gd II also has a slightly higher overall impurity content. Apparently to optimize the zone refining method the ultra high vacuum techniques used for SSE will need to be applied. However, the time advantage is then partially lost.

The mass spectrographic analysis results showing the metallic impurity levels for Gd II (Table 3) are most interesting. The major metallic impurity in Gd II was W, and this element was very successfully zoned to the end of the bar leaving < 1 ppma at the bar

beginning.  The 'end' value of 1433 ppma was visible as a precipitate on the bar surface near the end of Gd II.

Although O and N have been redistributed by zoning, the lowest levels of these impurities are considerably higher than has
been achieved by SSE on similar Gd starting material (2, 3).  However, the total refining time for electrotransport would be 500 -
1000 hours against 15 - 20 hours for the present zoning.  Even with
the additional time needed to introduce UHV techniques, zoning will
still be a more rapid refining technique.  To obtain metals with the
lowest levels of all impurities the best route may be to zone a metal to redistribute the metallic impurities, followed by electrotransport on the portion of the material low in metallic impurities, the
electrotransport being used to purify with respect to interstitial
impurities.

## ACKNOWLEDGMENTS

The authors would like to thank John Queen for measuring the
resistance ratios of Gd II.  They would also like to acknowledge
the financial support of the U. K. Science Research Council and
the U.S. Department of Energy, Office of Basic Energy Sciences,
Materials Sciences Division, for appropriate parts of these experiments.  The financial aid for travel granted by NATO which
helped coordinate this project is also gratefully acknowledged.

## REFERENCES

1.   D. W. Jones, D. Fort and D. A. Hukin, p. 309 in "The Rare
     Earths in Modern Science and Technology," G. J. McCarthy
     and J. J. Rhyne, eds. Plenum Press, New York (1978).

2.   O. N. Carlson and F. A. Schmidt, p. 460 in "The Proceed-
     ings of the 12th Rare Earth Research Conference," C. E.
     Lundin, ed. University of Denver, Denver (1976).

3.   D. W. Jones, S. P. Farrant, D. Fort and R. G. Jordan, p.
     11 in "Rare Earths and Actinides 1977," Inst. Phys. Conf.
     Ser. No. 37, Institute of Physics, Bristol (1978).

4.   B. J. Beaudry and K. A. Gschneidner, Jr., p. 173 in "Hand-
     book on the Physics and Chemistry of Rare Earths," Vol. 1,
     K. A. Gschneidner, Jr. and L. Eyring, eds. North-Holland
     Publishing Co., Amsterdam (1978).

# INTERMETALLIC COMPOUNDS IN R.E.(Al,Ga)$_2$ and R.E.(Cu,Ga)$_2$ ALLOYS

A. E. Dwight

Department of Physics, Northern Illinois University

DeKalb, Illinois 60115

## ABSTRACT

A series of alloys were made ranging from R.E.Al$_2$ to R.E.Ga$_2$, in which R.E. includes all trivalent rare earths and Ce. The only ternary compound has the CeCd$_2$-type structure instead of the AlB$_2$-type reported earlier. The unit cell constants and composition ranges of the ternary compounds were measured.

In the R.E.(Cu,Ga)$_2$ series, Ga substitutes for nearly half the Cu in the orthorhombic CeCu$_2$-type structure, for the heavy rare earths. Over a range of compositions centered around R.E.$_2$CuGa$_3$, a family of ternary compounds exists having the CaIn$_2$-type structure. This is significant in that previously reported ternary CaIn$_2$-type compounds exist at the equiatomic composition. All alloys were arc melted under argon, and annealed in vacuum. Crystal structure and unit cell constants were determined by the Debye-Scherrer method.

## INTRODUCTION

In the alloy series ranging from R.E.Al$_2$ to R.E.Ga$_2$, previous work[1] had established that a ternary compound exists near the equiatomic composition, and that the compound has a hexagonal unit cell. Further work was performed to delineate the composition range, and the structure type. A similar ternary compound was found in the alloy series ranging from R.E.Cu$_2$ to R.E.Ga$_2$, and the composition range and structure were investigated.

## EXPERIMENTAL

Alloys were made by arc melting on a water cooled copper hearth under argon, and homogenized in vacuum at 800°C. Debye-Scherrer patterns were taken with Cu and Fe radiation, and the patterns indexed by graphical methods. Unit cell constants were calculated from two high angle reflections, and indexing confirmed by comparing $I_o$ and $I_c$. For the $CaIn_2$-type structure the variable z was determined graphically from a plot of z vs $I_c$.

## THE R.E.$(Al,Ga)_2$ SYSTEM

The R.E.$Al_2$ binary compounds have the $MgCu_2$-type structure, for R.E. metals ranging from La thru Lu. The R.E.$Ga_2$ compounds have the $AlB_2$-type structure, for R.E. metals ranging only from La thru Er. Fig. 1 shows the phases present in alloys quenched from 800°C, in the ternary R.E.$(Al,Ga)_2$ alloys. One ternary phase is present, with the $CeCd_2$-type structure. This phase is stable for R.E. metals from Nd thru Tm. The $CeCd_2$-type structure is closely related to the binary $AlB_2$-type, in that both include a hexagonal ring of small atoms (Al,Ga in this example) above and below the large rare earth atom. In the $CeCd_2$-type, the ring is rippled; in the $AlB_2$-type the ring is flat. From Fig. 1, which is a plot of unit cell volume, it is seen that the packing efficiency is much greater for the $CeCd_2$-type. A relative size effect is apparent, in that the larger R.E. atoms, Pr, Ce and La form only the $AlB_2$-type structure. With the smaller R.E. atoms (e.g., Gd) the interatomic distance becomes reduced to the point where the more efficiently packed $CeCd_2$-type becomes stable.

## THE R.E.$(Cu,Ga)_2$ SYSTEM

The R.E.$Cu_2$ binary compounds have the orthorhombic $CeCu_2$-type structures, for R.E. metals from Ce thru Lu. When Ga is substituted for part of the Cu, the orthorhombic structure is retained, but unit cell volume is increased as shown in Fig. 2. The limit of solid solution is nearly the same as that reported[2] for the orthorhombic phase in the Ce$(Ni,Cu)_2$ system.

When the Ga substitution exceeds the amount that can be held in solid solution, a ternary phase with the $CaIn_2$-type structure becomes stable. Fig. 3 shows unit cell constants, from which it is noted that $a_o$ of the orthorhombic and hexagonal structures is nearly the same, but the hexagonal $c_o$ lies between the orthorhombic $b_o$ and $c_o$. This indicates that the $CeCu_2$-type is an orthorhombic distortion of the $CaIn_2$-type. The relation of the two types is shown in Fig. 4. GdCuGa, although orthorhombic, is shown as a pseudo-hexagonal array, for comparison with hexagonal $Gd_3Cu_2Ga_4$. The basic difference is that the chain of R.E. atoms is zig-zag in the

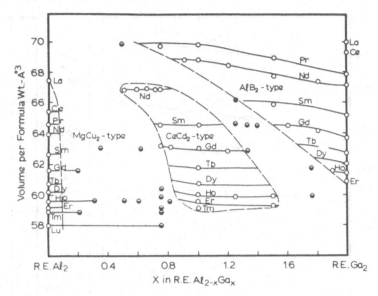

Fig. 1.   Volume per Formula Weight (V/M) of Intermetallic Compounds in Alloys from R.E.Al₂ to R.E.Ga₂, quenched from 800°C.

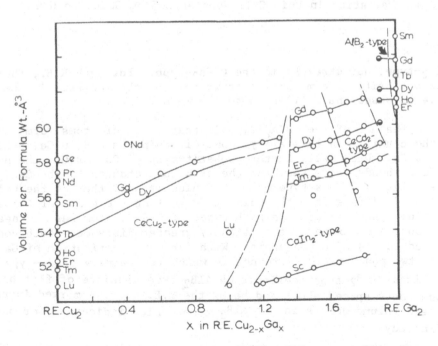

Fig. 2.   Volume per Formula Weight (V/M) of Intermetallic Compounds in Alloys from R.E.Cu₂ to R.E.Ga₂, quenched from 800°C.

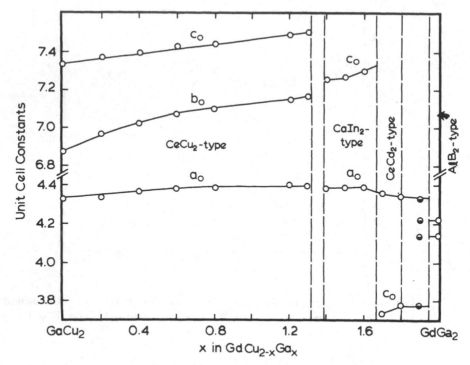

Fig. 3. Variation in Unit Cell Constants from GdCu$_2$ to GdGa$_2$.


CeCu$_2$-type, but straight in the CaIn$_2$-type. For each R.E., there exists a critical composition at which a slight increase in the Cu content causes the R.E. chain to become straight.

A feature of the CaIn$_2$-type is that pairs of atoms form parallel to the c axis. In this alloy, some of the pairs are Cu-Ga, others are Ga-Ga. With still further substitution of Ga for Cu, the number of Cu-Ga bonds decreases, and the structure changes to the CeCd$_2$-type (Fig. 4). This type has a c$_o$ which is half that of the CaIn$_2$-type, and differs from the CaIn$_2$-type in that the pairs of Cu-Ga atoms no longer exist. In both types, the R.E. atom has six nearest neighbors and six more at a slightly greater distance. Both contain rippled layers of Cu-Ga atoms. With further substitution of Ga for Cu, a two phase field is found, in which the ternary CeCd$_2$-type and the binary AlB$_2$-type coexist. The AlB$_2$-type consists of flat hexagonal nets spaced above and below the R.E. atom. A marked increase in cell volume occurs in the AlB$_2$-type, which indicates poor packing efficiency.

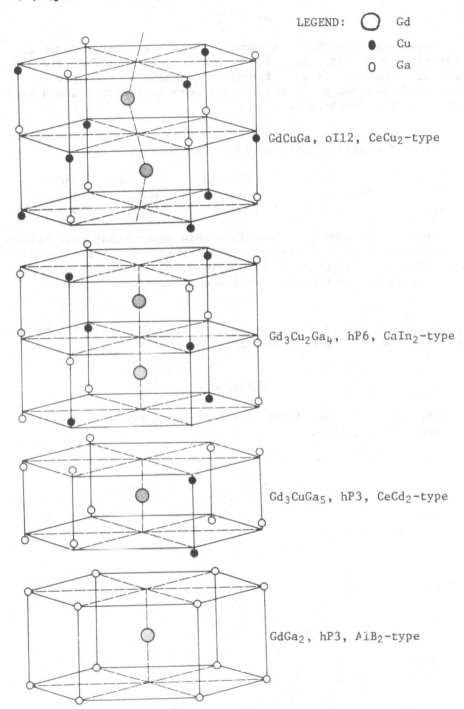

LEGEND: ◯ Gd
● Cu
◦ Ga

GdCuGa, oI12, CeCu$_2$-type

Gd$_3$Cu$_2$Ga$_4$, hP6, CaIn$_2$-type

Gd$_3$CuGa$_5$, hP3, CeCd$_2$-type

GdGa$_2$, hP3, AlB$_2$-type

Fig. 4. Variation in Crystal Structure with Decreasing Copper Content, from GdCuGa to GdGa$_2$.

## DISCUSSION

Compounds with the $CeCd_2$-type structure exhibit a lower c/a ratio than those with the $AlB_2$-type structure, therefore the c/a ratio is a useful criterion for distinguishing between the two phases.  Comparison of $I_O$ and $I_c$ is a more exact method, and will be discussed in a forthcoming paper.

The electron concentration e/a increases with increasing Ga content in the series $GdCu_2$ to $GdGa_2$, as follows:

| Alloy | GdCuGa | $Gd_3Cu_2Ga_4$ | $Gd_3CuGa_5$ | $GdGa_2$ |
|---|---|---|---|---|
| Structure Type | $CeCu_2$ | $CaIn_2$ | $CeCd_2$ | $AlB_2$ |
| e/a | 2-1/3 | 2-5/9 | 2-7/9 | 3 |

It will be shown in a sequel to this paper, that this sequence of crystal structures is present in other alloy systems, always in this order, but sometimes with omissions.  The relative size factor sometimes overrides the electron concentration.

---

A portion of this work was performed at Argonne National Laboratory under auspices of the U.S. Department of Energy.

## REFERENCES

1.  A.E. Dwight,  Proceedings, 7th Rare Earth Research Conf., Coronado, CA, 1968, p. 273.

2.  G.L. Olcese, J. Phys. Chem. Solids, 1977, 38, p. 1239.

# EQUILIBRIUM RELATIONS AMONG MAGNETIC PHASES IN MISCHMETAL COBALT SYSTEMS

E. M. T. Velu, E. C. Subbarao, H. O. Gupta, K. P. Gupta,
S. N. Kaul, A. K. Majumdar, R. C. Mittal,
T. A. Padmavathi Sankar, G. Sarkar, M. V. Satyanarayana,
K. Shankaraprasad and J. Subramanyam
Indian Institute of Technology, Kanpur 208016, India

## INTRODUCTION

The intermetallic compounds $RCo_5$ and $R_2Co_{17}$ (R = La, Ce, Nd, Sm or mischmetal MM) are important permanent magnetic materials whose magnetic properties ($H_c$ and $B_r$) are very sensitive to the phase composition (1-4). While the phase relations in pure binary R-Co systems are available (5-8), similar details are not known for a multicomponent system such as MM-Co. These are investigated in the present work.

## EXPERIMENTAL

Indian mischmetal (MM) and 99.3% pure Co were used. X-ray fluorescence and chemical analysis revealed MM to contain 52% Ce, 20.1% La, 15.7% Nd, 4.8% Pr, 6% Fe and the balance unidentified impurities. Batches of 20 g were arc-melted in a water-cooled Cu-crucible in a purified argon atmosphere. Repeated meltings for 20 sec were carried out to homogenize the composition. Weight loss during melting was 0.1 to 0.2%. Samples sealed in quartz capsules in a vacuum of $10^{-2}$ torr were annealed at 900°C for four days (C series) and 10 days (A, B and D series) and then quenched in water. The phase composition was determined by metallography, x-ray diffraction (Cr or Co radiation), thermomagnetic analysis (using PAR vibrating sample magnetometer) and elemental composition of some phases by electron probe microanalysis.

45

Table 1.   Compositions Selected in the MM-Co System and Phases
           Identified.

| | Composition (wt %) | | Phases (900°C, 9d) | |
| Alloy | Co | Fe | X-ray | Microstructure (No. of Phases) |
|---|---|---|---|---|
| A-1 | 55.15 | 2.72 | 1:3 + 2:7 | 2 |
| A-2 | 55.52 | 2.70 | 1:3 + 2.7 | 2 |
| A-3 | 55.90 | 2.69 | 2:7 | 1 |
| A-4 | 56.26 | 2.66 | 2:7 + 5.19 | 2 |
| B-2 | 57.66 | 2.58 | 2:7 + 5:19 | 2 |
| B-3 | 58.01 | 2.54 | 5:19 | 1 |
| B-4 | 58.33 | 2.53 | 5:19 + 1:5 | 2 |
| B-5 | 58.71 | 2.51 | 5:19 + 1:5 | 2 |
| C-5 | 64.54 | 2.16 | 5:19 + 1:5 | 2 |
| C-6 | 64.76 | 2.14 | 1:5 | 1 |
| C-7 | 65.22 | 2.11 | 1:5 | 1 |
| C-8 | 66.14 | 2.05 | 1:5 | 1 |
| C-9 | 66.48 | 2.04 | 1:5 + 2:17 | 2 |
| D-4 | 75.59 | 1.48 | 1:5 + 2:17 | 2 |
| D-5 | 75.80 | 1.47 | 2:17 | 1 |
| D-6 | 76.05 | 1.45 | 2:17 + β-Co | 2 |
| *E-1 | 63.80 | 3.90 | 1:5 + 5:19 | 2 |
| *E-2 | 55.90 | 11.80 | 1:5 + 5:19 | |

*E-1 and E-2 alloys were prepared from high purity metals (>99.9%
pure) and the phase analysis was made on as-cast alloys.


RESULTS AND DISCUSSION


Phase Relations in MM-Co System

     The phase analyses, based on x-ray diffraction and microstruc-
ture, of the alloys studied in the composition range 55 to 76 wt %
Co are shown in Table 1.   Six different phases (1:3, 2:7, 5:19, 1:5,
2:17 and β-Co) were identified in the microstructure.   Two phase
fields were detected in the intermediate compositions.   Since all
the strong x-ray diffraction lines of the 2:7 and 5:19 phases occur
nearly at the same position (2θ) with only a few weak lines characteristic of each phase, thermomagnetic analysis was used to clarify
this ambiguity.   The thermomagnetic curve of alloy A-1 (Fig. 1)
shows a major transition near 270°C, two weak transitions near 340
and 520°C and none near 60°C (Fig. 1).   The transition in A-1 at
60°C is attributed to the 2:7 phase, while the presence of 1:3 phase
could not be established since the $T_c$ of 1:3 phase lies below room
temperature.   The transition in B-5 at 270°C is attributed to the

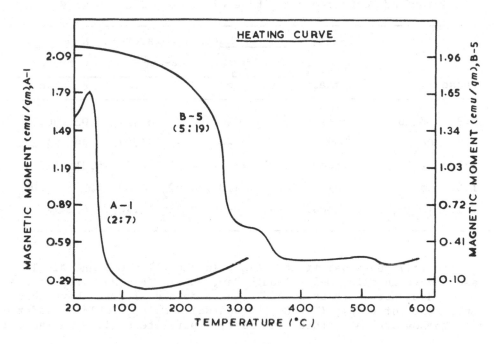

Figure 1.   Low field magnetic moment as a function of temperature
            for alloys A-1 and B-5, 900°C, 10 days, annealed.


5:19 phase and the other transitions to the minor phases present in
it (an unknown phase and 1:5 phase).  The unknown phase with a magnetic transition at 340°C lies between 5:19 and 1:5 phases and may
be related to the 1:4, 7:29, 4:17 and 3:19 phases indicated by
Strnat (9) on the basis of thermomagnetic evidence.  He did not present any corroborating x-ray or microstructural data for these
phases.

     The x-ray data for the 2:7 phase could be interpreted only on
the basis of a mixture of hexagonal and rhombohedral modifications.
Buschow (10) has reported that the two forms of $R_2Co_7$ could coexist,
while Khan (11) found the rhombohedral modification to be the stable
low-temperature form.  The lattice parameters of the 2:7 and 5:19
phases calculated from the present data are given in Table 2 and are
in good agreement with those given for CeMM-Co phases by Kahn (11).

     The 1:5 phase first appeared in alloy B-4 and increased in
amount with increasing cobalt, while composition C-6 becomes a single
phase material.  The alloys C-6, C-7 and C-8 are single phase (1:5),
indicating the homogeneity range of the 1:5 phase.  The lattice parameters of the hexagonal 1:5 phase are included in Table 2.

Table 2.  Comparison of Lattice Parameters of Stoichiometric Phases
of MM–Co System with Those of Ce–Co and Ce MM–Co Systems

| Phase | Structure | Crystal Symmetry | Lattice Parameters, Å | | | | | |
|-------|-----------|------------------|-----------------------|--|--|--|--|--|
|       |           |                  | Ce MM–Co[11] | | Ce–Co[11] | | MM–Co | |
|       |           |                  | a | c | a | c | a | c |
| 2.7 | $Gd_2Co_7$ | rhomb. | $5.02_3$ | $36.74_7$ | $4.95_6$ | $36.52_5$ | $5.02_4$ | $36.71_3$ |
|     | $Ce_2Ni_7$ | hex    | -- | -- | $4.95_9$ | $24.47_0$ | $4.99_0$ | $24.69_9$ |
| 5:19 | $Pr_5Co_{19}$ | rhomb. | $5.01_2$ | $48.71_6$ | $4.94_7$ | $48.74_3$ | $5.01_8$ | $48.65_7$ |
| 1:5 | $CaCu_5$ | hex. | $4.99_1$ | $4.00_4$ | $4.92_2$ | $4.03_0$ | $4.98_0$ | $4.00_8$ |
| 2:17 | $Th_2Zn_{17}$ | rhomb. | -- | -- | $8.38_1$ | $12.18_9$ | $8.47_2$ | $12.14_5$ |

The 2:17 phase was first noticed in the alloy C–9 and single
phase occurs in alloy D–5.  Though Ray, et al. (6), have reported a
rhombohedral structure below 400°C and a hexagonal type above that
temperature for $R_2Co_{17}$ (R = Ce, Pr and Nd), $MM_2Co_{17}$ could be indexed
as a rhombohedral structure with lattice parameters given in Table 2.

Compositions richer in cobalt than the 2:17 phase exhibit the
presence of β–Co (f.c.c.).  Electron microprobe analysis of the
β–Co phase showed that it contains predominantly Co with small
amounts of Fe and Ni and no detectable amounts of the rare earths
(Ce, La, Nd, Pr and Sm).  This is consistent with the lack of solubility of rare earths in Co in the R–Co system (6).

## Stability of MM $Co_5$

An alloy, C–6, which is a single phase 1:5 at 900°C, gave an
interesting microstructure on annealing at 700°C for 30 days (Fig. 2).
A bright phase (2:17) appeared along the grain boundaries of the 1:5
phase.  Buschow (12) has obtained similar microstructures for $NdCo_5$
phase annealed at 620°C for eight weeks.  He attributed this to an
eutectoid reaction in which the 1:5 phase decomposed into the two
neighboring (5:19 and 2:17) phases, the latter appearing as bright
grain boundary precipitate and the former as dark parallel lines inside the 1:5 phase with different orientations in each grain.  He
gave x-ray evidence only for the 2:17 product phase.  In the present
study, it was observed that the extent of decomposition of MM $Co_5$
even after 30 days of annealing was so small (<5%) that the x-ray
evidence could not be obtained for the product phases.  Since the
addition of Fe to the 1:5 phase has been reported to raise its eutectoid reaction temperature (13), two alloys (E–1 and E–2 of Table 1)

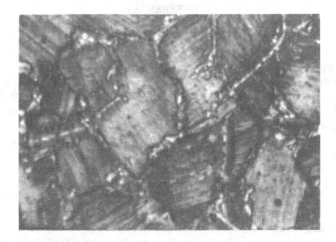

Figure 2.   Microstructure of alloy C-6, 700°C, 30 days annealed.
MMCo$_5$ (matrix) with MM$_2$Co$_{17}$ at the grain boundaries.
Dark parallel lines inside the grains (5:19?).   1250X

containing larger amounts of Fe were studied.  Examination of samples
annealed at 700°C for 30 days by x-ray diffraction, metallography
and thermomagnetic methods clearly establish the presence of 5:19 and
2:17 phases in the product of the eutectoid reaction of the 1:5 phase.

## CONCLUSIONS

The study of phase relations in the system MM-Co in the composition range of 55 to 84 wt % Co revealed six phases, namely, 1:3, 2:7,
5:19, 1:5, 2:17 and β-Co.  Thermomagnetic analysis was more useful
than x-ray diffraction to distinguish the structurally similar phases
(1:3, 2:7 and 5:19), since the Curie temperatures of these phases
are <20, 67 and 270°C respectively.  Evidence for a possibly new
phase between 5:19 and 1:5 phase with a $T_c$ of 340°C is presented.
The 1:5 phase exhibits a narrow but definite region of homogeneity.
The 1:5 phase undergoes an eutectoid decomposition into 5:19 and 2:17
phases, though the rate of decomposition is quite sluggish.  The
phase on the Co-rich side of 2:17 phase has a β-Co (f.c.c.) structure
with no solid solubility for rare earth elements.  The sequence of
occurrence of phases in multicomponent MM-Co system is essentially
similar to that observed in a pure binary Ce-Co system.

## ACKNOWLEDGEMENTS

We are grateful to Dr. V. S. Raghunathan, Reactor Research
Centre, Kalpakkam for his help in electron probe microanalysis.  This
work is supported by the Department of Electronics.

REFERENCES

1.  D. L. Martin and M. G. Benz, AIP Conf. Proc. 5:970 (1972).
2.  P. A. Nastepad, F. J. A. Den Broeder and R. K. J. Wassink. Powder Metall. Intern. 5:61 (1973).
3.  A. Benz and K. Bachmann, AIP Conf. Proc. 10:578 (1973).
4.  P. F. Weihrauch and D. K. Das, AIP Conf. Proc. 18:1149 (1974).
5.  K. H. J. Buschow and F. J. A. Den Broeder, J. Less Common Metals 33:191 (1973).
6.  A. E. Ray, A. T. Biermann, R. S. Harmer and J. E. Davison, Cobalt 4:103 (1973).
7.  Y. Khan, Phys. Stat. Sol. (a)21:69 (1174).
8.  Y. Khan, J. Less Common Metals 34:191 (1974).
9.  K. J. Strnat and A. E. Ray, AIP Conf. Proc. 24:680 (1975).
10. K. H. J. Buschow, Acta Cryst.  B26:1389 (1970).
11. Y. Khan, Acta Cryst.  B30:1533 (1974).
12. K. H. J. Buschow, J. Less Common Metals 29:283 (1972).
13. K. H. J. Buschow, J. Less Common Metals 37:91 (1974).

# MIXED HALIDE PHASES IN THE CHLORIDE-IODIDE SYSTEMS OF LANTHANUM AND GADOLINIUM

Susan Beda, University of Michigan, Ann Arbor, MI 48104;
John M. Haschke, Rockwell International, Golden, CO 80401;
Laura S. Quayle and Harry A. Eick, Michigan State
University, East Lansing, MI 48824

Numerous PbFCl-type mixed halogen halide phases, most of which contain the fluoride ion, have been reported for divalent lanthanides (1), and the existence of trivalent mixed-halogen halide phases has been postulated. An AlBrCl$_2$ phase, and poorly characterized mixed halogen halide phases of U(III) and U(IV), thought to be solid solutions, are also known (2,3). We present herein the first report of trivalent lanthanide mixed halogen halide phases.

The mixed halogen phases LnCl$_{3-x}$I$_x$, for which Ln = La and Gd, were prepared by sealing weighed masses of the pure trivalent chloride and iodide into evacuated quartz ampoules, and for Gd, by also reducing a mixture of HgCl$_2$ and HgI$_2$ with metal sponge. The reaction products, purified in the latter preparatory case by distillation of excess Hg and HgX$_2$, were examined by X-ray powder diffraction.

In the LaCl$_{3-x}$I$_x$ system one phase of the composition La$_4$Cl$_9$I$_3$ (LaCl$_{2.25}$I$_{0.75}$) was observed. This phase exhibits the structure type characteristic of LaCl$_3$ with the $a$ parameter doubled. The observed parameters are $a$ = 15.378 Å, $c$ = 4.459 Å (space group P6$_3$/m, Z = 2). A trial set of atomic coordinates was deduced on the basis of packing considerations and powder diffraction intensities were calculated with the program ANIFAC (4). For these calculations estimated isotropic thermal parameters were used. Atomic positional parameters were adjusted until the calculated intensities agreed acceptably with those observed. The atomic coordinates are listed in Table I.

In the GdCl$_{3-x}$I$_x$ system a phase of the approximate composition GdClI$_2$ was found. The X-ray powder diffraction pattern of this phase does not appear to be related to that of any pure trihalide,

Table I

Probable Atomic Coordinates of $La_4Cl_9I_3$

| Atom | x/a | y/a | z/c | B $(A^2)$ |
|---|---|---|---|---|
| La (1) | 2/3 | 1/3 | 1/4 | 0.4 |
| La (2) | 1/6 | 1/3 | 1/4 | 0.4 |
| Cl (1) | 0.191 | 0.048 | 1/4 | 2.5 |
| Cl (2) | 0.357 | 0.308 | 1/4 | 2.5 |
| Cl (3) | 0.191 | 0.548 | 1/4 | 2.5 |
| I (1) | 0.460 | 0.110 | 1/4 | 2.5 |

and the structure of this mixed-halogen halide has not been determined. This phase can be sublimed in vacuo to a sublimate of the same apparent composition (analyzed composition: $GdCl_{1.2(1)}I_{1.8(1)}$). Far IR spectra of the residue and sublimate are essentially identical, and are very different from those of the pure trihalides.

These significant differences observed between the lanthanum and gadolinium systems are to be expected. The coordination number difference between the $UCl_3(LaCl_3)$ and $PuBr_3(LaI_3)$ structure types is but one; the comparable value for the $UCl_3(GdCl_3)$ and $BiI_3(GdI_3)$ structure types is three (5).

At least one additional type of mixed halogen is expected in lanthanide chloride-iodide systems, and will be exemplified by Tb, whose chloride and iodide structures are of the $PuBr_3$ and $BiI_3$ types, respectively.

REFERENCES

1.  H. P. Beck, J. Solid State Chem. 23:213 (1978).
2.  L. H. Brixner, Mat. Res. Bull. 11:269 (1976).
3.  B. L. Clink and H. A. Eick, J. Solid State Chem. in press, and references contained therein.
4.  D. T. Cromer, private communication.
5.  J. D. Corbett and N. W. Gregory, J. Am. Chem. Soc. 75:5238 (1953).
6.  J. J. Katz and E. Rabinowitch, "The Chemistry of Uranium," Nat. Nucl. Energy Ser. Div. VIII, Vol. 5, McGraw-Hill, New York, NY (1951).
7.  J. M. Haschke, J. Solid State Chem. 18:205 (1976).

# PREPARATION AND PURIFICATION OF GdCl$_3$

Y. S. Kim, F. Planinsek,[*] B. J. Beaudry and
K. A. Gschneidner, Jr.
Ames Laboratory-DOE and Department of
Materials Science and Engineering
Iowa State University, Ames, IA  50011

## ABSTRACT

GdCl$_3$ has been prepared from high purity Gd$_2$O$_3$ with NH$_4$Cl as the chlorinating agent.  The optimum time, temperature, air flow rate, and the mixing ratio of Gd$_2$O$_3$ and NH$_4$Cl for chlorination were determined.  Almost complete conversion (99.8%) of Gd$_2$O$_3$ to GdCl$_3$ was obtained with 2.5 times or more the theoretical amount of NH$_4$Cl and with at least a 20 hr reaction time at 250°C under a dried air flow.  The excess NH$_4$Cl was removed from the GdCl$_3$ by sublimation at 350°C in vacuum.  The resulting GdCl$_3$ was purified to a dense crystalline GdCl$_3$ by distillation in vacuum using Ta crucibles and condensers.

## INTRODUCTION

The preparation of anhydrous chlorides of the rare earth elements is of considerable interest, since the chlorides are used widely for the production of metals both by electrolytic and metallothermic methods.

In general, the anhydrous rare earth chlorides can be obtained either by dehydration of the hydrated chlorides using a dehydrating reagent or by direct chlorination of the oxide using a chlorinating agent.  Although the dehydration of hydrated chloride using either HCl(g) (1-2) or NH$_4$Cl (3-6) yields a high purity product, the excessive length of time and careful temperature control for dehydration make this method less desirable for plant-scale use.  Direct conversion of oxide to chloride is obtained by passing a

---

*Present address:  Rockwell International, Rocky Flats Plant, Golden, Colorado  80401

stream of HCl(g) or $Cl_2$(g) over a hot, intimate mixture of oxide and carbon (7), or by reaction with $CCl_4$ (8), $COCl_2$ (9), $S_2Cl_2$ (10), $SOCl_2$ (11), $PCl_5$ (12), or $NH_4Cl$ (13). The presence of carbon has the advantage of improving the chlorination yields, but it is not convenient for the preparation of chlorides for metal preparation because of the carbon contamination problem.

Among the various chlorinating methods, Hopkins et al (13) first developed a rather simple method of preparation by heating a mixture of rare earth oxide ($R_2O_3$) and excess ammonium chloride to a temperature of 200°C or higher in accordance with equation (1).

$$R_2O_3(s) + 6NH_4Cl(s) = 2RCl_3(s) + 3H_2O(g) + 6NH_3(g) \qquad (1)$$

Excess $NH_4Cl$ is sublimed away in vacuum. This method has been used extensively by Lunex Corporation (14) to prepare rare earth chlorides for metal production and it has also been described by Reed et al (15). Since direct conversion of oxide to chloride using $NH_4Cl$ appeared quite efficient for producing anhydrous chlorides in quantity, this method was used in this investigation as a part of the process for the production of high purity gadolinium metal in large quantities. Since the optimum conditions for a complete conversion of oxide to chloride were not given in the previous work (13 - 15), the determination of these parameters as a function of the mixing ratio of $Gd_2O_3$ and $NH_4Cl$, time, temperature and flow rate was started. The optimum conditions for chlorination of the oxide and the purification of the crude $GdCl_3$ by distillation in vacuum are reported here.

### EXPERIMENTAL

GdCl$_3$ was prepared by heating a mixture of $Gd_2O_3$ and $NH_4Cl$ in a Pyrex boat (4cm wide x 13cm long x 2.5cm deep) which was inside a Pyrex tube used to contain the reactants. The Pyrex reaction tube measured 5cm in diameter by 56cm long and was closed at one end. During the reaction the Pyrex boat was placed near the closed end and the dried air, which had been passed through Drierite (granular $CaSO_4$) and molecular sieve (type 13x: alkali-metal alumino-silicate pellet), was introduced from a stopcock located 12.5cm from the open end to sweep away the $H_2O$ and $NH_3$ reaction products. Heating tapes were wound around the reaction vessel in addition to asbestos tape. Firebricks were also placed around the reaction vessel to obtain a more uniform temperature. After the reaction period, the system was evacuated in the presence of excess $NH_4Cl$ for 3 hrs at the reaction temperature to remove water condensed on the cool parts of the reaction vessel. The remaining $NH_4Cl$ was removed completely by heating in vacuum for 4 hrs at 350°C. The system was then cooled to room temperature while under vacuum prior to taking out the Pyrex boat and weighing. The product was obtained in the form of a finely divided

powder. It was extremely hygroscopic and hissed on addition of
water. The solution was clear when the chloride was pure with
respect to oxychlorides and oxides (hydrated chlorides also dissolve readily).

The conversion of Gd$_2$O$_3$ to GdCl$_3$ was determined by dissolving the entire reaction product in hot distilled water. The
solution was centrifuged and filtered. The residue was dissolved
in dilute HCl and oxalic acid was added to the solution to precipitate Gd$^{3+}$ as an oxalate. The solution was filtered and the oxalate
was ignited for 5 hrs at 850°C in a muffle furnace to convert to
Gd$_2$O$_3$. After ignition, the porcelain crucible containing Gd$_2$O$_3$
was moved into a desiccator, allowed to cool down, and weighed.
The percent conversion was obtained by

$$C(\%) = \frac{\text{Wt. initial Gd}_2\text{O}_3 - \text{Wt. of Gd}_2\text{O}_3 \text{ precipitated}}{\text{Wt. initial Gd}_2\text{O}_3} \times 100 \quad (2)$$

After determining important factors such as temperature,
time and mixing ratio of Gd$_2$O$_3$ and NH$_4$Cl for producing a 99.8%
conversion of oxide to chloride, the chloride was purified by distillation in vacuum using Ta crucibles and condensers.

## RESULTS AND DISCUSSION

The low yield of GdCl$_3$ when stoichiometric amounts of NH$_4$Cl
are used in the chlorination of Gd$_2$O$_3$ makes it necessary to add an
excess amount of NH$_4$Cl compared with the theoretical amount in
equation (1). The dependence on reaction time and excess NH$_4$Cl
for the conversion of Gd$_2$O$_3$ to GdCl$_3$ at 250°C is shown in Fig. 1,
in which the curves a, b, c and d are the results for the reactions
with the mixtures of 3.0, 2.5, 2.0 and 1.5 times the theoretical
amount of NH$_4$Cl in equation (1), respectively. As shown in Fig. 1,
almost complete conversion (97-99.8%) of Gd$_2$O$_3$ to GdCl$_3$ was obtained with 2.5 times or more the theoretical amount of NH$_4$Cl
(curves a and b) and with at least a 20 hr reaction time at 250°C.
The presence of excess NH$_4$Cl, by virtue of its acidic nature, pre-
vents the hydrolysis of the GdCl$_3$ in the reaction of Gd$_2$O$_3$ with
NH$_4$Cl. On the other hand, when only 1.5 times the theoretical
amount of NH$_4$Cl (curve d) was used, the percent conversion was
less than 73%. The reaction product contained the oxychloride
which might be formed according to the following reactions

$$GdCl_3(s) + H_2O(g) = GdOCl(s) + 2HCl(g) \quad (3)$$
$$GdCl_3(s) + \tfrac{1}{2}O_2(g) = GdOCl(s) + Cl_2(g) \quad (4)$$

The dependence of reaction temperature on the conversion of
Gd$_2$O$_3$ to GdCl$_3$ with 3.0 times the theoretical amount of NH$_4$Cl
is shown in Fig. 2, in which curves a, b, c, d, e and f are the results for the reactions at 350, 300, 250, 225, 200 and 150°C, respectively. In the reaction at 150°C, there was incomplete conversion of Gd$_2$O$_3$ to GdCl$_3$. The reactions at 350 and 300°C proceeded rapidly, but the yields of GdCl$_3$ decreased a little after a

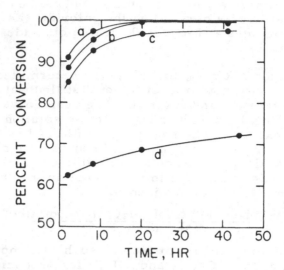

Fig. 1.  Dependence of reaction time on the conversion of $Gd_2O_3$ to $GdCl_3$ at 250° C; a: reaction of mixture containing 3.0 times of $NH_4Cl$, b: 2.5 times, c: 2.0 times, d: 1.5 times.

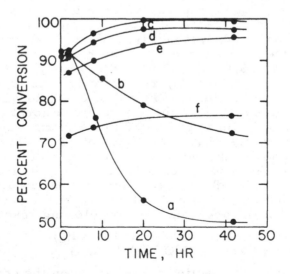

Fig. 2.  Dependence of the conversion of $Gd_2O_3$ to $GdCl_3$ on reaction temperature in reactions of mixtures containing 3.0 times of $NH_4Cl$; a: reaction at 350° C, b: 300° C, c: 250° C, d: 225° C, e: 200° C, f: 150° C.

longer reaction time than 2 hr. This tendency might indicate that the GdOCl was produced at these temperatures according to equations (3) and/or (4). This was also anticipated by Wendlant (16) from the results of thermal analysis for YCl$_3$ containing H$_2$O. At a reaction temperature around 250°C, however, almost complete conversion (97-99.8%) was obtained.

The effect of flow rate on degree of conversion of Gd$_2$O$_3$ to GdCl$_3$ is shown in Fig. 3, in which curves a and b are the results for the reactions with the mixtures of 3.0 times and 1.0 time the theoretical amount of NH$_4$Cl, respectively. In cases where a large excess of NH$_4$Cl (curve a) was used, the conversion % was not affected by changing the flow rate because a continuous amount of HCl(g) is available from the dissociation of the excess NH$_4$Cl. However, in cases where the exact amount of NH$_4$Cl was used, as shown by curve b in Fig. 3, the percent conversion increases, reaches a maximum with increasing flow rate and decreases again with further increasing flow rate. At slower flow rates, the H$_2$O formed in the reaction was not swept away and resulted in the formation of the oxychloride according to equation (3). At moderate flow rates, the degree of conversion reaches a maximum by appropriately sweeping away water vapor. At high flow rates the conversion was poor again due to a lower availability of HCl(g). This flow rate dependence shows that the reaction of Gd$_2$O$_3$ with NH$_4$Cl is a solid-gas reaction rather than a solid-solid reaction.

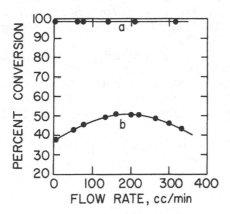

Fig. 3.  Effect of flow rate on degree of conversion of Gd$_2$O$_3$ to GdCl$_3$; a: reaction of mixture containing 3.0 times of NH$_4$Cl, b: 1.0 time.

The purification of GdCl₃ was done by distillation in vacuum using Ta crucibles and condensers. The temperatures of the distillation crucible and the condensation crucible were 860° C and 439° C, respectively. High yields (97. 9-98. 3%) were obtained after 33 hrs or more in the vacuum distillation of approximately 63g batches of crude chlorides.

## ACKNOWLEDGMENTS

This work was supported by the U. S. Department of Energy, Advanced Nuclear Systems and Projects Division. The authors would like to thank J. E. Powell, H. Burkholder and J. D. Corbett for their helpful comments and discussions.

## REFERENCES

1.   F. H. Spedding and A. H. Daane, J. Am. Chem. Soc. 74: 2783 (1952).
2.   H. J. Nolting, C. R. Simmons, and J. J. Klingenberg, J. Inorg. Nucl. Chem. 14:208 (1960).
3.   F. E. Block, T. T. Campbell, R. E. Mussler and G. B. Rodidart, U.S. Bur. Mines Rep. Invest. 5588 (1960).
4.   R. E. Musseler, T. T. Campbell, F. E. Block and G. B. Rodidart, U.S. Bur. Mines Rep. Invest. 6259 (1963).
5.   J. Mochinaga and K. Irisawa, Bull. Chem. Soc. (Japan) 47: 364 (1974).
6.   D. E. Cox and F. K. Fong, J. Crystal Growth 20:233 (1973).
7.   O. Pettersson, Z. Anorg. Chem. 4:1 (1893).
8.   J. F. Miller, S. E. Miller and R. C. Himes, J. Am. Chem. Soc. 81:4449 (1959).
9.   F. A. Schmidt and O. N. Carlson, U.S. Patent 3, 846, 121 (1974).
10.  F. Bourion, Ann. Chim. Phys. 21:49 (1910).
11.  G. Darzens and F. Bourion, Compt. rend. 153:270 (1911).
12.  L. A. Nisel'son, Yu. N. Lyzlov and K. V. Tret'yakova, Zhur. Neorg. Khim. 20:2362 (1975); Russ. J. Inorg. Chem. (Engl. transl) 20:1309 (1975).
13.  B. S. Hopkins, J. B. Reed and L. F. Audrieth, J. Am. Chem. Soc. 57:1159 (1935).
14.  J. L. Moriarty, J. Metals 20:41 (1968).
15.  J. B. Reed, B. S. Hopkins and L. F. Audrieth, in "Inorganic Syntheses," H. S. Booth, ed., McGraw-Hill, New York (1939).
16.  W. W. Wendlandt, J. Inorg. Nucl. Chem. 5:118 (1957).

SOLID-STATE REACTIONS OF SOME LANTHANIDE ORTHOPHOSPHATES WITH

SODIUM CARBONATE

M. Kizilyalli, Chemistry Department, Middle East Technical University, Ankara, Turkey;
and
A.J. E. Welch, Chemistry Department, Imperial College-
London, U.K.

## ABSTRACT

Solid-state reactions of lanthanum, gadolinium and cerium phosphates with sodium carbonate were investigated at 700° and 900°C.
If the stoichiometric amount of sodium carbonate was used, the reaction products were $Ln_2O_3$ and $Na_3PO_4$, but when the amount of sodium
carbonate was less than theoretical requirement, two intermediate
products, $Na_3Ln(PO_4)_2$ and $Ln_3PO_7$ were observed. These compounds
were found for the first time through these reactions. X-ray diffraction data for the intermediate phase is given.

## INTRODUCTION

The sintering reaction of monazite, which is an orthophosphate
of the rare earth elements and thorium, with sodium carbonate has
been known since 1932 (1). Kaplan and Uspenskaya (2) worked on
the sintering of monazite with sodium carbonate and lime. Kurup
and Moosath (3-7) claimed that the optimum amount of sodium carbonate to obtain lanthanide and thorium oxides was more than three
times the theoretical requirement. It has been found that by using
half of the theoretical amount of $Na_2CO_3$, it was possible to break
down monazite at 700°C (8). The separation of $CeO_2 + ThO_2$ from
other lanthanides was also achieved through these reactions. Solid-
state reactions of individual lanthanide phosphates with $Na_2CO_3$,
and the intermediate products formed, have not been described in
the literature.

The reaction between a lanthanide phosphate and sodium carbonate can be represented by the following equation:

$$2LnPO_4 + 3Na_2CO_3 \rightarrow Ln_2O_3 + 2Na_3PO_4 + 3CO_2$$

where Ln represents any rare earth element. If the stoichiometric
amount of sodium carbonate is used, the end product of the reaction
is expected to be the lanthanide oxide. The aim of the work described here was to investigate the reaction products when less
than the theoretical amount of sodium carbonate is used, and to
characterize intermediate products (8).


## EXPERIMENTAL


### Materials

Lanthanum, gadolinium and cerium orthophosphates were used in
these investigations. They were prepared in the laboratory by the
method of Buyers, et al. (9) and their purity was checked by determining their x-ray diffraction pattern and IR spectra. Full powder
data for LaPO$_4$ and GdPO$_4$ were published previously (10).


### Procedure

Several mixtures of each phosphate with sodium carbonate, in
compositions containing from 17.5% of the theoretical amount of
sodium carbonate to 10% excess of it were employed. The reactants
were weighed and ground together in an agate mortar. The sintering
operation was carried out in small platinum crucibles in an electric
muffle furnace around 700°C. In a few cases higher temperatures
were used to ensure a complete reaction. After the sintering operation the crucibles were cooled and weighed. The products were then
subjected to x-ray diffraction analysis. NaCl was used as an internal standard.

X-ray powder photographs were taken by cobalt or copper Kα
radiation. A Guinier type self-focusing camera (Guinier de Wolff
Camera No. II) was used. The results are summarized below.


## RESULTS


### Gadolinium and Lanthanum Orthophosphate Sintered with
### 10% Excess of Sodium Carbonate

The reaction proceeded in accordance with the equation given
above; the main products were lanthanide oxides and sodium orthophosphate. The γ-form of Na$_3$PO$_4$ was obtained when a reaction temperature higher than 700°C was employed (11). No unreacted

lanthanide phosphate was observed.  The experimental loss of weight
presumed to represent loss of carbon dioxide was in agreement with
the calculated value.  The products were extracted with cold water
and the insoluble residue filtered off, dried and heated to 700°C
again.  The x-ray diffraction film showed only lanthanide oxide
lines.  The loss of weight on extraction agreed well with the calculated weight of sodium orthophosphate required by the reaction.


   Sintering of Gadolinium Orthophosphate with 17.5, 25, 32.5,
40, 50, 60 and 75% of the Theoretical Amount of Sodium Carbonate

   Mixtures of reactants in several selected ratios were prepared
and sintered at 700°C.  In all cases there was a good agreement between experimental losses of weight and calculated values, assuming
complete expulsion of carbon dioxide from the reacting carbonate.

   An intermediate product was detected in all experiments employing the smaller proportions of sodium carbonate.  As the amount
of sodium carbonate employed was increased, the intermediate structure was observed to increase in quantity and the amount of unreacted gadolinium orthophosphate diminished.  With 40 to 50% of the
theoretical amount of sodium carbonate,the intermediate predomin-
ated, and very little unreacted gadolinium orthophosphate remained.
With 75% of the theoretical sodium carbonate, intermediate lines
were weaker, but gadolinium oxide and sodium orthophosphate lines
became very strong.  As before, no intermediate lines were observed
when a 10% excess of sodium carbonate was used.

   The product obtained by using 50% of the required sodium carbonate which had the largest content of the intermediate phase, was
extracted with water.  There was a small loss in weight after extraction of the product with water, but it was clear that the major part
of the sodium phosphate formed in the reaction of sodium carbonate
had remained combined in the insoluble intermediate phase.  As expected, the x-ray pattern of the water extraction residue after
heating at 700°C showed loss of the sodium orthophosphate lines,
but was otherwise identical with that of the product before extraction with water.  Evidently the intermediate phase was not appreciably soluble in water.  A more crystalline sample of the intermediate was obtained by heating the water extraction residue at
700°C overnight.  When this sample was heated to 950°C, some low
angle diffraction lines ascribed to the intermediate disappeared;
they were still present after heating at 900°C.  No detectable loss
of weight occurred in these heating operations at higher temperatures, indicating that the stoichiometry of the intermediate remained unchanged.  Table 1 contains the complete x-ray powder diffraction data of the intermediate phase at 900°C.  It contains
$Na_3Gd(PO_4)_2$, $Gd_3PO_7$ and another phase which is unknown (x phase).

Table 1.  The indexing of unidentified intermediate phase (x) in
          $GdPO_4$ + 50% $Na_2CO_3$ treatment, together with identified
          phase at 900°C.  Rad. $CuK_\alpha$, a = 14.547Å (for x-phase).

| $I/I_o$ | d | $\sin^2\theta_{obs.}$ | $\sin^2\theta_{calc.}$ | hkl | Remarks |
|---|---|---|---|---|---|
| 40 | 14.63 | 0.00278 | 0.00280 | 100 | x |
| 20 | 7.53 | 1098 | 1120 | 200 | x |
| B50 | 6.48 | 1412 | 1400 | 210 | x, $Gd_3PO_7$, $Na_3Gd(PO_4)_2$ |
| 50 | 4.59 | 2822 | 2800 | 310 | x |
| 5 | 4.40 | 3063 | 3080 | 311 | x, $Gd_2O_3$ |
| 100 | 4.14 | 3466 | -- | - | $Na_3Gd(PO_4)_2$, vaseline |
| B20 | 3.83 | 4043 | -- | - | $Na_3Gd(PO_4)_2$, $Gd_3PO_7$ |
| 10 | 3.54 | 4740 | 4760 | 410 | x |
| 25 | 3.47 | 4927 | -- | - | $Na_3Gd(PO_4)_2$ |
| 15 | 3.33 | 5359 | 5320 | 331 | x |
| 10 | 3.24 | 5637 | 5600 | 420 | x, $Na_3Gd(PO_4)_2$ |
| 10 | 3.165 | 5931 | 5880 | 421 | x, $Na_3Gd(PO_4)_2$ |
| 70 | 3.120 | 6109 | 6160 | 332 | $Gd_2O_3$, x, vaseline |
| 10 | 3.028 | 6478 | -- | - | $Na_3Gd(PO_4)_2$ |
| 25 | 2.936 | 6892 | -- | - | $Gd_3PO_7$ |
| 5 | 2.840 | 7363 | -- | - | $Gd_3PO_7$ |
| 70 | 2.770 | 7704 | -- | - | $Na_3Gd(PO_4)_2$ |
| 10 | 2.720 | 8024 | -- | - | $Na_3PO_4$ |
| 10 | 2.700 | 8143 | 8120 | 520 | x |
| 50 | 2.650 | 8456 | 8400 | 521 | x, $Na_3Gd(PO_4)_2$ |
| 5 | 2.570 | 8982 | 8960 | 440 | x |
| 10 | 2.459 | 9823 | 9800 | 531 | x |
| 10 | 2.300 | 11209 | -- | - | $Gd_2O_3$ |
| 20 | 2.255 | 11681 | -- | - | $Na_3Gd(PO_4)_2$ |
| 5 | 2.124 | 13172 | -- | - | $Gd_3PO_7$ |
| 20 | 2.107 | 13379 | -- | - | $Na_3Gd(PO_4)_2$ |
| 5 | 2.071 | 13858 | -- | - | $GdPO_4$ |
| 10 | 2.054 | 14082 | -- | - | $Na_3Gd(PO_4)_2$ |
| 10 | 1.997 | 14899 | 14840 | 720 | x, $Na_3Gd(PO_4)_2$ |
| 2 | 1.754 | 19313 | -- | - | $Gd_2O_3$ |
| 2 | 1.738 | 19659 | -- | - | $Gd_2O_3$, $Na_3Gd(PO_4)_2$ |
| 2 | 1.704 | 20463 | -- | - | $Gd_2O_3$, $Gd_3PO_7$ |
| B5 | 1.669 | 21314 | 21280 | 662 | $Gd_2O_3$, x |
| B5 | 1.650 | 21831 | 21840 | 752 | x, $Na_3Gd(PO_4)_2$ |
| 40 | 1.630 | 22360 | 22400 | 840 | x, $Gd_2O_3$ |
| 3 | 1.597 | 23305 | -- | - | $Gd_2O_3$ |
| 10 | 1.553 | 24631 | 24640 | 664 | x |
| 10 | 1.529 | 25402 | 25480 | 931 | x |
| w | 1.471 | 27472 | 27440 | 770 | x |
| Bw | 1.456 | 28019 | 28000 | 10.0.0 | x, $Na_3Gd(PO_4)_2$ |
| Bw | 1.209 | 40665 | 40600 | 12.1.0 | x |
| Bw | 1.177 | 42851 | 42840 | 12.3.0 | x |

*B:  Broad, w:  weak.

The x-phase can be indexed in the cubic system with a cell constant of 14.547Å.

The water extraction residue was also extracted with dilute hydrochloric acid and the insoluble residue was submitted to x-ray diffraction. Only gadolinium orthophosphate lines were observed, showing the solubility of the intermediate phase or phases in dilute hydrochloric acid. The appearance of the main intermediate phase at a composition providing only half the sodium carbonate required for complete reaction of the lanthanide phosphate, suggests that the important intermediate is a double phosphate possessing (at least approximately) the composition $GdPO_4 \cdot Na_3PO_4$ [or $Na_3Gd(PO_4)_2$]. The stoichiometry of the various reactions described is satisfactorily accounted for on this basis and later preparations of the double phosphates by solid state and precipitation reactions confirm the composition suggested (12,13). The additional lines in the powder photograph proved eventually to be those of $GdPO_4 \cdot Gd_2O_3$ ($Gd_3PO_7$)(14).

## Sintering of Lanthanum Orthophosphate with 50% of the Theoretical Amount of Sodium Carbonate

The x-ray pattern showed the presence of $Na_3La(PO_4)_2$ and $LaPO_4 \cdot La_2O_3$ as intermediates in the product of sintering at 700°C. The unidentified (cubic) phase which was observed in the corresponding gadolinium phosphate reaction was not present in this system.

## Sintering of the Cerium Orthophosphate with 50% of the Theoretical Amount of Sodium Carbonate

The sintering operation was carried out at 700°C for 24 hours. The x-ray diffraction pattern was very simple in comparison with those of the lanthanum and gadolinium products. The intermediate phase $Na_3Ce(PO_4)_2$ was present but no compound corresponding to the formula $CePO_4 \cdot Ce_2O_3$. Instead very strong $CeO_2$ lines were observed. Lines of cerous orthophosphate was also present. The loss of weight on sintering corresponded quite closely with the calculated value, adjusted for the oxidation of cerium to the tetravalent state. After heating at 900°C the product showed better crystallization of the intermediate phase and $CePO_4$ lines no longer appeared.

## GENERAL CONCLUSION

The occurrence of the double phosphates of the type $Na_3Ln(PO_4)_2$ as intermediate phases has been found for the first time in the sintering reactions of all three lanthanide phosphates with sodium carbonate. In the case of cerium the tendency of $Ce^{3+}$ to oxidize into

$Ce^{4+}$ favored the formation of cerium dioxide, so that the reaction tended to proceed further to the oxide.   The $LnPO_4 \cdot Ln_2O_3$ intermediate which was also prepared by solid-state reactions of $LnPO_4$ + $Ln_2O_3$ (14) did not occur at all in the case of cerium.

## REFERENCES

1.  O. Brunck and R. Höltje, Zeit. für. Angew. Chem. 45:331 (1932).
2.  G. E. Kaplan and T. A. Uspenskaya, Prod. 2nd U.N. Int. Conf. Geneva 3:378 (1958).
3.  K. N. N. Kurup and S. S. Moosath, Bull. Research Ins. Univ. of Kerala, Trivandurum Ser. A 5:9 (1957).
4.  K. N. N. Kurup and S. S. Moosath, Bull. Research Ins. Univ. of Kerala, Trivandurum Ser. A 5:15 (1957).
5.  K. N. N. Kurup and S. S. Moosath, Bull. Research Ins. Univ. of Kerala, Trivandurum Ser. A 5:21 (1957).
6.  K. N. N. Kurup and S. S. Moosath, Bull. Research Ins. Univ. of Kerala, Trivandurum Ser. A 6:1 (1958).
7.  K. N. N. Kurup and S. S. Moosath, Prod. of the Indian Acad. Science 48:76 (1958).
8.  M. Kizilyalli, PhD Thesis, University of London (1973).
9.  A. G. Buyers, E. Giesbrect and L. F. Audrieth, J. Inorg. Nucl. Chem. 5:133 (1957).
10. M. Kizilyalli and A. J. E. Welch, J. Appl. Cryst. 9:413 (1976).
11. M. Kizilyalli and A. J. E. Welch, J. Inorg. Nucl. Chem. 38:1237 (1976).
12. M. Kizilyalli and A. J. E. Welch, Rare Earths in Modern Science and Technology, Vol. 1, p. 209.  G. J. McCarthy and J. J. Rhyne (Eds.).  Plenum Press, NY (1978).
13. M. Vlasse, C. Parent, R. Salmon and G. LeFlem, Abst. 14th Rare Earth Res. Conf. North Dakota State University, Fargo, 25-28 June 1979.
14. M. Kizilyalli, METU Jour. of Pure Appl. Sci. 8:179 (1975).

# PREPARATION AND THERMAL STUDIES OF A CERIUM (III) SULFITE-SULFATE-HYDRATE

E. J. Peterson,* E. M. Foltyn, E. I. Onstott

Los Alamos Scientific Laboratory

MS 734, Los Alamos, New Mexico 87545, USA

Several thermochemical cycles for splitting water utilizing the Ce(III)-Ce(IV) couple have been demonstrated (1). Many of these cycles employ a Ce(III) compound which decomposes to gaseous products plus $CeO_2$ as the solid high temperature product. Recycle entails recombination of reagents except for one mol of $H_2O$ which is split into $H_2$ and $1/2 O_2$ in separate reactions. $CeO_2$ requires a reducing agent for formation of the Ce(III) compound which is subsequently decomposed. We investigated sulfurous acid as a reducing agent and report here some experimental results, and results on the characterization of the stable compound which is formed, $Ce_2(SO_3)_2SO_4 \cdot 4H_2O$. A method for synthesis of this new compound from Ce(III) carbonate and Ce(III) sulfate is reported also.

Ce(III) sulfite-sulfate compositions have not been reported in the literature previously. However, rare earth sulfite-sulfate-hydrates other than cerium are being investigated (2).

## EXPERIMENTAL SECTION

$CeO_2$ reacted slowly when stirred in sulfurous acid ($\sim 1M$) at 295K. At 375-385K, the reaction rate was much faster when stirred in a pressure vessel containing gaseous $SO_2$. Typically, 0.012 mol of $CeO_2$ was converted to $Ce_2(SO_3)_2SO_4 \cdot 4H_2O$ in 3 h in 97% yield on stirring in 1.1 mol of $H_2O$ with 0.02 mol of $SO_2$ (total) at 3.7 atm starting pressure. The final pressure was 2.0 atm. Over 90% of the pressure change took place in the time interval of one h, and is indicative of the reaction rate. The product was a white crystalline solid which was easily separated by filtration; it was washed with water, then air dried.

* Chemistry Division, Argonne National Laboratory, Argonne, IL 60439

For preparation of $Ce_2(SO_3)_2SO_4 \cdot 4H_2O$ from Ce(III), Ce(III) carbonate containing 0.200 mol of Ce (Apache Chemical Co.) was stirred in a flask containing one $\ell$ of sulfurous acid (6%, $\sim$ 1M) for 42 h at 295K.  A sweep gas of $SO_2$ at 0.75 atm was used.  Then Ce(III) sulfate (G. F. Smith Chemical Co.) containing 0.100 mol of Ce, dissolved in 0.5 $\ell$ of 1M sulfurous acid, was added and stirring continued for 16 h.  Excess $SO_2$ was removed by boiling at 356K for 5 h with $N_2$ substituted for the $SO_2$ sweep gas.  After cooling, the product was filtered, washed with one $\ell$ of water, then air dried.

Analysis for $Ce_2(SO_3)_2SO_4 \cdot 4H_2O$:  calc. wt.% Ce, 46.05; $SO_3^{2-}$ 26.32; $SO_4^{2-}$, 15.79; $H_2O$, 11.84; found, Ce, 46.18; $SO_3^{2-}$, 26.39; $SO_4^{2-}$, 14.56; $H_2O$ by difference, 12.87; by dynamic vac. to 630K, 11.81.  Ce was determined as $CeO_2$ after heating 1-2g samples in air at 1200K for 10 h.  A modified $I_3^-$ procedure was used for sulfite determinations.  The std. dev. for six samples was <0.5%.  Sulfate was determined by a barium sulfate procedure after oxidation of sulfite with aqueous $Br_2$.  Precision of this procedure was about 2%. A check of the $SO_4^{2-}$/Ce ratio in recrystallized $Ce_2(SO_4)_3 \cdot 8H_2O$ gave the value of 1.504.

## RESULTS AND DISCUSSION

$CeO_2$ reacts according to the stoichiometry:  $2 CeO_2 + 3 H_2SO_3 + H_2O \rightarrow Ce_2(SO_3)_2SO_4 \cdot 4H_2O$.  The product from this reaction gave the same X-ray diffraction pattern as the product from the alternate method of synthesis starting with Ce(III).  For the latter case, carbonate was converted first to bisulfite; then the product precipitated after adding sulfate and removing excess $SO_2$ to enhance sulfite formation.

Thermal decompositions of $Ce_2(SO_3)_2SO_4 \cdot 4H_2O$ to 1200K were done to establish reaction mechanisms and reaction stoichiometry.  Gaseous products were identified as $H_2O$, S, $SO_2$, and $O_2$.  $Ce_2O(SO_4)_2$, $Ce_2O_2SO_4$, $CeOSO_4$ and $CeO_2$ were identified as intermediate products. $CeO_2$ was the terminal solid product.  The quantitative results will be the subject of another communication.

Acknowledgements.  Funding was provided by the Department of Energy.  W. G. Witteman provided technical assistance, and M. G. Bowman and R. J. Bard provided valuable consultations.

## REFERENCES AND NOTES

(1)  E. J. Peterson, E. I. Onstott, M. R. Johnson and M. G. Bowman J. Inorg. Nucl. Chem. 40: 1357 (1978); C. M. Hollabaugh, E. I. Onstott, T. C. Wallace, Sr., and M. G. Bowman, Proceedings, 2nd World Hydrogen Energy Conf. 21-24 Aug. 1978, Zurich, Switzerland; E. J. Peterson and E. I. Onstott, J. Inorg. Nucl. Chem. 41: 517 (1979).
(2)  L. Niinisto, Helsinki Univ. Tech., Finland, Private Comm.

# SYNTHESIS AND CHARACTERIZATION OF $RE(Se_2O_5)NO_3 \cdot 3H_2O$,

## A NEW SERIES OF RARE EARTH COMPOUNDS

Lauri Niinistö, Jussi Valkonen and Paula Ylinen

Department of Chemistry, Helsinki University of Tech-

nology, Otaniemi, SF-01250 Espoo 15, Finland

## INTRODUCTION

Structure determinations by x-ray and neutron diffraction methods have shown that in metal selenites the anion may coordinate as a selenito, hydrogenselenito or diselenito ligand. Two rare earth selenite structures, $viz.$ $PrH_3(SeO_3)_2(Se_2O_5)$ and $Sc(HSeO_3)_3$ (1,2), have been determined so far and they illustrate the different forms of the selenite anion. The present work was initiated in continuation of the earlier studies (1-3), and it resulted in the synthesis and characterization of a hitherto unknown type of rare earth compounds.

## EXPERIMENTAL

The compounds were prepared at room temperature by slowly evaporating strongly acidic solutions which contained the trivalent rare earth, nitric acid and selenium dioxide. Chemical analysis (RE, Se, N, and $H_2O$) corroborated the formula $RE(Se_2O_5)NO_3 \cdot 3H_2O$ where RE = Y, Pr, Nd, Sm-Lu. The corresponding La, Ce or Sc compounds could not be prepared under the conditions investigated.

The crystalline precipitates were characterized by IR spectroscopy in the region 4000-300 $cm^{-1}$, by thermal analysis (TG, DTG, DTA) up to 1000°C, and by x-ray powder diffraction. The x-ray powder patterns were indexed and the unit cell parameters were calculated for all compounds.

The crystal structure of $Y(Se_2O_5)NO_3 \cdot 3H_2O$ was determined from single crystal data collected on a Syntex P2$_1$ diffractometer using graphite monochromatized MoK$\alpha$ radiation. The compound crystallizes in the orthorhombic space group $P2_12_12_1$ with four formula units in a cell of dimensions $a$ = 6.216(1), $b$ = 7.083(1), and $c$ = 20.616(5)Å. The structure was solved by direct methods and refined to an $R$-value of 0.051 for 1311 observed reflections, using anisotropic thermal parameters for all atoms. It was not possible to locate the hydrogen atom positions, however.

## RESULTS

Thermoanalytical studies indicate that the compounds are stable up to approximately 100°C where they begin to dehydrate in a single step. $RE(Se_2O_5)NO_3$ decomposes finally into an oxide through intermediate phases $RE_2O(SeO_3)_2$ and $RE_2O_2SeO_3$. The temperature of oxide formation depends on the ionic radii, being highest for praseodymium (>1000°C) and lowest for lutetium (920°C). The decomposition scheme is the same for all compounds.

X-ray diffraction studies confirm the isostructurality of the compounds. Structure determination of the yttrium compound shows that the coordination number is eight. Around yttrium there are four oxygens from the diselenite ligand at distances 2.324-2.359Å, three oxygens belonging to the water molecules at 2.369-2.442Å and one oxygen from the nitrate group at a distance of 2.459Å. The mean Y-O bond length is 2.38Å. The coordination polyhedron may be best described as a square antiprism.

The $YO_8$ polyhedra are connected into layers in the $xy$-plane through the bridging diselenite groups. Despite the tetradentate coordination the diselenite group maintains its symmetrical geometry and the bond distances and angles appear normal.

## REFERENCES

1.    M. Koskenlinna and J. Valkonen, Acta Chem. Scand., Sect. A
      31:638 (1977).
2.    J. Valkonen and M. Leskelä, Acta Crystallogr., Sect. B 34:1323
      (1978).
3.    E. Immonen, M. Koskenlinna, L. Niinistö and T. Pakkanen,
      Finn. Chem. Lett. 67 (1976).

# BINDING OF LANTHANIDES AND ACTINIDES BY POLYELECTROLYTES

Gregory R. Choppin

Department of Chemistry, Florida State University

Tallahassee, Florida 32306

## ABSTRACT

The complexation of lanthanide and actinide cations by polyelectrolyte ligands is reported. A synthetic carboxylic acid resin, BioRex-70, and a natural polyelectrolyte, humic acid, were
studied. A tracer method was used to avoid perturbation of the
polyelectrolyte structure by metal loading. The binding constants
for the metal-polyelectrolyte interaction are a function of the
degree of ionization of the polyelectrolyte. Comparison of the
analogous monomer ligand complexes show a considerable 'polyelectrolyte enhancement' which is discussed in relation to the enthalpy
and entropy of complexation.

## INTRODUCTION

The complexation of lanthanides and actinides in aqueous solution has been investigated for a wide variety of carboxylate ligands
ranging from acetate to diethylenetriaminepentaacetate. Interpretation of the thermodynamic parameters of complexation has provided
understanding of the role of ligand structure and basicity, hydration effects, metal ion charge density, etc. (1,2,3). We have
initiated similar studies with polyelectrolyte ligands with primary
emphasis on humic acid. As a principle component of soils throughout the world, humic acid is of major importance in the environmental behavior of metals. Humic acids are polymeric species whose
composition can be expected to vary somewhat with soil type. They
have at least two types of proton releasing groups, carboxylate
and phenolate. The cation exchange capacities vary between 3 and
5 meq per gm with the acid $pK_a$ values dependent somewhat on the

source of the humic acid and on the degree of ionization ($\alpha$).  Our earlier study of humic acid protonation indicated that the monomer analog of the principle carboxylate group is benzoate (4).  Perdue's similar study (5) is in agreement with this conclusion.

BioRex-70 is a synthetic acid resin with benzoate groups.  The thermodynamics of lanthanide and actinide binding to BioRex-70 resin was studied as a model polyelectrolyte system.

## Experimental

Humic acid.  Samples of humic acid were obtained from Aldrich Chemical Company, from sediment of a fresh water lake near Tallahassee, Florida and from A horizon (Chernozemic) soil near Joliot, Illinois.  The humic acid was isolated and purified by slight modification of standard procedures (6,7).  The BioRex-70 was obtained in the $Na^+$ form and converted to the $H^+$ form by washing with HCl solution.

Measurements.  Schubert's method (8) was used to obtain the binding constants for Eu(III) and Am(III) humate using Dowex-50 cation exchange resin as the counter phase.  For the other cations a solvent extraction technique was used with diethyhexylphosphoric acid dissolved in toluene as the organic phase.  The aqueous phase was adjusted to 0.1 M ionic strength with $NaClO_4$ and buffered with sodium acetate-acetic acid ($OAc_T$ = 0.05 M).  The binding constants to BioRex-70 were determined by Schubert's method.  The $^{152,4}Eu$ was counted in a NaI(Tl) well counter while $^{144}Ce$ and the alpha emitting nuclides $^{230}Th$, $^{233}U$, and $^{241}Am$ were counted in a liquid scintillation counter.

## Results

We can write the generalized reaction of a metal ion with humate anions as:

$$M + mX = MX_m \qquad \beta_m = (MX_m)/(M)(X)^m.  \qquad (1)$$

The distribution coefficient is:

$$\lambda = \frac{(M)_{org}}{(M)_{aq} + (MX_m)}  \qquad (2)$$

Defining

$$\lambda_0 = \frac{(M)_{org}}{(M)_{aq}} ,$$

$$\lambda = \frac{(M)_{org}}{(M) + \beta_m (M)(X)^{\overline{m}}} = \frac{\lambda_0}{1 + \beta_m (X)^{\overline{m}}} \qquad (3)$$

$$\log \frac{\lambda_0}{\lambda} - 1 = \log \beta_m + m \log(X). \qquad (4)$$

For humate systems, plots of log $(\lambda_0/\lambda-1)$ vs log (X) gave slopes between 1.3 and 1.8 ($\pm$0.1). Ardakani and Stevenson (9) have criti- cized the use of equation 4; however, their objections are valid for macroscopic quantitites of the metal ions but not for tracer concentrations. Analysis of our data by their modified method gave the same values of m. From these m values, we assume that the complexed species are MX and $MX_2$. This would not seem to be related to a stepwise sequence such as that found for monomers; but would seem to mean that the metal ions are associated simultaneously in 1:1 and 1:2 complexes in the polymer. So, in humic acid we inter- pret m>1 as indicating that some carboxylate sites bind singly to metal ions whereas other such sites are close enough together to bind in pairs.

Since we cannot a priori choose whether these two types of sites are identical or not, we designate the 1:1 sites as X and the 1:2 as Y. Then:

$$M + X = MX \quad \beta_x = (MX)/(M)(X) \qquad (5a)$$

$$M + 2Y = MY_2 \quad \beta_y = (MY_2)/(M)(Y)^2 \qquad (5b)$$

and equation 2 is modified to:

$$\lambda = \frac{(M)_{org}}{(M)_{aq} + (MX) + (MY_2)}$$

which leads to:

$$(\lambda_0/\lambda-1) = \beta_x (X) + \beta_y (Y)^2. \qquad (6)$$

Designating the total concentration of binding sites as [Z], we define $(X) = a(Z)$ and $(Y) = b(Z)$. Equation 6 becomes

$$(\lambda_0/\lambda - 1) = a\beta_x(Z) + b^2\beta_y(Z)^2 \qquad (7)$$

Plots of $(\lambda_0/\lambda-1)/(Z)$ vs $(Z)$ gives $a\beta_x$ as the intercept and $b^2\beta_y$ as the slope. The a and b values can be calculated from

$$a + 2b = m \qquad\qquad\qquad (8a)$$

$$a + b = 1 \qquad\qquad\qquad (8b)$$

The calculated values for $\beta_1(\equiv\beta_x)$ and $\beta_2(\equiv\beta_y)$ are listed in Table 1. These data are corrected for the effect of complexation by the acetate buffer (11) and are calculated based on the concentration of ionized humate. However, $\beta$ values are not corrected for the volume of the polymer. This is unnecessary for $\beta_1$ but according to Marinsky (12), for $\beta_2$ it may be as much as $10^3$. The thermodynamic parameters of binding of the metals to humate as determined from the temperature variation of log $\beta$ between 5° and 50° are listed in Table 2.

Table 1.   Binding Constants with Humate and BioRex-70.
$\mu = 0.10$ M; T = 25.0°C

a)   Humate

| Cation | $\alpha$ | log $\beta_1$ | log $\beta_2$ |
|---|---|---|---|
| Eu(III)[7] | 0.40[+] | 7.78±.04 | 10.70±.05 |
| Am(III)[7] | 0.40[+] | 7.26±.11 | 11.04±.02 |
| Th(IV)[10] | 0.25[+] | 11.14±.01 | 16.17±.02 |
| Th(IV)[10] | 0.43[+] | 12.03±.02 | 17.29±.04 |
| Th(IV)[10] | 0.54[+] | 13.18±.04 | 18.43±.17 |
| Th(IV)[10] | 0.37[‡] | 10.74±.01 | 15.79±.04 |
| Th(IV)[10] | 0.39[x] | 10.94±.02 | 16.43±.06 |
| $UO_2^{+2}$(11) | 0.48[x] | 7.28±.03 | 10.69±.04 |
| $UO_2^{+2}$(11) | 0.67[x] | 8.20±.03 | 11.55±.04 |

b)   BioRex-70

| Cation | $\alpha$ | log $\beta_3$ |
|---|---|---|
| Ce(III)[13] | 0.00[x] | 6.99±.09 |
| Eu(III)[13] | 0.00[x] | 7.25±.09 |
| Am(III)[13] | 0.00[x] | 7.46±.09 |

[+]Lake sediment humic acid
[‡]Illinois humic acid
[x]Aldrich humic acid

The sorption experiments with BioRex-70 were conducted at pH 3.5 so the reaction was:

$$M^{+3}_{(aq)} + 3HR_{(r)} = MR_{3(r)} + 3H^{+}_{(aq)} \qquad (9)$$

A third power dependence of extraction on pH confirmed this equation as expressing the reaction. The acid dissociation constant $pK_a$ at $\alpha=0$ was determined to be $5.90\pm0.1$ from which the binding constant for reaction 10 was calculated.

$$M_{(aq)} + 3R_{(r)} = MR_{3(r)} \qquad (10)$$

The binding constants and the corresponding thermodynamic parameters are listed in Tables 1 and 2, respectively.

Table 2.  Thermodynamic Parameters of Binding.
$\mu = 0.10$ M; T = 25.0°C

a)  Humate

| Cation | $\alpha$ | $-\Delta G_1^{+}$ | $\Delta H_1^{+}$ | $\Delta S_1^{x}$ |
|---|---|---|---|---|
| Am(III) | 0.40 | 41.41 | -19 ±3 | 73±11 |
| Th(IV) | 0.25 | 63.64 | 32.6±3.2 | 321±10 |
| $UO_2^{+2}$ | 0.48 | 41.56 | - 2.7±0.3 | 130±15 |

b)  BioRex-70

| | $\alpha$ | $-\Delta G_3$ | $\Delta H_3$ | $\Delta S_3$ |
|---|---|---|---|---|
| Ce(III) | 0.00 | 40.02 | 4.6±1.5 | 149±14 |
| Eu(III) | 0.00 | 41.50 | 10.7±2.4 | 175±18 |
| Am(III) | 0.00 | 42.67 | 13.3+3.3 | 187±22 |

$^{+}$kJ/m

$^{x}$J/m/K

## Discussion

The analogous monomer ligand for both humic acid and BioRex-70 is benzoic acid. For benzoate complexes, the values are:  Eu(III), log $\beta_1$=2.15 (14), log $\beta_3$ 4.6~(est.); Th(IV), log $\beta_1\approx4.0$ (est.) (10); $UO_2^{+2}$, log $\beta_1$=2.61 (15). The comparison of log $\beta$(Eu-BioRex-70) of 7.25 with log $\beta_3$ of 4.6 indicates an enhancement in the resin binding

of about a factor of 500. The apparent $pK_a$ ($\alpha=0$) of BioRex-70 is
5.9 compared to 4.1 for benzoic acid so the greater binding of the
resin would seem to be related directly to the greater basicity of
the carboxylate groups. Moreover, we can estimate that $\log \beta_1$(Eu)
for BioRex-70 would be ~3-3.5. To compare this value with the humate
values in Table 1 we must estimate $\log \beta(\alpha=0.5)$.

Marinsky has reviewed protonation and metal binding in poly-
electrolytes (12). He relates $\alpha$, the degree of ionization and $pK_a$
at $\alpha=0$ ($pK^o$) by:

$$pH - \log \frac{\alpha}{1-\alpha} = pK^o - \frac{2Z^- w}{2.303} \qquad (11)$$

where $Z^-$ is the charge on the polyelectrolyte and w is a complex
function involving the radius of the polymer, the distance of closed
approach, the dielectric constant, etc, w should not vary over small
ranges of $\alpha$. From equation 11 it is difficult to estimate the
variation of $Z^-$ with $\alpha$ but using series expansion, we find approxi-
mately $Z^- \propto \alpha$. We can develop an expression whereby $\delta \log \beta(\alpha)/\delta\alpha \propto Z^+ \cdot n$
where $Z^+$ is the cation charge and n is the slope of the $\log \beta(\alpha)$ vs
$\alpha$ curve. For protonation of BioRex-70, n-1.5. This leads to an
estimate of $\log \beta_1$~3.2 + 4.5 = 7.7 at $\alpha$ = 0.5 for Eu(III). This
value is reasonably close to the $\log \beta_1$ values for Eu(III) at
$\alpha$ = 0.5, reaffirming the identification of the primary binding site
in humic acid as a benzoate group. These estimates also reflect
the relation between the "polyelectrolyte enhancement" and the degree
of ionization which increases the charge of the polyelectrolyte.

The enthalpy and entropy data for humate complexing in Table 2,
within error limits, have a linear correlation with a slope of
about 200°. This suggests (2) that these values are dominated by
the cation dehydration upon complexation. Moreover, for Am(III)
the $\Delta S$ ($\alpha=.5$) for humate is comparable to the $\Delta S$ for Eu-fumarate
complexing in which it has been proposed that complexaton occurs
via one carboxylate with an effective charge of -2 (16).

This study has considered only one of the many complexing species in the environment. We have considered only soluble species
whereas the high binding constants for humic acid would indicate
that the related insoluble species (humin) in soils would retain
rather stronly lanthanide and actinide cations. By contrast, soluble humic acid as well as the even more soluble fulvic acids would
mobilize these cations by transport through solution. In solution,
hydrolysis and complexing by carbonate would be the principle competition.

The environmental behavior of plutonium is of major concern.
In Figure 1, the competition between humate complexation and

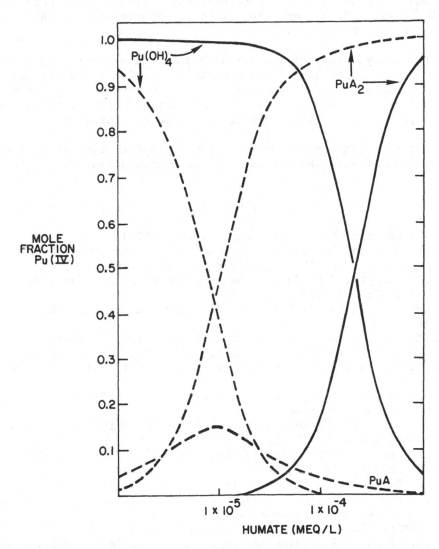

Figure 1. Variation of Pu(IV) species. Solid lines are pH 7 while dashed lines are pH 6.

hydrolysis for Pu(IV) is shown, using hydrolysis constants from
Baes and Mesmer (17) for mononuclear species only.  Humate concentrations in natural waters are often in the range $10^{-5}$ - $10^{-4}$ meq/l.
We see that for pH 6, humate complexing is dominant above [Hu] =
$10^{-5}$ meq/l whereas at pH 7 humate does not dominate until [Hu] $\geq$
2 x $10^{-4}$ meq/l.  In either case, it is the doubly bound (PuA$_2$) species which is significant.

In summary, study of lanthanide and actinide binding to poly-
electrolytes has provided thermodynamic data which is consistent
with the general model of complexing of these ions whereby ionic
intereaction is predominant and the enthalpy and entropy terms are
related primarily to cation dehydration upon complexation.

This research was supported by the U.S.D.O.E. through Contract
Ey-76-S-05-1797 with the Office of Basic Energy Sciences.

## REFERENCES

1.  G.R. Choppin, Pure Appl. Chem., 27:23 (1971).
2.  G.R. Choppin, Proc. 12th Rare Earth Res. Conf., C.E. Lundin,
    Ed., Vol. 1, pp 130-139, University of Denver, 1976.
3.  G.R. Choppin, M.P. Goedkin and T.F. Gritmon, J. Inorg. Nucl.
    Chem., 39:2025 (1977).
4.  G.R. Choppin and L. Kullberg, J. Inorg. Nucl. Chem., 40:651
    (1978).
5.  E.M. Perdue, Geochim. Cosmochim. Acta, 42:1351 (1978).
6.  M. Schnitzer and S.I.M. Skinner, Soil Sci., 105:392 (1968).
7.  E.L. Bertha and G.R. Choppin, J. Inorg. Nucl. Chem., 40:655
    (1978).
8.  J. Schubert, J. Phys. Coll. Chem., 52:340 (1948).
9.  Ardakani and Stevenson
10. K.L. Nash, Ph. Dissertation, Florida State University, Talla-
    hassee, Florida, 1979.
11. P.M. Shanbhag, Ph. Dissertation, Florida State University,
    Tallahassee, Florida, 1979.
12. J.A. Marinsky, Coord. Chem. Rev., 19:125 (1976).
13. F. Ohene-Aniapam, M.S. Thesis, Florida State University, 1979.
14. Y. Hasegawa, private communication.
15. J. Stary, Coll. Czech. Chem. Comm., 25:2630 (1960).
16. G.R. Choppin, A. Dadgar and R. Stampfli, J. Inorg. Nucl. Chem.
    35:875 (1973).

# THE TREND OF THE $Eu^{3+}$ - LIGAND BONDING IN COVALENT AND IONIC RARE-EARTH COMPOUNDS

F. Gaume, C. Linarès, A. Louat and M. Blanchard

Spectroscopy and Luminescence Laboratory, Lyon I University

43 Bd du 11 Novembre 1918, 69621 Villeurbanne, France

## ABSTRACT

This paper shows the influence of ionicity on the $e_\sigma$ and $e_\pi$ antibonding effects in some $Eu^{3+}$ doped compounds : $KY_3F_{10}$, $LaCl_3$ and $YMO_4$ (M=P,As,V).

In the ligand field approximation, the $e_\sigma$ and $e_\pi$ parameters are determined for each kind of ligands, using the angular overlap model of Jørgensen. The obtained $e_\sigma$ and $e_\pi$ must not only fit the crystal field energy values expressed in terms of experimental $B_n^m$ (deduced from fluorescence spectra) but also each $B_n^m$.

It is observed that the importance of the $\pi$ effects (compared to the $\sigma$ effects) increases with ionicity for all the $Eu^{3+}$ doped compounds whatever the ligand may be. Moreover, the contribution of the $\pi$ effects increases appreciably as $Eu^{3+}$ is replaced by $Er^{3+}$.

---

In order to follow the trend of the $Eu^{3+}$ - ligand bonding in ionic or covalent rare-earth compounds, it is interesting to compare the respective importance of the axial and the non axial antibonding effects $e_\sigma$ and $e_\pi$ . Indeed, starting from the hypothesis that the rare-earth experiences a ligand field, C.K. Jørgensen and al.(1)(2) have introduced the angular overlap parameters $e_\sigma$ and $e_\pi$ to describe the bonding. The $\sigma$ bonding is directed along the rare-earth ligand axis, while the $\pi$ bondings are perpendicular to this axis with two perpendicular components.

This paper presents the $e_\sigma$ and $e_\pi$ calculations in the serie: $LaCl_3$, $KY_3F_{10}$, $YMO_4$ (M=P,As,V) doped with $Eu^{3+}$ and occasionally with $Er^{3+}$.

Table 1

| Compounds and Eu$^{3+}$ site symmetry | Ligands | $R_j$ Å | $\Theta_j$ | $\varphi_j$ | | | |
|---|---|---|---|---|---|---|---|
| LaCl$_3$ D$_{3h}$ | 3 Cl$^-$ | 2,97 | 90 | - 60,4 | 59,6 | 179,6 | |
| | 6 Cl$^-$ | 2,99 | 43,13 | -118,1 | 1,9 | 121,9 | |
| | | | 136,87 | -118,1 | 1,9 | 121,9 | |
| KY$_3$F$_{10}$ C$_{4v}$ | 4 F$^-$ | 2,193 | 59,97 | 0 | 90 | 180 | 270 |
| | 4 F$^-$ | 2,331 | 130,81 | 45 | -135 | 135 | -45 |
| YPO$_4$ D$_{2d}$ | 4 O$^{2-}$ | 2,313 | 76,33 | -45 | 135 | | |
| | | | 103,67 | 45 | -135 | | |
| | 4 O$^{2-}$ | 2,374 | 30,22 | 45 | -135 | | |
| | | | 149,78 | -45 | 135 | | |
| YAsO$_4$ D$_{2d}$ | 4 O$^{2-}$ | 2,300 | 77,80 | -45 | 135 | | |
| | | | 102,20 | 45 | -135 | | |
| | 4 O$^{2-}$ | 2,412 | 31,88 | 45 | -135 | | |
| | | | 148,12 | -45 | 135 | | |
| YVO$_4$ D$_{2d}$ | 4 O$^{2-}$ | 2,291 | 78,10 | -45 | 135 | | |
| | | | 101,90 | 45 | -135 | | |
| | 4 O$^{2-}$ | 2,433 | 32,83 | 45 | -135 | | |
| | | | 147,17 | -45 | 135 | | |

To apply this model, it is necessary to know the site symmetry of the doping ion and to calculate the polar coordinates $(R_j, \Theta_j, \Psi_j)$ of each ligand $j$ (table 1).

The scheme of the calculation method is to equal the overlap angular energy expressed in terms of $e_\sigma$ and $e_\pi$ and the crystal field energy which depends on the phenomenological $B_n^m$ parameters deduced from the experimental $Eu^{3+}$ spectra. So, for each $f_i$ orbital, we have the following relation :

$$\sum_{n,m} \alpha_{(fi,n,m)} B_n^m = \sum_j a_{fi}(\Theta_j \Psi_j) e_{\sigma j} + b_{fi}(\Theta_j \Psi_j) e_{\pi j}$$

where the $\alpha_{(fi,n,m)}$ are numerical coefficients calculated from 3j symbols and the $a_{fi}(\Theta_j \Psi_j)$ and $b_{fi}(\Theta_j \Psi_j)$ are ligand angular functions, the seven wave functions of the rare earth ion being delocalized on the ligands.

In the $D_{3h}$, $C_{4v}$ or $D_{2d}$ symmetries of the presently studied compounds, the degeneracy of these f orbitals is not completely lifted ; two double representations are remaining (table 2).

From that, we get five equations with four unknowns $e_{\sigma 1}$, $e_{\pi 1}$, $e_{\sigma 2}$, $e_{\pi 2}$ since all the compounds are characterized by two rare earth - ligand distances $R_1$ and $R_2$. The $e_\sigma$ and $e_\pi$ values which solve the best these five equations are obtained by a fitting process. A good test of the validity of the method is to use these $e_\sigma$ and $e_\pi$ values to calculate again the original $B_n^m$ parameters ; if it happens that the original $B_n^m$ is not reobtained, we then adjoin to the five equations the Kibler equations (3)(4) giving each $B_n^m$ directly as a function of the different $e_{\sigma j}$ and $e_{\pi j}$.

We have observed that generally a correct fit is obtained only if the $B_2^0$ phenomenological parameter is modified. This looks to be due to the starting hypothesis which involves essentially a ligand field while the $B_2^0$ parameter is well known to be very sensitive to the contribution of more distant ions. Urland (5) has recently arrived to the same conclusion in studying $LaCl_3:Eu^{3+}$ but with the restrictive hypothesis $e_{\pi 1}/e_{\sigma 1}=e_{\pi 2}/e_{\sigma 2}$. The $B_2^0$ value is then chosen as being the nearest one from its experimental value which allows a good resolution of the system of equations. The experimental and modified values of this parameter are reported on table 3.

Table 2

$$f \longrightarrow \begin{cases} D_{3h} : & A_1' + A_2' + A_2'' + E' + E'' \\ C_{4v} : & A_1 + B_1 + B_2 + 2E \\ D_{2d} : & A_1 + A_2 + B_2 + 2E \end{cases}$$

F. GAUME ET AL.

Table 3

|  |  | $KY_3F_{10}$ | $LaCl_3$ | $YPO_4$ | $YAsO_4$ | $YVO_4$ |
|---|---|---|---|---|---|---|
| $Eu^{3+}$ | $B_2^0$ exp | - 264 | 89 | 150 | - 61 | - 54,5 |
| | $B_2^0$ modified | - 22 | - 20 | - 15 | - 40 | - 54,5 |
| $Er^{3+}$ | $B_2^0$ exp | | 94 | 141 | - 30,5 | -102,8 |
| | $B_2^0$ modified | | 6 | 0 | - 10 | -102,8 |

It can be seen that the more ionic is the compound, the more modified the $B_2^0$ must be ; for $KY_3F_{10}$ especially, no fit is possible without this modification.

The $e_\sigma$ and $e_\pi$ obtained are given in table 4. In order to discuss the results, we suggest to classify the compounds at first according to $R_1$, the first ligand - rare earth distance (table 5a), then according to their ionicity degree, roughly represented by the evolution of the $^5D_0$ level energy (table 5b).

From table 5a, it arises that whatever is the ligand, $F^-$, $Cl^-$ or $O^{2-}$, $e_{\sigma 1}$ is always decreasing as $R_1$ is increasing. We observe a high value of $e_\sigma$ for the ligand fluorin, in agreement with previous results on transition metal and actinide ions (6)(7)(8). For the second ligands, if the comparison is restricted to a serie of compounds having the same symmetry, such as the $YMO_4$ serie, $e_{\sigma 2}$ decreases as $R_2$ increases (table 4).

Table 4

| $Eu^{3+}$ | $e_{\sigma 1}$ | $e_{\pi 1}$ | $e_{\sigma 2}$ | $e_{\pi 2}$ | $R_1$ | $R_2$ |
|---|---|---|---|---|---|---|
| $LaCl_3$ | 229,2 | 51,2 | 224,4 | 0,3 | 2,97 | 2,99 |
| $KY_3F_{10}$ | 388 | 138 | 285 | 86 | 2,193 | 2,331 |
| $YPO_4$ | 261,9 | 44,2 | 234,9 | -19,5 | 2,313 | 2,374 |
| $YAsO_4$ | 269,6 | 4,9 | 195,9 | -26,1 | 2,300 | 2,412 |
| $YVO_4$ | 272,2 | - 5,5 | 191,5 | -35,3 | 2,291 | 2,433 |

Table 5a and 5b          5a                                                 5b

| Compound Eu$^{3+}$ | $R_1$ | $e_{\sigma 1}$ | Compound Eu$^{3+}$ | $^5D_{0_1}$ cm$^{-1}$ | $\dfrac{e_{\pi 1}}{e_{\sigma 1}}$ | $\dfrac{e_{\pi 2}}{e_{\sigma 2}}$ |
|---|---|---|---|---|---|---|
| KY$_3$F$_{10}$ | 2,193 | 388 | YVO$_4$ | 17 183 | - 2 % | - 19 % |
| YVO$_4$ | 2,291 | 272,2 | YAsO$_4$ | 17 210 | 1,5% | - 13 % |
| YAsO$_4$ | 2,300 | 269,6 | YPO$_4$ | 17 211 | 17 % | - 8 % |
| YPO$_4$ | 2,313 | 261,9 | LaCl$_3$ | 17 267 | 22 % | 0 % |
| LaCl$_3$ | 2,97 | 229,2 | KY$_3$F$_{10}$ | 17 269 | 36 % | 30 % |

On table 5b, a comparison of the $\pi$ effects relatively to the $\sigma$'s has been made. The ratios $e_{\pi 1}/e_{\sigma 1}$ as well as $e_{\pi 2}/e_{\sigma 2}$ clearly increase with ionicity. The percentage of the $\pi$ effects is particularly high for the fluoride as it is also observed on the transition metal ions (6)(7).

It seemed of interest to test these conclusions with an other trivalent doping ion, such as Er$^{3+}$. The results are reported on table 6. The observed discrepancy with LaCl$_3$ for the first ligands can be explained by the ion sizes. Indeed, those of Y$^{3+}$ and Er$^{3+}$ are very close to each other but that of La$^{3+}$ is much larger. It is not the case for Eu$^{3+}$, the ionic radius being then intermediate between that of La$^{3+}$ and Y$^{3+}$. In spite of that discrepancy, the values of $e_{\pi 1}/e_{\sigma 1}$ and $e_{\pi 2}/e_{\sigma 2}$ reported on tables 5 and 6 show a high percentage of $\pi$ antibonding effects in the rare earth - ligand bonding for ionic compounds.

At the end, a comparison between the tables 5b and 6 points out that the percentage of $\pi$ antibonding effects is always larger in the Er$^{3+}$ - ligand bonding than in the Eu$^{3+}$ - ligand bonding in agreement with the fact that the ion - ligand distances are shorter with Er$^{3+}$.

Table 6

| Compounds | YVO$_4$ | YAsO$_4$ | YPO$_4$ | LaCl$_3$ |
|---|---|---|---|---|
| $\dfrac{e_{\pi 1}}{e_{\sigma 1}}$ | 25 % | 36 % | 51 % | 40 % |
| $\dfrac{e_{\pi 2}}{e_{\sigma 2}}$ | 5 % | 15 % | 23 % | 28 % |

REFERENCES

1. C.E. Schäffer, C.K. Jørgensen, The Angular Overlap Model, Mol. Phys. $\underline{9}$, 401 (1965).

2. D. Küse, C.K. Jørgensen, Angular Overlap Treatment of erbium (III) in xenotime and yttrium Orthovanadate, Chem. Phys. Letters $\underline{1}$, 314 (1967).

3. M.R. Kibler, Comparison between the Point-charge Electrostatic Model and the Angular Overlap Model : Int. J. Quantum Chem. $\underline{9}$, 403 (1975).

4. M.R. Kibler, G. Grenet and R. Chatterjee, On the Interpretation of Crystal-field Parameters with additive Models, J. of Luminescence $\underline{18/19}$, 609 (1979).

5. W. Urland, The Interpretation of the Crystal-field Parameters for $f^n$ Electron Systems by the Angular overlap Model. Chem. Phys. Letters $\underline{53}$, 2, 296 (1978).

6. J. Glerup, O. Monsted, C.E. Schäffer, Nonadditive and additive Ligand Fields, Inorg. Chem. $\underline{15}$, 6, 1399 (1976).

7. J. Josephsen, C.E. Schäffer, The Position of 2,2'-Bypyridine and 1,10-Phenanthroline in the Spectrochemical Series, Acta Chem. Scandin., $\underline{A31}$, 813 (1977).

8. D. Brown, B. Whittaker, N. Edelstein, Spectral Properties of the Octahedral $(NEt_4)_2$ $PaF_6$, Inorg. Chem. $\underline{13}$, 1805 (1974).

# STUDY OF COMPLEX FORMATION OF THE LANTHANIDES BY MEANS OF FLUORESCENCE SPECTROSCOPY

B.A.Bilal

Hahn-Meitner-Institut für Kernforschung GmbH

D - 1000 Berlin 39, FRG

## INTRODUCTION

The absence of the covalent interaction between the rare earth ions and the ligands usually leads to no band shift in the electron absorption spectra due to complexation. Only small variation of the peak intensities is obtained, even at high concentration of the central ion. The determination of complex formation constants from such intensity difference is therefore very difficult, or even impossible for aqueous rare earth systems having low solubility products (e.g. fluoride, hydroxide, oxalate, carbonate). A concentration of not more than $10^{-5} - 10^{-6}$ M of the central ion has to be used to obtain the complex formation function of these systems.

The fluorescence intensity, in contrary, is sensitively reduced (or enhanced) by complex formation, even at such low concentration of the fluorescent species. This makes it possible to investigate the complex formation function up to a relatively high concentration of the ligand, and so to determine the stability constants $\beta_i$ of the different complex species formed in a successive complex formation system.

The complex formation function of the reaction

$$M + iA \rightleftharpoons MA_i \; ; \; \beta_i = [MA_i]/[M][A]^i \; ; \tag{1}$$

$\beta_o = 1$ ; ($i = 1,2....N$); M = metal; A = ligand, is given by the equation

83

$$\bar{n} = \sum_{i=o}^{i=N} i \ \beta_i [A]^i \ / \ \sum_{i=o}^{i=N} \beta_i \ [A]^i \qquad\qquad (2)$$

$\bar{n}$, the mean ligand number, is determined by means of the method of corresponding solutions, developed by Bjerrum(1) for evaluation of the formation function from spectrophotometric measurements. Applying this method to fluorescence measurement, the following premises must be fulfilled:

1.   The ligand must have no, or very low, fluorescence intensity at the emission range of the central ion.

2.   The measurement must be carried out in neutral salt medium of high ionic strength, in order to apply the mass action low in concentration terms.

3.   The fluorescence intensities of the individual complex species must contribute additively to the whole measured intensity according to

$$Q = I_o \ (2,3)d \sum_{i=o}^{i=N} c_i \ \epsilon_i \ \phi_o \qquad\qquad (3)$$

where Q = measured fluorescence intensity, $I_o$ = intensity of the excitation light, d = cell length, c = concentration, $\epsilon$ = molar absorptivity and $\phi$ = fluorescence efficiency.

To evaluate the formation function of the reaction (1), it is then only necessary to measure the fluorescence intensity of two sets of samples having different concentration of the central ion $C_M'$ and $C_M''$ in absence of the ligand (intensity = $Q_o$) and in presence of increasing concentration of the ligand (intensitiy = Q). Two samples having the same ratio $Q/Q_o$ have also the same percentual distribution of the various species containing M. That means the two samples have the same values of $\bar{n}$ and [A], since $\bar{n}$ (if no polynuclear complexes are formed) is only a function of [A]. $\bar{n}$ and [A] are then calculated according to

$$\bar{n} = \frac{C_A' - [A]}{C_M'} = \frac{C_A'' - [A]}{C_M''} = \frac{C_A' - C_A''}{C_M' - C_M''} \qquad\qquad (4)$$

and

$$[A] = \frac{C_M' \ C_A'' - C_M'' C_A'}{C_M' - C_M''} \qquad\qquad (5)$$

where $C_A^{'}$ and $C_A^{''}$ are the total concentrations of the ligand in the two samples.


## EXPERIMENTAL

The method was tested for determination of the stability constants of the fluoro complexes of $Eu^{3+}$ and $Tb^{3+}$. First the linear range of the fluorescence intensity as a function of the concentration of each element (in the absence of the ligand) was determined at the ionic strength 1.0 M (NaCl medium), pH = 5.4 and T = 25°C. Different sets containing each element at various concentration (within the linear range) were measured under the same conditions. Each set consisted of 20 samples (1 sample without fluoride and 19 samples with fluoride increasing in the concentration from $10^{-5}$M to $10^{-3}$ M). Eu was excited at 395 nm and Tb at 225 nm. For calculation of the ratio $Q/Q_0$ the intensities of the Eu peak at 590 nm and of the Tb peak at 545 nm were measured.


## RESULTS AND DISCUSSION

Figure 1 shows (as example) $\dfrac{Q}{Q_0}$ as a function of the concentration of fluoride $C_F$ for 4 sets of samples which contain Tb in the concentration of $1.5 \times 10^{-4}$M, $1.0 \times 10^{-4}$M, $0.5 \times 10^{-4}$M and $0.1 \times 10^{-4}$M. The curves descend gradually till there is a sudden fall at the point where the precipitation of fluoride takes place. At low Tb concentration ($10^{-5}$M) a plateau was found just before the precipitation occurred. This step indicates that the maximum ligand number obtainable till precipitation takes place is established.

The values of $\bar{n}$ and [F] were calculated according to the equations (4) and (5) from the $C_M^{'}$ and $C_M^{''}$ values of two curves and the $C_F^{'}$ and $C_F^{''}$ values at various points, at which they intersect different horizontal lines. The obtained [F] values are in agreement with those measured by means of fluoride membrane electrode within an error range of ±1%.

The stability constants $\beta_i$ of the fluoro complexes of the two lanthanides are calculated from the obtained formation function by means of a linear regression analysis using an ALGOL program. Table 1 presents the results and the corresponding values found by means of the potentiometric measurements.

The values of the stability constants of the mono- and difluoro complexes are quite in agreement with those obtained by means of the potentiometric and distribution methods.

B.A. BILAL

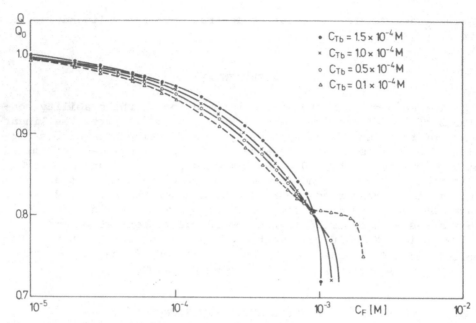

Figure 1.   The variation of the relative fluorescence intensity of
            Tb due to complex formation.


Table 1:  The stability of constants of the fluoro complexes of
          the lanthanides

| Central ion | $\beta_1$ $[M^{-1}]$ fluorimetric | potentiometric | $\beta_2$ $[M^{-2}]$ fluorimetric | potentiometric |
|---|---|---|---|---|
| $Eu^{3+}$ | 1210±90 | 1190±70 | $(3.2\pm0.3)\,10^5$ | $(3.2\pm0.2)\cdot10^5$ |
| $Tb^{3+}$ | 1910±120 | 1930±100 | $(2.3\pm0.2)\,10^5$ | $(2.5\pm0.2)\cdot10^5$ |

ACKNOWLEDGMENT


        The author is very thankful to Mr. E. Mueller for his accurate
assistance in the experimental work.

REFERENCE

1.  J. Bjerrum, A new optical principle for investigation of step
    equilibria.  Kgl. Danske Videnskab.  Selskab. math.-fys. Medd.
    21:  No. 4 (1944).

MAGNETIC CIRCULAR DICHROISM (MCD) OF LANTHANIDES.

STUDY OF THE $Eu^{3+}$ ION IN AQUEOUS SOLUTION

C. Görller-Walrand and Y. Beyens

Universiteit Leuven, Afdeling Anorganische en Analytische

Scheikunde, Celestijnenlaan 200F, 3030 Heverlee, Belgium

The nature of the complex species present in aqueous solution
of rare earth salts has been the subject of considerable discussion
in the literature for a number of years; yet no definite conclusion
has been reached. Questions still raise as whether the coordination
number is eight, nine or even less (1,2,3,4), whether only one (5)
or several species (6) are present in solution, whether the coordination geometry is trigonal or tetragonal (7).

It is the purpose of this paper to consider some of these aspects in the framework of the information provided by magnetic circular dichroism (MCD). We consider the example of the $^5D_1 \leftarrow ^7F_0$ and
$^5D_2 \leftarrow ^7F_0$ transitions in the MCD spectrum of the hydrated $Eu^{3+}$ ion.
The discussion about symmetry hypothesis will be based on coordination data proposed in the literature (8,9,10) and especially on a
coordination model developped in a recent review of high coordination complexes (11).

SYMMETRY OF EIGHT AND NINE COORDINATION COMPOUNDS

The two most frequently observed 8 coordinated polyhedra are
the dodecahedron (Dod) with symmetry $D_{2d}$ and the square antiprism

(SAP) with symmetry $D_{4d}$. Moreover another symmetry $C_{2v}$ or the bi-
capped trigonal prism (BCTP) seems often adequate to describe also
the arrangement of the eight coordination species.

Structures of 9 coordination complexes are always described
in terms of the tricapped trigonal prism (TCTP) with symmetry $D_{3h}$
or the capped square antiprism (CSAP) with symmetry $C_{4v}$.
Figure 1 illustrates the relevant coordination models as well as
the reaction paths between them.

EXPERIMENTAL

The spectrum of the hydrated $Eu^{3+}$ ion corresponds to an
aqueous solution of $Eu(ClO_4)_3$. The perchlorate anions are supposed
not to enter the first coordination sphere (12).
The absorption spectra are taken with a Cary 219 spectrophotometer.
The MCD spectra are registered on a Cary 61 dichrometer equipped
with an electromagnet providing a magnetic field of 7.5 kG. The
spectra are given in figure 2.

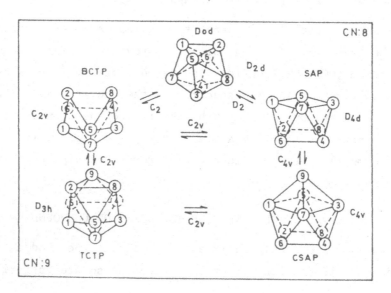

Fig. 1 Idealized coordination structures

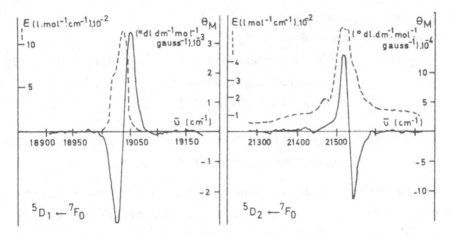

Instrumental variables : SBW (spectral bandwith)    0.2 nm
                          PP (penperiod) in ABS      1 sec
                                            MCD      3 sec
                          SR (scan rate)             0.1 nm/sec

Fig. 2 Absorption and MCD spectra of hydrated $Eu^{3+}$ (1.5 M)


MCD THEORY

Resulting from the interplay of matrix elements of the Zeeman operators $\mu_z$ and the electric or magnetic dipole operators
($\vec{0} = \vec{m}$ or $\vec{\mu}$ respectively), the MCD molar ellipticity is given by

$$[\theta]_M(J \leftarrow A) = -33,53 \, \bar{\nu}[\frac{A}{hc}f^1(\bar{\nu},\bar{\nu}_{JA}) + (B+C/kT)f^0(\bar{\nu},\bar{\nu}_{JA})] \quad [1]$$

with : A and J : ground state and excited state

    $hc\bar{\nu}_{JA}$ : zero-field energy of transition

    $f^0(\bar{\nu},\bar{\nu}_{JA})$ : zero-field shape function of transition

    $f^1(\bar{\nu},\bar{\nu}_{JA})$ : first derivative of $f^0(\bar{\nu},\bar{\nu}_{JA})$

    A,B and C : MCD parameters, defined by P.J. Stephens in (13)

For our purpose only the A term is explicitly written out :

$$A = -\frac{3}{2d_A} \sum_{a^0,j^0} [|<a^0|0_-|j^0>|^2 - |<a^0|0_+|j^0>|^2]$$

$$[<j^0|\mu_z|j^0> - <a^0|\mu_z|a^0>] \quad [2]$$

with : $a^o$ and $j^o$ : crystal field levels of the terms A and J

$d_A$ : zero-field degeneracy

$0_-$ and $0_+$ : dipole operators for left and right circularly polarised light where : $m_\pm = \mp 1/\sqrt{2}(m_x \pm im_y)$

and $\mu_\pm = \mp 1/\sqrt{2}(\mu_y \mp i\mu_z)$

APPLICATION TO THE $^5D_1 \leftarrow {}^7F_0$ (MAGNETIC DIPOLE) AND $^5D_2 \leftarrow {}^7F_0$ (INDUCED ELECTRIC DIPOLE) TRANSITIONS OF $Eu^{3+}$

From figure 2 it appears that the MCD signals can be approximately described by A terms, showing a sign inversion between the $^5D_1 \leftarrow {}^7F_0$ and the $^5D_2 \leftarrow {}^7F_0$ transition.

As $C_{2v}$ provides only B terms, this symmetry will be neglected in first approximation. Moreover the $D_{4d}$ symmetry can also be eliminated, because it is predicted that no electric dipole transition is allowed from the $^7F_0$ to the $^5D_2$ state, although a corresponding absorption signal is found.

In the retained symmetries a further simplification results from the fact that the MCD active crystal field components for the transitions under discussion can still be described by one quantum number $|M_J|$. In other words : $\mu = M_J$ with $\mu$ : crystal quantum number.

For the ground state $^7F_0$ with $M_J = 0$, the magnetic moment $<a_o|\mu_z|a_o> = <0|\mu_z|0>$ equals zero. Furthermore, the intensity matrix elements are equal : $|<0|0_-|M_J'>|^2 = |<0|0_+|M_J'^*>|^2$ so that the A parameter becomes :

$$A = 2 \times - \frac{3}{2d_A}<M_J'|\mu_z|M_J'> \times |<0|0_-|M_J'>|^2 \qquad [3]$$

By substituting [3] in [1] it is found that the sign of the MCD signal is thus determined by the sign of the Zeeman component (labeled by $M_J'$) that is active for left circular polarized light.

For example, for the magnetic dipole transition $^5D_1 \leftarrow {}^7F_0$, left circular polarized light $0_-$ will be absorbed between the $M_J = 0$ ground state and the $M_J' = +1$ Zeeman component of the excited state following the selection rule :

$$- M_J - 1 + M'_J = - 1 + M'_J = 0 \qquad\qquad [4]$$

This is the case for all the considered symmetries. The corresponding MCD signal is designed as a positive A term. The sign of the $^5D_2 \leftarrow {}^7F_0$ transition will now be defined with respect to this.

The $^5D_2 \leftarrow {}^7F_0$ transition is electric dipole induced. From the Judd-Ofelt theory it follows that the selection rules take into account a symmetry depended q-factor which appears in the expansion of the crystal field potential in odd k-terms. These q-values are given in Table 1.

For left circular polarized light $0_- = -1$ the active Zeeman component corresponds to a $M'_J$ value defined by the selection rule

$$- M_J - 1 + q + M'_J = - 1 + q + M'_J = 0 \qquad\qquad [5]$$

From table 1 it appears that accordingly to the symmetries the following signs of A terms are obtained for the $^5D_2 \leftarrow {}^7F_0$ transition :

| | | | |
|---|---|---|---|
| $D_{2d}$ | $q = + 2$ | $M'_J = - 1$ | negative A term |
| $D_{3h}$ | $q = + 3$ | $M'_J = - 2$ | negative A term |
| $C_{4v}$ | $q = 0$ | $M'_J = + 1$ | positive A term |

Comparison with the spectra shows that consequently two symmetries can be retained : $D_{2d}$ and $D_{3h}$.

On the basis of figure 1 one can try to find an explanation of these experimental features either within the "equilibrium" model (6,14) or within the "fraction coordination" model (15,16,17).

Table 1 : q-values for the appropriate coordination symmetries

| CF symmetry | $v^{even}$ | $v^{odd}$ | CF symmetry | $v^{even}$ | $v^{odd}$ |
|---|---|---|---|---|---|
| $D_{3h}$ | 0, ±6 | ±3 | $D_{2d}$ | 0, ±4 | ±2, ±6 |
| $D_{4d}$ | 0 | ±4 | $C_{4v}$ | 0, ±4 | 0, ±4 |

In the first one, two species should be present in solution with $D_{2d}$ and $D_{3h}$ symmetry respectively corresponding to a coordination number 8 and 9. In the second one, a single hydrated species is present with a fraction coordination number and a distortion symmetry of both $D_{2d}$ and $D_{3h}$ : the $C_{2v}$ symmetry.

Measurements on the relative value of the magnetic moments at different temperatures should enable us to distinguish between these two models.

## ACKNOWLEDGEMENTS

One of us (Y.B.) thanks the IWONL (Belgium) for a financial support. The laboratory is indebted to the IWONL and NFWO (Belgium) for the experimental equipment. The authors thank F. Morissens for technical aid.

## REFERENCES

1. R. Betts and O. Dahlinger, Can. J. Chem. 37: 91 (1959)
2. L. Morgan, J. Chem. Phys. 38, 2788 (1963)
3. L. Asprey, T. Keenan and F. Krise, Inorg. Chem. 3 :1137 (1964)
4. A.J. Graffeo and J.L. Bear, J. inorg. nucl. Chem. 30 :1577 (1968)
5. J. Reuben and D. Fiat, J. Chem. Phys. 51/11 :p. 4909 (1969)
6. F.H. Spedding, M.J. Pikal and B.O. Ayers, J. Phys. Chem. 70/8: 2440 (1966)
7. C. Görller-Walrand and J. Godemont, J. Chem. Phys. 66/1: 48 (1977)
8. J. Hoard and J.V. Silverton, Inorg. Chem. 2 :235 (1962)
9. E.L. Muetterties and C.M. Wright, Quart. Rev. 21: p. 109 (1967)
10. T. Moeller, D.F. Martin, L.C. Thompson, R. Ferrus, R. Feistel and W.J. Randall, Chem. Rev. 65 :1 (1965)
11. M.G.B. Drew, Coord. Chem. Rev. 24 :179-275 (1977)
12. L. Johansson, Coord. Chem. Rev. 12: p. 241 (1974)
13. P.J. Stephens, J. Chem. Phys. 52/7: (1970)
14. F.H. Spedding, J.A. Rard and A. Habenschuss, J. Phys. Chem. 81/11 :p. 1069 (1977)
15. J.L. Hoard, B. Lee and M.D. Lind, J. Amer. Chem. Soc. 87: p. 1612 (1965)
16. T. Mioduski and S. Siekierski, J. inorg. nucl. Chem. 37: p. 1647 (1975)
17. P. Geier, V. Karlen and A.v. Zelewsky, Helv. Chim. Acta 52/7 : p. 1967 (1969)

SOLVENT EFFECTS ON ERBIUM CHLORIDE COMPLEXATION REACTIONS*

Herbert B. Silber

Division of Earth and Physical Sciences, UTSA

San Antonio, Texas  78285, U.S.A.

ABSTRACT

Ultrasonic absorption measurements on 0.200 M $ErCl_3$ indicate a coordination number change accompanies inner sphere complexation in aqueous DMSO, but not in aqueous methanol.

INTRODUCTION

$ErCl^{2+}$ complexation proceeds via a multi-step mechanism (1):

$$Er^{3+}(solv) + Cl-(solv) \rightleftarrows Er^{3+}(solv)_xCl^- \rightleftarrows ErCl^{2+}(solv) \qquad (1)$$
$$\text{Step 12} \qquad\qquad \text{Step III}$$

Step 12 is the formation of the outer-sphere complex, $Er^{3+}(solv)_xCl^-$, and Step III is that of the inner-sphere complex, $ErCl^{2+}(solv)$.  The sound absorption, $\alpha/f^2$, is given by:

$$\alpha/f^2 = A_{12}/ [1+(f/f_{12})^2] + A_{III}/[1+(f/f_{III})^2]+B \qquad (2)$$

where $A_{12}$ and $A_{III}$ are the absorption amplitudes, $f_{12}$ and $f_{III}$ are the relaxation frequencies, f is the experimental frequency and B is the solvent (2) background.  Each relaxation can be converted to the excess absorption maximum, $\mu_{max_i}$, which at constant salt concentration is proportional to the square of the reaction volume change, by:

$$\mu_{max_i} = A_i \cdot f_{Ri} \cdot v/2 \qquad (3)$$

where v is the sound velocity in the mixed solvents (3,4).

With Er(III), the large ligands $NO_3^-$(5), $ClO_4^-$(6), $Br^-$(7) and

*Supported by a grant from the Robert A. Welch Foundation.

$I^-$(7) change coordination number during inner-sphere complexation, but not with the small $Cl^-$(8) ligand.  This study was initiated to study the effect of $ErCl_3$ concentration in methanol and to determine if the larger solvent DMSO could cause the coordination number change.


RESULTS


The preparation of the stock solutions (8) and the ultrasonic absorption apparatus have been described (7).  The absorption data are summarized in Tables I and II.  The relaxation calculations are shown in Table III.  Figure 1, is interpreted in the following manner.  For 0.200 M $ErCl_3$ in aqueous DMSO, $\mu_{max}$ increases as water is added, attributed to the solvation number change reaction with the larger DMSO molecules, consistent with the argument that the solvation number change is controlled by steric factors.

For 0.096 M $ErCl_3$ in aqueous methanol, $\mu max_{III}$ decreases as water is added due to decreased inner-sphere complexation caused by an increasing dielectric constant.  A more interesting variation occurs with 0.200 M $ErCl_3$ in aqueous methanol.

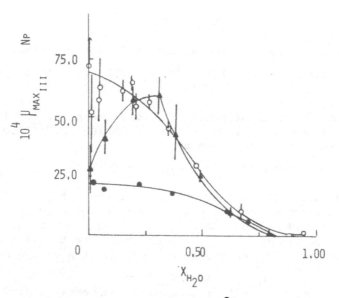

FIGURE 1.   The variation in $\mu_{max}$ for $ErCl^{2+}$ inner-sphere complexation as a function of water mole fraction, $X_{H_2O}$.

Key:   ● 0.096M in methanol; o 0.200M in methanol; ▲ 0.200M in DMSO

TABLE I.  The Experimental Absorption Data for 0.200M $ErCl_3$ in Aqueous Methanol at 25°C.

| $f$, MHz | $X_{H_2O}$ = 0.013 | 0.015 | 0.055 | 0.055 | 0.153 | 0.193 | 0.204 | 0.269 | 0.357 | 0.478 | 0.676 | 0.947 |
|---|---|---|---|---|---|---|---|---|---|---|---|---|
| 5.03 | - | 903.68 | 1056.13 | 1107.97 | 1212.98* | 1108.38 | - | 953.63 | 662.01 | 406.98 | 224.17 | - |
| 7.07 | 619.90 | 535.71 | 600.80 | 544.29 | 805.30 | 736.64 | 684.40 | 664.34 | 512.93 | 347.78 | 167.90 | - |
| 9.14 | 437.74 | 426.47 | 442.46 | 403.82 | 543.96 | 526.69 | 531.96 | 522.89 | 423.97 | 303.17 | 153.52 | 30.00 |
| 10.78 | 348.33 | - | 336.23 | - | - | - | 418.86 | - | - | 260.61 | 153.22 | 27.79 |
| 11.15 | 370.01 | 369.03 | 347.23 | 306.00 | 432.53 | 392.25 | 436.31 | 411.11 | 347.57 | 261.07 | - | 28.27 |
| 12.21 | 344.54 | 305.94 | 289.20 | 301.26 | 326.13 | 322.34 | 344.95 | 344.58 | 295.95 | 236.41 | 123.27 | 27.50 |
| 13.14 | 304.79 | 297.45 | 277.11 | 265.82 | 325.73 | 319.90 | 308.22 | 336.29 | 300.11 | 228.45 | 127.98 | 25.14 |
| 15.09 | 266.71 | 258.85 | 241.12 | 214.08 | 276.53 | 266.66 | 279.08 | 281.35 | 251.49 | 200.94 | 118.47 | 26.75 |
| 17.09 | 253.95 | 234.25 | 204.77 | 202.72 | 251.57 | 218.37 | 231.29 | 236.42 | 219.01 | 182.52 | 110.21 | 26.42 |
| 20.28 | 207.64 | 178.34 | 161.84 | 152.38 | 176.66 | 167.72 | 182.68 | 179.22 | 176.82 | 142.44 | 92.40 | 27.16 |
| 28.29 | 145.84 | 122.41 | 121.50 | 123.20 | 121.47 | 125.51 | 123.69 | 133.19 | 124.51 | 112.29 | 80.49 | 27.82 |
| 30.70 | 133.21 | 124.91 | 112.25 | 111.03 | 110.85 | 111.31 | 112.71 | 118.04 | 116.28 | 101.91 | 71.66 | 26.66 |
| 36.46 | 113.18 | 86.28 | 99.96 | 100.40 | 96.06 | 99.43 | 96.00 | 98.67 | 101.55 | 92.97 | 66.02 | 26.45 |
| 44.54 | 101.60 | 84.02 | 89.94 | 86.67 | 85.04 | 84.51 | 84.35 | 84.33 | 83.45 | 79.42 | 61.10 | 25.49 |
| 51.04 | 88.74 | 85.43 | 80.04 | 78.33 | 78.02 | 76.50 | 79.28 | 79.11 | 79.73 | 74.85 | 55.68 | 24.61 |
| 71.52 | 68.73 | 66.65 | 70.28 | 70.40 | 67.26 | 66.99 | 67.55 | 72.44 | 68.16 | 65.75 | 50.20 | 24.25 |
| 91.94 | 57.29 | 52.51 | 65.22 | 64.59 | 62.25 | 62.59 | 62.07 | 65.44 | 60.59 | 58.51 | 46.77 | 24.28 |
| 112.70 | 52.47 | 50.53 | 62.68 | 60.81 | 52.85 | 61.34 | 61.03 | 58.34 | 56.25 | 54.84 | 43.57 | 25.31 |
| 133.32 | 50.96 | 48.29 | 59.29 | 56.85 | 64.03 | 52.55 | 58.22 | 55.43 | 54.17 | 51.59 | 43.01 | 23.58 |
| 153.93 | 48.56 | 49.07 | 58.52 | 55.64 | 48.69 | 52.29 | 58.00 | 55.27 | 53.72 | 50.94 | 42.75 | 25.01 |
| 174.26 | 44.48 | 40.58 | 56.80 | 55.48 | 57.40 | 51.02 | 52.31 | 52.54 | 53.17 | 51.01 | 43.21 | 25.32 |
| 194.82 | 42.12 | - | 53.18 | 55.16 | 56.35 | 45.64 | 51.59 | - | 52.41 | 48.36 | 42.10 | 22.47 |
| 215.56 | 39.00 | - | 49.52 | 54.15 | - | - | - | - | - | - | - | 23.29 |
| 235.85 | 35.82 | - | - | - | - | - | - | - | - | - | - | - |

*f = 4.18 MHz
The units of the absorption data are:    $10^{17}$ Np $cm^{-1}$ $sec^2$

TABLE II.  The Experimental Absorption Data for 0.200M $ErCl_3$ in Aqueous DMSO at 25°C.

| f, MHz | $x_{H_2O}$ = 0.001 | 0.039 | 0.075 | 0.199 | 0.303 | 0.388 | 0.491 | 0.623 | 0.793 | 0.900 | 1.00 |
|---|---|---|---|---|---|---|---|---|---|---|---|
| 10.46±0.04 | 230.5 | - | 307.1 | 421.9 | 455.3 | 355.0 | 204.0 | 91.1 | 48.2 | - | N |
| 12.25±0.02 | 195.9 | 211.9 | 256.3 | 349.3 | 378.0 | 298.8 | 191.2 | 88.3 | 46.2 | 37.8 | O |
| 20.29±0.04 | 117.9 | 122.1 | 144.6 | 199.5 | 220.8 | 196.1 | 150.0 | 86.1 | 44.3 | 32.9 | |
| 28.27±0.07 | 91.4 | 90.7 | 106.0 | 142.3 | 158.9 | 145.1 | 123.0 | 80.3 | 43.1 | 29.4 | E |
| 30.31±0.13 | 79.0 | 89.2 | 90.7 | 125.3 | 154.0 | 149.9 | 111.7 | 76.6 | 40.4 | 29.9 | X |
| 36.45±0.02 | 80.0 | 75.5 | 84.9 | 110.5 | 128.8 | 119.8 | 105.5 | 74.8 | 40.2 | 27.8 | C |
| 44.48±0.03 | 72.2 | 67.6 | 75.0 | 94.7 | 108.1 | 106.3 | 96.9 | 72.5 | 40.7 | 27.9 | E |
| 50.89±0.19 | 63.7 | 67.7 | 63.0 | 84.3 | 101.1 | 100.1 | 92.8 | 72.7 | 41.9 | 28.8 | S |
| 71.40±0.27 | 57.1 | 62.4 | 61.1 | 71.1 | 81.9 | 85.3 | 81.0 | 67.7 | 40.9 | 28.5 | S |
| 91.96±0.16 | 53.2 | 54.0 | 52.6 | 61.9 | 69.1 | 73.0 | 72.9 | 61.1 | 38.2 | 25.9 | |
| 112.38±0.56 | 51.0 | 51.8 | 52.0 | 58.3 | 63.1 | 65.8 | 68.0 | 60.4 | 38.2 | 25.2 | A |
| 132.56±0.98 | 51.0 | 49.9 | 51.8 | 56.6 | 61.8 | 63.8 | 67.8 | 59.3 | 38.4 | 25.1 | B |
| 153.41±0.88 | 50.5 | 47.5 | 50.5 | 55.1 | 59.7 | 63.8 | 67.2 | 57.6 | 38.2 | 23.5 | S |
| 173.82±1.07 | 50.3 | 49.2 | 51.2 | 57.5 | 57.3 | 58.2 | 63.5 | 56.8 | 38.6 | 23.6 | O |
| 194.47 | - | - | - | - | - | - | - | 57.0 | - | - | R |
| 214.76 | - | - | - | - | - | - | - | 55.6 | - | - | P |
| | | | | | | | | | | | T |
| | | | | | | | | | | | I |
| | | | | | | | | | | | O |
| | | | | | | | | | | | N |

The units of the absorption are:   $10^{17}$ Np $cm^{-1}$ $sec^2$

TABLE III. The Relaxation Data for 0.200 M ErCl$_3$ in Mixed Solvents

| $x_{H_2O}$ | $A_{12}$,[a] | $f_{R_{12}}$, MHz | $\mu_{max_{12}}$,[b] | $A_{III}$,[a] | $f_{R_{III}}$, MHz | $\mu_{max_{III}}$,[b] | B,[a] Ref (2) | v, m/sec Ref (3) |
|---|---|---|---|---|---|---|---|---|
| **A. METHANOL** | | | | | | | | |
| 0.013 | 184.2±37.3 | 29.7± 4.3 | 30.5±10.6 | 7212 ±19500 | 1.74± 2.56 | 72 ±292 | 30.6 | 1114 |
| 0.015 | 202.0±42.4 | 23.1± 4.2 | 26.0±10.2 | 4623 ± 2285 | 2.04± 0.66 | 52.6± 43.1 | 30.4 | 1115 |
| 0.055 | 44.1±17.1 | 177.3±155.4 | 44.6±56.3 | 2993 ± 540 | 3.34± 0.47 | 57.0± 18.3 | 30.1 | 1140 |
| 0.055 | 56.7± 7.0 | 122.2± 29.2 | 39.5±14.3 | 3636 ± 290 | 2.99± 0.18 | 62.1± 8.6 | 30.1 | 1140 |
| 0.153 | 21.1± 7.8 | 1000 ± - | - | 1803 ± 59 | 5.70± 0.20 | 61.9± 4.2 | 29.8 | 1204 |
| 0.193 | 34.6± 3.8 | 177.4± 44.6 | 37.7±13.6 | 2362 ± 73 | 4.47± 0.12 | 64.8± 3.7 | 29.9 | 1227 |
| 0.204 | 25.2± 5.0 | 500 ±792 | 77.5±138 | 1385 ± 87 | 6.51± 0.33 | 55.5± 6.3 | 30.0 | 1231 |
| 0.269 | 35.1± 6.0 | 184.2± 76.7 | 40.8±23.9 | 1490 ± 46 | 6.02± 0.20 | 56.6± 3.6 | 30.3 | 1262 |
| 0.357 | 31.8± 4.5 | 190.1± 58.0 | 39.4±17.6 | 820.5± 17.0 | 8.07± 0.22 | 43.2± 2.1 | 31.6 | 1305 |
| 0.478 | 25.1± 2.4 | 184.4± 35.4 | 31.9± 9.2 | 423.7± 5.3 | 10.54± 0.21 | 30.8± 1.0 | 34.2 | 1380 |
| 0.676 | 57.1±16.7 | 40.2± 11.7 | 17.6±10.3 | 214.0± 26.5 | 6.23± 1.55 | 10.2± 3.8 | 32.8 | 1534 |
| 0.947 | 1.6±0.7 | - | - | 4.9± 0.9 | 38.55±16.96 | 1.5± 0.9 | 22.1 | 1545 |
| **B. DMSO** | | | | | | | Ref (2) | Ref (4) |
| 0.001 | 18.3± 2.7 | 243.2± 90.9 | 33.1±17.2 | 459.4± 62.8 | 8.22± 0.93 | 28.1± 7.0 | 35.5 | 1489 |
| 0.039 | 21.5± 2.2 | 186.1± 34.2 | 29.9± 8.6 | 672.9±150.2 | 6.69± 1.00 | 33.6± 12.6 | 34.6 | 1494 |
| 0.075 | 16.4± 2.3 | 509.8±579.1 | 62.6±79.9 | 684.3± 64.4 | 8.08± 0.61 | 41.5± 7.0 | 33.9 | 1499 |
| 0.199 | 28.7± 3.3 | 230.4± 63.0 | 50.5±19.7 | 862.0± 60.4 | 8.84± 0.54 | 58.1± 7.7 | 32.5 | 1526 |
| 0.303 | 50.8± 5.2 | 127.9± 17.0 | 50.5±11.9 | 903.8± 77.1 | 8.65± 0.67 | 60.7± 9.9 | 32.2 | 1554 |
| 0.388 | 54.0± 7.7 | 140.5± 26.9 | 59.9±20.0 | 543.2± 65.7 | 10.18± 1.32 | 43.6± 10.9 | 32.2 | 1578 |
| 0.491 | 39.8± 2.8 | 282.5± 53.5 | 90.7±23.6 | 177.4± 4.7 | 17.69± 0.97 | 25.3± 2.1 | 32.4 | 1614 |
| 0.623 | 27.1± 2.2 | 464.4±137.6 | 105 ±39.8 | 33.6± 2.1 | 36.78± 4.04 | 10.3± 1.8 | 32.0 | 1670 |
| 0.793 | 14.8± 1.0 | 476.8±216.9 | 60.0±31.4 | 12.6± 3.7 | 14.39± 6.43 | 1.5± 1.2 | 24.8 | 1703 |
| 0.900 | 2.7± 0.6 | - | - | 17.9± 3.4 | 19.04± 4.29 | 2.8± 1.2 | 22.1 | 1655 |

a) units = $10^{17}$ Np cm sec$^2$       b) units = $10^4$ Np

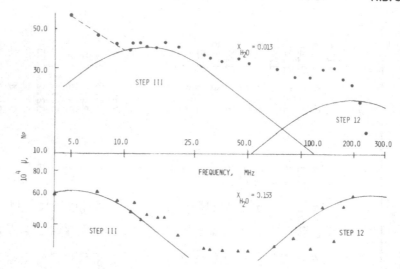

FIGURE 2. The excess absorption for 0.200M ErCl$_3$ in methanol.

(Figure 2). At low $X_{H_2O}$ the excess absorption below 9Hz can either
be the start of a new relaxation which cannot be characterized due
to its low $f_{Ri}$; a very high amplitude Step III relaxation introducing
computational difficulties within our double relaxation computer
program; or could represent extraneous sound scattering due to the
large path lengths required at low frequency. The extraneous scattering hypothesis can be eliminated from the $X_{H_2O}$ = 0.15 solution,
where the data are fit to a standard double relaxation program. If
a new relaxation is present, then $f_R$ rapidly increases with additional
water leading to the experimental scatter in $\mu_{III}$ below $X_{H_2O}$ ~ 0.2.
If this relaxation is real it may involve bulk solvent ordering by
the formation of inner-sphere complex. If a new relaxation is not
present, the reported $\mu_{III}$ for $X_{H_2O}$ = 0.013 is the minimum value and
the actual result must be higher. In any case, no increase occurs in
$\mu_{III}$ as water is added to 0.200 M ErCl$_3$ in aqueous methanol, and thus
no coordination number change accompanies complexation for ErCl$_3$ in
aqueous methanol.

REFERENCES

1.  H. Diebler and M. Eigen, Z. Phys. Chem. N.F. 20:  299(1959).
2.  M. Eigen and K. Tamm, Z. Electrochem. 66:  93, 107(1962).
3.  H. B. Silber, J. Inorg. Nucl. Chem. 39:  2284(1977).
4.  O. Nomoto, J. Phys. Soc. (Jpn) 8:  553(1953).
5.  D. E. Bowen, M. A. Priesand and M. P. Eastman, J. Phys. Chem.
    78:  2611(1974).
6.  J. Reidler and H. B. Silber, J. Inorg. Nucl. Chem 36:  175(1974).
7.  H. B. Silber, J. Phys. Chem. 78:  1940(1974).
8.  H. B. Silber and G. Bordano, J. Inorg. Nucl. Chem, in press.
9.  J. Reidler and H. B. Silber, J. Phys. Chem 78:  424(1974).

# COMPLEXES OF THE HEAVIER LANTHANOID NITRATES WITH CROWN ETHERS[1]

J.-C. G. Bünzli, D. Wessner, and B. Klein

Université de Lausanne, Institut de chimie minérale et

analytique, Place du Château 3, 1005 Lausanne, Switz.

## INTRODUCTION

Crown ethers form stable complexes with lanthanoid (III) ions (2), (3), (4), (5). We have shown, in previous papers of this series, that unsolvated complexes of the lighter lanthanoid nitrates (Ln = La-Gd) can be isolated with 15-crown-5 and 18-crown-6 ethers. The Ln(III)/crown ether ratio is 1:1 for the complexes with ligand 2 whereas complexes of ligand 3 may have two different stoichiometries,

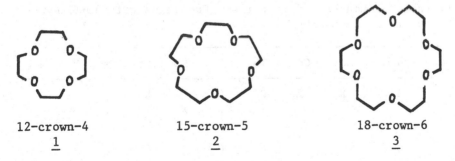

| 12-crown-4 | 15-crown-5 | 18-crown-6 |
| 1 | 2 | 3 |

1:1 for Ln = La-Nd and 4:3 for Ln = La-Gd (6). The strength of the Ln(III)/crown ether interaction depends on both the ratio of ionic diameter/ligand cavity diameter and the ligand flexibility. In this communication we discuss the interaction of ligands 2 and 3 with

___

1) Part 3 of the series "Complexes of Lanthanoid Salts with Macro-cyclic Ligands". For part 2 see reference (1).

the heavier lanthanoid(III) ions and we report the results of a
$^1$H-NMR study of the 18-crown-6 complexes with Ln = La, Pr, Nd, Eu,
Yb, as well as some preliminary results of our X-ray structural work
on the neodymium nitrate 1:1 and 4:3 complexes with 18-crown-6.

## COMPLEX FORMATION

Unsolvated and non-hygroscopic 1:1 complexes are obtained with
ligand 1 for the entire Tb-Lu series (7). Isolation of 1:1 complexes
with ligands 2 and 3 is more difficult and the resulting compounds
form readily tri- or tetrahydrates on exposure to air. Table 1 summarizes the composition of the compounds for which we have obtained
an elemental analysis. Depending on the particular lanthanoid ion
and on the drying procedure, mono- di- or trihydrates can be characterized.

Thermogravimetric measurements on the 15-crown-5 complexes show
a single and complete decomposition around 250-300 °C for the lighter
lanthanoid complexes. For Ln = Eu, the decomposition starts at a
lower temperature and an intermediate of limited thermal stability
is formed. For the Gd-Lu ions, the thermograms first indicate the

Table 1. Isolated Lanthanoid Trinitrate Complexes with 15-Crown-5
and 18-Crown-6 Ethers[a].

| L | Ln:L | ·nH₂O | La | Gd | Tb | Dy | Ho | Er | Tm | Yb | Lu |
|---|------|-------|----|----|----|----|----|----|----|----|----|
| 2 | 1:1 | 0 | X | X | C | C | C | X | | | |
|   |     | 2 |   | X |   | X | X | | | | |
|   |     | 3 |   |   |   |   |   | C | X | X | X |
|   | 4:3 | 0 |   |   |   | C |   | C | C | X | C |
| 3 | 1:1 | 0 | X |   | C | C | X | X | C | C | C |
|   |     | 1 |   |   | X | X |   | X | X | X | X |
|   | 4:3 | 0 | X | X | X | X | X | X[b] | C | C | C |

[a] Key : X = complete elemental analysis obtained, C = complexometric
analysis of Ln only.  [b] Typical example : Er 30.53 % (calc. 30.23%)
C 19.51 % (19.60 %), H 3.51 % (3.29 %), N 7.38 % (7.62 %).

formation of anhydrous 1:1 complexes around 60–100 $^\circ$C. These complexes are thermally unstable and they soon decompose with loss of
ligand into a compound which is thermally stable up to $\sim$ 280 $^\circ$C and
the composition of which corresponds to a 4:3 salt:ligand ratio.
These complexes are similar to those isolated with ligand 3 and the
lighter lanthanoid nitrates (6) and they can easily be obtained in
larger quantities by heating the corresponding 1:1 hydrated complexes.
Similarly, all the heavier lanthanoid nitrates form 4:3 complexes
with the 18-crown-6 ether, which can either be obtained by thermal
decomposition of the 1:1 complexes or, in some cases, directly by
crystallization out of an equimolar solution of lanthanoid nitrate
and ligand in acetonitrile.

## STUCTURAL STUDY

The 4:3 complexes are rather unexpected and new species for
f-elements. To our knowledge, only one other 4:3 complex is known,
$[U(BH_4)_3]_4 \cdot (3)_3$, the synthesis of which has been recently reported
(8). In order to understand better why such complexes form, we have
undertaken an X-ray study on both the 1:1 and 4:3 complexes of neodymium nitrate with the 18-crown-6 ether.

X-Ray diffraction measurements have been performed on single
crystals and the structure of the 1:1 complex was solved by Patterson
and Fourier methods. The 1:1 complex is orthorhombic and belongs to
the $P_{bca}$ space group. The unit cell measures 12.23 x 15.62 x 21.80 Å
and contains 8 molecules. The neodymium ion is 12-coordinated, being
bound to the 6 oxygen atoms of the crown ether and to 3 bidentate
nitrato groups. It lies approximately in the center of the macrocycle and above the mean plane defined by the 6 oxygen atoms. The
crown ether adopts a "cup" conformation in order to equalize the
Nd–O bond lengths. The mean C–C and C–O distances amount to 1.507
and 1.450 Å, respectively, and two types of C–O–C angles are present:
109–110$^\circ$, and 112–113$^\circ$. The mean Nd–O bond lengths amount to 2.70
(8) Å (crown), 2.60(1) Å (nitrate), and 2.65(4) Å (overall). One
nitrato group is pointing inside the cup and the two other ones are
positioned on the other side of the macrocycle.

The crystal structure of the 4:3 complex has not yet been
completed. The compound is monoclinic and belongs to the C2/m
space group. Its unit cell contains two 4:3 complex molecules and
is needle-like : 11.26 x 12.04 x 29.00 Å, with $\beta$ = 113.5$^\circ$. The
4:3 complex could well have a "club sandwich" structure similar to
that of the 2:3 complex of cesium with dibenzo-18-crown-6 ether (9).

## SOLUTION STUDY

The Ln(III) ion/18-crown-6 ether interaction has been further studied in acetonitrile by means of $^1$H-NMR spectroscopy in order to determine the formation constant of the complexes. The spectra were measured with a FT-spectrometer (Bruker WP-60) since the solubilities of the complexes are usually much smaller than 0.02 M. Solutions of 1:1 and 4:3 complexes (Ln = La, Pr, Nd, Eu, Yb) were prepared under inert atmosphere in 99.9 % deuterated acetonitrile. The water content of the solvent was less than 0.02 M (0.05 %) so that the water/Ln ratio usually did not exceed 2. In the case of neodymium, we have shown that water does not influence the formation constant, at least up to a water/Nd ratio of 10.

At room temperature, the exchange reaction between bulk and coordinated crown is completely frozen and the complexed ligand shows only one singlet well separated from the free ligand resonance. Solutions of 1:1 and 4:3 complexes give essentially the same spectrum, except for Ln = Nd; in this case the chemical shift of the complexed species varies from −0.5 ppm (1:1 complex) to +0.5 ppm (4:3 complex).

Table 2. $^1$H-NMR Shifts (ppm from TMS)[a] of Acetonitrile Solutions of 18-Crown-6 Complexes, and Formation Constants at 25°C.

| Ln | $10^2 \cdot [Ln^{3+}]$ mol·l$^{-1}$ | Ln:L | $\delta_L$ ppm | $\delta_C$ ppm | C/L[b] | log $K_f$ | log $K_f$[c] |
|----|------|------|------|------|------|------|------|
| La | 0.86 | 1:1 | 3.62 | 3.85 | 14.3 | 4.4 | 3.29 |
|    | 0.46 | 4:3  | d)   | 3.85 |      |     |      |
| Pr | 1.34 | 1:1  | 3.62 | −8.05 | 7.9 | 3.7 | 2.63 |
| Nd | 1.70 | 1:1  | 3.57 | −0.53 | 6.9 | 3.5 | 2.44 |
|    | 1.14 | 4:3  | 3.62 | 0.47  | 9.4 | 3.4 |      |
| Eu | 1.45 | 1:1  | 3.55 | 8.57  | 2.0 | 2.6 | 1.84 |
|    | 0.84 | 4:3  | 3.62 | 8.60  | 3.0 | 2.9 |      |
| Yb | 0.81 | 1:1  | 3.62 | 35.95 | 0.9 | 2.3 | e) |
|    | 0.71 | 4:3  | 3.62 | 35.87 | 1.3 | 2.5 |      |

a  Resonances with positive shifts are at lower field than TMS.
b  Ratio of the complexed to the free ligand.
c  Data of reference (10) for LnCl$_3$/18-crown-6 in CH$_3$OH/H$_2$O.
d  Too small to be observed.
e  No interaction observed.

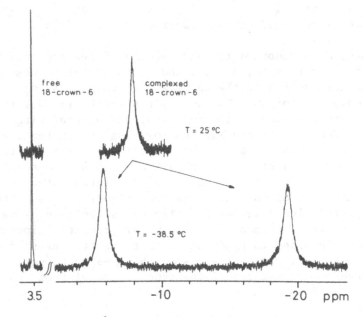

Figure 1.    60 MHz $^1$H-NMR Spectrum of $Pr(NO_3)_3 \cdot \underline{3}$ 0.013 M in
            99.9 % $CD_3CN$.

We have calculated the overall formation constant $K_f$ assuming that
only the 1:1 complex forms in these diluted solutions :

$$Ln(NO_3)_3 + L \rightleftharpoons Ln(NO_3)_3 \cdot L \qquad K_f \ (298 \ K)$$

Solubility problems prevented us from studying the influence of an
excess of ligand. The results are summarized in Table 2. As expected
from our synthetic work, the formation constant decreases when the
atomic number increases. Izatt et al. have studied the $LnCl_3$/18-
crown-6 systems in a methanol/water mixture by titration calorimetry.
They have found that no interaction takes place between the crown
ether and the heavier lanthanoid ions (Tb-Lu). Working with a poorer
Lewis base than water or methanol as solvent allowed us to observe
an appreciable interaction between ligand $\underline{3}$ and Yb(III), formation
constant of the corresponding complex being around 200. On average,
the formation constants we found are 10 times larger than in the
$LnCl_3$/$CH_3OH$/$H_2O$ system. These results give an additional proof of
the importance of the ring flexibility on complex formation, since

no interaction at all is observed between the heavier lanthanoid ions and the more rigid dibenzo-18-crown-6 ether (3).

We have also recorded the NMR spectra of the above solutions at −38.5 °C. For Ln = La, Eu, Yb, the only changes we observe are larger chemical shifts of the paramagnetic complexes ($\delta_C$ = +10.9 ppm and +49.4 ppm for Ln = Eu and Yb, respectively), due to the increased paramagnetism of the solution, and a broadening of all the resonances. For Ln = Pr and Nd, the lowering of the temperature leads to the observation of two broad resonances arising from distinct paramagnetic species : $\delta_C$ = −5.9 and −19.3 for Ln = Pr (cf. Fig. 1), and +0.1 and −7.1 for Ln = Nd. The intensities of these two resonances are approximately in the ratio 1:1. The ratio of the total complexed ligand to the free ligand is the same as measured at 25 °C, within experimental error. We are currently investigating this problem in order to find out whether the two resonances arise from non equivalent hydrogen atoms on the same macrocycle, from two different conformations of the crown ether or from two differently complexed species.

## ACKNOWLEDGMENTS

Support from the Swiss National Science Foundation through grant No 2.150-0.78 is gratefully acknowledged.

## REFERENCES

1. J.-C. G. Bünzli, D. Wessner, and Huynh Thi Tham Oanh, Inorg. Chim. Acta 32, L33 (1979).
2. A. Cassol, A. Seminara and G. de Paoli, Inorg. Nucl. Chem. Letters 9, 1163 (1973).
3. R.B. King and P.R. Heckley, J. Am. Chem. Soc. 96, 3118 (1974).
4. J.F. Desreux, A. Renard, and G. Duyckaerts, J. Inorg. Nucl. Chem. 39, 1587 (1977).
5. M. Ciampolini and N. Nardi, Inorg. Chim. Acta 32, L9 (1979).
6. J.-C. G. Bünzli and D. Wessner, Helv. Chim. Acta 61, 1454 (1978).
7. J.-C. G. Bünzli and D. Wessner, to be submitted for publication.
8. D.C. Moody, R.A. Penneman, and K.U. Salazar, Inorg. Chem. 18, 208-9 (1979).
9. J.J. Christensen, D.J. Eatough, and R.M. Izatt, Chem. Rev. 74, 351 (1974).
10. R.M. Izatt, J.D. Lamb, J.J. Christensen, and B.L. Haymore, J. Am. Chem. Soc. 99, 8344 (1977).

# SPECTRA AND REDOX PHOTOCHEMISTRY OF CROWN ETHER COMPLEXES

## OF LANTHANIDES

Terence Donohue

Laser Physics Branch
Naval Research Laboratory
Washington, DC 20375

## ABSTRACT

Complexation of lanthanides by crown ethers is shown to cause remarkable spectroscopic changes in charge-transfer, f-d and f-f transitions. Furthermore, the quantum yields for photoreduction of Eu(III) in ethanol are increased dramatically in the presence of 18-crown-6. The possibilities for using crown ethers to stabilize unstable oxidation states of lanthanides and actinides is discussed.

## INTRODUCTION

The lanthanides (Ln) normally occur in their only common oxidation state of +3. It is possible to change the oxidation state of two of them, cerium to +4 and europium to +2 by chemical (1) or photochemical (2,3) means, and such processes have been used in separation schemes for these elements. However, it is much more difficult to cause photoredox processes in the other lanthanides (Figure 1), due to the instability of the non-III state (4). While Yb(II) and Sm(II) can be produced as transient species in solution or stabilized in molten salts, they cannot survive as stable species in common solvents for any appreciable time. But it may be possible to stabilize these unstable states using complexing agents which bind Ln(II) ions much more strongly than Ln(III) ions. One such class of complexes include the macrocyclic ligands, in particular crown ethers.

The crown ethers are cyclic polyethers, the first of which was synthesized by Pedersen in 1967 (5). They have been found to bind all types of metal ions, including alkalis and alkaline earths (6). More recently, it has been determined, using chemical analysis (7)

and titration calorimetry (8), that crown ethers can bind Ln ions as
well. This paper is the first report that complexation between lanthanide ions and crown ethers can be detected solely through spectroscopic means.

The crown ethers best suited for studies with lanthanide ions
are 15-crown-5 and 18-crown-6, due to the similarity of the crown
cavity (ring) size and radius of the Ln ion. Small perturbations
can be made in the effective cavity size by substituting cyclohexyl
or benzo groups on a ring, but these effects will not be considered
here.

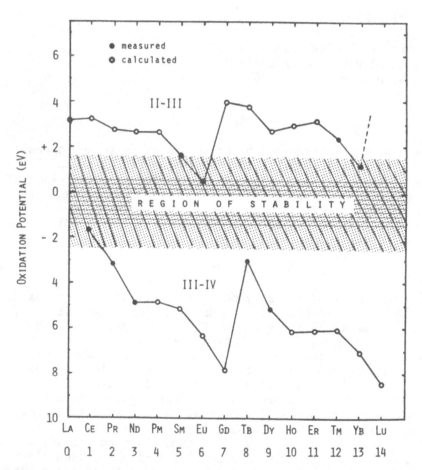

Figure 1. Oxidation potentials for those lanthanide oxidation
states that fall directly below the III state in stability. Calculated values are from Ref. 4; references for the measured values
can be found in Ref. 4. Limits for the "Region of Stability" are
given by the degree of resistance to oxidation and/or reduction
typically found in common solvents.

## EXPERIMENTAL

Since Ln ions tend to form only weakly bound complexes, non-aqueous solvents must be used in studies of crown complexes, because of the strong tendency for water to compete for binding sites on the central metal ion. Ethanolic solutions of Ln(III) chlorides were prepared by dissolving the normally hydrated salts (6-7 moles of water per mole of salt) directly in ethanol. Ethanolic solutions of Ln perchlorates were prepared by evaporating aqueous solutions of Ln perchlorates (prepared from the appropriate lanthanide oxide and 60% perchloric acid) almost to dryness and then adding ethanol. This process was repeated twice, and then the solutions diluted to give solutions 0.01 M in Ln(III). 0.1-1.0 M solutions of crown ethers were added dropwise to the Ln solutions, using a calibrated pipette, and absorption spectra were measured following the mixing of each drop.

For the photochemical experiments, ethanolic solutions of EuCl$_3$ were used, after thorough deoxygenation with argon. Concentrations were determined spectrophotometrically, using previously determined spectra for the various complexes of Eu(III) and Eu(II). Photolysis

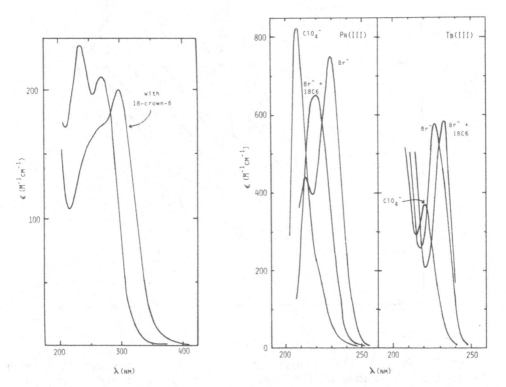

Figure 2.  (a)  Eu(III) chloride charge-transfer spectra in ethanol. (b)  f-d spectra of several Ln(III) ions in ethanol.

sources used included an argon fluoride laser (193 nm), a low pressure mercury lamp (254 nm), a xenon chloride laser (308 nm) and an argon ion laser (351/363 nm).

## RESULTS AND DISCUSSION

### Spectra

Distinct changes in absorption spectra were observed when 18-crown-6 was added to solutions of the lighter lanthanides.  These changes were observed for all three fundamental types of electronic transitions found in lanthanides (9); charge-transfer, f-d (Figure 2) and f-f.  In the case of charge-transfer transitions, the crown ether invariably shifted the band to lower energies.  For f-d tran-

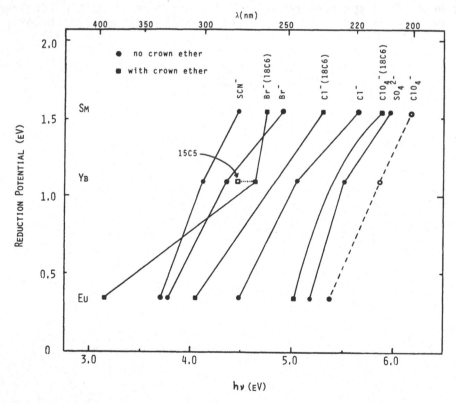

Figure 3.   Peak of the longest wavelength (lowest energy) lanthanide charge-transfer bands with a number of different ligands.  The two curves on the right ($SO_4^{2-}$) and $ClO_4^-$) are in aqueous solution; all others are in ethanol.

sitions, including Eu(II), the band(s) were usually shifted to higher energies, except in the case of Tb(III), where the band was shifted to lower energy.  The 4f intrashell transitions display minor shifts in energies, but a number of them change intensities by up to a factor of 10 upon crown complexation, and, furthermore, the bands become narrower and/or split upon addition of the crown.

The data for the effect of 18-crown-6 on all Ln charge-transfer transitions (except Ce(IV); the poor stability of this species in ethanol precludes this study) is collected in Figure 3.  Note the consistent shift to lower energies when complexed by the crown ether. Apparently Yb(III) is too small to interact with 18-crown-6.  However, a complex with 15-crown-5 was observed.  This fact correlates well with the well known lanthanide contraction in ion size.  In all cases studied, the complexes are 1:1 Ln(III) to crown ether, as deduced by titration spectrophotometry.

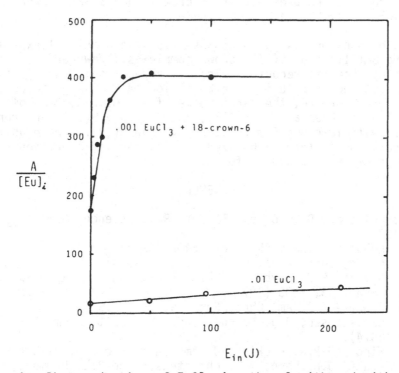

Figure 4.  Photoreduction of $EuCl_3$ in ethanol with and without 18-crown-6.  The photolysis source is an $Ar^{2+}$ laser (351/363 nm).  The absorbance is measured at a maximum in the Eu(II) absorption, at about 320 nm ($\varepsilon = 6000M^{-1}cm^{-1}/1080M^{-1}cm^{-1}$, with/without 18-crown-6).

## Photochemistry

The photochemical and chemical reactions of lanthanide halides in alcohols have been reported earlier (10). Basically there are two competing reactions, photoreduction of Eu(III) and photooxidation of Eu(II). These rates can become equal (depending upon concentrations) at a point called the photostationary state. The significant observation here is that in all the experiments attempted, photoreduction of Eu(III) is enhanced in the presence of 18-crown-6. The effect is most dramatic at the longest wavelengths studied, where photoreduction quantum yields are normally quite low indeed. Figure 4 plots the relative concentrations of the product, Eu(II), as a function of time, with and without 18-crown-6. The dramatic differences observed are a result of several effects: (i) the Eu(III) crown complex absorbs at longer wavelengths; (ii) the Eu(II) complex has smaller absorption at the photolysis wavelength; and (iii) the Eu(II) complex is more strongly bound than the Eu(III) complex. The first two effects explain why the photostationary state is achieved at a much larger [Eu(II)]/[Eu(III)] ratio than without crown complexation. The third effect explains why quantum yields for photoreduction are always higher in the presence of 18-crown-6 and always lower for photooxidation of Eu(II).

Photoreduction has not been observed in the initial experiments involving Sm(III) and Yb(III) crown complexes. However, in experiments using chemical reduction (Mg), it appears that the lifetime of Sm(II) in ethanolic solution can be prolonged in the presence of 18-crown-6. Furthermore, the first successful photoreduction of uranium(IV) has been achieved (11), only because U(III) apparently complexes with 18-crown-6 more strongly than U(IV). These unstable states could be further stabilized if a closer match between ion size and crown size could be found.

## REFERENCES

1.  E.K. Hulet and D.D. Bode, MTP Int. Rev. Science, Ser. 1, 7, 1 (1972).
2.  T. Donohue, J. Chem. Phys. 67, 5402 (1977).
3.  T. Donohue, Chem. Phys. Lett. 61, 601 (1979).
4.  L.J. Nugent, R.D. Baybarz, J.L. Burnett and J.L. Ryan, J. Inorg. Nucl. Chem. 33, 2503 (1971).
5.  D.J. Pedersen, J. Am. Chem. Soc. 89, 2945 (1967).
6.  J.J. Christensen, D.J. Eatough and R.M. Izatt, Chem. Rev. 74, 351 (1974).
7.  R.B. King and P.R. Heckley, J. Am. Chem. Soc. 96, 3118 (1974).
8.  R.M. Izatt, J.D. Lamb, J.J. Christensen and B.L. Haymore, J. Am. Chem. Soc. 99, 8344 (1977).
9.  C.K. Jørgensen, Prog. Inorg. Chem. 12, 181 (1970).
10. T. Donohue, Opt. Eng. 18, 181 (1979).
11. T. Donohue, unpublished results.

# NEW APPROACHES TO THE STUDY OF LANTHANIDE/ACTINIDE CHLORIDE -

# ALUMINUM CHLORIDE VAPOR PHASE COMPLEXES*

E. J. Peterson, J. A. Caird, W. T. Carnall, Jan P.

Hessler, H. R. Hoekstra, and C. W. Williams

Chemistry Division, Argonne National Laboratory,

9700 South Cass Avenue, Argonne, Illinois  60439, U.S.A.

Recent interest in vapor phase complexes of lanthanide and actinide halides with aluminum halides has been focused on their dynamic optical properties, since these systems have potential as optical gain media.  In addition, their thermodynamic properties indicate that elemental separation by chemical transport is potentially viable.  The reaction for gas complexation is generally written as follows for lanthanide chlorides and aluminum chloride

$$m LnCl_y(c) + n Al_2Cl_6(g) \rightleftarrows Ln_m Al_{2n} Cl_{6n+my}(g). \tag{1}$$

Vapor phase studies over a condensed solid phase (in all cases $LnCl_3$ or $AnCl_y$) have been made for the $Nd^{+3}$ (1), $Sm^{+3}$ (2), $Ho^{+3}$ (3), $U^{+4}$ and $U^{+5}$ (4) systems.  Vapor phase studies over a condensed liquid phase are more complicated and have been discussed for the $Eu^{+2}$ system (5).  The present discussion will be limited to the pressure-temperature region of solid-vapor equilibrium.

Equilibrium studies of reaction (1) are made to establish the temperature and $Al_2Cl_6$ pressure dependence of the vapor density of complexed lanthanide or actinide ions (1-5).  The stoichiometric coefficient, n, is then derived from the isothermal pressure dependence of the complexed metal ion density.  If n is temperature dependent, then more than one vapor species exist in the measurement region.  The value of m cannot be calculated from data of this type, but a value of 1 is in agreement with mass spectro-

*Work performed under the auspices of the Office of Basic Energy
 Sciences of the Department of Energy.

scopic studies of transition metal chloroaluminate vapor complexes (6) and is consistent with our results.

A variety of direct and indirect methods have been used to determine vapor densities of complexed metal ions in transition metal systems (6), but the spectrophotometric technique has been used exclusively for the lanthanides and actinides. Unfortunately, the stoichiometries inferred from the vapor density information derived from these experiments are contradictory for the lanthanides (1-3). In the $NdCl_3(Al_2Cl_6)_n$ system n varies from 1.85 to 1.65, while in the $SmCl_3(Al_2Cl_6)_n$ system n is reported to be invariant with temperature and equal to 1.5. This type of deviation in chemistry is unusual for two similar lighter lanthanides. These and other discrepancies have prompted us to re-evaluate the application of the spectrophotometric technique to these systems and to develop a technique for vapor density measurements which circumvents some of the problems associated with optical experiments.

## SPECTROPHOTOMETRIC TECHNIQUE

This method relates the optical absorbance of the system to the vapor density of complexed metal ions. The implicit assumption in this analysis is that the optical absorption of the measured bands follows Beer's law over the entire range of temperatures and group IIIB halide pressures used to study the system. Unfortunately, for many lanthanide systems this is not the case (2,3,7). Several authors have treated the temperature dependence of the molar absorptivity by empirically deriving an equation to express the dependence (2,5). This approach is fundamentally flawed. Three effects, which are considered to contribute to the deviations (from Beer's law), are (1) the temperature dependence of the natural line width, (2) the temperature dependence of the population distribution within the ligand field components of the ground J-manifold and (3) changes in the absorption spectrum due to the existence of several complex molecular species whose vapor density distributions are functions of temperature and complexing agent pressure (8). To overcome these effects, integration over the entire absorption band is performed.

The fundamental equation for the spectrophotometric analysis is

$$\frac{\ln 10}{\ell} \int_{band} A(\sigma)d\sigma = \frac{\pi e^2}{mc^2} \sum_n \rho_n \overline{f_n} ,$$

(2)

in which $\ell$ is the optical path length, $A(\sigma)$ is the observed optical absorbance at wavenumber $\sigma$, e and m are the charge and mass of the electron respectively, c is the velocity of light, $\rho_n$ is the vapor density of the nth complexed species and $\overline{f_n}$ is the average

oscillator strength of the nth complexed species for transitions
between two J-manifolds (11). Eq. (2) should be used to relate
optical absorbance data to the vapor density of complexed metal
ions. In essence this approach relates the vapor density to the
area under the peak due to the particular f-f transition being
probed instead of to the peak height at a specific wavelength. In
practice the optical experiments are done in a manner similar to
that reported in the literature (1,2). The pressure and temperature dependence of the effective oscillator strength, $f*(T,P_D)$, is
determined in experiments with a known total vapor density of complexed metal ions (unsaturated cell). Then the pressure and temperature dependence of the vapor density is determined in cells
with excess metal halide (saturated cell)

$$\rho_T(T,P_D) = \sum_n \rho_n(T,P_D) = \frac{mc^2}{\pi e^2} \frac{1}{f^*(T,P_D)} \frac{\ln 10}{\ell} \int_{band} A(\sigma)d\sigma. \quad (3)$$

Because this technique measures the product of the vapor density
and the oscillator strength, two assumptions were necessary. The
oscillator strengths of the various species must be independent of
temperature and the relative vapor densities of multiple species
for saturated and unsaturated cells must be identical. To test
these assumptions and to provide information when the spectrophotometric technique is not applicable, a technique for direct measurement of vapor densities of complexed metal ions was developed.


RADIOACTIVE TECHNIQUE

This method of complexed metal ion density measurement involves labelling the metal ions of interest with a radioactive
isotope whose decay leads to γ-ray emission. The vapor density is
conveniently monitored by γ-counting (usually >0.4 MeV γ-radiation)
in a specified energy range (the energy integral of the γ-ray
spectrum). The applicability of this technique was evaluated
during studies of the $TbCl_3$-$AlCl_3$ system using tracer [160]$TbCl_3$.
The sample vessel was loaded with $TbCl_3$ and $AlCl_3$ and was placed
in a furnace surrounded except for a detection slot by a shield of
lead bricks. The end of the tube opposite to that monitored by
the detector was kept at a lower temperature by ca. 20 K. The non-
volatilized portion of $TbCl_3$(s) thus remained out of the detection
field. Vapor equilibrium was established over the condensed phase.
A suitable detector was used to monitor the γ-activity in the hot
end of the sample vessel. The signals were fed through an ampli-
fier/discriminator to a pulse height analyzer. Thus, in principal
more than one characteristic γ-ray in a mixture could be monitored.
The measured count rate was proportional to the density of Tb
atoms in the vapor plus any background noise. The equation which
relates vapor density to count rate is

$$\rho = \frac{4\pi R^2}{\bar{\varepsilon}\, A^* V_D\, f_o}\, \frac{e^{(\ln 2)t/T_{\frac{1}{2}}}}{I_\gamma}\, P \tag{4}$$

in which R is the distance from sample to detector, A* is the effective detector area, $V_D$ is the volume of the vessel sampled by the detector, $\bar{\varepsilon}$ is the effective detector efficiency, $f_o$ is the initial amount of radioactive isotope, $e^{(\ln 2)t/T_{\frac{1}{2}}}$ considers the isotopes radioactive decay with time, $I_\gamma$ is the probability per unit time of radioactive decay, and P is the count rate (9). Experiments are carried out in a manner similar to optical experiments. Unsaturated cells of known total complexed metal ion density are examined to calibrate the counting geometry (R, $\bar{\varepsilon}$, $f_o$, A* and $V_D$). Then saturated cells are studied to determine the temperature and complex agent pressure dependence of the vapor density. The greatest strength of this technique is that it is a direct measure of the complexed metal ion density and is not dependent upon the chemistry of the system. Accurate measurements of multi-species vapor systems can be readily accomplished. The accuracy and precision of the data are superior to data obtained by optical experiments (9). The tracer method is applicable to systems, such as $TbCl_3$–$AlCl_3$, which are not amenable to study by the spectrophotometric technique. In situations where both methods are available complementary data will be available to aid in characterization of the vapor system. A complete thermodynamic analysis of the $TbCl_3$–$AlCl_3$ system is currently underway using the tracer technique.

## ANALYSIS OF THERMODYNAMIC PROPERTIES

From the temperature and pressure dependence of vapor density data, the identity and apparent thermodynamic formation parameters for the vapor complex species are inferred. The apparent equilibrium quotient for reaction (1) is written

$$K^* = P_C / P_D^{n*} \tag{5}$$

where the condensed phase is a solid. If n* is a non integer temperature dependent variable, then multiple vapor species exist. The prior method of analysis (1,5,10) involved fitting vapor pressure information to an appropriate two species model using a nonlinear least squares program. The data was not considered adequate for consideration of more complex models, but this assumption was not tested. The drawback of this approach is that it forces the data into a preconceived picture, the validity of which cannot be reliably tested. Instead of relying on this approach, a systematic method of data analysis has been developed (11).

Taking the natural logarithm of eq. (5) yields

$$\ln P_C = \ln K^* + n^* \ln P_D . \tag{6}$$

A series of isothermal plots of $\ln P_C$ vs. $\ln P_D$ data yield $K^*$ and
$n^*$ values as a function of temperature. In a multiple species
system the vapor pressure can be expressed as

$$P_C = \sum_{m=0}^{B} K_m P_D^{m/2} \qquad\qquad (7)$$

where B is twice the oxidation state of the metal ion in the halide
and $m = 2n$. Possible m values can be chosen by considering the
range of n values derived in the measurement region from $\ln P_C$ vs.
$\ln P_D$ plots with the largest m value being derived by extrapolation to low temperature. For a two species model eq. (7) is
written

$$P_C = K_{m_1} P_D^{m_1/2} + K_{m_2} P_D^{m_2/2} \qquad\qquad (8)$$

where $m_1 > m_2$. Dividing both sides by $P_D^{m_1/2}$ yields

$$P_C/P_D^{m_1/2} = K_{m_1} + K_{m_2}/P_D^{(m_1-m_2)/2} \qquad\qquad (9)$$

with the choice of $1/P_D^{(m_1-m_2)/2}$ as independent variable being
arbitrary. Isothermal plots based on eq. (9) indicate the temperature dependence of the equilibrium quotients $K_{m_1}$ and $K_{m_2}$. If
the natural logarithm of these quotients are well behaved straight
line functions of $1/T(K)$ ($K_m = A_m e^{-B_m/T}$), then the two species
model will adequately reproduce the vapor pressure data. If there
are deviations from straight line behavior for all possible values
of $m_1$ and $m_2$, then a three species model can be tested with the
proposed third species being dependent upon the nature of the deviation of the previous model.

$$P_C = K_{m_1} P_D^{m_1/2} + K_{m_2} P_D^{m_2/2} + K_{m_3} P_D^{m_3/2} \qquad\qquad (10)$$

For example, in the $HoCl_3(AlCl_3)_x$ system (3), the data indicate
that a previously unreported higher temperature species, $HoAl_2Cl_9$,
makes a 20% contribution to the total vapor density at 750 K and
2 atm of dimer pressure. Re-analysis of the $Nd^{+3}$ system also led
us to infer the existence of this lower molecular weight species
at higher temperature (11). Examination of temperature dependent
optical spectra indicates that spectral changes of the $^5I_8$ to $^5G_6$
transition of $Ho^{+3}$ are consistent with this interpretation (3).


CONCLUSIONS


    The spectrophotometric technique for vapor density measure-
ments of complexed metal ions has been reformulated to account for

temperature dependent effects and multi-species systems. Analysis of vapor pressure information indicates that the $NdCl_3$-$AlCl_3$ and $HoCl_3$-$AlCl_3$ systems are adequately explained by the existence of three vapor species. The two higher molecular weight complexes $LnAl_4Cl_{15}$ and $LnAl_3Cl_{12}$ were first proposed by Øye and Gruen. The newly identified higher temperature species, $HoAl_2Cl_9$, contributes significantly to the vapor density above 750 K and below 3 atm of dimer pressure. In view of the consistency of the $Nd^{+3}$ (1) and $Ho^{+3}$ (3) chemistry the data for the $Sm^{+3}$ system (2) should be viewed with reservation. A new method for vapor density measurements involving use of radioactive tracers has been discussed in terms of its applicability to the study of $(Ln,An)Cl_3(AlCl_3)_x$ systems.

## REFERENCES

1.   H. A. Øye and D. M. Gruen, J. Amer. Chem. Soc. 91:2229 (1969).

2.   G. N. Papatheodorou and G. H. Kucera, Inorg. Chem. 18:385 (1979).

3.   E. J. Peterson, J. A. Caird, W. T. Carnall, Jan P. Hessler, H. R. Hoekstra and C. W. Williams, to be published.

4.   D. M. Gruen and R. L. McBeth, Inorg. Chem. 8:2625 (1969).

5.   M. Sørlie and H. A. Øye, J. Inorg. Nucl. Chem. 40:493 (1978).

6.   H. A. Øye and D. M. Gruen, 10th Materials Research Symposium, Gaithersberg, MD, Sept. 1978, in press.

7.   Jan P. Hessler and C. W. Williams, unpublished results.

8.   Jan P. Hessler, J. Phys. Chem., in press.

9.   E. J. Peterson, J. A. Caird, Jan P. Hessler, H. R. Hoekstra and C. W. Williams, J. Phys. Chem., in press.

10.  A. Anudskas and H. A. Øye, J. Inorg. Nucl. Chem. 37:1609 (1975).

11.  E. J. Peterson, Jan P. Hessler and J. A. Caird, J. Phys. Chem., submitted.

EFFECT OF TEMPERATURE ON THE STABILITY OF FLUORO COMPLEXES OF THE

RARE EARTH ELEMENTS IN FLUORITE BEARING MODEL SYSTEM

B.A. Bilal and P. Becker

Hahn-Meitner-Institut für Kernforschung GmbH

D - 1000 Berlin 39, FRG

## INTRODUCTION

Thermodynamic processes like complex formation and liquid/ solid distribution equilibria are assumed to be responsible for the fractionation of the rare earth elements (REE) during the mineralization of Ca minerals, particularly fluorite.

The formation of the REE fluoro complexes in a fluorite bearing model system has, therefore, been studied at various ionic strengths. Bilal, Herrmann and Fleischer (1) and Bilal and Koß (2) determined the stability constants of the fluoro complexes of trivalent La, Ce, Nd, Eu, Tb, Er, Yb and Lu at pH $\leq$ 5, T = 25$^{\circ}$C and the ionic strength 1.0 M. Bilal and Becker (3) further extended these studies a maximum in the stability of the monofluoro complexes was found at Tb and a lower stability for the La than the Lu monofluoro complex. The stability constants of the difluoro complexes, on the other hand, increase gradually through the series. For the evaluation of a model for the fractionation of the REE in hydrothermally formed fluorite, it is, of course, necessary to carry out the investigation under a pressure of $\sim$ 300$^{\circ}$C in special equipment. However, it was possible to apply the described (1) potentiometric method up to 75$^{\circ}$C to obtain the initial temperature effect on the stability variation within the series.

## EXPERIMENTAL

The formation function of the fluoro complexes of five selected REE (La, Nd, Tb, Er and Lu) was determined at the ionic strengths 0.1, 0.5, 1.0 and 3.0 M (NaCl), at pH = 3 and temperatures of

117

25, 40, 60 and 75°C by means of potentiometric measurements of the free fluoride and free hydrogen ions in system. All experimental details are described in (1). The stability constants $\beta_i$ were obtained from the formation functions

$$\sum_{i=0}^{i=N} (\bar{n} - i)\beta_i [F]^i = 0 \qquad \begin{array}{l} (\bar{n}, N = \text{average and maximum} \\ \text{ligand number}) \end{array} \qquad (1)$$

by means of a linear regression analysis using an ALGOL program.


## RESULTS AND DISCUSSION

The variation of the stability constants $\beta_1$ of the monofluoro complexes within the REE series was determined at the four ionic strengths at 25°C (3). The best agreement between this variation and the fractionation tendency of the REE found in fluorite seems to be obtained at the ionic strength I = 0.5 M, which corresponds to the most commonly assumed salinity of fluorite bearing solutions.

Fig. 1 shows (for example) the variation of $\beta_1$ and $\beta_2$ across the REE series at the ion strength 0.5 M and at temperatures of 25, 40, 60 and 75°C. A relatively large deviation of the $\beta_1$ values of La and Nd complexes was obtained with increasing temperature. No reasonable values resulted for $\beta_2$ of the complexes of these two REE.

This deviation may probably be due to a drift of the fluoride membrane electrode used for measurement of the free fluoride, at elevated temperatures. An increasing exchange of the $REE^{3+}$ ions between the $LaF_3$ crystal of the electrode and the solution is expected with increasing temperature. This exchange, which is more probable for La and the neighboring REE, may lead to a large error of the electrode potential. Moreover, the $LaF_3$ crystal may accelerate the precipitation of La in solution, so that a relatively small range is available for the evaluation of the formation function.

The relatively strong temperature dependence of the stability constants of the REE fluoro complexes is due to the electrostatic character of the bond between ligand and central ion, which increases as the dielectric constant of water becomes smaller with elevating temperature. Considering the relative variation of $\beta_1$ with increasing temperature (figure 1) it is remarkable that the order

$$(\Delta\beta_1/\beta_1)_{La} > (\Delta\beta_1/\beta_1)_{Lu} > (\Delta\beta_1/\beta_1)_{Tb}$$

was obtained. This leads to a flattening of the maximum of $\beta_1$ at Tb as the temperature increases. Figure 2, in which log $\beta_i$ is plotted versus 1/T, illustrates this tendency more clearly. The curves of

Table I.    $\Delta H_i$ and $\Delta S_i$ of the REE Fluoro Complexes at Various Temperatures and Ionic Strengths

| REE | I [M] | $\Delta H_1$ [k J/mol] | $\Delta S_1$ [J/degree/ mol] | $\Delta H_2$ [k J/mol] | $\Delta S_2$ [J/degree/ mol] |
|-----|-------|------------------------|------------------------------|------------------------|------------------------------|
| Nd  | 0.1   | 22.9                   | 138                          |                        |                              |
|     | 0.5   | 21.6                   | 124                          |                        |                              |
|     | 1.0   | (13.3)                 | (96)                         |                        |                              |
| Tb  | 0.1   | 12.1                   | 109                          | 34.2                   | 230                          |
|     | 0.5   | 9.6                    | 96                           |                        |                              |
|     | 1.0   | (9.2)                  | 92                           | 31.1                   | 109                          |
|     | 3.0   | 8.6                    | 92                           |                        |                              |
| Er  | 0.1   | 13.9                   | 113                          | 23.5                   | 197                          |
|     | 0.5   | 13.4                   | 106                          |                        |                              |
|     | 1.0   | 11.6                   | 100                          |                        |                              |
|     | 3.0   | 6.7                    | 84                           |                        |                              |
| Lu  | 0.1   | 15.7                   | 121                          | 34.1                   | 230                          |
|     | 0.5   | 14.1                   | 108                          |                        |                              |
|     | 1.0   | 16.9                   | 120                          | 20.3                   | 176                          |
|     | 3.0   | 16.4                   | 113                          |                        |                              |

FIGURE 1.  The variation of $\beta_1$ and $\beta_i$ within the REE series at ionic strength 0.5M and different temperatures.

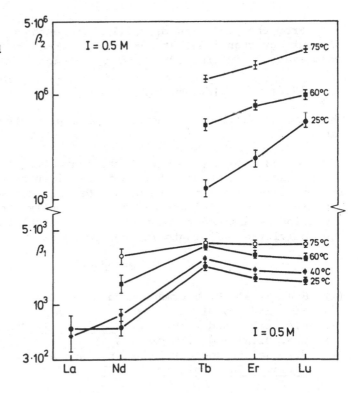

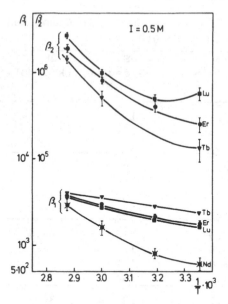

FIGURE 2.   log $\beta_{1,2}$ of the fluoro complexes as a function of $1/T$
the stability constant $\beta_1$ decrease  more strongly in the order
Tb < Lu < Nd.


From the plots log $\beta_i$ = f(1/T) the enthalpy change $\Delta H_i$ and
the entropy change $\Delta S_i$ attending the formation of the REE fluoro
complexes are determined· according to the fundamental equation

$$\ln \beta_i = -\ \Delta H_i/RT + \Delta S_i/R \tag{2}$$

The results obtained at the four ionic strengths are summarized in
Table 1.

The $\Delta H_1$ and $\Delta S_1$ values at the ionic strength 1.0 M are lower
by the factor of about 0.5 than those given by Walker and Choppin (4)
for the same ionic strength but in $NaClO_4$ medium.  Since the participation of both $ClO_4^-$ and $Cl^-$ to the complexation is negligible, this
deviation is assumed to be mainly attributable to the difference of
the activity coefficients in the two media.

REFERENCES

(1)  B.A. Bilal, F. Herrmann, and W. Fleischer, J. Inorg. Nucl. Chem.,
     41:  347 (1979).
(2)  B.A. Bilal and V. Koß, J. Inorg. Nucl. Chem., (in press)
(3)  B.A. Bilal and P. Becker, J. Inorg. Nucl. Chem., (in press)
(4)  J.B. Walker and G.R. Choppin, Lanthanide/Actinide Chemistry,
     Advances in Chemistry 71, 127 (1967).

# STABILITY CONSTANTS OF RARE-EARTH COMPLEXES WITH SOME ORGANIC LIGANDS

Y. Suzuki, S. Yokoi, M. Katoh, M. Minato, and
N. Takizawa
Department of Industrial Chemistry, Faculty of
Engineering, Meiji University
Ikuta, Tama-ku, Kawasaki 214, Japan

## INTRODUCTION

Although stability constants of many rare-earth complexes
have been determined by various methods, the elucidation of some
problems such as the relatioship between the structure and stability trends has not yet been satisfactorily given.  The present
work was intended to accumulate stability data of high accuracy,
as well as to improve the simplified methods of determining stability constants.  Since no thorough stability data have been reported
on the β-hydroxypropionate and malate complexes, the complex formation equilibria were investigated by the potentiometric method.

During the last decade, ion-selective electrodes have had
extensive applications to various aqueous systems.  Determining
stability constants of the chelates with complexans such as ethyl-
enediaminetetraacetic acid ($H_4$edta) may require the utilization of
displacement equilibria.  Among the ion-selective electrodes commercially available, copper-, lead-, and cadmium-ion electrodes
have been examined, and the results with copper- and lead-ion
selective electrodes are reported.

## THEORETICAL

The method employed for the β-hydroxypropionato and malato
complexes was based on the conventional potentiometry developed
by Bjerrum (1), Sonesson (2), and others, and has been used in
previous work (3,4) by one of the authors.

The method involving ion-selective electrodes is based upon

the displacement equilibrium

$$MA + Ln \rightleftharpoons LnA + M \tag{1}$$

where Ln and A are the rare-earth cation and the complexan anion, respectively, and M is the cation to be measured by the specific ion-selective electrode. All the ionic charges are omitted hereafter. The equilibrium constant of reaction 1, $K_I$, can be given by

$$K_I = [LnA][M]/[MA][Ln]$$

$$= [M]^2/([MA]_0 - [M])([Ln]_0 - [M])$$

where $[MA]_0$ and $[Ln]_0$ denote the initial concentrations of MA and Ln, respectively. Since $K_I$ can be rewritten by using the stability constants of the M- and rare earth-chelates, $K_{MA}$ and $K_{LnA}$, respectively,

$$K_I = K_{LnA}/K_{MA}$$

thus the relationship

$$K_{LnA} = K_I K_{MA} \tag{2}$$

shows that the values of $K_{LnA}$ can be obtained from the values of $K_I$ and $K_{MA}$.


## EXPERIMENTAL

Rare-earth perchlorate and nitrate standard solutions were prepared from the respective rare-earth oxides of 99.9% or higher purity, supplied by Shin-etsu Chemical Industries Co., Ltd. Other reagent solutions were prepared from reagent grade chemicals.

The pH and electromotive force measurements were made with a Corning model 101 digital electrometer, a Corning model 12 pH meter, and a Beckman model 3500 pH meter. The electrodes used were Beckman 39000, Corning 476022 and 476024 glass electrodes, an Orion 94-29A copper-ion selective electrode, a Denki Kagaku Keiki 7180-1P(S) lead-ion selective electrode, and an Orion 90-02 double junction reference electrode. All measurements were performed at 25.0 $^\circ$C and at an ionic strength of 0.10 with either sodium perchlorate or potassium nitrate as the supporting electrolyte.

Since ion-selective electrodes require adequate solution conditions such as pH and concentration, a careful preliminary experiment has to be carried out to find the optimum experimental conditions. A copper-ion electrode has to be used in a consistent brightness, and a lead-ion electrode should be conditioned in a pH 6 buffer solution.  During the measurements using copper- and lead-ion electrodes, the pH values of the solutions were kept at 4 and 5, respectively.

The stability constants of the β-hydroxypropionato complexes were calculated using a FACOM 230-75 digital computer at the Japan Atomic Energy Research Institute, by the revised computational programs of the previous work (4).  The constants of the malato complexes were calculated by successive approximation from the potentiometric titration data.

## RESULTS AND DISCUSSION

### β-Hydroxypropionato and Malato Complexes

The values of the acid dissociation constants of β-hydroxypropionic acid were determined as $4.001 - 4.616 \times 10^{-5}$ within a concentration range of $4 \times 10^{-4}$ to $5 \times 10^{-2}$ mol dm$^{-3}$ at $25.0^\circ$C and at an ionic strength of 0.10.  The values are in good agreement with those by Crutchfield, et al. (5) and Cefola, et al. (6).  The acid dissociation constants of malic acid, $pK_1$ and $pK_2$, were obtained by titration method as 3.28 and 4.79, respectively.  The values are also in good agreement with those by Cefola, et al. (6) and of Davidenko (7).  The stability constants of the β-hydroxypropionato and malato complexes are given in Table 1 and in Fig. 1.

Table 1.  Stability Constants of the β-Hydroxypropionato and Malato Complexes at $25.0^\circ$C and at I = 0.10 (NaClO$_4$ and KNO$_3$, respectively)

| Element | β-Hydroxypropionates | | Malates | |
|---------|----------|----------|----------|----------|
|         | log $K_1$ | log $K_2$ | log $K_1$ | log $K_2$ |
| La | 1.910 | 1.38 | 4.60 | 2.9 |
| Ce | 2.072 | 1.19 | 4.88 | 3.4 |
| Pr | 2.049 | 1.26 | 4.87 | 3.3 |
| Nd | 2.232 | 1.46 | 4.93 | 3.2 |
| Sm | 2.232 | 1.49 | 5.04 | 3.3 |
| Eu | 2.080 | - | 5.03 | 3.3 |
| Gd | 1.846 | - | 5.00 | 3.3 |
| Tb | 1.987 | 1.21 | 5.08 | 3.4 |
| Dy | 2.034 | 1.31 | 5.12 | 3.4 |
| Ho | 1.932 | 1.41 | 5.18 | 3.5 |
| Er | 1.836 | 1.18 | 5.20 | 3.6 |
| Tm | 1.959 | 1.00 | 5.27 | 3.6 |
| Yb | 1.986 | 1.00 | 5.36 | 3.8 |
| Lu | 1.969 | 1.23 | 5.42 | 3.9 |
| Y | - | - | 5.03 | 3.3 |

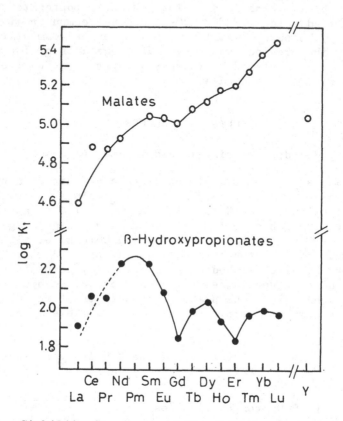

Figure 1.   Stability Constants of the β-Hydroxypropionates and
            Malates.

Cefola, et al. (6) determined the stabilities of malates for the
first four elements of the lighter rare earths.  Their values at
$30^{\circ}C$ are higher than those of the present work, while those by
Davidenko (7) and by Březina (8) are lower.  Any of the previous
results do not cover the entire rare earths.

   The stability trend of the β-hydroxypropionates shown in

Fig. 1 is quite peculiar; it has some similarity to the result of
Jones and Choppin (9), particularly in the heavier rare earths.
Higher stability values of the present work may be due to the
difference in the experimental conditions.  A simple steric effect
appears not to be the reason of low stability constants in the
heavier rare-earth complexes.  The presence of at least two breaks
at gadolinium and erbium, which can also be found in the other
complexes studied, is appreciable.  The malato complexes, on the
other hand, yield a smoothly rising stability curve, although the
breaks again interrupt the curve at gadolinium, erbium, and also
at neodymium.


## The Ethylenediaminetetraacetato Chelates

Stability constants of the rare earth-edta chelates have
already been determined by Wheelwright, et al. (10) and by
Schwarzenbach, et al. (11)  As mentioned, the present method of
utilizing ion-selective electrodes requires the knowledge of the
accurate stability data of the edta chelates of copper and lead.
Unfortunately, however, these stability constants were too high to
be determined directly by the respective electrodes.  We thus
selected the standard logarithmic values of 18.80 and 18.04 for
copper and lead chelates, respectively (both at $20.0^{\circ}C$ and at
$I = 0.10$ with potassium nitrate (12)).  The results are shown in


Table 2.  A Comparison of the log $K_1$ values of the Rare Earth-edta
Chelates at $I = 0.10$ $(KNO_3)$

| Element | This work* | | Wheelwright, et al.**(10) | Schwarzenbach, et al.**(11) |
| | With Cu electrode | With Pb electrode | | |
|---|---|---|---|---|
| La | 15.70 | 15.08 | 14.72 | 15.50 |
| Ce | – | 15.61 | 15.39 | 15.98 |
| Pr | 16.46 | 16.14 | 15.75 | 16.40 |
| Nd | 16.71 | 16.31 | 16.06 | 16.61 |
| Sm | 17.22 | 16.76 | 16.55 | 17.14 |
| Eu | 17.38 | 16.93 | 16.69 | 17.35 |
| Gd | 17.41 | 16.95 | 16.70 | 17.37 |
| Tb | 17.96 | 17.54 | 17.38 | 17.93 |
| Dy | 18.38 | 17.82 | 17.75 | 18.30 |
| Ho | 18.64 | 18.09 | 18.31 | – |
| Er | 18.97 | 18.34 | 18.55 | 18.85 |
| Tm | 19.23 | 18.99 | 19.07 | 19.32 |
| Yb | 19.47 | 19.19 | 19.39 | 19.51 |
| Lu | 19.64 | 19.52 | 19.65 | 19.83 |
| Y | 18.12 | 17.78 | 17.56 | 18.09 |

*$25.0^{\circ}C$.     **$20.0^{\circ}C$.

Table 2. Any error in the pK of the copper- or lead-edta chelates leads to a systematic deviation in the data from the appropriate electrode. The stability differences between 20 and 25°C are not appreciable, and the relative relationships among the constants are consistent within the rare-earth metals. The present stability data may thus be considered reasonably reliable. A comparison of the results may be made by plotting the values of $K_I$, instead of the stabilities, versus atomic number.

It is now apparent that the ion-selective electrodes can be efficiently used for determining stability constants, with sufficient accuracy. The rare earth-ion sensitive electrodes, on which we have been working, could also be effective sensors for solutions containing trivalent rare-earth ions.

## ACKNOWLEDGEMENTS

The authors wish to express their gratitude to Shin-etsu Chemical Industries Co., Ltd. for supplying the rare-earth oxides, and to Messrs. H. Umezawa and Y. Nakahara of the Japan Atomic Energy Research Institute for computational assistance. The present work was partially supported by a Grant-in-Aid for Developmental Scientific Research from the Institute of Science and Technology, Meiji University.

## REFERENCES

1.  J. Bjerrum, "Metal Ammine Formation in Aqueous Solution," P. Haase and Son, Copenhagen, 1957.
2.  A. Sonesson, Acta Chem. Scand. 12: 165 (1958).
3.  J. E. Powell and Y. Suzuki, Inorg. Chem. 3: 690 (1964).
4.  Y. Suzuki and J. E. Powell, Bull. Chem. Soc. Japan 49: 2327 (1976).
5.  C. A. Crutchfield, Jr., W. M. McNabb, and J. F. Hazel, J. Inorg. Nucl. Chem. 24: 291 (1962).
6.  M. Cefola, A. S. Tompa, A. V. Celiano, and P. S. Gentile, Inorg. Chem. 1: 290 (1962).
7.  N. K. Davidenko, Zhur. Neorgan. Khim. 7: 2709 (1962); Russ. J. Inorg. Chem. 9: 859 (1964).
8.  F. Brézina, Monatsh. 94: 772 (1963).
9.  A. D. Jones and G. R. Choppin, J. Inorg. Nucl. Chem. 31: 3523 (1969).
10. E. J. Wheelwright, F. H. Spedding, and G. Schwarzenbach, J. Amer. Chem. Soc. 75: 4196 (1953).
11. G. Schwarzenbach, R. Gut, and G. Anderegg, Helv. Chim. Acta 37: 937 (1954).
12. G. Anderegg, Helv. Chim. Acta 47: 1801 (1964).

CRYSTAL FIELD PARAMETERS FOR Eu(III) IN LANTHANIDE PERCHLORATES

WITH HEXAMETHYLPHOSPHORAMIDE (HMPA)

Sérgio Maia Melo * and Osvaldo Antonio Serra**

*Centro de Ciências - UFC - 60.000 - Fortzleza-Brasil

** Fac. Filosofia, Ciências e Letras - USP

Coordination compounds of hexamethylphosphoramide (HMPA) with lanthanide ions have been reported in the literature (1,2), but no doped organic complexes with lanthanide ions are known. The preparation and the crystal field calculations of the doped lanthanide perchlorates with HMPA are described. The emission spectra were measured and the probable assignments of the excited levels of the trivalent ions were made. The analysis of the low temperature fluorescence spectra of the prepared complexes shows that the local symmetry around the trivalent lanthanide ion is nearly $D_{2d}$. With this symmetry, the values of the crystal field parameters have been calculated.

PROCEDURES

Starting Materials

The rare earth oxides were 99.9% pure as obtained from Ventron Corporation - Alfa Products. The hexamthylphosphoramide (HMPA), obtained from Aldrich Chemical Co., was also used as received. All other chemicals were reagent grade, and the solvents treated to be anhydrous by known procedure.

Preparation of compounds

The lanthanide perchlorates were obtained by the reaction of the perchloric acid with the corresponding oxides (La, Eu, Gd and Y). These were dissolved in acetonitrile and HMPA was added. The compounds formed after 48 hours, were washed with cold acetronitrile

and stored in a vacuum desiccator. The coordination compounds of
HMPA with perchlorates of La, Gd and Y were "doped" with Eu(III)
and after recrystallization in acetonitrile, compounds were obtained
with the general formula $\{Ln(HMPA)_6\}(ClO_4)_3$. Powdered samples were
introduced in a quartz tube and immersed in a glass dewar flask with
a quartz tail and filled with liquid nitrogen.

Physical Methods

Infrared spectra were recorded in the region 1400- 400 $cm^{-1}$
with a Perkin-Elmer 337 spectrophotometer using KBr windows.

Visible spectra (fluorescence) were recorded in the region
5200 - 7000 $\overset{\circ}{A}$ on a GCA/McPherson RS-10 spectrophotometer.

Analytical Procedures

Elemental analysis (C,N and H) was performed in a Perkin-Elmer
240. The analytical procedure for the determination of the lanthanide ions were performed by EDTA titration (3).

RESULTS AND DISCUSSION

The $(2J + 1)$ degenerate levels that arise in the spherical
symmetry (free ion) are split in crystal forms. This occurs because
the neighbors create an electrical field which acts on the central
ion and has the same symmetry as the point group. In the fluorescence spectra of $Ln_{.99}Eu_{.01}(HMPA)_6(ClO_4)_3$; Ln = La, Eu, Gd, and
Y the observed transitions have been assigned by comparison with the
paper of De Shazer and Dieke (4) for Eu(III) in $LaCl_3$. The low
concentration of europium ion gives rise to a small number of lines,
all arising at the $^5D_0$. In a free ion transitions are forbidden by
the parity rule for electric dipole transitions, but in a crystal
forced electric transition they become allowed due to the coupling
of odd electronic wave functions. Due to the weakness of the
transition $^5D_0 \rightarrow {}^7F_0$ ther is no indication of centro-symmetric systems. For the transition $^5D_0 \rightarrow {}^7F_1$ we observe two intense lines, one
of which is slightly in the matrixes of gadolinium, europium and
lanthanum. The transition $^5D_0 \rightarrow {}^7F_2$ presents four bands, two strong
and two weak. In this way, the level $^7F_2$ would be split into four
levels, among which could occur two allowed transitions and if in the
other two levels we would have two other transitions then the possible symmetries would be $C_{4v}$ or $D_{2d}$.

The study of compounds of europium (100%) shows the existence
of two or three lines that appear in the region of the transition
$^5D_0 \rightarrow {}^7F_3$ in agreement with $C_{4v}$ (2 lines) and $D_{2d}$ (3 lines) symmetries since in the other cases we would expect a greater number
of bands in this region. The symmetry is lower than $C_{4v}$ or $D_{2d}$

since in some spectra (La and Gd) the double degenerate line for the transition $^5D_0 \rightarrow {}^7F_1$ indicates a tendency for splitting. For calculation purposes we will hold to symmetry $D_{2d}$ considering also the two weak emission in the infrared region as belonging to the $F_2$ level since they are approximately 230 cm$^{-1}$ apart from other transitions at the same level. We would emphasize that in the compounds $Cs_2NaEuCl_6$ (5) and $Eu(HMPA)_5Cl$ (6) this separation is also of the order of 220 cm$^{-1}$. Fig. 1 shows the spectrum of the gadolinium compound.

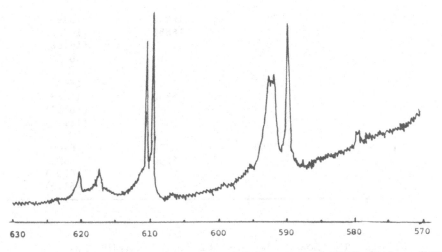

Figure 1. The Fluorescence Spectrum of $Gd_{.99}Eu_{.01}(HMPA)_6(ClO_4)_3$(77K)

The values of the observed transition in the fluorescence spectra of the prepared compounds, and the lowest energy levels observed are indicated in Tables I and II, respectively.

The crystal field parameters of these systems were determined using a FORTRAN program appropriate for $D_{2d}$ symmetry which calculates at the base $^7F_J$ energy levels for which we observe transitions. In this program, the matrixes for each level $^7F_J$ are calculated separately, not considering the effect of J-mixing. In order to find the CF parameters, a set of values $B_q^k$ are automatically introduced, one by one, in the matrices and by the method of minimum chi-squared, the program searches for that value which best simulates reality. This set of parameters will be that one which reproduces a better fitting with the observed energy levels. For our compounds the values found are shown in Table III.

Table 1. Transitions observed by Fluoresence for
$Ln_{.99}Eu_{.01}(HMPA)_6(ClO_4)_3$ Compounds.

| Compound of | \multicolumn{6}{c}{Transitions ($cm^{-1}$)} |
|---|---|---|---|
|  | $^5D_0 \rightarrow {}^7F_0$ | $^5D_0 \rightarrow {}^7F_1$ | $^5D_0 \rightarrow {}^7F_2$ |
| La | 17244 | 16955 16895 | 16407 16383 (16202) (16126) |
| Eu | 17247 | 16952 16883 | 16396 16380 (16186) (16124) |
| Gd | 17247 | 16955 16880 | 16409 16383 (16197) (16121) |
| Y | 17250 | 16958 16880 | 16409 16383 (16194) (16116) |

Table 2. The Lowest Levels for $Ln_{.99}Eu_{.01}(HMPA)_6(ClO_4)_3$

| Level | Irreducible Representation | \multicolumn{8}{c}{Energy ($cm^{-1}$)} |
|---|---|---|---|---|---|---|---|---|---|
|  |  | \multicolumn{2}{c}{La} | \multicolumn{2}{c}{Eu} | \multicolumn{2}{c}{Gd} | \multicolumn{2}{c}{Y} |
|  |  | obs | cal | obs | cal | obs | cal | obs | cal |
| $^7F_0$ | $A_1$ | 0 | 0 | 0 | 0 | 0 | 0 | 0 | 0 |
| $^7F_1$ | $A_2$ | 289 | 287 | 295 | 297 | 292 | 292 | 292 | 292 |
|  | E | 349 | 350 | 364 | 363 | 367 | 367 | 370 | 370 |
| $^7F_2$ | $B_2$ | 837 | 848 | 851 | 851 | 838 | 847 | 841 | 849 |
|  | E | 861 | 860 | 867 | 864 | 864 | 865 | 867 | 869 |
|  | $A_1$ ou $B_1$ | 1042 | 1049 | 1061 | 1058 | 1050 | 1054 | 1056 | 1060 |
|  | $B_1$ ou $A_1$ | 1118 | 1115 | 1123 | 1126 | 1126 | 1128 | 1134 | 1137 |

Table III Crystal field parameters of $Eu^{+3}$ in $Ln_{.99}Eu_{.01}(HMPA)_6(ClO_4)_3$

| $B_q^k$ (cm$^{-1}$) | Compound of | | | |
|---|---|---|---|---|
| | La | Eu | Gd | Y |
| $B_0^2$ (cm$^{-1}$) | -210 | -230 | -250 | -260 |
| $B_0^4$ (cm$^{-1}$) | -1340 | -1380 | -1390 | -1414 |
| $B_4^4$ (cm$^{-1}$) | ± 960 | ± 984 | ± 991 | ±1009 |

The energy levels calculated with the parameters indicated in Table III are in reasonable agreement with the experimental results, especially considering the small number of transitions obtained and the fact of not having carried out calculations with "J-mixing". The compounds are well removed from the point $O_h$, for which the ratio $B_0^4/B_4^4 = 1.67$ (07); for our compounds this ratio was constant and equal to 1.40.

## ACKNOWLEDGMENT

We are grateful to the Institute de Química da Universidade de São Paulo for the use of their laboratories and to Conselho Nacional de Desenvolvimento Científico e Technólogico (CNP$_q$) and Organization of American States (OAS) for financial support. We wish to thank Dr. L.C. Thompson for the fluorescence spectra and Dr. P. Porcher for helpful comments.

## REFERENCES

1. S.M. Melo and O.A. Serra, Proc. 12th Rare Earth Res. Conf., 180. (1976).
2. E. Giesbrecht, and L.B. Zinner., Inorg. Nucl. Chem. Lett., 5: 575, (1969).
3. I.M. Kolthoff, and P.S. Elving. "Treatise on Analytical Chemistry" Part. II, vol. 8, Interscience, New York. (1963).
4. L. De Shazer and G.H. Dieke., J. Chem. 38: 2190 (1963).
5. O.A. Serra and L.C. Thompson, Inorg. Chem. 15: 504 (1976).
6. P. Porcher, P. Caro and O.A. Serra, Unpublished results.
7. P. Caro "Structure Électronique des Élements de Transition; Presses Universitaires de France (1976).

# SOLVATION OF Eu(III) AND Tb(III) IONS IN ANHYDROUS $CH_3CN$ AND DMF

J.-C. G. Bünzli, J.-R. Yersin, and M. Vuckovic

Université de Lausanne, Institut de chimie minérale et analytique, Place du Château 3, 1005 Lausanne, Switz.

## INTRODUCTION

Lanthanoid(III) ion solvation has been investigated by many different techniques. Few studies, however, have been performed in strictly anhydrous solvents, and the coordination number of these ions is not always established with certainty. We report here some results we obtained on the solvation of Eu(III) and Tb(III) ions in anhydrous acetonitrile, dimethylformamide, and in binary solvent mixtures. We have studied solutions of nitrate and perchlorate (0.0001-0.2 M) by means of conductimetry, vibrational spectroscopy, and fluorescence measurements which are very sensitive to changes in the first solvation sphere (1). The long fluorescence lifetimes reported in Table 1 reflect the very low water content of the salts (2) and solvents (3) used (no water quenching).

Table 1. Fluorescence lifetimes (ms) of $Ln(NO_3)_3$ and $Ln(ClO_4)_3$ [a].

| Solvent | $Eu(NO_3)_3$ | $[Eu(NO_3)_5]^{2-}$ | $Eu(ClO_4)_3$ | $Tb(NO_3)_3$ | $Tb(ClO_4)_3$ |
|---|---|---|---|---|---|
| $CH_3CN$ | 1.35 0.77[b] | 1.26 | 2.10 1.22[b] | 1.87 | 2.42 |
| $CD_3CN$ | | 1.35 | 2.50 2.00[b] | | |
| DMF | 1.47 | | 1.44 | 1.86 | 2.35 |

[a] $[Eu^{3+}] = 0.05$ M, $^5D_0 \rightarrow {}^7F_2$ transition; $[Tb^{3+}] = 0.02$ M, $^5D_4 \rightarrow {}^7F_5$ transition.
[b] Reference (4).

### Eu(III) SOLVATION

Conductometric measurements on 0.0001–0.01 M solutions of europium nitrate in strictly anhydrous acetonitrile indicate that essen-
tially no dissociation occurs ($\Lambda \sim 15$–$20 \Omega^{-1}.cm^2.mol^{-1}$). The vibrational
spectra show the presence of coordinated acetonitrile, with bands
at 395, 2270 and 2310 $cm^{-1}$, compared to 380, 2240 and 2290 $cm^{-1}$ for
bulk solvent. Nitrato group frequencies, as well as Raman band
polarizations are consistent with bidentate (or bridging) anions of
$C_{2v}$ local symmetry (2). No band appears at 830–835 $cm^{-1}$, which precludes the presence of uncoordinated (ionic) nitrate. When the total
concentration of nitrate is increased, the pentanitrate species
$[Eu(NO_3)_5]^{2-}$ forms. The IR spectrum of the solution is then practically identical with the spectrum of the solid pentanitrate and
since crystal structures of such salts demonstrate the presence of
five bidentate nitrato groups (5), one may conclude that the Eu(III)
ion is decacoordinated in such solutions. This conclusion is further
demonstrated by the following experimental facts. (i) The fluorescence lifetime of a 0.05 M solution of $[Eu(NO_3)_5]^{2-}$ only increases
a few percent when acetonitrile is replaced by deuterated acetonitrile (cf. Table 1). A more pronounced change would be expected
if solvent molecules were involved in the first solvation sphere.
(ii) When more nitrate is added, no further change occurs, neither
in the fluorescence spectrum, nor in the fluorescence lifetime.
This is also illustrated by Fig. 1 which shows the influence of the
progressive replacement of acetonitrile by nitrato groups in the
first solvation sphere of europium perchlorate upon the intensity
ratio of the forced electric dipole transition $^5D_o \rightarrow {}^7F_2$ to the
magnetic dipole transition $^5D_o \rightarrow {}^7F_1$ : this ratio increases linearly
up to a nitrate/Eu ratio equal to 5 and then remains constant.
(iii) When water is added to an acetonitrile solution of europium
nitrate, the fluorescence lifetime experiences a drastic drop from
1.35 ms to 0.58 ms when the water/Eu ratio, R, reaches 1. If nitrate
is added to such a solution (R = 0.9), the fluorescence lifetime
of 0.72 ms first decreases to 0.49 ms (cf. Fig. 2), reflecting the
replacement of acetonitrile by a nitrato group and then increases
again sharply to 1.23 ms until the nitrate/Eu ratio reaches 5. That
is, the water molecule is expelled from the first solvation sphere
and the Eu(III) ion is 10-coordinated.

The strong nitrate/europium interaction is evidenced when water
is added to acetonitrile solutions of europium nitrate. The first
two acetonitrile molecules are easily and quantitatively replaced by
water molecules. The replacement of the remaining solvent molecules
is more difficult to achieve and requires R = 8. FT-IR difference
spectra of solutions having R = 18 show that all the nitrato groups
are still coordinated; they can only be displaced at higher water

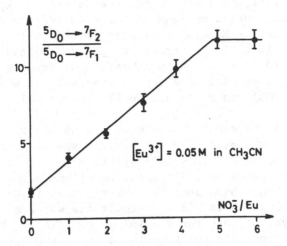

Figure 1.   $^5D_o \rightarrow {}^7F_2$ / $^5D_o \rightarrow {}^7F_1$ intensity ratio for Eu³⁺ in anhydrous acetonitrile versus total nitrate concentration.

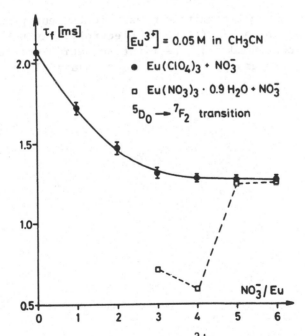

Figure 2.   Fluorescence lifetime of Eu³⁺ in anhydrous acetonitrile
versus total nitrate concentration.

concentration. Interpretation of this re-solvation experiment leads
to the conclusion that a maximum of 4 acetonitrile molecules are
directly bound to the Eu(III) ion; that is, the Eu(III) ion is again
10-coordinated since the 3 nitrato groups are bidentate. It seems
therefore that 10 is a common coordination number for Eu(III), even
in water where, according to Horrocks & Sudnik, an equilibrium occurs
between nona- (40%) and deca-aquo (60%) species (6).

The weak fluorescence band around 580 nm allows one to observe
the formation of mono-, di- and trinitrato species. This transition does not undergo ligand-field splitting and thus each of its
components corresponds to a different europium-containing species.
We have studied this band under high resolution (0.08 nm) for europium nitrate solutions in mixed acetonitrile/water mixtures. Figure
3 displays a typical spectrum showing an equilibrium between 3
species : the mono-, di- and trinitrato species occurring at 579.47,
579.78, and 580.17 nm, respectively. The position and width at half
height of these bands remain constant when the nitrate concentration
is varied, whereas their varying relative intensities reflect the
equilibrium displacement. Such experiments show that di- and trinitrato species form almost simultaneously with the mononitrato
species.

According to conductometric measurements, europium nitrate is
a 2:1 (<0.02 M), or a 1:1 (>0.02 M) electrolyte in DMF solutions.
Krishnamurthy concluded that the first solvation sphere contains
one or two nitrato groups (7). Abrahamer deduced later from the

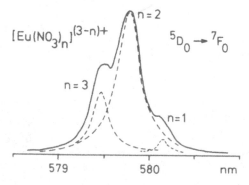

Figure 3.    $^5D_0 \rightarrow {}^7F_0$ fluorescence band of $Eu(NO_3)_3$ 0.05 M in
              $CH_3CN$, with $H_2O/Eu$ = 38.

study of absorption spectra that the Ln(III)/nitrate interaction in
DMF was an <u>outer</u>-sphere one (8). We have now found conclusive facts
to support the idea of an <u>inner</u>-sphere interaction. (i) The IR and
Raman spectra of a 0.04 M solution clearly shows the presence of
ionic (831, 1350 cm$^{-1}$) and coordinated (818, 1305, 1490 cm$^{-1}$) nitrate.
(ii) The fluorescence spectra of europium perchlorate and nitrate in
DMF are quite different; in the latter the forced electric dipole
(hypersensitive) transition is much enhanced, reflecting a lowering
of the symmetry around the Eu(III) ion due to the penetration of
nitrato groups into the first solvation sphere. (iii) When DMF is
added to a 0.05 M solution of europium nitrate in acetonitrile, a
close scrutiny of the $^5D_0 \rightarrow {}^7F_0$ band reveals that the component
corresponding to the trinitrato species almost vanishes whereas
the component corresponding to the dinitrato species becomes the
more intense.

## Tb(III) SOLVATION

Conductometric measurements on strictly anhydrous solutions lead
to the same conclusions as found for Eu(III) salts : the perchlorate
is a 3:1 electrolyte in both acetonitrile and DMF, the nitrate is a
2:1 (<0.02 M) or a 1:1 (>0.02 M) electrolyte in DMF; it is completely
undissociated in acetonitrile with extremely low equivalent conducti-
vities ($\Lambda \sim 10-15 \Omega^{-1}.cm^2 .mol^{-1}$). Fluorescence lifetimes ($^5D_4 \rightarrow {}^7F_5$
transition) of perchlorate solutions in both acetonitrile and DMF
are almost identical, and so are the fluorescence lifetimes of tri-

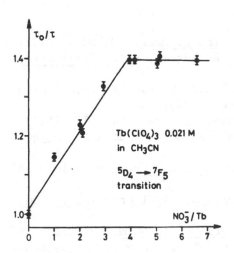

Figure 4. Fluorescence lifetime
versus nitrate concentration.

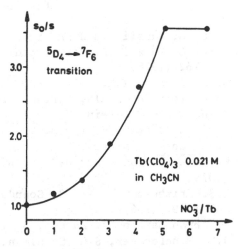

Figure 5. Relative band intensities versus nitrate concentration.

nitrate solutions, although they are much smaller due to the quenching effect of the nitrato groups. This clearly reflects an inner-
sphere nitrate/Tb(III) interaction and it is also indicated in the
fluorescence spectra. We have tried to determine the coordination
number of Tb(III) by adding nitrate to 0.02–0.05 M solutions of
perchlorate. Stern-Völmer plots (9) of $\tau_0/\tau$ are linear up to a
nitrate/Tb ratio R equal to 4 (acetonitrile, cf. Fig. 4), respectively 3 (DMF). They both show a break for R = 4. The variation of
the relative fluorescence band intensities (band areas) cf, for
instance, the $^5D_4 \rightarrow {}^7F_4$ and $^5D_4 \rightarrow {}^7F_6$ transitions gives a different
result. For acetonitrile solutions, both $s_0/s$ plots have breaks at
R = 5 (cf. Fig. 5) whereas for DMF solutions the break occurs
between R = 3 and 4 and it is less pronounced, which means that
the DMF molecules are not replaced easily by the nitrato groups.
A re-solvation experiment of terbium nitrate in acetonitrile by
water allows us to conclude that 3 acetonitrile molecules belong
to the first solvation sphere. These experimental facts point to a
smaller coordination number for Tb(III) compared to Eu(III); this
number is probably 9. We are now trying to support this conclusion
by studying the IR spectra of terbium salts in different solvents.


ACKNOWLEDGMENTS

Support from the Swiss National Science Foundation (grant
2.150–0.78) is gratefully acknowledged.


REFERENCES

1. J.-C. G. Bünzli and J.-R. Yersin, Inorg. Chem. 18: 605 (1979).
2. J.-C. G. Bünzli, E. Moret, and J.-R. Yersin, Helv. Chim. Acta
   61: 762 (1978).
3. D. D. Perrin, W.L.F. Armarego, and A.R. Perrin, Purification of
   Laboratory Chemicals, Pergamon Press 1966.
4. Y. Haas and G. Stein, J. Phys. Chem. 75: 3668 and 3677 (1971).
5. J.-C. G. Bünzli, B. Klein, G. Chapuis and K.J. Schenk, Inorg.
   Nucl. Chem., in press.
6. W. de W. Horrocks and D.R. Sudnik, J. Amer. Chem. Soc. 101: 334
   (1979).
7. S.S. Krishnamurthy and S. Soundararajan, J. Inorg. Nucl. Chem.
   28: 1689 (1966).
8. I. Abrahamer and Y. Marcus, J. Inorg. Nucl. Chem. 30: 1563 (1968).
9. J.D. Winefordner, S.G. Schulman, and T.C. O'Haver, Luminescence
   Spectrometry in Analytical Chemistry, Wiley-Interscience 1972,
   p. 71.

# YTTRIUM PURIFICATION BY SOLVENT EXTRACTION

Y. Minagawa, K. Kojima, T. Kaneko, F. Yajima,
K. Yamaguchi, T. Miwa, T. Yoshitomi
Mitsubishi Chemical Industries Limited
5-2, Marunouchi 2-chome
Chiyodaku, Tokyo, JAPAN

## INTRODUCTION

Solvent extraction of yttrium from the heavier rare earths can be significantly improved by the use of a chelating agent in the slightly alkaline aqueous phase. In particular, we report the results for yttrium extraction using DTPA(diethylene triamine penta acetic acid) or EDTA (ethylene diamine tetra acetic acid) as the chelating agent in the aqueous phase with a pH of ~ 8.7 where D2EHPA (di-2-ethyl-hexyl-phosphoric acid) was used as an extraction solvent.

Among these two chelating agents, DTPA brought much better outcome. Up to this work, the usage of chelating agents had been aimed to mask unwanted rare earth ions to prevent to be extracted under the equilibrium state (1), (2). In this study, however, the role of chelating agent has been found for the first time to control the extraction velocities of various rare earth ions, and this system is under the non-equilibrium state. Utilizing this controlled difference among the extraction velocities of rare earth ions and yttrium, a yttrium purification process has been devised and tested in a batch mode.

## THE PART OF EXPERIMENT

(I) Fundamental investigation:
(1) Feed solutions were prepared containing $YCl_3$ and one of the rare earth chlorides $RECl_3$ where RE=Yb, Er, Ho, Dy, Nd and the concentration of each chloride was adjusted to 0.05 mol/l. The

pH was adjusted to 0.1.  A 50 ml of the feed solution was extracted
with 50 ml of kerosene containing 0.5 mol/1 of D2EHPA for more
than 5 minutes.  The results are shown in Fig. 1. β´ indicates the
calculated separation coefficient β during extraction.  $\beta_{eq}$ is the
separation coefficient under the extraction equilibrium.

(2) An aqueous solution (feed solution) containing 0.05 mol/1
of $YCl_3$ and 0.05 mol/1 of $DyCl_3$ (Er, Ho) and 0.10 mol/1 of EDTA
was used.  The pH was adjusted to 8.7 with ammonia water.  A 100 ml
of the feed solution was adjusted to 8.7 with ammonia water.  A
100 ml of the feed solution was extracted with 100 ml of kerosene
containing 0.5 mol/1 of D2EHPA.  The relation between the extraction
time and percent extraction is shown in Fig. 2.

(3) As a feed solution, aqueous solutions containing 0.05 mol/1
$YCl_3$, 0.05 mol/1 of $RECL_3$ (RE=Dy, Ho, Er, or Yb) and 0.11 mol/1 of
DTPA were used.  The pH was adjusted to 8.6 with ammonia water.  A
100 ml of the feed solution was extracted with 100 ml of kerosene
containing 1.0 mol/1 of D2EHPA for more than 5 minutes.  In Fig. 3,
the resulting relation between percent extraction of Y ions
(abscissa) and percent extraction of rare earths (RE=Dy, Ho, Er, Yb)
(ordinate) is shown.

1: The all reagents used in this experiment were guaranteed as
special reagent grade and rare earth compounds had 99.99% purity.
2: The analytical methods used to determine the rare earth
were spectrophotometry, fluorescent x-ray, liq. chromatography,
optical emission spectroscopy, and chelating titration.

(II) Discussion:
From Fig. 1, it should be understood that the order of the
extracting velocity of rare earths to the D2EHPA solution is as
follows:   (La)> Nd > Dy > Ho > Y > Er ···· > Yb ⟵ faster.

However, Fig. 2 shows that yttrium is extracted to D2EHPA
faster than dysprosium when the feed solution contains the
chelating agent EDTA.  According to the stability constant (3) of
rare earths with EDTA, the position of yttrium is as follows:
Y < Dy < Ho < Er ···< Yb ⟶ stable.

Then the free ion concentration of yttrium under the presence
of EDTA is greater than that of dysprosium.  Supposing that the
extracting velocity is written; V=k C , where k is the velocity
constant depending on the rare earth and C is the free ion concentration of the rare earth.

We speculate from Fig. 1 that the order of constants k with
respect to the rare earths are $k_{Nd}> k_{Dy}> k_{Ho}> k_{Y}> k_{Er}> k_{Yb}$.

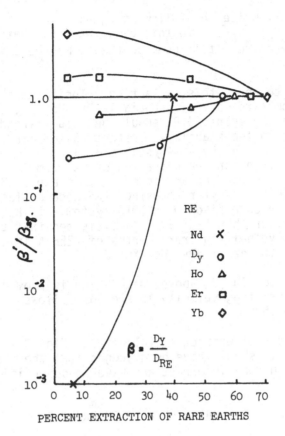

Figure I.

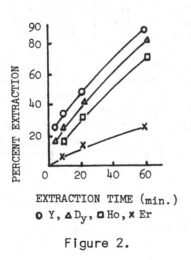

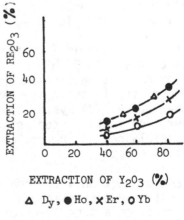

Figure 2.                              Figure 3.

Now, we can write the following equation:
$V_Y = K_Y \cdot [C_Y]$ , $V_{Dy} [C_{Dy}]$.  In this case, we find that $V_Y$ is greater than $V_{Dy}$.  Therefore, it is guessed that C may be a very important factor.

In order to apply this extracting velocity to the yttrium separation from rare earths, especially Dy, Ho, Er, it was thought that the chelating agent, DTPA should be used instead of EDTA. Now, according to the stability constant (3) of rare earth with
Y< Yb< Tm <Tb< Er< Ho< Dy ⟶ stable
Therefore, the free ion concentration of yttrium is much greater than that of Dy, Ho and Er and so on.  It is expected that the extracting velocity of yttrium using DTPA becomes faster than that of Dy, Ho, Er in comparison with EDTA system.  If $k_{Yb}$ is extra-ordinarily smaller than $K_{Er}$, it is anticipated that the order of the extracting velocity of rare earths of D2FHPA is as follows:
Y >Dy> Ho> Er> Yb  or  Y >Dy> Ho> Yb> Er.

According to Fig. 3, above speculation is **proved** to be correct and extracting velocity is seen as follows:
Y >Dy> Ho> Er> Yb.

If the order of extracting velocity in Fig. 3 were as follows: Y  Er  Ho  Dy, it should have been judged that the above mentioned order of k's would be incorrect and have no scientific meaning.

## THE BATCH EXTRACTION PROCESS

The high yttrium extraction velocity in the above extraction system is applied to yttrium purification by the authors using a batch operation.

This process comprises mainly three steps, a light rare earth separation step, a heavy rare earth separation step, and a DTPA recovery step.  The flow diagram of this process is shown in Fig. 4.

The first step is the light rare earth separation step.  This step takes place in D2EHPA HCl system, 6-stage mixer settlers having a 3-stage extraction section and a 3-stage scrub section were used. The second step is the heavy rare earth separation step.  This step uses the D2EHPA-DTPA system and is accomplished by a batch operation. The operation in this step is shown in Fig. 5.  The details of this step is explained in the following using the example of the n-th extraction stage.

Each extraction stage comprises three operations, the feed adjustment, the extraction, where the difference of the extracting velocity causes a separation of yttrium from other rare earth, and

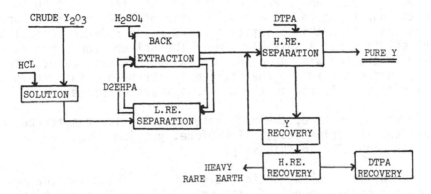

Figure 4.

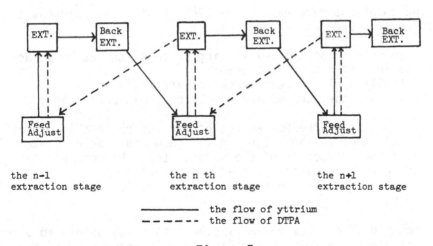

the n-1                    the n th                   the n+1
extraction stage           extraction stage           extraction stage

——————— the flow of yttrium
— — — — — the flow of DTPA

Figure 5.

the back extraction.

The feed solution is made from the raffinate solution in the n+1 th extraction stage and the stripping solution in the n-1 th back extraction stage.

Ammonia water and DTPA are added to this feed solution in order to adjust the pH and the mole ratio of DTPA to yttrium and rare earths. Then this solution is placed in intimate contact with D2EHPA solution for a suitable time. The solvent concentrated with yttrium is treated with an acid such as hydrochloric acid, nitric acid or sulfuric acid and the yttrium is stripped. This stripping solution is transferred to the n+1 th extraction stage.

On the other hand, the raffinate solution is transferred to the n-1 th extraction stage. Therefore, yttrium flows counter-currently with DTPA in this step.

The raffinate solution in the first stage contains the heavy rare earths, small amount of yttrium and DTPA. This solution is treated with D2EHPA to recover yttrium using the extracting systems. Then the pH of the solution contianing DTPA and heavy rare earths is adjusted to 1.0 such that DTPA loses its chelating ability. This solution is then treated with D2EHPA solvent in order to separate the heavy rare earths from the solution.

The pH of the rare earth free DTPA solution is adjusted to 1.8 to recover DTPA as its precipitate.

In the course of this extracting process, the authors faced the following difficult requirements.
(1)  Controlling of the percentage of extraction.
(2)  Decreasing the consumption of the back extraction acid.

The control of the percentage of extraction in this process is difficult because the system is in a non-equilibrium state. In advance of the extraction of the second step, the concentration of DTPA and total rare earths in the feed solution is determined from titrarion in order to adjust the mole ratio of DTPA to total rare earths.

Then the pH and the volume of the feed solution are adjusted to a suitable value. The extraction time is determined by the concentration of the total rare earths in the feed solution. The percentage of extraction is controlled by this method.

The second requirement is achieved by using about 50% sulfuric acid in back extraction. In this process the back extraction should be repeated many times. Sulfuric acid is used in order to decrease the consumption of the back extraction acid. In sulfuric acid back extraction, yttrium is stripped into sulfuric acid as sulfate precipitate. Yttrium sulfate is filtered to recover the excess sulfuric acid and it is recycled to next back extraction step.

Although yttrium purification process using the difference of the extracting velocity between yttrium and rare earths has the difficulty that it is in a non-equilibrium state, nevertheless, it has the highest selectivity between yttrium and heavy rare earths ever known.

The authors have solved above the difficulty and obtained the yttrium oxide having 99.9% purity at 90% recovery.

## ACKNOWLEDGEMENT

We are indebted to Dr. M. Yamaguchi, Mr. T. Kurata and Mr. H. Fukutani for their valuable discussions and encouragement.

## REFERENCES

1.  D. J. Bauer et al, Bureau of Mines RI 6396 (1964).
2.  R. B. Brown et al, Proceedings of the 7th Rare Earth Conference Vol. 1, 385-396 (1968).
3.  F. H. Spedding and A. H. Daane, The Rare Earths, Chap. 55.55, (1961).

# TRIGONAL PRISMATIC VERSUS OCTAHEDRAL COORDINATION: THE CRYSTAL AND MOLECULAR STRUCTURES OF $Dy\{S_2P(C_6H_{11})_2\}_3$ AND $Lu\{S_2P(C_6H_{11})_2\}_3$

A. Alan Pinkerton

Institut de Chimie Minérale et Analytique, Université
de Lausanne, CH-1005 Lausanne, Switzerland

## INTRODUCTION

Simple arguments concerning ligand-ligand repulsion are often used to predict coordination polyhedra. This approach has been quantified over the past few years by Kepert et al. for a number of different ligand types and coordination numbers (1). For the specific case of M(bidentate)$_3$ the structures are found to be intermediate between octahedral and trigonal prismatic if the ligands are not of the dithiolene type (2). The ideal coordination polyhedron may be described by the twist angle, $\theta$, between opposite triangular faces where $\theta = 0°$ for a trigonal prism and $\theta = 60°$ for an octahedron. Calculations of ligand-ligand repulsions predict small values of $\theta$ for ligands with a small "bite", b, and more octahedral geometry for ligands with larger values of b. This approach was extended by Avdeef et al. (3) who presented a large number of examples to illuminate the discussion which spanned a number of ligand, central ion and charge types. We thus thought it worthwhile to examine a series of compounds which are iso-electronic (excluding f electrons), have the same ligand, the same total charge and only differ in the radius of the central ion.

The preparation of the series $[Ln\{S_2P(C_6H_{11})_2\}_3]$ for Ln = Pr-Lu was recently reported (4) and the structures of the Pr and Sm compounds have been determined (5). The complexes for Sm-Lu are isomorphous, hence the structural changes along the series should be small. This offers the possibility of studying the effect of systematically changing the ionic radius of the central ion, and hence the ratio of ligand bite to bond length, on the twist angle $\theta$.

147

The almost perfect $D_3$ symmetry of this series of compounds is an
added advantage in that they are well described in the above way,
whereas many similar compounds would be better described in terms
of rotations about the individual ligand two-fold axes (6).

A further stimulus to carry out this investigation was the fact
that few examples of structures with $\theta$ between 30 and 0° i.e. the
midpoint and the trigonal prismatic extreme of the $D_{3d}$-$D_{3h}$ reaction
coordinate, exist, as pointed out by Muetterties et al. (7).

## RESULTS AND DISCUSSION

Suitable crystals for X-ray studies were prepared by a diffusion
controlled reaction between the metal ions and the ligand,
$S_2P(C_6H_{11})_2^-$ (8), in ethanol solution over a period of 2-3 weeks.
X-ray measurements were carried out with a Syntex $P2_1$ automatic
four-circle diffractometer using Nb filtered Mo radiation ($\lambda$ =
0.71069 Å).

The structures were solved using the published atomic coordinates for the analogous samarium compound (5) as the starting model
and refining these by block diagonal least squares to a final residual of R = 0.042 (Dy) and R = 0.039 (Lu). Calculated bond lengths
and angles for the coordination polyhedra are reported in Table 1.

In both complexes, $[Ln\{S_2P(C_6H_{11})_2\}_3]$ where Ln = Dy and Lu, as
in the samarium analogue, the metal atoms are bonded to six sulfur
atoms as shown in the Figure. As expected, the metal-sulfur bond
lengths (Table 1) decrease with the ionic radius of the metal ion,
the bond to lutecium being the shortest Ln-S bond observed to date
(2.692 Å). This bond shortening is accompanied by a concomitant
reduction in the SPS angles of the chelate rings (112.8 to 111.8
to 110.7°) whilst the angles at sulfur remain essentially constant.
The metal atom lies in the plane described by the three phosphorus
atoms of the ligands. The four membered chelate rings defined by
the metal, one phosphorus and two sulfur atoms are planar and are
tilted with respect to the metal-phosphorus plane (Table 2). The
similarity of these three angles in each case is a manifestation of
the almost perfect threefold axis to be discussed below.

The coordination polyhedron may best be described as intermediate
between an octahedron and a trigonal prism, having nearly $D_3$ symmetry.
As noted above, in previous discussions of the expected geometry
of M(bidentate)$_3$ compounds where the ligands are not dithiolene in
character, calculation of the minimum ligand-ligand repulsion energy
has been used to predict the deviation from octahedral symmetry

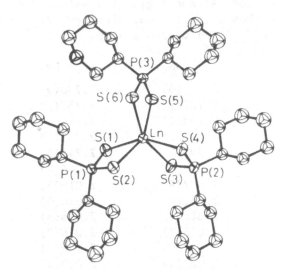

Figure. [Ln{S$_2$P(C$_6$H$_{11}$)$_2$}$_3$] molecule viewed along the threefold axis.

with reasonable accuracy (2,3). The most important parameter to determine the distortion is the ratio of the ligand "bite", b, to the M-S bond length, a. This is quoted as b/a in Table 3 where a number of other parameters describing the polyhedron are presented. The distortion is expressed as the trigonal-twist angle, θ, measured as the angle between opposite triangular faces of edge s. For the range of values of b/a studied here, the theoretical correlation with θ may be considered as linear. Examination of the experimental values obtained (Table 3) shows that they are also linear and have essentially the same slope as the theoretical curve.

We have also calculated the theoretical twist-angle, θ$_{calc}$, corresponding to the minimum repulsion energy for our experimental values of b/a using the expression for the repulsion energy given by Avdeef et al. (3) (Table 3). We thus observe that the experimental value for θ is about 12° lower than the theoretical value in each case. This is a large discrepancy, being double the previously observed maximum deviation from the theoretical curve for non-dithiolene ligands (2,3). It is significant that deviations of this type have previously been used as evidence for the existence of supplementary π bonding. For example, the 1,2-dithiolene complex [Zr(S$_2$C$_6$H$_4$)$_3$]$^{2-}$ (9) is not trigonal prismatic where strong π bonding is to be accepted (10), but rather, distorted towards an octahedron.

Table 1. Bond Lengths (Å) and Angles (°) with standard deviations
in parentheses for the coordination polyhedra of
$[Ln\{S_2P(C_6H_{11})_2\}_3]$ where Ln = Sm, Dy, Lu.

| (a) Distances | Ln = Sm | Ln = Dy | Ln = Lu |
|---|---|---|---|
| Ln-S(1) | 2.781(6) | 2.730(2) | 2.681(2) |
| Ln-S(2) | 2.787(7) | 2.733(2) | 2.685(2) |
| Ln-S(3) | 2.785(6) | 2.747(2) | 2.696(2) |
| Ln-S(4) | 2.790(6) | 2.743(2) | 2.694(2) |
| Ln-S(5) | 2.790(7) | 2.745(3) | 2.698(2) |
| Ln-S(6) | 2.796(6) | 2.746(2) | 2.697(2) |
| P(1)-S(1) | 2.007(9) | 2.019(3) | 2.021(3) |
| P(1)-S(2) | 2.032(8) | 2.020(3) | 2.018(3) |
| P(2)-S(3) | 2.018(8) | 2.016(3) | 2.020(3) |
| P(2)-S(4) | 2.005(8) | 2.016(3) | 2.017(3) |
| P(3)-S(5) | 2.033(8) | 2.022(3) | 2.021(3) |
| P(3)-S(6) | 2.007(9) | 2.018(4) | 2.016(3) |
| S(1)...S(2) | 3.371(9) | 3.345(3) | 3.326(3) |
| S(3)...S(4) | 3.346(8) | 3.340(3) | 3.320(3) |
| S(5)...S(6) | 3.349(9) | 3.341(3) | 3.321(3) |
| (b) Angles | | | |
| S(1)-Ln-S(2) | 74.5(2) | 75.52(7) | 76.60(6) |
| S(1)-Ln-S(3) | 89.9(2) | 89.45(7) | 89.19(6) |
| S(1)-Ln-S(4) | 152.4(2) | 153.54(7) | 154.97(7) |
| S(1)-Ln-S(5) | 90.6(2) | 89.78(7) | 89.42(7) |
| S(1)-Ln-S(6) | 111.7(2) | 110.98(7) | 110.03(6) |
| S(2)-Ln-S(3) | 111.8(2) | 111.27(8) | 110.69(7) |
| S(2)-Ln-S(4) | 90.8(2) | 90.18(7) | 89.69(6) |
| S(2)-Ln-S(5) | 152.5(2) | 153.29(7) | 154.68(6) |
| S(2)-Ln-S(6) | 90.2(2) | 89.44(7) | 88.94(6) |
| S(3)-Ln-S(4) | 73.8(2) | 74.94(6) | 76.03(6) |
| S(3)-Ln-S(5) | 90.6(2) | 90.38(7) | 89.78(7) |
| S(3)-Ln-S(6) | 153.0(2) | 154.37(7) | 155.60(6) |
| S(4)-Ln-S(5) | 111.3(2) | 111.15(7) | 110.27(6) |
| S(4)-Ln-S(6) | 91.1(2) | 90.64(6) | 90.23(6) |
| S(5)-Ln-S(6) | 73.7(2) | 74.95(7) | 75.99(6) |
| Ln-S(1)-P(1) | 86.5(2) | 86.4(1) | 86.30(9) |
| Ln-S(2)-P(1) | 85.8(3) | 86.3(1) | 86.26(9) |
| Ln-S(3)-P(2) | 86.8(2) | 86.55(9) | 86.59(8) |
| Ln-S(4)-P(2) | 86.9(2) | 86.66(9) | 86.70(8) |
| Ln-S(5)-P(3) | 87.0(3) | 86.7(1) | 86.6(1) |
| Ln-S(6)-P(3) | 87.3(2) | 86.73(9) | 86.72(9) |
| S(1)-P(1)-S(2) | 113.1(3) | 111.8(1) | 110.8(1) |
| S(3)-P(2)-S(4) | 112.5(3) | 111.8(1) | 110.7(1) |
| S(5)-P(3)-S(6) | 112.0(4) | 111.6(1) | 110.7(1) |

Table 2. Angles ($^o$) between the Ln-S-P-S Rings and the LnP₃ Plane.

|                    | Ln = Sm | Ln = Dy | Ln = Lu |
|--------------------|---------|---------|---------|
| Ln, S(1), P(1), S(2) | 72.7    | 72.2    | 71.4    |
| Ln, S(3), P(2), S(4) | 71.2    | 70.9    | 70.3    |
| Ln, S(5), P(3), S(6) | 71.5    | 70.9    | 70.0    |

Table 3. Structural parameters for $[Ln\{S_2P(C_6H_{11})_2\}_3]$

| Ln | a     | b     | s     | b/a   | $\theta_{obs}$ | $\theta_{calc}$ |
|----|-------|-------|-------|-------|------|-------|
| Sm | 2.788 | 3.355 | 3.965 | 1.203 | 26.4 | 37.6  |
| Dy | 2.741 | 3.342 | 3.875 | 1.219 | 27.7 | 39.3  |
| Lu | 2.692 | 3.322 | 3.791 | 1.234 | 29.2 | 40.9  |

However, the trigonal-twist angle is less than expected from ligand-ligand repulsions by 11°. This was interpreted as evidence for some residual π contribution to the M-S bonds as previously proposed for $[Mo\{S_2C_2(CN)_2\}_3]^{2-}$ and $[W\{S_2C_2(CN)_2\}_3]^{2-}$ (11). While the above reasoning may indeed be correct, we must point out that the same situation exists in the present case where π-bonding is not allowed if we accept the normal M.O. description of the π orbitals of 1,1-dithiolato ligands (12).

It might be suggested that the 12° discrepancy observed here is due to steric interactions between adjacent cyclohexyl rings, however molecular models show that deviations with opposite sign would be more likely. It is tempting to attribute the distortion to the effect of packing forces on the disc like molecules tending to flatten the molecule as a whole. This process would produce the desired effect and should not require excessive energy as the minimum in the repulsion potential is rather flat for small values of b/a, especially on the low θ side. However, no excessively close contacts between neighbouring molecules were observed.

It is interesting to comment at this point on the previously published structure of $[Pr\{S_2P(C_6H_{11})_2\}_3]$ (5). The structure may be discussed in terms of the average twist angle and be considered as part of the same series as above, however, it is better considered in terms of the individual ligand distortions from a trigonal prism. The formation of a regular octahedron requires a twist of 35.3° around the two-fold axis of each ligand. In the molecules above with effective D₃ symmetry, the ligands are twisted on average by 18.2, 18.7, and 19.4° for Sm, Dy and Lu respectively. However, in the Pr compound one of the ligands is twisted by 30.7° and the other two are twisted by only 9.7 and 11.1° respectively, values close to the trigonal prismatic limit of 0°.

We are currently investigating the structures of similar complexes of the lanthanides and actinides and their early transition metal analogues in the hope of finding the origin of the rather unexpected geometry found for the above series of complexes.

## ACKNOWLEDGMENTS

We thank the Swiss National Science Foundation for financial support.

## REFERENCES

1. D.L. Kepert, Progr. Inorg. Chem. 23, 1 (1977).
2. D.L. Kepert, Inorg. Chem. 11, 1561 (1972).
3. A. Avdeef and J.P. Fackler, Inorg. Chem. 14, 2002 (1975).
4. A.A. Pinkerton, Y. Meseri and C. Rieder, J.C.S. Dalton, 85 (1978).
5. Y. Meseri, A.A. Pinkerton and G. Chapuis, J.C.S. Dalton, 725 (1977).
6. J.L. Martin and J. Takats, Inorg. Chem. 14, 1358 (1975).
7. E.L. Muetterties and L.J. Guggenberger, J. Amer. Chem. Soc. 96, 1748 (1974).
8. M.M. Rauhut, H.A. Currier and V.P. Wystrach, J. Org. Chem. 26, 5133 (1961).
9. M. Cowie and M.J. Bennett, Inorg. Chem. 15, 1595 (1976).
10. E.I. Stiefel, R. Eisenberg, R.C. Rosenberg, and H.B. Gray, J. Amer. Chem. Soc. 88, 2956 (1966); G.N. Schrauzer and V.P. Mayweg, J. Amer. Chem. Soc. 88, 3236 (1966).
11. G.F. Brown and E.I. Stiefel, Inorg. Chem. 12, 2140 (1973).
12. R. Eisenberg, Progr. Inorg. Chem. 12, 295 (1970).

# CRYSTAL CHEMISTRY OF THE EUROPIUM PNICTIDES

F. Hulliger[*] and R. Schmelczer[+]

[*]Solid State Physics Laboratory ETH, CH-8093 Zürich

[+]Crystallogr. Institute, University, CH-1015 Lausanne

Europium occurs in the pnictides with oxidation number two and three, depending on concentration and electronegativity of the anions. The crystal chemistry of trivalent Eu resembles that of the normal rare-earth elements, whereas the divalent Eu compounds are closely related to their alkaline-earth analogs. The alkaline earths as well as divalent Eu have a relatively low electronegativity. Nonmetallic properties are therefore most likely to occur in the phosphides and arsenides, possibly also in the antimonides. The discussion of the crystal chemistry of these pnictides thus will turn out to be an illustration of the Mooser — Pearson rule (1,2). This rule correlates the electronic properties of a compound to its crystal structure or short-range order. Nonmetallic properties require that in an idealized crystal (i.e. a crystal without the unavoidable imperfections) at T=0 the valence electrons occupy all states in some low-lying energy bands (valence band) while the next higher available states (in the conduction band) are well separated energetically. The generation of free charge carriers therefore requires an activation energy. From a chemical point of view this means that in a nonmetallic compound all the chemical bonds have to be saturated while non-bonding electrons (such as 4f-electrons) have to be localized on the ions. In normal valence compounds $M_m^{\mu+} X_x^{\chi-}$ the Mooser — Pearson formula

$$n_C' + n_A = 8 N_A$$

(where $n_C'$ ($n_A$) = number of bonding cationic (anionic) valence electrons per unit cell; $N_A$ = number of anions per unit cell) is equivalent to the octet rule or the neutrality condition $m\mu = x\chi$.

153

In polycompounds valence electrons are engaged not only in the heteropolar M — X bonds but also in homopolar M — M and/or X — X
bonds. In the europium pnictides, however, we cannot expect localized cation — cation bonds, so that we restrict our considerations
to polyanionic compounds. The additional covalent X — X bonds reduce the effective valence of the anions with respect to the cations. For the heteropolar part we still have the neutrality condition $m\mu = x\chi'$, where now $\chi' = \chi - B_A$ with $B_A$ = number of anion —
anion bonds per anion. Since in non-metallic compounds these
X — X bonds are single bonds we can replace $B_A$ by $C_A$, the coordination number of the anion towards other anions. This leads to the
generalized 8-N rule for the anion partial structure

$$C_A = 8 - \frac{n_A + n'_C}{N_A} ,$$

i.e. the anions form a pseudo-element structure with a reduced coordination number $C_A = 8 - N - n'_C/N_A$ . This analogy even includes
semimetals like As $\leftrightarrow$ $CaSi_2$.

In Table 1 we have listed the Eu pnictide phases in the sequence of increasing anion contents. Compounds which obey the
Mooser — Pearson formula (the Mooser — Pearson phases) are underlined. Broken lines point to transitional cases (semimetals); metals are not underlined. We have divided the table into two parts.
The upper group contains all compounds which according to their
chemical formula must exhibit metallic properties. Mainly two structure types occur: With decreasing cation to anion radius ratio the
$Mn_5Si_3$ type (the $Ca_5Pb_3$ type is a slightly distorted $Mn_5Si_3$ type)
is replaced by the anti-$U_3S_5$ type. The $Mn_5Si_3$ structure type is a
strongly distorted defect $Ni_2In$ type (which in turn is a filled-up
NiAs type) and the anti-$U_3S_5$ structure is a $Rh_5Ge_3$ superstructure,
i.e. a defect $Fe_2P$ structure (which itself is a filled-up CoSn
type). Since in $Eu_5\square As_3$ the charges are almost balanced we suspect
that these phases are stabilized by oxygen (23).

The valence compounds comprise the trivalent NaCl-type phases
EuN and EuP, which could be semiconductors but up to now appear to
be semimetals. Divalent Eu is found in the anti-$Th_3P_4$-type derivatives. On filling up the empty anion sites in the anti-$Ce_2S_3$ structure
of $Eu_3P_2$ ($=Eu_4P_{8/3}\square_{1/3}$) and $Eu_3As_2$ the structure suffers a rhombohedral distortion (7-9) if trivalent Eu is created, but appears to
remain cubic if a chalcogen anion is introduced, as in $Eu_4P_2S$ (24).
It is to be expected that ternary phases $Eu_2^{2+}Ln^{3+}P_3$ and $Eu_3^{2+}Ln^{3+}As_3$
are feasible, all the more as (cubic) $Eu_3GdSb_3$, $Eu_3DySb_3$ and $Eu_3GdBi_3$
(10) are known already.

Table 1   The binary europium pnictides

| M : X | Eu — P | Eu — As | Eu — Sb | Eu — Bi |
|---|---|---|---|---|
| 5:3(?) | | rt:$Mn_5Si_3$(3)<br>ht:disord.$Ca_5Pb_3$(3) | anti-$U_3S_5$(4) | anti-$U_3S_5$(4) |
| 3+x:2 | | tetragonal(5) | | |
| 3:2 | anti-$Ce_2S_3$(6) | anti-$Ce_2S_3$(6) | tetrag.(?)(7) | |
| 4:3 | *rhombohedral<br>anti-$Th_3P_4$(7,8) | *rhombohedral<br>anti-$Th_3P_4$(9) | | anti-$Th_3P_4$(10) |
| 5:4 | | $Eu_5As_4$(5,11) | | |
| 11:10 | "$Eu_{11}P_{10}$"(12) | $Ho_{11}Ge_{10}$deriv.(5) | $Ho_{11}Ge_{10}$(4,7) | $Ho_{11}Ge_{10}$(4) |
| 1:1 | *NaCl | $Na_2O_2$(13,14) | | |
| 3:4 | $Sr_3As_4$(7,25) | $Sr_3As_4$(13,15) | | |
| 2:3 | | $Ca_2As_3$(4,16) | $Eu_2Sb_3$(17,25) | |
| 2:3+x | | $Eu_2As_{3+x}$(13) | | |
| 1:1.82 | "$EuP_{1.82}$"(12) | | | |
| 1:2 | $BaAs_2$(12,25) | $EuAs_2$,orth.(13,25) | $CaSb_2$(18,4,25) | $YbBi_2$(4) |
| 1:2.25 | $EuP_{2.25}$(8) | | | |
| 1:3 | $SrAs_3$(19)<br>ht:$SrP_3$(25) | $SrAs_3$(13,19,21) | | $Cu_3Au$(22) |
| 1:7 | $EuP_7$(20) | | | |

— Mooser — Pearson phase        --- border-line case (semimetal?)
*containing trivalent europium

The compounds with higher anion contents will probably all contain divalent europium. At least in the phosphides and arsenides
(that are most likely Mooser — Pearson phases) the excess valence
electrons have to be bonded in covalent anion — anion bonds. It may
be possible only in some simple cases to predict the crystal struc-

ture of a polyanionic semiconductor, but we can guess possible anion arrays. Assuming semiconductor properties in $Eu_5As_4$ its "ionic" formula will be $Eu_5^{2+}(As_2)^{4-}As_2^{3-}$. In other words, half the anions have to be bonded in pairs. This is indeed the case in the actual structure (11) which is a more symmetrical version of the $Sm_5Ge_4$ structure (which also contains these X —— X pairs, although the isomorphous $Ln_5^{3+}X_4$ phases with X = Si, Ge, Sn, Pb, cannot be semiconductors).

In $Eu_{11}P_{10}$ 22 cation valence electrons are opposed to 30 anion electrons. If we allow for P rings or chains and P pairs in addition to the single P anions we end up with $Eu_{10}^{2+}P_x^{-1}P_y^{-2}P_{10-x-y}^{-3}$ and various possibilities fulfilling the condition 2x+y=8, namely x=0, y=8 as the most simple combination, x=1, y=6 (which requires Z=2n), x=2, y=4 (realized in the $Ho_{11}Ge_{10}$ type), x=3, y=2, (Z=2n) and x=4, y=0 as the next simple solution with possibly one P square plus 6 single P anions per formula unit. In $Ho_{11}Ge_{10}$-type $Eu_{11}Sb_{10}$ the Sb —— Sb distance within the pairs is reasonable for a single bond: 2.93 Å (7); within the Sb squares, however, it is definitely too large: 3.04 Å (7). Bond saturation, therefore, is impossible, and indeed the $Eu_{11}Sb_{10}$ samples showed a very low resistivity and Seebeck coefficient. In $Eu_{11}As_{10}$ the orthorhombic deformations (5) may act towards bond saturation. It is a puzzle why a structure with such a complicated anion arrangement as the tetragonal $Ho_{11}Ge_{10}$ type at all forms, when the resulting phases nevertheless are metallic as eg. $Ca_{11}Bi_{10}$, $Ln_{11}Ge_{10}$, $Ln_{11}Sn_{10}$, $Ln_{11}Pb_{10}$.

A nonmetallic $Eu^{II}X$ pnictide may contain as many ring or chain members $X^{-1}$ as it contains single anions $X^{-3}$ besides the paired anions $X^{-2}$. The simplest case is realized in EuAs where all As atoms occur in pairs. Its $Na_2O_2$ structure derives from the NiAs structure (where Ni chains run along the trigonal axis). EuAs is nonmetallic like the isomorphous CaP, CaAs,.. We hoped to find $Li_2O_2$-type (anti-TiAs derived) EuSb, but obtained only $Eu_{11}Sb_{10}$.

$Eu_4P_5$ (12) is most likely $Eu_3P_4$, but if it really exists as a separate phase it must contain more than one kind of anion units, e.g. with Z=2 two 4-membered chain fragments and one P pair per unit cell. An 8-membered chain fragment plus 2 single P atoms are less likely.

EuAs may be considered as the first member of a series of structures containing all anions in the form of chain fragments. If we take phosphides as examples their formula must be of the form $M_{n+2}^{2+}P_{2n}$:SrP with pairs, $Eu_5P_6$, $Eu_3P_4$, $Eu_7P_{10}$, $Eu_2P_3$ and $EuP_2$ with 3- to 6-membered and infinite chains, resp. The structure of $Eu_3As_4$, $Eu_2Sb_3$ and $EuSb_2$ do indeed adhere to this family. The semiconductors $Eu_3P_4$ and $Eu_3As_4$ have a defect $\alpha$-$ThSi_2$ structure (15,16,18):

$Eu_3As_4\square_2$. The structures of $Eu_2Sb_3$ and $Eu_2As_3$ are closely related to each other. Both can be derived from a more symmetrical structure containing anion spiral chains (17). Slight variations of the anion site parameters lead to 6-membered chain fragments in $Eu_2Sb_3$ and to 4- and 8-membered fragments in $Eu_2As_3$. The $CaSb_2$-type structure of $EuSb_2$ can be derived from the $ZrSi_2$ structure. Distortions transform the square anion nets of the $ZrSi_2$ structure (present in metallic $YbSb_2$) into infinite chains of the $CaSb_2$ structure as required for a Mooser — Pearson phase (18). The structure of $EuP_2$ ($BaAs_2$ type) contains infinite $P^-$ chains that can be derived from As double layers while in $EuAs_2$ the $As^-$ anions form two kinds of "half-spiral" chains (25). $EuP_2$ might as well be built up from infinite chains in which each chain member is bonded (and thus becomes $P^o$) to an additional $P^{-2}$, according to the formula $EuP^oP^{-2}$. The X-ray diagram given for $EuP_2$ by Mironov et al.(12) points indeed to a second modification.

The structures of $EuP_3$, $EuAs_3$ and $EuP_7$ can be derived from black P or As layers. If in these layers we substitute 1/4 of the anion pairs by two cations, we can generate the idealized structures of $CaP_3$, $SrP_3$, $EuP_3$ and $BaP_3$ (20). Only 1/8 of the P pairs has to be replaced in order to create the structure of $EuP_7$ (20). By eliminating 1/3 of the anions, all parallel in a row, we end up with $\square_2^{4+}P_2^oP_2^{-2}$ appropriate for the $EuP^oP^{-2}$ modification proposed above. If the $\square_2$ holes form a hexagonal pattern, the remaining anions form infinite saw-tooth chains corresponding to $EuP_2^{-1}$ (reminiscent of a pseudo-tellurium anion partial structure).

## REFERENCES

1  E. Mooser and W.B. Pearson, "The chemical bond in semiconductors", Progr. Semicond. 5:103 (1960).

2  E. Mooser, "The electrical properties of the solids, a problem of the chemical bond", Chimia 23:169 (1969).

3  Yu Wang, L.D. Calvert, E.J. Gabe and J.B. Taylor, "Structure of two forms of europium arsenide $Eu_5As_3$", Acta Cryst. B34:2281(1978).

4  J.B. Taylor, L.D. Calvert, T. Utsunomiya, Yu Wang, "Rare-earth arsenides: The metal-rich europium arsenides", J. Less-Common Met. 57:39 (1978).

5  J.B. Taylor, L.D. Calvert and Yu Wang, "Powder data for some new europium arsenides", J. Appl. Cryst. 10:492 (1977).

6  F. Hulliger and O. Vogt, "New ferromagnetic europium compounds", Solid State Commun. 8:771 (1970).

7  F. Hulliger and R. Schmelczer, unpublished.

8  M. Wittmann (Stuttgart), private commun. (1979).

9  F. Hulliger, "$Eu_4As_3$, a new trigonal anti-$Th_3P_4$ type compound", Mat. Res. Bull. 14:33 (1979).

10  R.J. Gambino, "Rare-earth-Sb and -Bi compounds with the $Gd_4Bi_3$ (anti-$Th_3P_4$) structure", J. Less-Common Met. 12:344 (1967).

11  Yu Wang, L.D. Calvert, E.J. Gabe and J.B. Taylor, "The crystal structure of $Eu_5As_4$: A more symmetrical version of the $Sm_5Ge_4$-type structure", Acta Cryst. B34:1962 (1978).

12  K.E. Mironov, G.P. Brygalina and V.N. Ikorskii, "Magnetism of europium phosphides", Proc. 11th Rare Earth Res. Conf., Traverse City:105 (1974).

13  S. Ono, F.L. Hui, J.G. Despault, L.D. Calvert and J.B. Taylor, "Rare-earth pnictides: The arsenic-rich europium arsenides", J. Less-Common Met. 25:287 (1971).

14  A. Iandelli and E. Franceschi, "On the crystal structure of the compounds CaP, SrP, CaAs, SrAs and EuAs", J. Less-Common Met. 30:211 (1973).

15  M. Smart, L.D. Calvert and J.B. Taylor, private communic. (1977).

16  F. Hulliger, "Pnictides", in: Hdb. Phys. Chem. Rare Earths, Vol. IV, K.A. Gschneidner and L. Eyring, eds., North-Holland (1979).

17  G. Chapuis, F. Hulliger and R. Schmelczer, "The crystal structure and some properties of $Eu_2Sb_3$", J. Solid State Chem. (in press).

18  F. Hulliger and R. Schmelczer, "Crystal structure and antiferromagnetism of $EuSb_2$", J. Solid State Chem. 26:389 (1978).

19  H.G. v. Schnering, W. Wichelhaus and M. Wittmann, "New polyphosphides of the rare-earth metals", 5th Conf. Trans. Elem. Compds., Uppsala, PII 68 (1976).

20  H.G. v. Schnering, "Catenation of phosphorus atoms", in: "Homoatomic rings, chains and macromolecules of main-group elements", A.L. Rheingold, ed., Elsevier, Amsterdam (1977).

21  J.F. Brice and A. Courtois, "On the existence of $EuAs_3$ and $SrAs_3$", C.R. Acad. Sci. Paris 283C:479 (1976).

22  B. Kempf, B. Elschner, P. Spitzli and Ø. Fischer, "Superconductivity and magnetic ordering in $Bi_3Sr_{1-x}Eu_x$", Phys. Rev. B17:2163 (1978).

23  F. Hulliger, "On the usefulness of bond considerations in phase characterization: The 2:1 alkaline-earth pnictides", Z. Krist. (in press).

24  F. Hulliger, "New ternary anti-$Th_3P_4$-type europium compounds", Mat. Res. Bull. 14:259 (1979).

25  M. Wittmann, W. Schmettow, D. Sommer, W. Bauhofer and H.G. v. Schnering, "Phosphides, arsenides and antimonides of divalent europium", VI. Int. Conf. Solid Compds. Trans. Elements, Stuttgart 1979, PO III/20.

# ABSORPTION SPECTROPHOTOMETRIC AND X-RAY DIFFRACTION STUDIES OF

# $NdI_3$, NdOI AND $SmI_3$[*]

R. G. Haire, J. P. Young,[†] and J. Y. Bourges[††]

Transuranium Research Laboratory, Oak Ridge National

Laboratory, Oak Ridge, TN 37830

## INTRODUCTION

A systematic study of the first five transplutonium halides and oxyhalides, using spectrophotometry and X-ray diffraction, has been in progress in our laboratory. One of the goals was to establish the trends and differences in the properties of these 4f and 5f compounds. During the investigation of americium and curium tri-iodides and comparison of their X-ray data with literature values [1] for the lanthanide tri-iodides, it became apparent: (1) that the lattice parameters for samarium tri-iodide were not in accord with those of other lanthanide tri-iodides; and (2) that a second crystal form of neodymium and samarium tri-iodide may exist. Thus, neodymium, samarium and other selected lanthanide tri-iodides were examined with the same investigative tools that have been used for the transplutonium halides.

Described here are the results of examining the anhydrous tri-iodides of neodymium and samarium, and the oxyiodide of neodymium, by X-ray powder diffraction and absorption spectrophotometry.

---

*Research sponsored by the Division of Nuclear Sciences, Office of Basic Energy Sciences, U. S. Department of Energy under contract (W-7405-eng-26), with Union Carbide Corporation.
†Analytical Chemistry Division, ORNL.
††Guest Scientist from Centre d'Etudes Nucléaires, B.P. No. 6, 92260 Fontenay-aux-Roses, France.

## EXPERIMENTAL

The lanthanide tri-iodides were prepared from doubly-sub-
limed iodine and commercial lanthanide metals (99.5%, with oxygen,
nitrogen, hydrogen being major impurities) using material cut
from the center portion of metal ingots.  The reactants were
sealed in quartz capillaries under high ($10^{-7}$ torr) vacuum and
then heated at 300°C for ~14 hours.  Oxyhalides were prepared by
sealing the above reactants together with a small quantity of
moist air in the capillaries and then proceeding as with the tri-
iodide preparation.  A variable gradient, micro-type furnace was
employed to sublime the excess iodine to the cooler end of the
capillary and the portion of the capillary containing the lanthanide product was then flame-sealed to provide a sample suitable
for both absorption spectroscopy and X-ray diffraction analyses.

X-ray powder diffraction data were obtained from the capillaries using modified Debye-Scherrer type cameras and Ni-
filtered Cu radiation.  Lattice parameters were refined by the
LCR-2 program [2] and theoretical line intensities and positions
used for indexing the experimental data were generated by the
POWD program [3].

Absorption spectrophotometric data were obtained from the
same samples as used for X-ray analyses using a single-beam,
microscope spectrophotometer (quartz optics).  A description of
the spectrophotometer is available in the literature [4].  Both
systems are capable of examining the samples at room or elevated
temperatures.

## RESULTS AND DISCUSSION

Analysis of the X-ray powder diffraction data on the neodymium and samarium tri-iodide samples after reaction at 300°C
showed that neodymium tri-iodide existed in an orthorhombic form
(PuBr$_3$ type), while samarium tri-iodide exhibited the BiI$_3$ structure type, usually reported in terms of hexagonal axes (Wyckoff
[5]).  Our lattice parameters for the orthorhombic neodymium
tri-iodide agreed well with the literature values [1], but the
hexagonal parameters we obtained for samarium tri-iodide were
larger than those previously reported by Asprey et al. [1]
(Table I).  By plotting the molecular volume of the lanthanide
tri-iodides as a function of the lanthanide, Asprey et al. [1]
noted that the lighter and heavier members of the series fell into
two distinct groups.  The lighter lanthanide tri-iodides (lanthanum through neodymium) exhibited the orthorhombic crystal
type while the remaining tri-iodides crystallized in the BiI$_3$
type structure.  Each grouping in their plot of the tri-iodides
formed a smooth curve with the exception of the samarium compound,

Table 1.  X-Ray Data for Selected Lanthanide Tri-Iodides

| Compound | Structure Type | Lattice Parameters ($\overset{o}{\text{A}}$) $a_o$ | $b_o$ | $c_o$ | Molecular Volume | Ref. |
|---|---|---|---|---|---|---|
| LaI$_3$ | PuBr$_3$ | 4.37 | 14.01 | 10.04 | 153.6 | 1 |
| LaI$_3$ | " | 4.377(8) | 14.01(2) | 10.04(1) | 153.9 | * |
| PrI$_3$ | " | 4.309 | 13.98 | 9.958 | 150.0 | 1 |
| PrI$_3$ | " | 4.309(2) | 13.969(10) | 9.978(9) | 150.2 | * |
| NdI$_3$ | " | 4.284 | 13.979 | 9.948 | 148.9 | 1 |
| NdI$_3$ | " | 4.293(2) | 13.986(11) | 9.952(8) | 149.4 | * |
| NdI$_3$ | BiI$_3$ | 7.610(7) | - | 20.89(3) | 174.6 | * |
| SmI$_3$ | " | 7.49 | - | 20.80 | 168.4 | 1 |
| SmI$_3$ | " | 7.590 | - | 20.88(6) | 173.6 | * |
| GdI$_3$ | " | 7.539 | - | 20.83 | 170.8 | 1 |
| GdI$_3$ | " | 7.551(2) | - | 20.85(8) | 171.6 | * |
| TbI$_3$ | " | 7.526 | - | 20.838 | 170.4 | 1 |
| NdOI | PbFCl | 4.015(4) | - | 9.197(6) | - | * |

*This work

whose cell volume was smaller than expected from extrapolation. Promethium tri-iodide and europium tri-iodide were not reported.

The molecular volume of samarium tri-iodide calculated from our hexagonal lattice parameters, falls on an extension of a molecular volume vs. element curve for the heavier lanthanides. The reason for this difference between our lattice parameters for samarium tri-iodide and those of Asprey et al. [1] is not known.  Although samarium tri-iodide thermally decomposes to yield the di-iodide, the Sm(II) ion should have a larger radius than Sm(III) and any replacement of Sm(III) in the tri-iodide structure by Sm(II) ions would be expected to yield larger lattice parameters, and a larger molecular volume for the samarium tri-iodide.

It is possible to obtain a hexagonal (BiI$_3$ type) form of the neodymium tri-iodide by heating the orthorhombic form above 600°C and quenching from elevated temperatures.  The lattice parameters and molecular volume for this structure of neodymium tri-iodide are in accord with our values for samarium tri-iodide and the published values for the heavier lanthanide tri-iodides [1] and supports our larger parameters for samarium tri-iodide [see Table I].

The transition temperature for the orthorhombic to hexagonal conversion of neodymium tri-iodide was estimated to be 600 ± 30°C, by both X-ray diffraction and absorption spectrophotometric analyses. A phase change at 570°C was also noted by Druding and Corbett [6] in their thermal analysis work with neodymium tri-iodide, but the phase was not identified. In our work, we found this phase transition to be reversible; the hexagonal form could be retained at room temperature by quenching from above ∿650°C but reheating at 300-550° converted it back to the orthorhombic form.

A large volume change accompanies the neodymium tri-iodide transition and these effects can be visually noted in crystals by heating and observing them with the aid of a microscope. Visually, the orthorhombic form of neodymium tri-iodide is a pale yellow, which becomes more colored on heating, turning dark yellow-green near the transition temperature. After conversion, the hexagonal form appears white up to its melting point (785°C, in agreement with the literature value [1]) where it turns into a dark, red-orange liquid.

Neodymium oxyiodide was also prepared and studied in this work. This compound appeared a dull, off-white material, whose lattice parameters (see Table I) are in accord with those of other lanthanide oxyiodides [7].

In Figure 1 are shown absorption spectra of the two crystal forms of neodymium tri-iodide and neodymium oxyiodide at room temperature. The assignments of each spectrum were first made on the basis of identification by X-ray diffraction of the exact sample used for the spectral study. These assigned spectra could then be used to analyze other samples of neodymium tri-iodide or oxyiodide. It was possible to determine the transition temperature by successively heating the samples to higher temperatures in the spectrophotometer and obtaining spectral data after quenching. Spectra were also obtained at these elevated temperatures but these data are complicated by two things: (1) in general, the absorption peaks lose detail and become broader; and (2), certain transitions are enhanced due to changes in the population of different energy levels due to the thermal energy being supplied.

The major differences between the absorption spectra (Fig. 1, f-f transitions) of the two neodymium tri-iodides are: (1) increased splitting of bands in the high temperature form (hexagonal structure); (2) the change in the relative peak heights of the bands in the 760-900 nm region as compared to the peaks at ∿600 nm; and (3) the presence of two smaller peaks at 690 and 640 nm in the low-temperature, orthorhombic form. The spectrum of neodymium oxyiodide is characterized by weak absorption in the

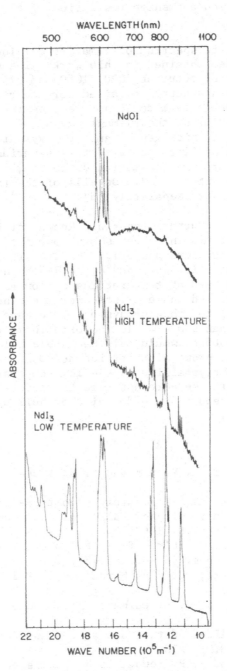

Figure 1. Solid State Absorption Spectra of Neodymium Tri-iodide and Oxyiodide.

680 to 900 nm region and sharp resolution of the peaks around 600 nm.

The solid state spectrum of samarium tri-iodide (hexagonal structure) was also obtained in this work. The major observable f-f absorption peaks occur at 1090, 1080 and 975 nm. This was the only absorption spectrum obtained for samarium tri-iodide, since an orthorhombic form could not be prepared by low temperature reaction (250°C) of the elements or by storage of the hexagonal form in liquid nitrogen. Samarium oxyiodide was not prepared in this work. On heating, we found that samarium tri-iodide decomposed (~750°C) to yield samarium di-iodide, in accord with the findings of Asprey et al. [1]. Details of the thermal decomposition will be published separately [8].

Although two structure types are known for the lanthanide tri-iodides, only neodymium has been shown to exist in both forms (high and low temperature phases). The transition temperature for neodymium tri-iodide is about 600°C, while the hexagonal samarium tri-iodide is obtained at temperatures as low as 250°C. This implies that if a second phase existed for the other lanthanide tri-iodides, those lighter than neodymium would have transition temperatures higher than 600°, while the orthorhombic form for the transsamarium compounds would only be stable below 250°C. We would predict that promethium tri-iodide would have a hexagonal form and may also crystallize in the low-temperature, orthorhombic form. In this work, we were not able to prepare a second phase of lanthanum, praseodymium, gadolinium or holmium tri-iodides by thermal treatment.

## REFERENCES

1. L. B. Asprey, T. K. Keenan and F. H. Kruse, Inorg. Chem. 3, 1137 (1964).
2. D. E. Williams, Ames Laboratory Report IS-1052 (1964).
3. D. K. Smith, University of California, Lawrence Radiation Lab. Rept. UCRL-7196 (1963).
4. J. P. Young, R. G. Haire, R. L. Fellows and J. R. Peterson, J. Radioanal. Chem. 43, 479 (1978).
5. R. W. G. Wyckoff, "Crystal Structures," Vol. II, Wiley, New York, 1963.
6. L. F. Druding and J. D. Corbett, J. Am. Chem. Soc. 83, 2462 (1961).
7. D. Brown, "Halides of the Lanthanides and Actinides," Wiley, New York, 1968, p. 226.
8. R. G. Haire and J. P. Young, to be published.

# SOLID STATE CHEMISTRY IN THE FLUORITES AND ITS INFLUENCE

# ON THE SPECTROSCOPY

John C. Wright and D. S. Moore

Department of Chemistry, University of Wisconsin

Madison, WI  53706

Fluorite materials occupy a prominent position in the understanding of solid state chemistry and in the control defects exert on that chemistry.  Consequently, the fluorites are one of the world's best studied materials and the knowledge gained from them penetrates much of solid state science.  Despite this work, the present understanding of defect chemistry in at least some of the fluorites (alkaline earth fluorides) is probably wrong and the fundamental phenomena controlling defects in fluorites remain poorly understood.

The details of the solid state defect equilibria can be determined by doping the fluorites with optically active trivalent lanthanide ions and using a tuneable dye laser to selectively excite fluorescence from specific defect sites (1).  This technique permits one  to determine how many different kinds of sites are present, to classify them according to whether there are several cations clustered or not, and to determine the absolute concentration of each one (2).  We have followed the changes in the defect equilibria in the fluorites as a function of dopant concentration and annealing temperature.  We have found that defect clustering is important, even at concentrations as small as 0.01 mole %.  In addition, the changes in defect concentrations do not follow the behaviors expected for any of the traditional models of defect equilibria in fluorites.

In order to explain these observations, a new model for the defect equilibria has been proposed.  It is sketched below for $CaF_2$ using the Kroger-Vink notation.

perfect $CaF_2$ lattice $\rightleftharpoons F_i' + V_F^{\cdot}$

$$+$$
$$Er_{Ca}^{\cdot}$$
$$\Updownarrow$$
$$(Er_{Ca}^{\cdot} \cdot F_i)^X + (Er_{Ca}^{\cdot} \cdot F_i)^X \rightleftharpoons (2Er_{Ca}^{\cdot} \cdot 2F_i)^X$$
$$+$$
$$F_i'$$
$$\Updownarrow$$
$$(2Er_{Ca}^{\cdot} \cdot 3F_i)'$$

The important addition in this picture is the equilibrium between $(2Er_{Ca}^{\cdot} \cdot 2F_i)^X$ and $(2Er_{Ca}^{\cdot} \cdot 3F_i)'$ which represents scavenging of free fluorite interstitial ions by clusters. In order to be consistent with experimental observations, the scavenging of fluoride interstitials must be very efficient. The equilibria between defect structures with one $Er_{Ca}$ and those with more than one is established and quenched at a high temperature because these equilibria require cation mobility while the relative concentrations between structures containing different numbers of $F_i$ are established and quenched at lower temperatures where only fluoride interstitual mobility is required.

The observed temperature and concentration dependences of the individual defect structures is in agreement with this model. The model predicts that higher temperatures will cause dissociation of the $(2Er_{Ca}^{\cdot} \cdot 3F_i)'$ structure and the consequent release of the scavenged fluoride interstitial and formation of a $(2Er_{Ca}^{\cdot} \cdot 2F_i)^X$ structure. The increased fluoride interstitial concentration forces an increase in the $(Er_{Ca}^{\cdot} \cdot F_i)^X$ concentration and a decrease in the $Er_{Ca}'$ concentration. All of these changes must be correlated with each other. At higher temperatures still, the clusters themselves must dissociate and cause an increase in both the $(Er_{Ca}^{\cdot} \cdot F_i)^X$ and $Er_{Ca}'$ concentrations. All of these predictions are in agreement with experiment.

## REFERENCES

1.  D.R. Tallant and J.C. Wright, Proceedings of the 11th Rare Earth Research Conference (1974).
2.  D.R. Tallent and J.C. Wright, *J. Chem. Phys.* 63:468 (1975); D.R. Tallant, M.P. Miller, and J.C. Wright, *J. Chem. Phys.* 65 510 (1976); D.R. Tallant, D.S. Moore, and J.C. Wright, J. Chem. Phys. 67:2897 (1977).

# NEW ANION-EXCESS, FLUORITE-RELATED SUPERSTRUCTURE PHASES IN $LnF_2$-$LnF_3$, $CaF_2$-$YF_3$, AND RELATED SYSTEMS

Ortwin Greis

Department of Chemistry, Texas A&M University

College Station, Texas 77843

## ABSTRACT

Cubic solid solutions in yttrofluorites $(Ca,Y,RE)F_{2+\delta}$ with $0.0 < \delta < 0.5$ are often described as some of the most representative examples for nonstoichiometry. The classic, so-called Goldschmidt-Zintl model with statistical distribution of cations and interstitial anions, however, is nowadays substituted by short-range order models based on neutron powder diffraction analysis. In contrast, long-range order has been revealed by means of single crystal electron diffraction. The observed microdomains are of the same nature as the corresponding superstructure phases. These can be obtained by longtime annealing of the solid solutions, in most cases only at lower temperatures. Earlier, they have been characterized only by means of X-ray powder methods and a systematic procedure for indexing superstructure powder patterns. In the meantime, however, the superstructure geometry of all investigated phases has been confirmed also by single crystal electron diffraction. Presently, six different superstructure types are revealed; all belong to one of the two homologous series, $M_mF_{2m+5}$ or $M_mF_{2m+6}$, or to both in the case of a phase width. The atomic arrangement of cations and anions is already known for rhβ and rhγ. Only one model exists for the cation distribution in t, rhα, and rhδ. The phenomenon of ordering can be described by two borderline models. In the first and dominant case, the superstructure is determined by a fully ordered cation lattice, and in the second and less important case by anion clusters, which are built up from both interstitial and shifted parent structure anions. These anion clusters, however, are significantly more extended than the so-called Willis cluster.

## INTRODUCTION

The fluorite structure is well-known for its almost unique feature to accommodate high concentrations of chemical defects as substitution in the cation sublattice and interstitial anions or anion vacancies (1). These defects can be accommodated in a dis-ordered or ordered way depending on the annealing conditions during preparation. At higher temperatures or after quenching, cubic solid solutions $MX_{2+\delta}$ or $MX_{2-\delta}$ are revealed, while super-structure phases are detected at lower temperatures and in most cases only after longtime annealing. The $LnO_{2-\delta}$ systems are good examples for anion deficiency, while yttrofluorites $(Ca,Y,RE)F_{2+\delta}$ demonstrate anion excess very well. This paper deals with yttro-fluorite-related binary and ternary rare earth fluorides of the systems $LnF_2-LnF_3$ and $MF_2-REF_3$ with M = Ca, Sr, Ba. The recent literature on preparation and phase investigations is reviewed briefly. Furthermore, new structural information from X-ray and electron diffraction will be discussed leading to a better under-standing of the nature of ordering in fluorite-related, anion-excess superstructures. (A poster with the same title presented at the 14th Rare Earth Research Conference at Fargo/USA is com-plementary to this paper. A reprint of all tables and figures is available from the author on request).

## $LnF_2-LnF_3$ SYSTEMS

Up to about 1970, only six rare earth fluorides with unusual oxidation states were known: $CeF_4$, $PrF_4$, $TbF_4$, and $SmF_2$, $EuF_2$, $YbF_2$ (2). Several phase studies on $LnF_2-LnF_3$ systems, however, indicated the existence of one or two intermediate phases in the composition range $LnF_x$ with $2.25 < x < 2.50$ (3-5). The formulae of these fluorite-related superstructure phases could not be revealed, since their X-ray powder patterns could not be indexed. Therefore, reinvestigation of these systems seemed to be appropriate with special emphasis on chemical and structural characterization.

Water and oxygen-free rare earth trifluorides were prepared by means of a combination of the so-called wet and dry methods (6). Metal and fluorine analysis (6) showed compositions within the limits of $LnF_{3.00\pm0.01}$. Water and oxygen-free rare earth difluo-rides $LnF_{2.00\pm0.01}$ were prepared using the so-called double cell method (7). Intermediate fluorides $LnF_x$ with $2.00 < x < 3.00$ were prepared by several methods: (a) annealing of mixtures of $LnF_2$ and $LnF_3$ in sealed platinum tubes, (b) partial reduction of $LnF_3$ with Ln (g) or $H_2$, and (c) thermal decomposition of $LnF_2$. Details for the individual systems are described in the corresponding phase studies: $SmF_2-SmF_3$ (8), $EuF_2-EuF_3$ (9), "$TmF_2$"-$TmF_3$ (10), and $YbF_2$-$YbF_3$ (11).

Table 1.   Phase Designation and Formulae of Thirteen New
Binary Rare Earth Fluorides with Mixed-Valency

| t | rhα | cβ | rhβ |
|---|---|---|---|
| $Sm_3F_7$ | $Sm_{14}F_{33}$ | $Sm_{27}F_{64}$ | $Sm_{13}F_{32-\delta}$ |
| $Eu_3F_7$ | $Eu_{14}F_{33}$ | $Eu_{27}F_{64}$ | $Eu_{13}F_{32-\delta}$ |
|  |  |  | $Tm_{13}F_{32-\delta}$ |
| $Yb_3F_7$ | $Yb_{14}F_{33}$ | $Yb_{27}F_{64}$ | $Yb_{13}F_{32-\delta}$ |

Each sample has been characterized by X-ray powder diffraction using the very precise Guinier technique (6,11).   The phase analysis resulted in four new types of fluorite-related super-structures and 13 new binary rare earth fluorides with mixed-valency.   Their powder patterns were indexed by means of a systematic procedure (12).   The formulae have been derived from the superstructure geometry on the basis of the crystal chemistry of anion-excess fluorites (Table 1).   In the case of $TmF_x$, only one II/III-fluoride could be prepared, while the other superstructure phases with smaller F/Ln ratios and $TmF_2$ seem not to exist.

## $CaF_2$-$YF_3$ AND RELATED SYSTEMS

The radii of $Ln^{2+}$ and $Ca^{2+}$, $Sr^{2+}$, $Ba^{2+}$ are very similar and their mutual substitution in crystals is known for a long time. It was, therefore, obvious to reinvestigate the systems $MF_2$-$REF_3$, with M = Ca, Sr, Ba and RE = La, Y, Ln, with particular regard to low-temperature phase relationships.   The high-temperature phase relationships in these systems were already explored by Russian scientists, mainly Sobolev, Fedorov et al.   In some cases, super-structure phases were also detected in the composition region of interest $(M,RE)F_x$ with $2.3 < x < 2.5$ (13-15).   We cannot agree in any case with the crystallographic characterization of these phases, even though in some case single crystals have been used. Of the 48 systems, we examined not all but some of the most significant borderline cases.   Several of our results are already published as for example $CaF_2$-$YF_3$ (16,17), $CaF_2$-$YbF_3$ (18), $SrF_2$-$EuF_3$ (9).   The systems $BaF_2$-$REF_3$ were investigated by Kieser (19); manuscripts are to be submitted for publication (20).

Table 2.   Superstructure Types in $LnF_2$-$LnF_3$ and $MF_2$-$REF_3$

| I-II | m = 19 | rhγ | $M^{II}_{14-\delta}RE^{III}_{5+\delta}F_{43+\delta} = M^{II}_{13+\delta}RE^{III}_{6-\delta}F_{44-\delta}$ |
|---|---|---|---|
| I | m = 15 | t | $M^{II}_{10}RE^{III}_5F_{55}$  $(= M^{II}_2 RE^{III} F_7)$ |
| II | m = 14 | rhδ | $M^{II}_8 RE^{III}_6 F_{34}$ |
| I | m = 14 | rhα | $M^{II}_9 RE^{III}_5 F_{33}$ |
| I | m = 13+14 | cβ | $M^{II}_{17}RE^{III}_{10}F_{64}$  (rhα + rhβI) |
| I-II | m = 13 | rhβ | $M^{II}_{8-\delta}RE^{III}_{5+\delta}F_{31+\delta} = M^{II}_{7+\delta}RE^{III}_{6-\delta}F_{32-\delta}$ |

All samples of interest were prepared by longtime annealing or quenching of mixtures of $MF_2$ and $REF_3$ in sealed platinum tubes. Powder patterns of superstructures were indexed analogous to the $LnF_{2+\delta}$ phases using a systematic procedure (12). Besides t, rhα, cβ, rhβ two further superstructure types were revealed, rhδ in $BaF_2$-$REF_3$ and rhγ in tveitite, the only known yttrofluorite mineral with a superstructure (17,21). The superstructure phases in $MF_2$-$REF_3$ show a significant higher thermal stability than the $LnF_2$-$LnF_3$ phases. Therefore, electron diffraction from single crystals could be carried out (AEI-EM 802, $10^{-5}$ torr, 100 kV). The superstructure geometry could be confirmed for t, rhα, rhβ, rhγ, rhδ. In the case of cβ, the lattice parameters could be confirmed. Systematic extinctions, however, indicate that the true symmetry is probably not cubic but orthorhombic (work in progress). Table 2 shows all superstructure types presently known in $LnF_2$-$LnF_3$ and $MF_2$-$REF_3$ systems. They belong to one of the two homologous series $M_mF_{2m+5}$ or $M_mF_{2m+6}$, or to both in the case of phase width. Single crystal electron diffraction were carried out also on high-temperature solid solutions of $(Ca,Yb)F_x$ with $2.25 \le x \le 2.40$ (cα). In contrast to X-ray powder diffraction, microdomains of the low-temperature superstructure phase could be observed frequently besides the expected undistorted fluorite structure of cα. It is obvious that partial ordering takes place in cα, but the microdomains are too small to be revealed with X-ray diffraction. Analogous behavior has been observed in the case of tysonites (22).

## DISCUSSION

The superstructure rhβ is isostructural to $Na_7Zr_6F_{31+\square}$ (23), as pointed out in (8-11,16,18) with respect to the second formula (see Table 2). Tveitite rhγ is related to rhβ and $Na_7Zr_6F_{31+\square}$ by the equation rhγ = rhβ + $6CaF_2$ (17). Therefore, both structures are built up from the same fluorine cuboctahedrons surrounded by six face-sharing $(REF_8)^{5-}$ square antiprism in an octahedral array. For both superstructures these clusters are centered on 000, 2/3-1/3-1/3, and 1/3-2/3-2/3. Their concentrations in the fluorite matrix, however, are different in rhβ and rhγ in view of different sizes of their unit cells. Expanding these principles further hypothetical superstructures and their unit cells can be predicted: $M_{1+\delta}^{II}RE_{6-\delta}^{III}F_{20-\delta}$ ($\sqrt{7}a,c$), $M_{19+\delta}^{II}RE_{6-\delta}^{III}F_{56-\delta}$ (5a,c), $M_{25+\delta}^{II}RE_{6-\delta}^{III}F_{68-\delta}$ ($\sqrt{31}a,c$), and so on. It is questionable, however, if they really exist. In the first case, the clusters are probably too concentrated, while in the other cases too diluted. Most likely the member with Z=25 could exist (composition range $(M,RE)F_x$ with $2.20 \leq x \leq 2.24$). In contrast to the rare earth oxideflourides (1), all superstructure phases in $LnF_2$-$LnF_3$ and $MF_2$-$REF_3$ systems have different cations, and fully ordered cation distribution is obviously the dominant structure principle. Anion clustering is then only a consequence of local charge balance. Full cation order is, therefore, also expected in the other superstructure phases. Indeed, only one model of ordered cation arrangement is possible for t and rhα (16), as well as for rhδ where $RE^{3+}$ cations are only in every second cation layer along the unique rhombohedral axis. These are symmetrically arranged around the centered $M^{2+}$ cations.


## ACKNOWLEDGMENT

The author is very much indebted to D. J. M. Bevan, J. Strähle, M. Kieser, and A. Clearfield for their support or interesting discussions.


## REFERENCES

1.  D.J.M. Bevan, Non-Stoichiometric Compounds: An Introductory Essay, in: "Comprehensive Inorganic Chemistry", Vol. IV:453, Pergamon Press, Oxford & New York (1973).

2.  Gmelin-Handbuch der Anorganischen Chemie, 8th Edition, No. 39, Part C3, Springer Verlag, Berlin/Heidelberg/New York (1976).

3.  E. Catalano, R.G. Bedford, V.G. Silveira, and H.H. Wickman, Nonstoichiometry in Rare-Earth Fluorides, J. Phys. Chem. Solids 30:1613 (1969).

# STRUCTURAL CHEMISTRY OF TERNARY METAL BORIDES: RARE EARTH METAL-NOBLE METAL-BORON

Peter Rogl, Institute of Physical Chemistry,
University of Vienna, A-1090 Wien, Austria and
Hans Nowotny, Institute of Materials Science
University of Connecticut, Storrs, Connecticut 06268, USA

## INTRODUCTION

Recent work on new ternary superconducting compounds has mainly concentrated on ternary rare earth borides (1-6). Their interesting physical properties, such as reentrant superconductivity and/or magnetic ordering have thus stimulated a general interest in the structural chemistry of combinations: rare earth metal-noble metal-boron.

## EXPERIMENTAL

Impurity level of starting materials, details of sample preparation (arc melting) of ternary alloys: RE-T-B (T=noble metal), X-ray techniques applied, as well as a general description of magnetic measurements (Faraday pendulum compensation method) can be found in earlier publications on ternary rare earth borides (7,8).

## RESULTS AND DISCUSSION

A study of the phase relationships within an isothermal section (1100°C, quench) of the systems RE - {Ru,Os}-B, revealed the existence of several ternary compounds of formula: $RETB_4$, $RET_4B_4$, $RET_3B_2$.

RE{Ru,Os}B$_4$-borides were found to crystallize with the $YCrB_4$-type structure (7,8), which according to the high boron/metal ratio, groups among typical boron net type structures such as $AlB_2$,

TABLE 1

CRYSTALLOGRAPHIC AND MAGNETIC DATA OF THE NEW TERNARY BORIDES
$MM'_4B_4$ (M=Y,RE,Th; M'=Os,Ir). $NdCo_4B_4$-TYPE STRUCTURE, $P4_2/n$, Z=2.

| Compound | a ±0.003 Å | c ±0.002 Å | $\mu_{eff}(RE^{3+})$ theor. | $[\mu_B]$ exp. | $\theta_p[K]$ |
|---|---|---|---|---|---|
| $LaOs_4B_4$ | 7.611 | 3.998 | 0 | weak TDP | - |
| $CeOs_4B_4$ | 7.538 | 4.005 | 2.56 (0.0 $Ce^{4+}$) | weak TDP | - |
| $PrOs_4B_4$ | 7.567 | 4.002 | 3.62 (2.54 $Pr^{4+}$) | 3.76 | -9 |
| $NdOs_4B_4$ | 7.552 | 4.003 | 3.68 | 3.88 | -6 |
| $SmOs_4B_4$ | 7.526 | 4.009 | 1.60* (3.45* $Sm^{2+}$) | 1.65* | - |
| $ThOs_4B_4$ | 7.579 | 3.999 | - | - | - |
| $YIr_4B_4$ | 7.547 | 3.980 | 0 | 0 | - |
| $LaIr_4B_4$ | 7.672 | 3.974 | 0 | weak TDP | - |
| $CeIr_4B_4$ | 7.642 | 3.970 | 2.56 | 2.51 | -16 |
| $PrIr_4B_4$ | 7.629 | 3.974 | 3.62 | 3.70 | -18 |
| $NdIr_4B_4$ | 7.616 | 3.974 | 3.68 | 3.78 | +19 |
| $SmIr_4B_4$ | 7.590 | 3.976 | 1.60* | 1.75* | - |
| $GdIr_4B_4$ | 7.571 | 3.979 | 7.94 | 8.07 | +35 |
| $TbIr_4B_4$ | 7.557 | 3.979 | 9.72 | 9.89 | +6 |
| $ThIr_4B_4$ | 7.668 | 3.972 | - | - | - |

*) values at room temperature

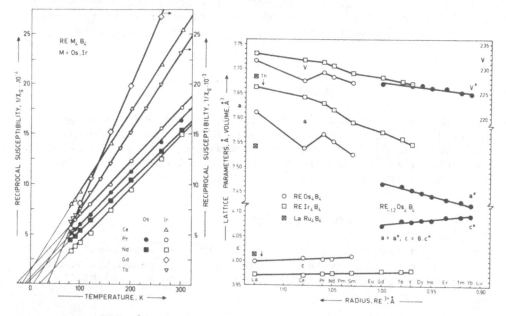

FIG. 1
Reciprocal gramsusceptibility
versus temperature for
$RE\{Os,Ir\}_4B_4$-borides; values
for $CeIr_4B_4$ are reduced by 2/3.

FIG. 2
Lattice parameters and volumes
of $\{RE,Y\}\{Os,Ir\}_4B_4$-borides
versus $R_{RE}^{3+}$.

TABLE 2

CRYSTALLOGRAPHIC DATA FOR THE TERNARY BORIDES $UM_4B_4$
(M=Ru,Os). $LuRu_4B_4$-TYPE STRUCTURE, $I4_1/acd$, Z=8

| | | | | |
|---|---|---|---|---|
| $URu_4B_4$ | a = 7.459(2)Å<br>c = 14.986(3)Å<br>c/a = 2.009 | | $UOs_4B_4$ | a = 7.512(2)Å<br>c = 15.053(4)Å<br>c/a = 2.004 |

$RE_2ReB_6$ and $ThMoB_4$. Due to a decisive radius ratio $R_{RE}/R_T$ compound formation was observed for Y- and the smaller rare earth metals (9). No isotypic Sc-containing compounds were found to exist.

{RE,U,Th}$Ru_3B_2$-borides: From single crystal X-ray analysis hexago-nal symmetry was established (Weissenberg photographs, axis  001 ), indicating a superstructure (10) of the $CeCo_3B_2$-type structure (13) conforming to the following relations:

$$a(RERu_3B_2) \approx 2a(RECo_3B_2); \qquad c(RERu_3B_2) \approx 2c(RECo_3B_2).$$

{RE,Y,Th,U}$T_4B_4$-borides (T=noble metal): $RERh_4B_4$ (Nd → Lu,Th; $CeCo_4B_4$-type) as well as $RERu_4B_4$ (Ce → Lu,Y,Th; $LuRu_4B_4$-type) have been thoroughly investigated by Matthias and coworkers (1-5).

The larger rare earth members of the osmium and iridium series, however, crystallize in a different structure type (Table 2). Precise atomic parameters have been refined for $LaIr_4B_4$ from single crystal diffractometer data (312 independent observations, $|F_o|>2$ ) leading to a final R-value of 3.9%. Virtually identical parameters were found from powder intensity data refinement for $NdIr_4B_4$. Slightly different values were obtained for the compounds of the osmium series from Weissenberg photographs ($LaOs_4B_4$, axis  001 ). In all other cases the symmetry 4/m and space group $P4_2/n$ were con-firmed from powder data. Among actinide metals: $ThOs_4B_4$ and $ThIr_4B_4$ show isotypic behavior ($NdCo_4B_4$-type) with rather large cell param-eters, comparable to those of the largest RE members, whereas the uranium compounds $URu_4B_4$ and $UOs_4B_4$ crystallize with $LuRu_4B_4$ type structure.

Magnetic measurements were performed for all members of the RE{Ru,Os}$B_4$ -, RE{Os,Ir}$_4B_4$- and $RERu_3B_2$-series of compounds within the temperature range 80-300 K. In all cases typical Curie Weiss be-havior  was obvious from the linear temperature dependency of the reciprocal gram susceptibility χg emu/g . The paramagnetic moment per formula unit $\mu_{eff}$ was in very close agreement with calculated theoretical values $g\sqrt{J(J+1)}$ $\mu_B$ on the basis of trivalency of the rare earth component (see Fig. 1, Table 1). Accordingly the paramag-netic moment of the samarium-containing compounds Sm(Os,Ir)$_4B_4$ (room temperature values) clearly indicates Sm to be also in a trivalent state (Tab. 1, Ref. 14). With respect to the multivalent behavior of cerium it is of interest to note, that from paramagnetic moments a $Ce^{4+}$ state is derived for the $CeOs_4B_4$ compound, whereas in combination with the larger T-metal component iridium, an almost 100% $Ce^{3+}$ state is observed. These results are in perfect agreement with the linear dependency of lattice parameters versus radii $R_{RE}3+$ (see Fig. 2). Earlier magnetic measurements (1,3) indicate paramagnetic $RE^{3+}$ behavior for all Ru as well as Rh compounds (80-300 K, except Ce,Th) of the two structural series $RERu_4B_4$ and $RERh_4B_4$. Confirmation of this linear correlation is found from corresponding plots of lattice parameters (L.P.) versus $RE^{3+}$ radius. Furthermore, for the $RERu_4B_4$ phases two different series with slightly different c/a ratios are shown on the graph. This fact might indicate the formation of super-structure cells for the larger RE members of $LuRu_4B_4$-type borides (Ce → Eu). Similarily, formation of superstructure cell was observed from L.P.-$RE^{3+}$ plots for the smaller rare earth compounds (Gd → Yb,Y) of the $REOs_4B_4$ structural series (Fig. 2). Weissenberg photographs confirmed the $NdCo_4B_4$-superstructure cell conforming to the relation $a=a(NdCo_4B_4)$, $c=8c(NdCo_4B_4)$.

The slightly different c/a ratio reflects the higher RE/T ratio $(Y_{1.25}Os_4B_4$-type structure (10)) found from microprobe analysis. Thus, at present ternary rare earth-noble metal-borides of formula 1/4/4 were found to crystallize in at least four different structure types ($CeCo_4B_4$, $LuRu_4B_4$, $NdCo_4B_4$ and $Y_{1+x}Os_4B_4$). Furthermore, for rhenium compounds (La,Ce,Y)$Re_4B_4$ a new structure type was suggested by Mihalenko (15).

Formation of $T_4$-metal clusters (tetrahedra) appears to be the general structural unit common to all $RET_4B_4$-type phases. Inter-atomic distances showed boron-boron interaction (1.81-1.85 A) to be rather weak compared to typical boron pair formation (1.76-1.82A); B-T distances, however, prove strong metal boron interaction. In-volving two rather remote RE atoms, boron coordination most resem-

TABLE 3

GEOMETRICAL RELATIONSHIP BETWEEN THE DIFFERENT STRUCTURE
TYPES $RET_4B_4$ (T=Ru,Rh,Os,Ir).

| $LuRu_4B_4$ - $CeCo_4B_4$ | | $NdCo_4B_4$ - $CeCo_4B_4$ | |
|---|---|---|---|
| $a(LuRu_4B_4)$ | $2a(CeCo_4B_4)$ | $a(NdCo_4B_4)$ | $2a(CeCo_4B_4)$ |
| $c(LuRu_4B_4)$ | $2 \ 2a(CeCo_4B_4)$ | $c(NdCo_4B_4)$ | $1/2c(CeCo_4B_4)$ |

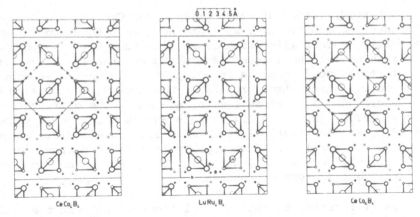

FIG. 3

Comparison of the crystal structures of $CeCo_4B_4$ and $LuRu_4B_4$; projection along 001 and 100 respectively. The atomic arrangement of $LuRu_4B_4$ consists of alternating slabs of $CeCo_4B_4$ type structure and its mirrored counterpart. For detailed description see also Fig. 4.

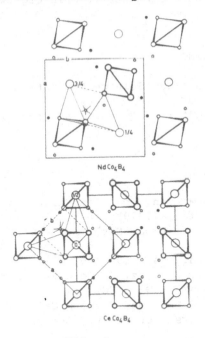

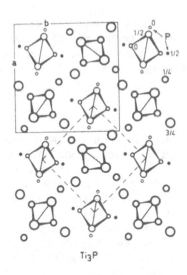

FIG. 4

Comparison of the crystal structure $NdCo_4B_4$; projection along 001 . Largest circles are RE; tetrahedra are formed by noble metal atoms. Smallest circles are boron atoms. Size of noble metal atoms (Co) depends on height in projection.

FIG. 5

Crystal structure of $Ti_3P$; projection along 001 . For comparison with the structure types of $NdCo_4B_4$, $CeCo_4B_4$ See Fig. 4.

bles a distorted triangular metal prism. For $CeCo_4B_4$ and $LuRu_4B_4$-type borides one additional T atom and no direct RE-RE contacts indicate a better space filling with respect to $NdCo_4B_4$-type, which in turn is favored for the largest RE-members and thorium. Confirmation of this fact is found from the recently reported $LaRu_4B_4$-compound with the $NdCo_4B_4$-type structure (11). Considering the packing of the $T_4$-tetrahedral units, a close resemblance among the different structure types $RET_4B_4$ (Fig. 3,4) is obvious and lattice geometry can be expressed by simple relations (see Table 3). Indeed close agreement is found for extrapolated lattice parameters derived from observed $NdCo_4B_4$-type parameters of $LaRu_4B_4$ ($R_{Rh} \approx R_{Ru}$) with the hypothetical $LaRu_4B_4$ and $CeCo_4B_4$ type structures.

It is of interest to note the relationship between the $NdCo_4B_4$ type structure and that of $Ti_3P$ ($Fe_3(B,P)$-type, Ref. 16). Within the same space group ($P4_2/n$) substitution of the eight fold position for Ti(II) and P by one large atom (Nd) leads to the $NdCo_4B_4$ type structure $(Ti_8^{II}P_8).Ti_8^{I} Ti_8^{III}= Nd_2Co_8B_8$ (see Fig. 5). Both the eightfold positions form almost spherical structure elements (two intergrown tetrahedra $Ti_4+P_4$).

ACKNOWLEDGEMENT

..   This work was sponsored by the Austrian Science Foundation (Osterr.Forschungsrat) under grant 3620.
        The authors thanks are due Dr. Endter, Degussa, Hanau for kindly supplying us the noble metals.

REFERENCES

1.  B. T. Matthias, E. Corenzwit, J. M. Vandenberg and H. E. Barz,
    Proc. Natl. Acad. Sci. USA, 74:  1334 (1979)
2.  J. M. Vandenberg and B. T. Matthias, Proc. Natl. Acad. Sci.
    USA, 74:  1336 (1977)
3.  D. C. Johnston, Solid State Commun., 24:  699 (1977)
4.  B. T. Matthias, C. K. N. Patel, H. E. Barz, E. Corenzwit and
    J. M. Vandenberg, Phys. Lett., 68A:  119 (1978)
5.  W. A. Fertig, D. C. Johnston, L. E. DeLong, R. W. McCallum,
    M. B. Maple and B. T. Matthias, Phys. Rev. Lett., 38:  989
    (1977)
6.  R. N. Shelton, Less Common Met. 62:  191 (1978)
7.  P. Rogl, Mater. Res. Bull. 13:  519 (1978)
8.  R. Sobczak and P. Rogl, J. Solid State Chem. 27:  343 (1979)
9.  H. Nowotny and P. Rogl, in Boron and Refractory Borides,
    ed. M. V. Matkovich, N.Y. Springer (1977)
10. P. Rogl, to be published
11. A. Gruttner and K. Yvon, Acta Cryst. B35, 451 (1979)

12.  P. Rogl, Mh. Chem. <u>110</u>:  235 (1979)

13.  Yu. B. Kuz'ma, P. I. Kripyakevich and N. S. Bilonishko,
     Dopov. Akad. Nauk.  URSR Ser. A, <u>10</u>:  939 (1969)

14.  B. Rupp, P. Rogl and R. Sobczak, to be published

15.  S. I. Mihalenko, Yu. B. Kuz'ma and A. S. Sobolev
     Poroschkov. Met. <u>169(1)</u>:  48 (1977)

16.  S. Rundqvist, Acta Chem. Scand. <u>16</u>:  1 (1962)

# THE CRYSTAL STRUCTURE OF PrOHSO$_4$*

James A. Fahey, Department of Chemistry, Bronx Community
College, W181 St. & University Ave. Bronx, N.Y. 10453;
Graheme J. B. Williams, Chemistry Department
Brookhaven National Laboratory, Upton, N.Y. 11973;
J. M. Haschke, Rockwell International, Rocky Flats Plant
Golden, CO, 80401

## INTRODUCTION

The phase equilibria and structural relationships of the anion
substitution phases of the lanthanide hydroxides are both interesting and complex. The consequences of anion substitution have been
most thoroughly investigated for monovalent anion systems of the
type Ln(OH)$_{3-x}$X$_x$ (X=Cl$^-$, Br$^-$, I$^-$, NO$_3^-$). Equilibria, structures and
structural correlations of these systems are described and reviewed
in recent reports.(1-3)

Similar studies of the anion accommodation processes of common
divalent anions by lanthanide hydroxides have not been described,
and an effort to investigate the sulfate-containing systems of the
lighter lanthanides has been initiated.(4) A determination of the
structure of praseodymium monohydroxide monosulfate has been undertaken as part of this effort.

## EXPERIMENTAL SECTION

Single crystals of PrOHSO$_4$ were selected from products of the
hydrothermal reaction of Pr(OH)$_3$ with Pr$_2$(SO$_4$)$_3$·xH$_2$O in gold capsules
at 450-500 °C and 1,200 ± 200 atm water pressure. The preparative
procedures were similar to those outlined previously.(5) Weissenberg
data for several multifacet crystals showed 2/m diffraction symmetry
and systematic absences consistent with space group P2$_1$/c. The
crystal used for data collection measured 0.09X0.07X0.06 mm. The
final unit cell parameters refined from data for 28 reflections were
in excellent agreement with those obtained from Guinier data. The

Table I. Atomic Coordinates and Thermal Parameters of PrOHSO$_4$

| Atom | Coordinate[a] | | | B-Value X10$^4$ | | | | | |
|------|------|------|------|------|------|------|------|------|------|
| | x | y | z | 11 | 22 | 33 | 12 | 13 | 23 |
| Pr | 0.1590(1) | 0.43505(5) | 0.30128(9) | 79 | 7 | 32 | -3 | 40 | -2 |
| S | 0.4022(6) | 0.1456(2) | 0.3888(4) | 77 | 7 | 32 | 0 | 40 | 1 |
| O(1) | -0.305(2) | 0.3348(8) | 0.060(1) | 61 | 20 | 36 | 6 | -5 | 9 |
| O(2) | 0.225(2) | 0.2457(7) | 0.319(1) | 188 | 8 | 80 | 21 | 63 | 5 |
| O(3) | 0.475(3) | 0.0954(8) | 0.233(1) | 265 | 15 | 50 | 6 | 88 | -2 |
| O(4) | 0.195(2) | 0.0703(7) | 0.450(1) | 136 | 6 | 59 | -2 | 63 | 0 |
| OH | -0.239(2) | 0.4149(7) | 0.461(1) | 102 | 10 | 60 | 8 | 78 | -2 |
| H[b] | 0.375 | 0.063 | 0.167 | - | - | - | - | - | - |

a. The uncertainty in the last digit is given in parentheses.

b. The fixed coordinates of H used in the final refinement are listed; $B_{11}$ was also invariant at 2.0 $Å^2$. Anisotropic terms are defined by the expression $\exp(-2\pi^2(h^2B_{11}\cdots + 2hkB_{12} + \cdots))$.

crystal data follow: a=4.488 (1), b=12.495 (2), c=7.091 Å (3), β=111.08° (2), Z=4, $d_x$=4.49 g/cm³.

Intensity data were collected with an Enraf Nonius CAD4 auto-diffractometer using MoKα radiation (λ=0.7107Å). A data set of 1890 2θ≤100° was corrected for polarization and absorption. The Pr and S atoms were positioned in 4e sites using the Patterson method. Oxygen positions (also 4e) were located by Fourier synthesis. After the data had been weighted (w=$\sigma^2 F_o^2$+12X10⁻⁴$F_o^2$) and corrected for secondary extinction, the final least squares refinement converged at R=0.072. In this cycle, the H position was fixed at the coordinates obtained from difference Fourier data. The atomic scattering factors for the neutral atoms were those of Cromer and Waber[6], and the anomolous-dispersion corrections for Pr and S were those of Cromer and Liberman.[7] The refined positional coordinates and anisotropic thermal parameters for PrOHSO₄ are given in Table I. Selected values of bond distances and angles appear in Table II.

Table IIa. Distances in the coordination spheres of Pr and S

| Pr-O(1) | 2.504 Å* | Pr-O(4)[d] | 2.556 | S-O(1) | 1.455(8) |
|---|---|---|---|---|---|
| Pr-O(2) | 2.383 | Pr-OH | 2.541 | S-O(2) | 1.469(9) |
| Pr-O(3)[a] | 2.840 | Pr-OH[e] | 2.537 | S-O(3) | 1.466(9) |
| Pr-O(3)[b] | 2.636 | Pr-OH[f] | 2.451 | S-O(4) | 1.493(9) |
| Pr-O(4)[c] | 2.547 | | | | |

*esd's in Pr-O distances are all 0.009 Å.

[a] x,½-y,½+z, [b] 1-x,½+y,½-z, [c] x,½-y,z-½, [d] -x,½+y,½-z, [e] 1+x,y,z, [f] -x, 1-y, 1-z

Table IIb. Unique angles in the coordination sphere of S

| O(1)-S-O(2) | 110.3(6)° | O(2)-S-O(3) | 110.1(6)° |
|---|---|---|---|
| O(1)-S-O(3) | 110.6(6) | O(2)-S-O(4) | 108.5(5) |
| O(1)-S-O(4) | 109.2(6) | O(3)-S-O(4) | 107.8(6) |

## DISCUSSION

A stereoscopic projection of the cation coordination sphere of the PrOHSO₄ structure is shown by Fig. 1. The nine-fold coordination frequently observed for the lighter lanthanides is achieved by oxygens from three hydroxides and six sulfate ions. The coordination geometry is best described as a slightly distorted mono-capped square antiprism. The Pr-OH and Pr-O distances listed in Table II are consistent with those of similar lanthanide compounds. (2) The structural data for the sulfate ion (cf. Table II) provides an internal standard for evaluating the refinement and shows that the ion is not significantly distorted in the hydroxide sulfate. The average S-O distance (1.47 ± 0.01Å) is in good agreement with

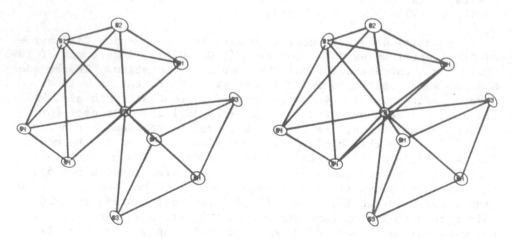

Figure 1.   The nine-fold coordination geometry of Pr in PrOHSO₄

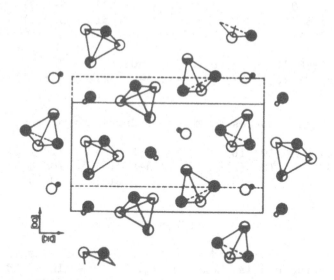

Figure 2.   The idealized structure of PrOHSO₄ projected on (011).
            (Small circles represent Pr atoms and large circles
            represent O atoms and OH; S atoms centered in the tetra-
            hedra are not shown.  Dark circles are near 1/4, open
            circles are near 3/4 and half-darkened circles are at an
            intermediate distance near 1/2.)

the 1.49Å value reported for sulfate.(8)  The average O-S-O angle
(109.4 ± 1.1°) is close to the ideal tetrahedral value.

The $PrOHSO_4$-type structure is characterized by a framework of
double chains of $PrOH^{2+}$ which are parallel to [100] and staggered
by half a repeat unit.  These chains are encircled by six sulfate
ions which link a given double chain to six adjacent chains and
thereby provide strong binding interaction normal to [100].  This
extended structure is shown by the idealized projection onto (011)
in Fig. 2.  The contents of only a single unit-cell-thick layer is
projected in this view, and each successive layer normal to [100]
is displaced along [001] by 1.61Å.  The staggered feature of the
double chains is clearly seen in Fig. 2.  The Pr in a given $PrOH^{2+}$
chain is sandwiched between two hydroxides which are almost coin-
cident with the projected metal position.  In the adjacent chain,
the Pr and OH positions relative to (011) are reversed.  The presence of hydrogen bonding is indicated by small contact distances
(2.617 ± 0.012 and 2.631 ± 0.013Å) between OH and O3.  In Fig. 2,
the O3 atoms are indicated by the half-darkened circles at the
sulfate apices pointing toward the $PrOH^{2+}$ chains.  Since neither
oxide is clearly favored for hydrogen bonding to a given hydroxide,
a random site distribution is believed to be most probable.

The structures of several hydrated basic sulfates with 1:1
ratios of metal to hydroxide have been determined, but reports for
anhydrous salts are limited.(8)  Both the $FeOHSO_4$- and $CuOHIO_3$-type
structures are classified by Wells as one-dimensional hydroxy-metal
complexes because single metal-hydroxide chains are joined by the
oxy-anions to form the three dimensional structure.  Two- and three-
dimensional lanthanide hydroxide frameworks have been identified for
$Ln(OH)_2X$ and $Ln_7(OH)_{18}X_3$ phases (X=Cl⁻, Br⁻, I⁻, NO₃⁻), respectively.
(2,3)  Although similarities to the one-dimensional hydroxy-metal
structures are obvious, $PrOHSO_4$ has an unusual type of structure
which is known to exist for La and Nd,(4) and which is probably
stable for the basic sulfates of additional lanthanides and
actinides.

## REFERENCES

*Research performed, in part, under the auspices of the U.S. Depart-
ment of Energy.

1.  E. T. Lance-Goméz and J. M. Haschke, J. Solid State Chem. 23:
    275 (1978).

2.  E. T. Lance-Goméz, J. M. Haschke, W. Butler and D. R. Peacor,
    Acta Crystallogr. B34:758 (1978).

3.  J. M. Haschke, J. Solid State Chem. 18:205 (1976).

4.  J. M. Haschke, unpublished results.

5.  J. M. Haschke and L. Eyring, Inorg. Chem. 10:2267 (1971).

6.  D. T. Cromer and J. T. Waber, Acta Crystallogr. 18:104 (1965).

7.  D. T. Cromer and D. Liberman, J. Chem. Phys. 53:1891 (1970).

8.  A. F. Wells, "Structural Inorganic Chemistry," Fourth Ed.
    Clarendon Press, Oxford (1975).

# OXIDATION OF PRASEODYMIA IN TERNARY RARE EARTH OXIDE PHASES

G. K. Brauer and B. Willaredt

Laboratorium für Anorganische Chemie, Universität Freiburg

D-7800 Freiburg, F.R. Germany

In the binary oxide systems Ce-O, Pr-O, and Tb-O there are two-phase regions at the oxygen richest parts where a mixed valence phase $CeO_{1.81}$, or $PrO_{1.833}$, or $TbO_{1.81}$ coexists with the corresponding dioxide $CeO_2$, $PrO_2$, or represents the phase richest in oxygen under normal laboratory conditions of temperatures and pressures. Both coexisting or neighbouring phases are very similar in structure; they have an oxygen-deficient or a sound fluorite type lattice. As the two phases (i.e., $CeO_{1.81}$ and $CeO_2$) are so similar, the area between them appears as a sort of miscibility gap, and one wonders if the gap may close at suitable conditions, forming a coalescent (critical) point.

Authors such as Eyring, Bevan, Hyde, Faeth, Clifford and others have shown that in addition to the parameters x and T (as usually shown in the published diagrams), a third variable quantity, the pressure of oxygen, must by no means be neglected. Thus a three-dimensional p,T,x diagram has to be considered where the phase boundaries are represented by faces, more or less bent, and sometimes even bent in a strange manner as has been described (1).

In the case of the gap $CeO_{1.81}$-$CeO_2$ it was observed (2) that the coalescence can simply be realized by increasing the temperature of a mixture of the two phases. sealed in a capsule. The "critical temperature" lies at ∿685°C. This experimentally favourable situation is due to the very low corresponding $O_2$-pressures (about $10^{-10}$-$10^{-30}$ Torr).

In contrast to the ceria case, the coalescence of the $PrO_{1.83}$ and the $PrO_2$ phases is to be expected only at an extremely high $O_2$-pressure. We have investigated in the praseodymia case how far the transition from the oxygen–deficient to the oxygen–sound phase may be realized by preparing mixed crystals with other RE oxides using the extremely high pressures.  Mixed crystals of praseodymia with $Dy_2O_3$, $Ho_2O_3$ or $Er_2O_3$ favour the oxidation of $Pr^{3+}$ to $Pr^{4+}$.  The oxidation performable under relatively simple and mild conditions (350–400°C, $O_2$–pressure 0.2 and 1 atm), achieves, indeed, a complete oxidation of Pr, but only in the way of forming mixed crystals:  No pure $PrO_2$ is produced but rather ternary phases containing $Pr^{4+}$. The crystal structure of the mixed oxide phases passes from pseudo-cubic ($Pr_6O_{11}$) continuously to C-sesquioxide types.

For the system $Pr_6O_{11}$ + $Dy_2O_3$, about 60 at.% Dy in the mixed crystal phase is needed to complete the Pr oxidation.  The lattice parameters parallel the transition of $Pr^{3+}$ to $Pr^{4+}$ combined with the change due to the gradual transition from pseudo-fluorite type to C-sesquioxide type.  For $Pr_6O_{11}$ + $Ho_2O_3$, about 50–70 at.% Ho is needed to achieve the $Pr^{3+}$ to $Pr^{4+}$ oxidation.  The percentage depends on the preparation temperature and oxygen pressure in a reasonable direction.  Again, the lattice parameters parallel the oxidation state of Pr and the transition from a pseudo-fluorite to C-sesquioxide type.  The system $Pr_6O_{11}$ + $Er_2O_3$ is quite similar except that there occurred some confusing zones on the way from $Pr_6O_{11}$ to $Er_2O_3$, probably due to non-equilibrium samples.  Complete tetravalency of Pr was already reached at 50–55% Er in the mixed phase. A small area of two coexisting phases of the same C-sesquioxide type is also observed in this system.

Though $Dy_2O_3$, $Ho_2O_3$, and $Er_2O_3$ behave quite similarly in favouring the oxidation of $Pr^{3+}$ to $Pr^{4+}$ insofar as it is always completed at 50–70 at.% of the tetravalent RE there are appreciable differences with regard to the phases occurring in the three mixed oxide systems. We suggest that the smaller ionic radii of the admixed trivalent RE's are generally responsible for favouring the small $Pr^{4+}$.  However, we do not yet have an explanation for the differences in the observed phase relationships comparing the three closely related systems of mixed crystal oxides.  It is to be hoped that the understanding of such a situation may be evolved by using the concept of microdomains in phases (3).

REFERENCES

1.    G. Brauer, in Progress in the Science and Technology of the Rare Earths, L. Eyring, Ed., Vol. 2, p. 312.  Pergamon Press (1966).
2.    G. Brauer, K. A. Gingerich and U. Holtschmidt, J. Inorg. Nucl. Chem. 16:77, 87 (1960).
3.    D. J. M. Bevan, private communication.

THE TRIVALENT BEHAVIOUR OF CERIUM.

IMPLICATIONS FOR COSMOCHEMISTRY AND THERMODYNAMICS

Long-Den Nguyen

Institut de Physique Nucléaire

B.P. N° 1, 91406 – Orsay, France

## INTRODUCTION

Several investigations have indicated the existence of a non-stoichiometric single phase of cerium oxide $CeO_{2-x}$ $(0 \leqslant x \leqslant 0.5)$ (1). These experimental results were obtained using either the study of condensed phase (x-ray measurements, thermogravimetry...) or of the gaseous phase (particularly with the mass-spectrometry coupled with effusion from a Knudsen cell or free vaporization according to Langmuir...). These studies principally concerned the binary systems Ce-O and in some cases the ternary systems: Ce-M-O (where M represents another element). The different interpretations of these results are not completely satisfactory and sometimes conflicting, which shows the difficulty of obtaining a clear understanding of the high temperature behaviour of cerium oxides.

Due to its high sensitivity to oxygen fugacity, it is interesting to use the behaviour of cerium oxides as a probe of the high temperature evolution of the systems studied. These studies carried out in correlation with other rare earths (REs), particularly those with two valence states: Pr and Tb (III and IV) and Eu and Yb (II and III) provide a means of investigating thermodynamic behaviour of the whole system.

We have studied not only binary or ternary systems but also complex systems involving more than three elements, in order to approach as nearly as possible real systems. Furthermore we present in this paper evidence for the existence of the trivalent state in cerium oxide and the use of this property to study the mechanisms of the solar system formation (cosmochemistry and planetology) through vaporization and condensation of the elements.

189

## IMPLICATIONS IN COSMOCHEMISTRY

One of the most striking features of the RE geochemistry of
lunar samples is the negative europium anomaly (Figure 1) (2).  The
majority of geochemists have agreed that this phenomenon is probably due to the presence of divalent europium as the major chemical state in the particularly reducing conditions of the lunar
magmas in partial melting and/or fractional crystallization (3).
In 1970 (4), we proposed a mechanism by which the depletion of
europium may be explained by the differences in volatilization of
REs from their oxides at a high temperature and in vacuum presumably
during an early high temperature stage of the sun, the Hayashi
phase (5) before and during accretion of lunar materials.  Furthermore we indicated that according to our hypothesis, the negative
anomaly of cerium was not expected in spite of the high volatility
of $CeO_2$ (more than $Eu_2O_3$) because of the reduction of $CeO_2$ to
$Ce_2O_3$ (less volatile).

Our theory assumes that the moon was formed by an accretion of
small grains of solid in vacuum.  We can represent the relative concentration q% after a heating at the temperature T during the time
t by the formula:  $q\% = 100\ Ct/Co$ with $Ct = C_o exp(-3Gt/rd)$.  The
value for $G = p\sqrt{M/2\pi RT}$, where p is the vapor pressure, M is the
molecular weight, d is the density of the grain, r is its radius
and R is the gas constant.  We have calculated q% of elements remaining after heating grains of radius:  0.01, 0.05 and 0.1 cm at
1400°K for 100 years (5) (Figure 2) (2,4).  We note that for r =
0.05 cm, the calculations give the same RE distribution as that of
the rock 12035 which is supposed to have a closer RE composition to
the "original" lunar material (6).

In order to verify our theory, we simulate the Hayashi phase
(7).  The experiments are carried out with an equimolecular mixture
of RE oxides.  After heating them in the vacuum (simulation of the
case of lunar formation) or in the atmosphere (simulation of the
case of formation of the earth) we analysed by mass-spectrometry,
using the method described in our previous work (2), the composition of REs in different phases:  condensed phase, remaining phase
and residue on the crucible surface.  The results of the analysis
of the remaining phase are reproduced (continuous lines) in Figure
3 (volatilization in the vacuum:  simulation of the conditions of
lunar formation) and in Figure 4 (volatilization in the air:  simulation of the conditions of the earth's formation).

For the theoretical values (dotted lines) we calculated the
relative concentration of REs in the remaining phase q% for the
following conditions:

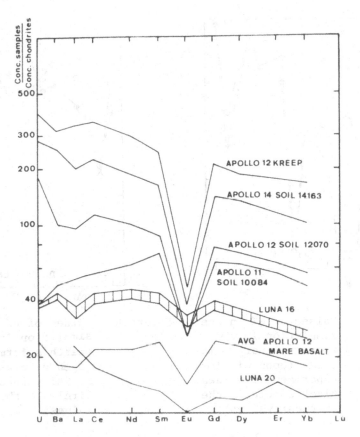

Figure 1.  RE concentrations normalized to chondites in lunar
samples.

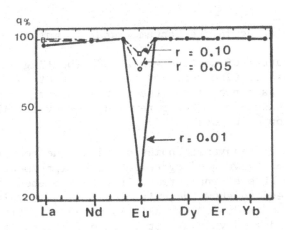

Figure 2.  Theoretical calculations of relative RE concentration
q% with different radius r(cm) of grains.

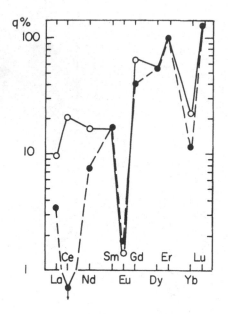

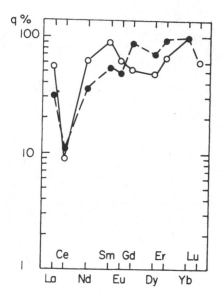

Figure 3.  Values of q% in the
          simulation in the
          vacuum (lunar forma-
          tion) (open circles:
          experimental values;
          black circles:  the-
          oretical values).

Figure 4.  Values of q% in the
          simulation in the air
          (earth's formation)
          (open circles:  experi-
          mental values; black
          circles:  theoretical
          values).

Condition 1:  Heating in the Vacuum (lunar formation simulation)

     The temperature of heating was 2273°K (heating at this tem-
perature for a short time replaced the heating at 1400°K during 100
years in the Hayashi phase) and the partial pressures of different
RE oxides were taken from Shchukarev and Semenov's works (8).  We
notice the following points:

     (i) From a thermodynamic point of view, the results of theore-
tical calculations give good agreement with experimental results
for the studied RE's except cerium.  For this element the results
of simulation tend to show the existence of a quite different behaviour of cerium oxide $CeO_2$ when it is mixed with trivalent RE
oxides and heated in a vacuum.  It is then far less volatile than
when it is alone; its vapor pressure approaches that of neodymium.

One could suppose that, under these special conditions, it would be reduced to trivalent state.

(ii) From the cosmochemistry point of view, these results confirm our theory.  In lunar samples we observed an important depletion of europium, and a relative concentration of cerium either analogous to or slightly higher than (about 7%) (9) other REs. These observations agree very well with our results obtained by simulation.  Our theory based on the volatilization of REs during the Hayashi phase, and on the interactions in complex systems with the ambient environment provides a coherent explanation for europium and cerium concentrations as well as for those of the other REs.


Condition 2:  Heating in the air (earth formation simulation)

Our calculations, performed with vapor pressures of the RE oxides under the atmosphere given by Benezech and Foëx (10) agree very well with our results obtained in simulation.  We observe, then, a smaller volatility of europium and a quite high volatility of cerium in the tetravalent oxide form.  These observations tend to confirm our theory for the explanation of the relative concentrations of REs in terrestrial rocks.

These two sets of experiments seem to confirm our theory for explanation of the relative concentrations of RE in lunar and terrestrial samples by volatilization of RE oxides under vacuum or in the air during the Hayashi phase.  These studies of the behaviour of REs and particularly of cerium could be extended to the mechanisms of supernova formation by condensation.  They explain perfectly the differences of RE distributions in solar and supernova condensates reported by Boynton (11).


CONCLUSIONS

Due to the high sensitivity of cerium to oxygen fugacity, the study of the behaviour of cerium oxides in some complex systems allows us to better understand the evolution of these systems.  Thus cerium as europium stand as a probe of the evolution of the system to which they belong.  This ability offers many applications in different fields.


REFERENCES

1.  R. J. Ackermann and E. G. Rauh, J. Chem. Therm. 3:609 (1971);
    J.R. Sims, Jr. and R. N. Blumenthal, High Temp. Sci. 8:99,
    111, 121 (1976).
2.  L. D. Nguyen, M. de Saint Simon, G. Puil and Y. Yokoyama,

Geochim. Cosmochim. Acta, Suppl. 4, Vol. 2:1415 (1973).

3.  See different proceedings of "Lunar Science Conference."

4.  L. D. Nguyen and Y. Yokoyama, 33rd Annual Meeting of The
    Meteoritical Society, Skyland, VA, USA, October 27-30, 1970;
    Meteoritics 5:214 (1970).

5.  J. W. Larimer and E. Anders, Geochim. Cosmochim. Acta 31:1239
    (1967).

6.  A. Masuda and T. Tanaka, Contr. Mineral. and Petrol. 34:336
    (1972).

7.  L. D. Nguyen, M. de Saint Simon, J. P. Coutures and M. Foëx,
    Journées d'Odeillo (France) B111:1 (1975).

8.  S. A. Shchukarev and G. A. Semenov, Dokl. Akad. Nauk. 141:652
    (1961).

9.  A. Masuda and N. Nakamura, Geoch. J. 7:179 (1973).

10. G. Benezech and M. Foëx, C. R. Ac. Sc. (France) 268:2315 (1969).

11. W. V. Boynton, Lunar and Plan. Sc. IX, Part 1:120 (1978).

# THE STRUCTURES OF $Na_3Ln(XO_4)_2$ PHASES (Ln = RARE-EARTH, X = P, V, As)

Marcus Vlasse, Claude Parent, Roger Salmon and Gilles
Le Flem
Laboratoire de Chimie du Solide du C.N.R.S., Université
de Bordeaux I, 351, cours de la Libération, 33405,
Talence Cedex, France

## ABSTRACT

A complete and detailed structural description of sodium rare-earth orthophosphates, orthovanadates and orthoarsenates, with general formula $Na_3Ln(XO_4)_2$ (Ln = rare-earth, X = P, V, As), is presented. Their structural evolution has been studied not only as a function of the $Ln^{3+}$ ion size and the $(XO_4)^{3-}$ group, but also as a function of temperature.

A number of allotropic forms have been observed for each pair $Ln^{3+}-(XO_4)^{3-}$. The different structure types found during this study have been identified due to the complete structure determinations of the low temperature varieties of $Na_3Nd(PO_4)_2$, $Na_3La(VO_4)_2$, $Na_3Nd(VO_4)_2$ and $Na_3Er(VO_4)_2$, and the high temperature form of $Na_3Yb(PO_4)_2$. The first four structures can be considered as derivated structures of the glaserite type, while the last one corresponds to phases having an extended non-stoichiometry and is related to the structure of $NaZr_2(PO_4)_3$.

## INTRODUCTION

The great interest in new luminophors has motivated recently considerable research concerning materials susceptible to containing rare-earth ions. A precise knowledge of the crystal structure of such materials is essential in the study of the luminescence mechanisms of the activator ions.

Keeping this in mind, we have undertaken a study of the compounds with general formula $Na_3Ln(XO_4)_2$ where Ln = rare-earth and

Table 1.   Structure types of the $Na_3Ln(XO_4)_2$ phases.

| | Structure types | Space group | Relationships between the cell parameters and those of the glaserite hexagonal cell($a_h, c_h$) | z | Cationic ordering in the($\beta$) sublattice along the $[210]$ direction |
|---|---|---|---|---|---|
| I | $K_3Na(SO_4)_2$ glaserite | $P\bar{3}m1$ | $a_h$ ; $c_h$ | 1 | |
| II | $\beta\text{-}K_2SO_4$ | Pnma | $a_{orth.} = c_h$ ; $b_{orth.} = a_h$ ; $c_{orth} = a_h\sqrt{3}$ | 4 | |
| III | $Na_2CrO_4$ L.T. | Cmcm | $a_{orth.} = a_h$ ; $b_{orth.} = a_h\sqrt{3}$ ; $c_{orth.} = c_h$ | 4 | |
| IV | $Na_3La(VO_4)_2$ L.T. | $Pbc2_1$ | $a_{orth.} = a_h$ ; $b_{orth.} = 2c_h$ ; $c_{orth.} = 2a_h\sqrt{3}$ | 8 | 2-2 |
| V | $Na_3Nd(VO_4)_2$ L.T. | Cc | $a_{mon.} = 3a_h\sqrt{3}$ ; $b_{mon.} = a_h$ ; $c_{mon.} = 2c_h$ | 12 | 2-1 |
| VI | $Na_3Nd(PO_4)_2$ L.T. | $Pbc2_1$ | $a_{orth.} = 3a_h$ ; $b_{orth.} = 2c_h$ ; $c_{orth.} = 2a_h\sqrt{3}$ | 24 | 2-2 |
| VII | $Na_3Er(VO_4)_2$ L.T. | $P2_1/n$ | $a_{mon.} = a_h$ ; $b_{mon.} = a_h\sqrt{3}$ ; $c_{mon.} = c_h$ | 2 | 1-1 |
| VIII | $Na_3Tm(PO_4)_2$ I.T. (5) | Pnmm, $Pnm2_1$ | $a_{orth.} = 2a_h\sqrt{3}$ ; $b_{orth.} = c_h$ ; $c_{orth.} = 3a_h$ | 12 | 2-2 |
| IX | $Na_3Yb(PO_4)_2$ H.T. | $R\bar{3}c$ | $a_{hex.} = a_h\sqrt{3}$ ; $c_{hex.} = 3c_h$ | 9 | |

X = P, V, As.  For each Ln-X couple, all the compounds are characterized by similar crystal structure (1-4).  It seems, therefore, important to give an overall description of the $Na_3Ln(XO_4)_2$ phases and discuss the relationships which exist between the various structural
types.

## EXPERIMENTAL

The nine structural types characterizing the different phases
$Na_3Ln(XO_4)_2$ can be classified using Table 1.  They were identified
and characterized either using high temperature x-ray diffraction,
or by structural determination of the room temperature form of
$Na_3La(VO_4)_2$ (IV), $Na_3Nd(VO_4)_2$ (V), $Na_3Nd(PO_4)_2$ (VI), $Na_3Er(VO_4)_2$ (VII)
and the high temperature form of $Na_3Yb(PO_4)_2$ (IX).

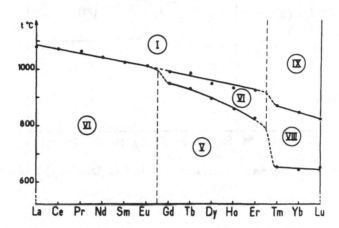

Figure 1.   Thermal behavior of the Na$_3$Ln(PO$_4$)$_2$ phases.

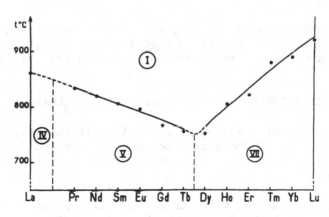

Figure 2.   Thermal behavior of the Na$_3$Ln(VO$_4$)$_2$ phases.

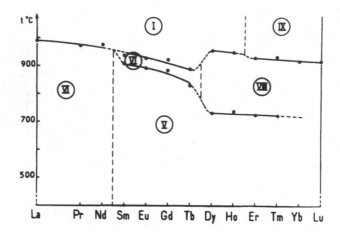

Figure 3.   Thermal behavior of the $Na_3Ln(AsO_4)_2$ phases.


## VARIATIONS OF THE $Na_3Ln(XO_4)_2$ PHASES AS A FUNCTION OF TEMPERATURE

In Figs. 1, 2 and 3 are given the transformation temperatures of the $Na_3Ln(XO_4)_2$ phosphates, vanadates and arsenates.  All trans-formations are reversible except those between the forms V and VIII in the case of the phosphates and arsenates (may be the result of slow kinetics).

For the vanadates $Na_3Ln(VO_4)_2$ (Ln = Dy to Lu) a fast quench of a type I phase gives a metastable type III variety, which when an-nealed at a low temperature transforms into the type VII form.


## RELATIONSHIP BETWEEN THE STRUCTURES OF THE $Na_3Ln(XO_4)_2$ PHASES

The structure of $K_3Na(SO_4)_2$ (glaserite) is the simplest of all nine structures and is the parent compound from which derive the eight remaining structures.  It is characterized by two basic struc-tural features:

(a)  a central octahedron with its 3-fold axis normal to the (001) plane linked at the corners to three down pointing and three up pointing tetrahedra (Fig. 4).

(b)  In the (210) plane there are found rows of $(XO_4)$ tetrahedra and K atoms (A) alternating with rows of Na and K atoms (B), all running along the $\vec{c}$ axis (Fig. 5).

All the prototype structures (except IX) derive from the

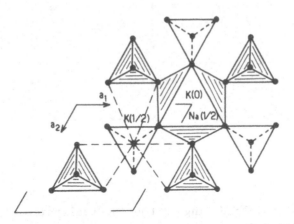

Figure 4.  The glaserite structure projected down the 3-fold axis.

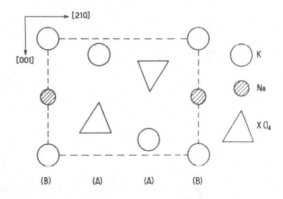

Figure 5.  Rows (A) and (B) in glaserite.

the glaserite through the intermediate $\beta-K_2SO_4$ or the orthorhombic $Na_2CrO_4$ type structure, depending on the size of the rare-earth ion. They can be described as being ordered superstructures made up of rows of alternating Na and $(XO_4)$ (A) and Na and Ln (B).  In the (B) sublattice the Na–Ln ordering is 1-1 along the direction of the rows but leads to different sequences of planes perpendicular to the [210] direction of the glaserite structure.  The sequence is 2-2 for IV, VI and VIII types, 2-1 for the V type and 1-1 for the VII types.  As the temperature and the ionic rare-earth size decreases, the order increases (Fig. 6).

Contrary to the other $Na_3Ln(XO_4)_2$ phases, the particular

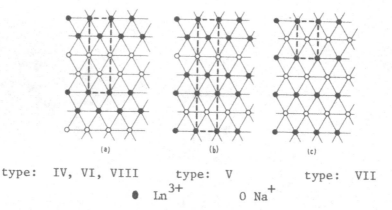

type:  IV, VI, VIII      type:  V          type:  VII

● $Ln^{3+}$          ○ $Na^+$

Figure 6.    Cationic ordering in the (B) sublattice along the [210] direction.

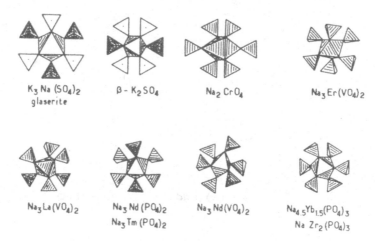

Figure 7.    Basic building-blocks found in the $Na_3Ln(XO_4)_2$ structures.

structure of Na$_3$Yb(PO$_4$)$_2$ (IX) derives directly from the NaZr$_3$(PO$_4$)$_2$
structure type (6).  It can be described as a three-dimensional
framework made up of central [(Na,Yb)O$_6$] octahedra and their six associated [PO$_4$] tetrahedra.  The rigid 3D skeleton forms a stable lattice while the remaining Na atoms sites can be empty, partially or
fully occupied.  This explains the non-stoichiometry of these phases
and their ionic conductivity (4).

The parenthood of the glaserite structure with respect to all
other lattices is demonstrated also by Fig. 7 where are shown the
basic building blocks found in all the phases:  these blocks are
made up of a central octahedron [MO$_6$] sharing common corners with
six surrounding [XO$_4$] tetrahedra capable of rotating themselves to
accommodate the various coordinations of the rare-earth ions.

## REFERENCES

1.   M. Vlasse, R. Salmon and C. Parent, Inorg. Chem. 15:1440 (1975).
2.   R. Salmon, C. Parent, G. Le Flem and M. Vlasse, Acta Cryst.
     B32:2799 (1976).
3.   R. Salmon, C. Parent, M. Vlasse and G. Le Flem, Mat. Res. Bull.
     13:439 (1978).
4.   R. Salmon, C. Parent, M. Vlasse and G. Le Flem, Mat. Res. Bull.
     14:85 (1979).
5.   S. K. Apinitis, Y. Ia Sedmalis, Izv. Akad. Nauk Latv. SSR,
     3:373 (1978).
6.   L. O. Hagman and P. Kierkegaard, Acta Chem. Scand. 22:1822
     (1968).

# INVESTIGATION OF THE REACTIONS OF RE + RE$_2$O$_3$

## UNDER HIGH PRESSURES

J. M. Leger, N. Yacoubi, and J. Loriers

C.N.R.S., ER 211 - Lab. Bellevue

92190 Meudon, France

## INTRODUCTION

Rare earth monoxides in the solid state have attracted consider-able attention. Except for the well documented case of europium the existence of all the reported monoxides obtained at normal pressure has beeen questioned and it is now established on good grounds that these compounds were in fact oxinitrides or hydrides (1,2). Detailed calculations have been performed in order to understand why SmO and YbO could not be obtained in contrast with EuO. In particular the variation of the Gibbs energy $\Delta G$ has been calculated for the reduc-tion of the sesquioxide by the pure metal along the reaction RE + RE$_2$O$_3$ → 3 REO. It has been shown (3) that $\Delta G$ is positive for Sm and Yb but negative, as expected, for Eu. However, the values found are small (+ (8-13) kcal/mole for Yb, + 7 for Sm and -10 kcal/mole for Eu) and are of the same order as the exeprimental errors so that they do not really allow us to draw any definite conclusion. If the small positive values found for Yb and Sm can explain that under nor-mal conditions the reaction has not been observed they do not pre-clude that once the monoxide is made under other circumstances it can be kept metastably in the normal conditions.

In the case of Yb the volume of the pure metal is so large that the volume of the expected monoxide is much less than the volume of the starting reactants. In these conditions the term P$\Delta$V is negative and no longer negligible in the Gibbs energy variation provided the reaction is made under pressure. If the applied pressure is suffi-ciently high, this term could even reverse the sign of the normal pressure value of $\Delta G$ thus making the reaction possible. It has been demonstrated for Yb (4) that effectively such a reaction occurs under high pressures but at pressures much lower than those estimated from the initial value of $\Delta G$. This would indicate that the value of $\Delta G$ at normal pressure is less positive than calculated in ref. (3).

In the case of Sm the variation of volume $\Delta V$ calculated along the reduction process is very small so that pressure is not expected to appreciably favour the appearance of a monoxide with the rare earth in the divalent state. However, it has been shown (5) that Sm would not be in the divalent state but that the monoxide would be metallic. In effect when the cell parameters of the chalcogenides decrease with the size of the anion the d band widens and in the case of SmO overlaps the f level thus making this monoxide metallic. This one would be unstable and Sm would contract towards a state of higher valence. In these conditions there is again an important volume contraction during the reaction and pressure could favour the appearance of such a monoxide.

We have investigated the systems $RE + RE_2O_3$ under high pressures in order to determine which reactions were possible.


RESULTS

The samples are made of mixed powders. The metal powder is obtained by filing an ingot (nominal purity 99.9%); the sesquioxide (nominal purity >99.9%) is calcinated above 1000°C in order to reduce nonmetallic impurities. All the handling is carried out under argon. An excess of metal with respect to the stoichiometric ratio is generally added. The experiments are conducted in a compressible gasket apparatus of the belt type. After unloading the samples are crushed and an x-ray diffractogram is recorded. The sensitivity to second phases is estimated to be about 5%. Chemical analyses for C,H,N were performed on different samples by acidimetric or catharometric methods.


Reactions $RE + RE_2O_3$, RE = La to Sm

The experiments with La were performed at 40 kbar and 900°C. In these conditions a golden yellow compound with a metallic luster was obtained. The structure is face centered cubic (fcc) and the parameter is $5.144 \pm 0.005$Å. In open air it rapidly transformed into a compound with a hexagonal structure which was identified as being $La(OH)_3$.

In the case of Ce the reaction of the metal with the dioxide, but not with the sesquioxide, was investigated. After an eight-hour run at 15 kbar and 700°C we obtained a mixture of two compounds: the first one has a hexagonal structure and the parameters are those of

$Ce_2O_3$; the second one has a fcc structure with a parameter of 5.089 ± 0.005Å. This last one was stable in open air while $Ce_2O_3$ converted slowly into $CeO_2$. After a four-hour reaction at 50 kbar and 600°C we did not find any trace of the fcc phase but only two phases which can be attributed to $Ce_2O_3$ (H) and $CeO_2$.

The reaction for Pr was investigated at 50 kbar and 800°C. After treatment a golden yellow compound with a metallic luster was recovered. Its structure is fcc and the parameter is 5.032 ± 0.005Å. No foreign lines could be seen in the diffractogram. In open air it converted to the hexagonal form of the hydroxide $Pr(OH)_3$ as in the case of La but more slowly.

The reaction for Nd was also studied at 50 kbar but at a higher temperature, 1000°C. Again, in this case only a golden yellow compound with a metallic luster was obtained. The structure is fcc and the parameter is 4.994 ± 0.005Å. No foreign lines could be seen in the diffractogram. In open air it transformed very slowly into hexagonal $Nd(OH)_3$.

Reactions for Sm were performed at 1000°C at different pressures. Above 50 kbar, after a four-hour treatment, the reaction was complete and a compound with the same aspect as in the case of La, Pr, Nd was obtained. The structure is fcc and the parameter is 4.943 ± 0.005Å. No foreign line could be seen in the diffractogram. This compound appeared to be stable in open air. When the reaction was performed at 40 kbar the reaction was not complete.

## Reactions RE + $RE_2O_3$, RE = Gd to Lu

Only the reactions with RE = Gd, Dy, Tm and Yb were studied under high pressure. For Gd and Dy no reaction could be detected after runs at 50 kbar and 1000°C. The metal was recovered non-oxidized and the sesquioxide was found in the monoclinic high pressure form. Various conditions of pressure and temperature were applied for the Tm reaction: 10-80 kbar, 600-1200°C, but as in the two previous cases the metal was recovered non-oxidized and the sesquioxide was found in the cubic or monoclinic phase according to the pressure and temperature conditions.

For Yb the reaction was investigated in a large range of pressures and temperatures (4). At lower temperatures a dark coloured sample with a fcc structure is obtained. The parameter is 4.877 ± 0.005Å. At higher temperatures a compound with an orthorhombic structure was found which was attributed to $Yb_3O_4$ because of the close similarity of the x-ray pattern with those of $CaYb_2O_4$ and $Eu_3O_4$.

DISCUSSION

Rare earths are well known to easily form binary and ternary compounds with hydrogen, carbon, nitrogen and oxygen. Many of these have a fcc structure but their cell parameters are larger than the values we report here for the fcc compounds we obtain under high pressure.

The impurity contents of the reaction products have been determined for H, C and N. The values found do not show large variations for the different compounds and a typical analysis is (ppm in weight): $C \simeq 900$ ppm, $N \simeq 200$ ppm, $H < 200$ ppm. So the principal impurity is H (less than 3 at %), followed by C ($\simeq 1$ at %); there is considerably less N ($\sim 0.2$ at %). These values are not very different from those of the starting products as measured just before introduction into the high pressure cell. From these analyses, it can be inferred that the fcc compounds we obtain are really the RE monoxides; however, their exact stoichiometry has not yet been exactly established. This conclusion is supported by our experiments with Gd, Dy, Tm in which no fcc compound is found although the same hydrides, oxynitrides, etc., also exist for these RE's and similar conditions of environment, pressure and temperature are applied in all cases.

The cell parameters of the monoxides obtained for RE = La to Sm are much lower than the values which can be calculated for ionic compounds with the rare earth in the 2+ state, but are higher than the values found when taking the trivalent ionic radius of the rare earth. This discrepancy can be accounted for in the metallic character of these compounds demonstrated by the low resistivity of the samples (of the order of 1000 $\mu\Omega$cm). In order to determine the valence state of the rare earth in these monoxides it is necessary to calculate the cell parameters which they would have if the RE were in the 2+ and 3+ states. A way to determine the parameter corresponding to the RE in the 3+ state is to extrapolate the lattice constant of the chalcogenides, which are metallic, as a function of the anionic radius. From La to Nd these extrapolations give values which are in excellent agreement with our experimental results, the uncertainty on the reported parameters of the chalcogenides being larger than the possible discrepancy. This shows that the rare earths are really trivalent in the monoxides which explains their metallic properties.

The case of Sm is more difficult to assess as there are no trivalent samarium chalcogenides. However, three different ways can be used to estimate the parameter of $Sm^{3+}O$. The parameters of hypothetical trivalent Sm chalcogenides can be calculated by interpolation between the neodymium and terbium chalcogenides; these values are then extrapolated as a function of the anionic radius to obtain the lattice constant of $Sm^{3+}O$. This procedure gives 4.917Å. Another way is to proceed by comparison with neighboring compounds: the difference between the ionic radii of $Nd^{3+}$ is 0.066Å). The comparison

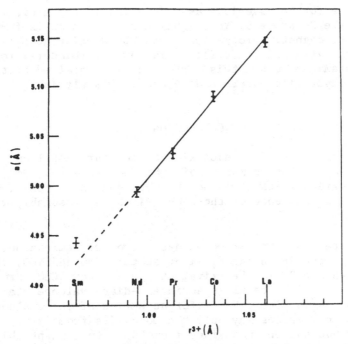

Fig. 1.  Lattice parameter of monoxides as a function of the trivalent radius of the rare earth.

with the cell parameter of NdO (4.994Å), given then 4.924Å.  A third
way is to extrapolate the cell parameters of the now existing monoxides as a function of the 3+ ionic radius of the RE.  This yields
4.922Å (Fig. 1).  The three values are in good agreement but the experimental value, 4.943Å, is definitely higher.  The difference cannot be explained by impurities; this would imply more than 10 at %
of H and still larger contents of C or N which is mich higher than
the values found by chemical analysis.  So, we have to admit that Sm
is probably not in a pure 3+ but in a state of intermediate valence.
Such a situation is well known for SmS above 7 kbar where this compound exhibits the same metallic golden yellow aspect as our SmO
samples.  For calculating the exact valence state of Sm in SmO we
have to know the cell parameter of hypothetical Sm$^{2+}$O.  Extrapolation
of the lattice constant of Sm chalcogenides leads to a value of 5.23Å
which seems too large as compared to the experimental value for Eu$^{2+}$O
(5.142Å); a value of 5.15Å seems more reasonable.  The valence state
of Sm is then found to be 2.9 which is close to the value determined
for SmS above transition to the metallic state (2.8-2.9)(6).

Ytterbium monoxide is different from the others:  its aspect is
dark instead of golden yellow and appears to be an insulator, or a
semiconductor, instead of being metallic.  The cell parameter is
equal to the value calculated by extrapolation of the lattice con-

stants of the Yb chalcogenides as a function of the ionic radius. This shows the 2+ state of Yb in this monoxide which is further confirmed by its magnetic properties. The $Yb^{2+}O$ oxide should be diamagnetic (the 4f shell is full); it is however found (4) to be slightly paramagnetic but this difference is explained by the amounts of impurities determined by chemical analysis.

## CONCLUSION

Reduction of the RE sesquioxide by the pure metal under high pressure leads to the formation of LaO, CeO, PrO, NdO, SmO and YbO, six new monoxides which could not be obtained up to now at normal pressure. The existence of those monoxides is reasonably well established.

There are now seven RE monoxides: two are semiconductors or insulators with the RE in the divalent state: EuO and YbO: four are metallic with the RE in the trivalent state: LaO, CeO, PrO and NdO; one is metallic with the RE in an intermediate valence state: SmO. These compounds complete the different series of the chalcogenides. The question now arises why only the monoxides from La to Sm could be obtained and not the monoxides from Gd to Lu (except Yb). Pressure favours the appearance of divalent or trivalent RE monoxides provided that $\Delta V$ is negative and large. For divalent RE monoxides only YbO can be expected in addition to EuO. For trivalent RE monoxides $\Delta V$ is large and nearly constant all over the series and so is the additional $P\Delta V$ energy term. It is thus likely that use of higher pressures would allow the synthesis of the still missing RE monoxides.

## REFERENCES

1.   G. Brauer, H. Bärnighausen and N. Schultz, Z. Anorg. Allg. Chem., 536:46 (1967).
2.   T. L. Felmlee and L. Eyring, Inorg. Chem. 7:660 (1968).
3.   G. J. McCarthy and W. B. White, J. Less-Common Met. 22:409 (1970).
4.   J. M. Léger, J. Maugrion, L. Albert, J. C. Achard and C. Loriers, C.R. Acad. Sc. Paris 286C:201 (1978).
5.   B. Batlogg, E. Kaldis, A. Schegel and P. Wachter, Phys. Rev. B 14:5503 (1976).
6.   G. Günterodt, R. Keller, P. Grünberg, A. Frey, W. Kress, R. Merlin, W. B. Holzapfel and F. Holtzberg, in Narrow Instabilities and Related Narrow-Band Phenomena, R. D. Parks, Ed., Plenum Press, NY (1977).

# POLYMORPHISM OF THE RARE EARTH DIGERMANATES AND DISILICATES UNDER HIGH PRESSURE AND INFORMATION GIVEN BY THE $Eu^{3+}$ STRUCTURAL PROBE

G. Bocquillon, C. Chateau, J. Loriers

Centre National De La Recherche Scientifique
Laboratoires De Bellevue - 92190 Meudon, France

## INTRODUCTION

The structural relationships between the compounds of general formula $M_2A_2O_7$ have been the subject of numerous investigations. In particular, Brown [1] and Shannon [2] have shown that the crystalline structure of these compounds depends mainly on the respective volumes of the cations M and A : most of the compounds having a relatively large A cation (ionic radius >0.60 Å) crystallize in the pyrochlore structure, whereas the others (phosphates, silicates and germanates) exhibit a large variety of structure types, with frequent analogies between the series.

Recent studies have shown that under pressure the polymorphism of this last class of compounds is still more extended.

The present report summarizes the results of a comparative study of the new phases obtained at high pressure in the rare earth disilicates and digermanates series $Ln_2A_2O_7$ (where Ln designates a lanthanide, yttrium, indium or scandium). These series are interesting for polymorphism studies because of their extended range of ionic radii, $r_{Ln}$. Moreover, as will be seen below, the europium ion $Eu^{3+}$ can be conveniently used in these compounds as a structural probe for spectroscopic studies giving additionnal informations about the symmetry of the crystalline site occupied by the $Ln^{3+}$ ion in the different structures.

## EXPERIMENTAL

The $Ln_2Si_2O_7$ and $Ln_2Ge_2O_7$ compounds are prepared at atmospheric pressure by thermal treatments between 900°C and 1500°C of a stoichiometric mixture of the rare earth hydroxide and finely powdered $SiO_2$ or $GeO_2$.

The high pressure experiments are performed in a belt-type apparatus [3]. The samples, surrounded by a gold foil, are put into a

boron nitride container, heated under pressure for 1/2-2 h, and
quenched before releasing the pressure. The products are identified
by their powder X-ray diffraction patterns.

The fluorescence spectra of the compounds, activated by 5 at.%
of europium, are recorded at ambiant temperature with a grating
monochromator using 2537 Å U.V. excitation.

POLYMORPHISM UNDER PRESSURE

The analogies in the polymorphism, which have been emphasized
between the silicates and germanates series, are present only in a
limited extent in the compounds reported here. Specifically, seven
phases are known in the $Ln_2Si_2O_7$ series (4) and five phases in the
$Ln_2Ge_2O_7$ series (5), but among these only two forms are found in both
the disilicates and digermanates.

In the disilicates, the most significant effect of pressure is
observed on the compounds with the smaller cations (Ln=Tm,Yb,Lu,In,Sc)
which are only obtained in the thortveitite, monoclinic structure
(C phase) at atmospheric pressure (6).

For example, figure 1 shows the pressure-temperature phase diagram
of $Lu_2Si_2O_7$. Starting from the C-form three new varieties are obtained at increasing pressure, called D (monoclinic) X (tetragonal) and
B (triclinic). At atmospheric pressure the D and B phases are known
in the disilicates of the larger rare earths, and the X phase in the
$Ln_2Ge_2O_7$ compounds with Ln = Lu to Gd. For $Yb_2Si_2O_7$ and $Tm_2Si_2O_7$ the
same phases are observed in roughly the same relative positions but

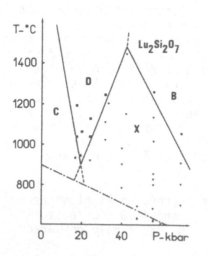

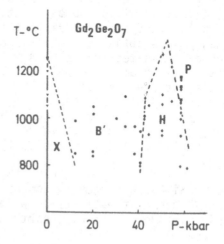

Figure 1 : P - T phase
diagram of $Lu_2Si_2O_7$

Figure 2 : P - T phase
diagram of $Gd_2Ge_2O_7$

with transformation pressures for C→X and C→B lower than in $Lu_2Si_2O_7$.
    The C-phase of $In_2Si_2O_7$ transforms to the D phase (7) between
40 and 80 kbar, i.e. at pressures much higher than for $Lu_2Si_2O_7$. The
same behavior is observed for the C→X and C→B transformations, and
it can be noted that generally the pressure needed to obtain a high
pressure phase is greater the smaller the ionic radius of Ln.
$Sc_2Si_2O_7$ is the first compound in which direct transformation from
the C form to the pyrochlore cubic structure (P phase) has been obtained. The volume change at this transformation is 21 % which is
much greater than the change observed in the other transformations
in the $Ln_2Si_2O_7$ series.
    For $Y_2Si_2O_7$, $Gd_2Si_2O_7$ and $La_2Si_2O_7$, experiments at 80 kbar and
1000°C with water added as mineralizer, lead to new structures presently unidentified.
    In the digermanate series the $Ln_2Ge_2O_7$ compounds which have been
studied under pressure are the Sm, Gd, Y and Lu analogs. Only
$Gd_2Ge_2O_7$ exhibits an important polymorphism (8) with three high pressure transformations (X→B,H and P) as shown in the figure 2.The H
phase is a new form, and has a X-ray pattern which is different from
those known for the other phases of disilicates and digermanates.Its
measured density is similar to that of the P phase and it is possible
that its structure is related to the pyrochlore one.
    The triclinic phase of $Sm_2Ge_2O_7$, prepared at atmospheric pressure,
is stable at least up to 40 kbar; between 58 and 76 kbar, the H-phase
is formed but up to now it has not been possible to transform it into
the pyrochlore structure.
    $Y_2Ge_2O_7$ and $Lu_2Ge_2O_7$ crystallize at atmospheric pressure in the
X-form; for $Lu_2Ge_2O_7$ the C phase is also obtained below 1000°C but
only as mixtures with the X-form. The two compounds transform under
pressure into the pyrochlore structure previously obtained by Shannon
(9). This P-form appears near 850°C at about 20 kbar for $Y_2Ge_2O_7$,
and 10 kbar for $Lu_2Ge_2O_7$; the P phase of $Lu_2Ge_2O_7$ might be stable
at atmospheric pressure above 1300°C.
    From our investigations under pressure it must be concluded that
two of the new phases obtained are common to the disilicates and
digermanates : the X phase which is observed in $Ln_2Si_2O_7$ (Ln=Lu,Yb,
Tm) and the pyrochlore structure found in $Sc_2Si_2O_7$ and $Ln_2Ge_2O_7$
(Ln=Sc,Lu to Gd). However, large differences remain between the two
series. In the disilicates the polymorphism is important for
$(Lu,Yb,Tm)_2Si_2O_7$, whereas in the digermanates it is $Gd_2Ge_2O_7$ for
which several phases are obtained. So it is not only the size or the
nature of the Ln cation which are important in the polymorphism of
$Ln_2A_2O_7$ compounds but also the relative size of the Ln and A cations.
In geophysics the germanates are often used as a model for the behavior of silicates at high pressure; it can be seen that this is not
valuable in the present series. For example $Lu_2Ge_2O_7$ transforms
directely at moderate pressure from X phase to P phase, whereas
X-$Lu_2Si_2O_7$ transforms into B-$Lu_2Si_2O_7$ at 60 kbar, and is not converted into the pyrochlore structure at pressures up to 120 kbar (10).

An additional remark can be made concerning the pyrochlore structure. The pressure at which the pyrochlore phase appears in a $M_2A_2O_7$ series is greater when the ratio $(r_M - r_A)/r_M$ increases, i.e. for a given A cation, when the ionic radius of the M cations increases. This last behavior is at the contrary to that observed for D, X and B phases and it could result from the fact that the transformation into the P phase involves a change in the coordination number of the A cation (from IV to VI), which is not the case for the other transformations.

## SPECTROSCOPIC STUDY

The luminescence emission lines of the $Eu^{3+}$ ion originate mainly from the transitions between the $^5D_0$ excited level and the $^7F_J$ manifold. The $^5D_0 \rightarrow {}^7F_0$ and $^5D_0 \rightarrow {}^7F_2$ transitions are electric dipole transitions and are strictly forbiden in a non-vibrating site having a center of symmetry. On the other hand, the $^5D_0 \rightarrow {}^7F_1$ transition is allowed by the magnetic dipole mechanism and is therefore insensitive to the symmetry. The ratio of the intensities of $^5D_0 \rightarrow {}^7F_2$ and $^5D_0 \rightarrow {}^7F_1$ (the most fluorescing transitions of $Eu^{3+}$)therefore may be considered as an indication of the degree of deviation from a centro-symmetric environment around the $Eu^{3+}$ ion. Moreover, for the $^5D_0 \rightarrow {}^7F_0$ transition, since the J=0 values indicate that the levels are not subject to splitting by the crystal field, only one emission line should be observed. The appearence of more lines then indicates that $Eu^{3+}$ is present in more than one kind of site in the host lattice, assuming that each site has a resolved J=0→J=0 transition.

The results given by the $Eu^{3+}$ spectra are in good agreement with the data obtained by the X-ray analysis.

Figure 3a shows the $Eu^{3+}$ spectrum in the pyrochlore phase of $Gd_2Ge_2O_7$. There are only two lines in the $^5D_0 \rightarrow {}^7F_1$ transition,which is in agreement with the $D_{3d}$ symmetry of the rare earth site in the structure (11).

In the figures 3b,c,d are given the $Eu^{3+}$ spectra in X-$Lu_2Ge_2O_7$, D-$Lu_2Si_2O_7$ and C-$Lu_2Si_2O_7$ respectively. The presence of only one line for the $^5D_0 \rightarrow {}^7F_0$ transition indicates the existence of only one site for the rare earth in the structures. Moreover, the appearence in the spectrum of the C-phase of only four lines for the $^5D_0 \rightarrow {}^7F_2$ transition indicates a probably $C_{3v}$ symmetry for the rare earth site in this thortveitite structure.

Figure 3e shows the $Eu^{3+}$ spectrum in the B-phase of $Gd_2Si_2O_7$. In this structure the rare earths occupy four different sites, which is confirmed by the observation of four lines in the $^5D_0 \rightarrow {}^7F_0$ transition.

Figure 3f gives the $Eu^{3+}$ spectrum in the H-phase of $Gd_2Ge_2O_7$. For this unknown structure the four lines for the $^5D_0 \rightarrow {}^7F_0$ transition indicate at least the existence of four different sites for the rare earths.

The examination of the spectra of all the other compounds brings out the similarities in the emission of the europium ion in one type

of structure; the number of lines and their relative intensities
remain unchanged, only their positions are slightly shifted. Going
from C-phase to P-phase, an increase of the relative intensity of
the magnetic dipole transition $^5D_0 \rightarrow {}^7F_1$ with regard to the electric
dipole transition $^5D_0 \rightarrow {}^7F_2$ is observed; it seems that the rare earth
environment tends towards centrosymmetry (strictly achieved in the
pyrochlore structure) as the crystal density increases in going from
the C-phase to the P-phase.

　　　Compared with the $Eu^{3+}$ spectra published by McCauley (12) for
titanates and stannates, the spectra of our high pressure pyrochlore
phases appear sharper. This fact seems linked to the electronegativity of the A cations of $M_2A_2O_7$ compounds, which increases along the
sequence titanates-stannates-silicates-germanates.

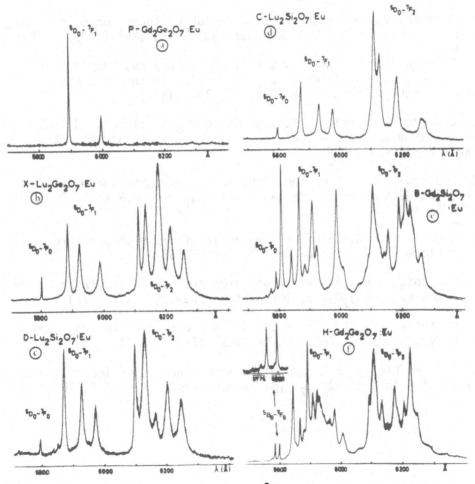

Figure 3  :  Examples of $Eu^{3+}$ fluorescence spectra

REFERENCES

1.I.D.Brown, C.Calvo, The crystal chemistry of large cation dichromates, pyrophosphates and related compounds with stoichiometry $X_2Y_2O_7$, J. Solid State Chem. 1, 173 (1970).

2. R.D.Shannon, C.T.Prewitt, Synthesis of pyrosilicates and pyrogermanates having the thortveitite structure, J.Solid State Chem.2, 199 (1970).

3. G.Bocquillon, C.Susse, B.Vodar, Allotropie de l'oxyde d'hafnium sous haute pression,Rev.Int.Hautes Tempér.et Réfract.5,247 (1968).

4. J.Felsche, Polymorphism and crystal data of the rare earth disilicates of type R.E.$_2$Si$_2$O$_7$, J.LessCommon Metals 21, 1 (1970).

5. G.Jouhet-Vetter, F.Queyroux, Etude de quelques composés $Ln_2Ge_2O_7$ (Ln = La, Nd, Sm, Eu, Gd) Mat. Res. Bull. 10, 1201 (1975).

6. G.Bocquillon, C.Chateau, C.Loriers, J.Loriers, Polymorphism under pressure of the disilicates of the heavier lanthanoids $Ln_2Si_2O_7$ (Ln=Tm,Yb,Lu)J.Solid State Chem. 20, 135 (1977).

7. C.Chateau, J.Loriers, Polymorphisme des disilicates d'indium et de scandium sous haute pression et à haute température, C.R.Ac. Sc. Paris 288C, 421 (1979).

8. G.Bocquillon, J.Maugrion, J.Loriers, Polymorphisme sous haute pression du digermanate de gadolinium $Gd_2Ge_2O_7$ C.R.Ac.Sc.Paris 287C,5 (1978).

9. R.D.Shannon, A.W.Sleight, Synthesis of new high-pressure pyrochlore phases, Inorg. Chem.7, 1649 (1968).

10.A.F.Reid, C.Li, A.E.Ringwood, High pressure silicate pyrochlores $Sc_2Si_2O_7$ and $In_2Si_2O_7$, J. Solid State Chem.20, 219 (1977).

11.M.Faucher, C.Caro  Ordre et désordre dans certains composés de type pyrochlore, J. Solid State Chem. 12, 1 (1975).

12.R.A. Mc Cauley, F.A. Hummel, Luminescence as an indication of distorsion in $A_2^{3+}$  $B_2^{4+}$ $O_7$ type pyrochlores J.Luminesc.6, 105 (1973).

# CORRELATION OF OPTICAL AND PARAMAGNETIC PROPERTIES OF NEODYMIUM

# $(4f^3)$ IN THE SOLID STATE

P. Caro, J. Derout, L. Beaury and E. Antic Fidancev

C.N.R.S. - ER. 210 - Laboratoires de Bellevue
1 Place A. Briand - 92190 Meudon Bellevue - France

## INTRODUCTION

The optical absorption spectra of neodymium compounds at 4 K, can be reproduced using, on the complete basis of 364 $|SLJM_J>$ kets in the configuration, an hamiltonian which involves at least 12 free-atom parameters (Racah's, Trees $\alpha$ and $\beta$ , Judd's six $T^k$ three-body interaction parameters and the spin orbit coupling constant) and the crystal field hamiltonian written in Wybourne's formalism. Recent examples in the literature are to be found in references (1) to (3).

The wave-vectors which are obtained as a result of the energy calculation can be used to compute other properties. Among those is the paramagnetic susceptibility, which is a function (3) of the diagonal and non diagonal matrix elements of the operator $\vec{L} + 2\vec{S}$ applied on the basis formed by the eigenvectors which correspond to the eigenvalues which are solution of the secular equation formed from the matrix of the crystal field and free-atom hamiltonians applied on the basis of a selection of $|SLJM_J>$ states within $4f^3$. The selection can for example be limited to the 10 $|SLJM_J>$ kets of the ground state $^4I_{9/2}$. In that case the summation on the energies of the levels in the Van Vleck equation will have of course to be limited to the 5 lowest energy states (which are Kramers doublets) of the $4f^3$ configuration. However if ones want to check the paramagnetic susceptibility at high temperature (i.e. 1000 K) it is necessary to use the wave vectors from a complete calculation of the energy levels. Use of 18 energy levels (that is including $^4I_{9/2}$, $^4I_{11/2}$ and $^4I_{13/2}$) seems sufficient.

The present paper reports an overview of the results of calculations of paramagnetic susceptibilities according to the principal crystal axis for four compounds for which the optical absorption spectra were thoroughly interpreted that is with a mean square deviation of 15 to 20 cm$^{-1}$ for 80 to 120 Stark levels in $4f^3$. The results are compared to experiment when available. The paramagnetic susceptibility is extremely sensitive to the values of the coefficients of the kets in the wave vectors which in turn are extremely dependent on the crystal field parameters. It is apparent that the paramagnetic susceptibility is very useful to check the values of the crystal field parameters and especially to determine, as it was suspected (1)(2)(4), if the set of crystal field parameters which fits the overall manifold of Stark levels, is also convenient for the $^4I_{9/2}$ ground state alone.

## I - A-Nd$_2$O$_3$

The site symmetry of Nd$^{3+}$ in A-Nd$_2$O$_3$ (3) is exactly C$_{3v}$. Several sets of the six crystal field parameters (Table 1) were defined in this structure. One set (set A) fits the 88 Stark levels observed with a mean square deviation of 17 cm$^{-1}$, however it is not very good for the ground $^4I_{9/2}$ state. Another set (set B) will fit the 88 Stark levels to a much larger standard deviation (25 cm$^{-1}$) but gives a very good reproduction of the ground state splitting. Set C and D when applied on the basis of the ten $^4I_{9/2}$ |SLJM$_J$> kets give an excellent ground state splitting, but not if those sets are used in the complete matrix. In every case C$_{3v}$ double group (S$_1$ + S$_3$) symmetry is associated with the Stark ground state. When the magnetic susceptibility is computed from the wave vectors the results in Table 2 are obtained.

It can be seen that the paramagnetic susceptibilities are best approximated by sets of crystal field parameters C and D which fit exactly the ground state alone for the energies. Set B which fits well the energy splitting of the ground state under the complete calculation is inadequate both for the overall manifold of 88 Stark levels and for the paramagnetic susceptibility. Set A fits badly the paramagnetic susceptibility (although the curve at high temperature (see reference 3) follows perfectly the experimental one, whereas the one with set C does not so well).

The data indicate the extreme sensitivity of the paramagnetic susceptibility to the real values of the crystal field parameters. It is consequently advisable to have magnetic measurements as precise as possible. The result also suggest that the crystal field parameters are different for the ground state and for the excited levels, a fact which has already being suggested (1)(2)(4). The difference can be here appreciated in a quantitative way.

Table 1 – Sets of crystal field parameters for A-Nd$_2$O$_3$
          (see text for explanation)

| cm$^{-1}$ | A* | B | C | D |
|---|---|---|---|---|
| B$_0^2$ | – 836 | – 804 | – 818 | – 645 |
| B$_0^4$ | 634 | 970 | 1196 | 992 |
| B$_3^4$ | – 1606 | – 1367 | – 1338 | – 1141 |
| B$_0^6$ | 752 | 748 | 557 | 585 |
| B$_3^6$ | 237 | 183 | 246 | 311 |
| B$_6^6$ | 672 | 804 | 847 | 962 |

\* see reference (3) for error margin's

Table 2 – Parallel paramagnetic susceptibility for
          A-Nd$_2$O$_3$ (computed on the five Kramers doublets
          of $^4$I$_{9/2}$)

| T°K | $1/\chi_{//}$ for 2 at. Nd u.e.m. cgs | | | | |
|---|---|---|---|---|---|
| B$_q^k$'s set | A | B | C | D | EXP (5) |
| 10 | 8.84 | 12.84 | 16.56 | 25.24 | 22. |
| 50 | 38.57 | 45.25 | 54.19 | 54.24 | 52. |
| 100 | 65.93 | 71.59 | 77.59 | 74.04 | 75.7 |
| 200 | 102.20 | 105.52 | 108.28 | 103.4 | 106.5 |

## II – $(NdO)_2S$

The oxysulfide is isomorphous to the oxide $A-Nd_2O_3$. However the crystal field parameters are very different. They read

| $B_q^k$ | $B_0^2$ | $B_0^4$ | $B_3^4$ | $B_0^6$ | $B_3^6$ | $B_6^6$ |
|---|---|---|---|---|---|---|
| $cm^{-1}$ | 194 | 912 | -925 | 512 | 301 | 257 |

Work in progress in this laboratory indicates that the difference between the two materials can be explained on the basis of a complete electrostatic field calculation including the dipole moments on the anions. The ground state is also of $(S_1 + S_3)$ symmetry, but as the wave vectors are very different it is predicted that the paramagnetic anisotropy is reversed, that is we have now
$$1/\chi_\perp > 1/\chi_{//}$$
Computed typical values (on the basis of the 10 $^4I_{9/2} \mid SLJM_J \rangle$ ) are :

| T°K | $1/\chi_{//}$ for 2 at. Nd u.e.m c.g.s. | $1/\chi_\perp$ for 2 at. Nd u.e.m c.g.s. |
|---|---|---|
| 10 | 2.00 | 25.20 |
| 50 | 13.00 | 28.48 |
| 100 | 29.53 | 41.52 |
| 200 | 63.34 | 70.97 |

## III – $NdF_3$

The optical absorption spectrum of hexagonal $NdF_3$ was interpreted in the framework of a $D_{3h}$ point site symmetry and also in a $C_{2v}$ distortion with respect to the ternary axis. (the real site symmetry is $C_2$, the $C_2$ axis being perpendicular to the ternary axis). The coordination is eleven. The sets used are :

| $B_q^k(cm^{-1})$ | $B_0^2$ | $B_2^2$ | $B_0^4$ | $B_2^4$ | $B_4^4$ | $B_0^6$ | $B_2^6$ | $B_4^6$ | $B_6^6$ |
|---|---|---|---|---|---|---|---|---|---|
| $D_{3h}$ | 201 | | 1249 | | | 1573 | | | 809 |
| $C_{2v}$ | 204 | – 179 | 1081 | -34 | -17 | 1531 | 110 | -298 | 931 |

The ground state is of 2 $S_1$ symmetry in $D_{3h}$ equivalent to $(S_1 + S_3)$ in $C_{3v}$ that is involving again only the kets with $M_J = 9/2$ and $- 3/2$.

The computed and experimental data for the paramagnetic susceptibilities are given below on the basis of the extended calculation (18 levels). Detailed results will be published elsewhere.

| T°K | $1/\chi_{\parallel}$ 1 at. Nd u.e.m cgs | | $1/\chi_{\perp}$ 1 at. Nd u.e.m cgs | |
|-----|-------|---------|-------|---------|
|     | calc. | exp (6) | calc  | exp (6) |
| 10  | 6.87    | 15.5 | 42.50  | 14   |
| 50  | 29.95   | 44.  | 60.96  | 57   |
| 100 | 54.77   | 71.  | 91.58  | 96.5 |
| 200 | 104.84  | 128. | 145.47 | 155  |

## IV - NdAlO₃

The optical absorption spectrum of $NdAlO_3$, was interpreted in $C_{3v}$ symmetry with the following set of crystal field parameters.

| $B_q^k$ | $B_0^2$ | $B_0^4$ | $B_3^4$ | $B_0^6$ | $B_3^6$ | $B_6^6$ |
|---------|---------|---------|---------|---------|---------|---------|
| $cm^{-1}$ | −481 | 481 | −390 | −1700 | −950 | −1080 |

The ground state is, this time, of $D_{1/2}$ symmetry and the following results were obtained for the paramagnetic susceptibility on the basis of the 10 $^4I_{9/2}$ $|SLJM_J\rangle$ kets

| T°K | $1/\chi_{\parallel}$ 1 at. Nd u.e.m cgs | $1/\chi_{\perp}$ 1 at. Nd u.e.m cgs |
|-----|--------|--------|
| 10  | 31.60  | 11.76  |
| 50  | 100.57 | 48.49  |
| 100 | 140.87 | 85.65  |
| 200 | 208.02 | 154.07 |

The anisotropy is quite large. The measurement has not been made, to our knowledge, but may be difficult because of the easy twinning, or microtwinning, observed in crystals of this material (7).

## CONCLUSION

The present work shows that the wave vectors obtained from the complete calculation of the optical absorption spectrum of neodymium compounds may be used to compute the single crystal paramagnetic susceptibility up to 1000 K. This data is so sensitive to crystal field parameters that it may be possible, in this way, to know if the ground state is best fitted with a set of crystal field parameters different from the set which fits the overall manifold of states in the configuration.

## REFERENCES

1.   H.M. Crosswhite, H. Crosswhite, F.W. Kaseta and R. Sarup, J. Chem. Phys., $\underline{64}$, 1981 (1976).
2.   P. Caro, D-R. Svoronos, E. Antic and M. Quarton, J. Chem. Phys., $\underline{66}$, 5284 (1977).
3.   P. Caro, J. Derouet, L. Beaury and E. Soulie, J. Chem. Phys, $\underline{70}$, 2542 (1979).
4.   B.R. Judd, Phys. Rev. Lett., $\underline{39}$, 242 (1977).
5.   R. Tueta, A.M. Lejus, J.C. Bernier and R. Collongues, C.R. Acad. Sci. Ser C., $\underline{274}$, 1925 (1972).
6.   Measured at the University of Bordeaux by J. Chaminade and M. Pouchard.
7.   M. Marezio, P.D. Dernier and J.P. Remeika, J. Solid State Chemistry, $\underline{4}$, 11 (1972).

# DISORDERED CATIONS IN PSEUDO SPINELS $FeR_2S_4$ (R = Yb, Lu, Sc)

Alain Tomas, Luc Brossard, Micheline Guittard and
Pierre Laurelle
Laboratoire de Chimie Minérale Structurale, associé au
CNRL – Faculté de Pharmacie, 4, av. de l'Observatoire,
75270 Paris Cedex 06, France

## ABSTRACT

The $FeR_2S_4$ compounds (R = Yb, Lu and Sc) were previously described in terms of the direct spinel structure. In the case of R = Yb, two varieties are known: orthorhombic of the $MnY_2S_4$-type at low temperature and cubic of the spinel type at high temperature. There is a reversible transition at about 1000°C. In the orthorhombic structure, Fe has only an octahedral coordination.

In a previous crystallographic study, Tomas and Guittard [1] have shown that, in addition to the normal tetrahedral (8a) and octahedral (16d) metal sites of the spinel, a (16c) octahedral site is partially occupied by metals. Moreover, Riedel, Karl and Rackwitz [2] from Mössbauer study have shown that iron is partially in the octahedral (16d) site, as $Fe^{3+}$.

In order to determine the distribution of Fe and R atoms among these 3 metal sites, cubic $FeR_2S_4$ compounds were studied by single crystal x-ray diffraction and Mössbauer spectroscopy.

The x-ray refinement of the structures in combination with the Mössbauer information give the following distributions of metals in the sites:

For $FeYb_2S_4$: $0.80Fe^{2+}$ in (8a) site, $(0.88Yb^{3+} + 0.045Fe^{2+})$ in (16d) site, $(0.12Yb^{3+} + 0.055Fe^{2+})$ in (16c) site. The R factor is 0.034.

For $FeLu_2S_4$: $0.71Fe^{2+}$ in (8a) site, $(0.84Lu^{3+} + 0.095Fe^{2+})$ in (16d)

site, $(0.16Lu^{3+} + 0.050Fe^{2+})$ in (16c) site.  The R factor is
0.022.

In conclusion, the tetrahedral (8a) site contains only Fe; the
two series of octahedral (16d) and (16c) sites contain mixtures of
Fe and R atoms.  All the sites are partially filled.  Fe is exclusively divalent.  Finally, the overall composition is exactly
$FeR_2S_4$.

For $FeSc_2S_4$, the structure is a true direct spinel (R factor =
0.041).  Cubic solid solutions are obtained from this compound by
addition of $Sc_2S_3$ or by addition of FeS.  The crystal structures of
the two compositions $Fe_{0.85}Sc_{2.10}S_4$ and $Fe_{2.29}Sc_{1.14}S_4$ were solved
by Tomas, Guittard, Rigoult and Bachet (3).  In both cases, the same
structural characteristics as obtained previously for $FeYb_2S_4$ and
$FeLu_2S_4$ were observed.

This work will be published in the Journal of Solid State Chemistry.

REFERENCES

1.    A. Tomas and M. Guittard, Mat. Res. Bull. 12:1043 (1977).
2.    E. Riedel, R. Karl and R. Rackwitz, Mat. Res. Bull. 12:599
      (1977).
3.    A. Tomas, M. Guittard, J. Rigoult et B. Rachet, Mat. Res. Bull.
      14:249 (1979).

# TWINS IN $LaSe_2$ AND $LaSe_{2-x}$ AND A PHASE TRANSITION IN $LaS_2$

Simone Bénazeth, Jérôme Dugué, Daniel Carré, Micheline
Guittard and Jean Flahaut
Laboratoire de Chimie Minérale Structurale, associé au
CNRS – Faculté de Pharmacie, 4, av. de l'Observatoire,
75270 Paris Cedex 06, France

## ABSTRACT

The rare earth dichalcogenide structures are derived from a
common type, the tetragonal anti $Fe_2As$-type ($P4/_n$ mm). All of these
structures have layers of $(LaX)_n^{n+}$ (X = S, S3) that alternate with
chalcogen layers (S or Se)$_n$. They are assembled with a pseudo-
tetragonal symmetry. In the layers of S or Se, the atoms form covalent pairs.

Lanthanum diselenide can be stoichiometric or Se-deficient up to
a composition of $LaSe_{1.86}$. The stoichiometric form of $LaSe_2$ crystallizes as a monoclinic cell with: a = 8.51Å; b = 8.58Å; c = 4.25Å;
$\gamma$ = 90.12°; space group $P2_1/a$; Z = 4.

This cell is related to the anti-$Fe_2As$ type by: a $\simeq$ $2a_0$; b $\simeq$
$c_0$; c $\simeq$ $a_0$.

All crystals are twinned because of the nearly tetragonal aspect
of the $LaSe_2$ cell. The twins are built in accordance with the following laws. First, we observe a (100) twin plane. This twin is
possible because the $\gamma$ value is close to 90°. Secondly, a twin is
formed by a reflection across the (201) plane; it is possible because
the a/c value is very close to 2. Despite the presence of twins,
the structure was refined in the two cases to R values of 0.06 and
0.08 respectively. It is shown that the stoichiometry is preserved
in both cases.

A third kind of twinning results from packing of the anti-$Fe_2As$
structure along the fourfold axis, and the crystals exhibit super-
structure with a cell of volume V = $10V_0$ (a' $\simeq$ b' $\simeq$ $a_0$ $\sqrt{5}$, c' $\simeq$ $2c_0$).

Analysis with Castaing's probe demonstrates non-stoichiometry while crystallographic observations indicate that all vacancies are ordered, though the structure cannot be determined. Evidence is given that vacancies lie in the selenium plane.

The low temperature form of $LaS_2$ is isostructural with the stoichiometric diselenide, but only twinning with the (100) twin plane is observed. The parameters of the monoclinic $P2_1/a$ cell are $a = 8.27Å$; $b = 8.21Å$; $c = 4.21Å$; $\gamma \simeq 90°$

The high temperature form of $LaS_2$ is orthorhombic, space group Pnma, with parameters: $a = 8.13Å$; $b = 16.34Å$; $c = 4.14Å$.

Its structure was solved by Dugué, Carré and Guittard (1). The structures of the two forms differ from one another in the arrangement of the $(LaS)_n^{n+}$ layers. Two consecutive layers are related by a translation parallel to $\vec{b}$ in the low temperature form, and by reflection across the mirror plane (010) in the high temperature form.

The temperature of this reversible transition is about 750°C. Similar dimorphism has not been observed for the other rare earth disulfides: only the monoclinic structure is observed for the $RS_2$ composition with Ce, Pr and Nd.

This work will be published in Acta Crystallographica.


REFERENCE

1.   J. Dugué, D. Carré and M. Guittard, Acta Cryst. B34:403 (1978).

MÖSSBAUER SPECTROSCOPY ON SUPERCONDUCTING AND MIXED VALENT

EUROPIUM COMPOUNDS

Frank Pobell

Inst.f.Festkörperforschung, Kernforschungsanlage Jülich

D-5170 Jülich, W. Germany

## ABSTRACT

Mössbauer spectroscopy has been applied to investigate $Eu_zSn_{1-z}Mo_6S_8$ and $(ScEu)Al_2$. The former compound is superconducting for $z < 0.9$. Its $^{151}Eu$ Mössbauer spectra show a magnetic hyperfine splitting at $T < 0.5$ K, and in addition a single line at $T > 0.1$ K. The behavior is considered to result from spin-glass type magnetic ordering. The linewidth of the single line increases by about a factor of three from 300 K to 0.15 K, and its isomer shift indicates a very low conduction electron density at the Eu site. The $^{151}Eu$ Mössbauer spectra of $(ScEu)Al_2$ show a line whose position changes from -5.8 mm/sec at 325 K to -1.9 mm/sec at 4.2 K. This change results from a change of the valence of the mixed valent Eu impurity ions in $ScAl_2$ from $(2.34 \pm 0.05)$ to $(2.73 \pm 0.07)$ as the temperature is decreased. For the valence fluctuations we obtain a fluctuation temperature of $(50 \pm 10)$ K and an excitation energy of $(280 \pm 40)$ K.

## INTRODUCTION

Recoilless nuclear resonance is a powerful method for investigating solid state porperties. The sharpness of the Mössbauer resonance line allows in many cases detailed investigation of weak electro-nuclear interactions. These interactions give rise to shifts or splittings of the nuclear transitions. The main limitation of Mössbauer spectroscopy is the limited number of atoms containing suitable Mössbauer isotopes. Fortunately, most rare earths have isotopes with nuclear transitions allowing Mössbauer spectroscopy. In this paper, I will concentrate on the behavior of two groups of unusual Europium compounds: magnetic superconductors of the

Chevrel phase type and compounds containing Eu ions as impurities
with an unstable valence.


## MAGNETIC CHEVREL PHASE SUPERCONDUCTORS

There is a long standing interest in the investigation of a
possible coexistence of superconductivity and magnetic order.
Former examinations concentrated on superconducting compounds containing low concentrations of rare earth impurity ions. In these
substances the magnetic interactions lead to clustered or spin-
glass type magnetic behavior resulting from the random distribution
of RE ions (1). Recently, a new class of ternary superconductors
containing RE ions has attracted much interest because of their
unusual properties. These are the Chevrel phase superconductors (2)
RE $Mo_6S_8$, (3) and RE $Mo_6Se_8$, (4) as well as the rare earth rhodium
borides RE $Rh_4B_4$ (5). Here, we have for the first time superconductors containing a regular lattice of RE ions which interact only
weakly with the superconducting state. The periodic distribution
of the magnetic ions gives rise to a sharp magnetic ordering temperature. In the following, I will discuss the RE-Chevrel phase
materials.

Empirical arguments, (2) as well as band structure calculations
(6) show that the superconducting properties of these compounds are
essentially determined by narrow 4d-bands of molybdenum at the Fermi
energy. The particular structure of the Chevrel phase results in a
very weak overlap of these 4d electrons and the RE ions. (2,3). This
weak exchange interaction between the conduction electrons and the
localized magnetic moments is only of order 0.01 eV (7). Spin flip
scattering of conduction electrons from the magnetic ions is therefore very weak, and the two series of ternary molybdenum chalcogenides are superconductors (with the not quite understood exception
of the Ce and Eu compounds), although they contain about 7 at %
magnetic RE ions (3,4).

In these compounds, superconductivity is not even destroyed
when the RE ions order antiferromagnetically, whereas ferromagnetic
order seems to suppress superconductivity (7). This behavior was
convincingly confirmed by magnetic neutron scattering experiments
performed at Brookhaven (8). We have here a unique opportunity to
study the coexistence of superconductivity and various types of
magnetism.

The first Mössbauer investigation of the magnetic behavior of
Chevrel phases was performed with the 21.6 keV $\gamma$-line of $^{151}Eu$ in
$Eu_zSn_{1-z}Mo_6S_y$, (9,10) which is superconducting for $z < 0.9$ (3). As
an example, four spectra for $Eu_{0.94}Mo_6S_{6.61}$ which is non-superconducting at least above 0.05 K, (3,7) are shown in Fig. 1. Spectra
for other z, like the superconductor $Eu_{0.63}Sn_{0.38}Mo_6S_{6.93}$,

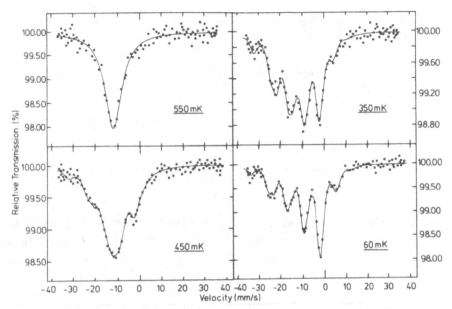

Figure 1. Mössbauer spectra of the 21.6 keV γ-line of [151]Eu in
Eu$_{0.94}$Mo$_6$S$_{6.61}$ at the indicated temperatures. The solid
lines are fits to the data points.

($T_c$ = 8.0 K), are very similar.

Below 0.1 K, the spectra show a magnetic hyperfine splitting
resulting from an internal field of 27.3 T. Above 0.5 K we see a
single, broadened line typical for slow electronic relaxation in a
paramagnetic compound. For the intermediate range 0.1 K < T < 0.5 K,
the spectra consist of a superposition of the static hyperfine
spectrum observed for T < 0.1 K, and of the broadened single line
observed for T > 0.5 K. The temperature range where we see a
superposition of the two spectra is wider for the pseudoternary
compounds with z ≠ 1. The intensity of the hyperfine spectrum increases with decreasing temperature at the expense of the single
line. These features are shown in Fig. 2.

The width Γ of the single line in the spectra is strongly
temperature dependent. It increases from 5 mm/sec at 300 K to
18 mm/sec at 0.15 K, (see Fig. 3). This increase arises from paramagnetic relaxation effects due to exchange scattering from the
conduction electrons. It follows a Korringa law Γ ∝ 1/T for
T > 15 K (10,11). The slow electronic relaxation of the Eu moments
responsible for the line broadening confirms the weak interaction
between conduction electrons and magnetic ions. It is due to a
small density of states at the Eu site and/or a small exchange
coupling (10,11). At T < 1 K the line broadening can again be

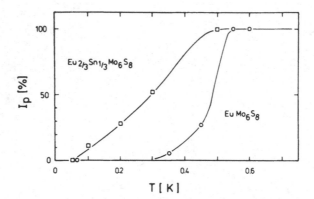

Figure 2. Relative intensities of the single line in the Mössbauer
spectra of $Eu_{0.63}Sn_{0.38}Mo_6S_{6.93}$ (□) and of $Eu_{0.94}Mo_6S_{6.61}$ (○)

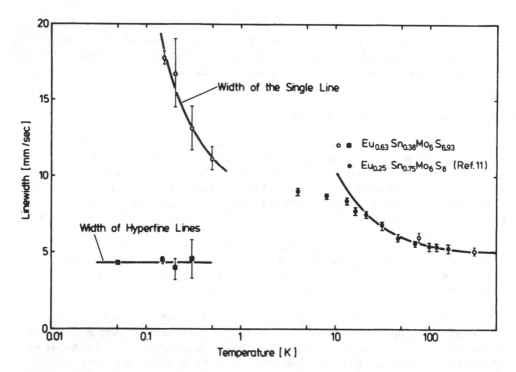

Figure 3. Mössbauer linewidth of the single line and of the hyperfine lines of $Eu_{0.63}Sn_{0.38}Mo_6S_{6.93}$, and of $Eu_{0.25}Sn_{0.75}$
$Mo_6S_{6.61}$. The lines through the single-line data represent
$\Gamma = 4.9 + 56.4/T$ (from Ref.11), and $\Gamma = 8 + 1.5/T$, respectively.

described by a Korringa law but now the relaxation rate has changed
considerably. At low temperatures dipolar interactions among Eu
spins which eventually lead to magnetic order are becoming important.
As is shown in Fig. 3, magnetic ordering is accompanied by a dramatic
sharpening of the resonance lines.

The isomer shift is shown in Fig. 4. It is $(-12.9 \pm 0.3)$mm/sec,
(relative to a $^{151}Sm_2O_3$ source), for the hyperfine spectra. The
negative isomer shift for the single, broadened line has a slightly
larger value of $(-13.5 \pm 0.4)$mm/sec at higher temperatures but decreases strongly with decreasing temperature. In Ref. 10 a value of
$(-14.0 \pm 0.1)$mm/sec has been observed for $Eu_{0.5}Sn_{0.5}Mo_6S_8$. These
large negative isomer shifts are typical for $Eu^{2+}$ in ionic compounds
but they are anomalously large for conducting systems (13). There
is therefore a very low conduction electron density at the Eu site
in these compounds. This result again confirms the weak overlap of
the conduction electrons and the Eu ions.

In Ref. 10 it was shown that the internal field in these
compounds is negative. Fully polarized $Eu^{2+}$ ions in insulators
have a saturation hyperfine field due to core polarization of
-34 T (12). Therefore, the contribution from the conduction-
electron polarization is about +6.7 T at the Eu site. This is a

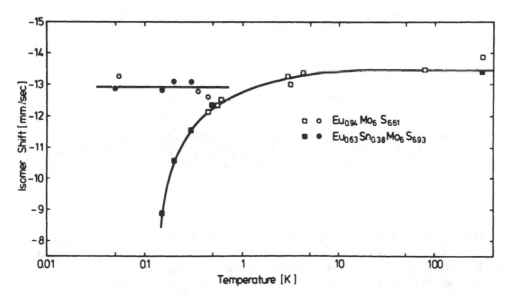

Figure 4.    Isomer shift of the single lines ($\square,\blacksquare$), and of the
hyperfine spectra (o, $\bullet$) of $Eu_{0.94}Mo_6S_{6.61}$ and of
$Eu_{0.63}Sn_{0.38}Mo_6S_{6.93}$.

rather low value for the conduction electron contribution in intermetallic Eu compounds, and indicates a weak conduction electron
spin polarization at the Eu site.

The results for linewidth, isomer shift, and hyperfine field
show that the conduction electron density as well as the exchange
interaction at the Eu site are small. These features have their
origin in the structure of the Chevrel phase. The large Eu-Mo and
Eu-Eu distances, and their shielding due to the sulphur atoms make
their superconductivity possible. Another unusual feature of these
compounds is the temperature dependence of the internal magnetic
field (9). It is temperature independent for T < 0.35 K, and col-
lapses steeply in a small temperature region (see Fig. 5). This
behavior is quite different from a Brillouin function. The temperature dependence of the internal field as well as of the intensity

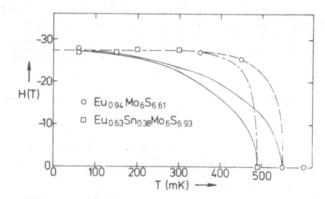

Figure 5. Temperature dependence of the hyperfine fields in
$Eu_{0.94}Mo_6S_{6.61}$ and in $Eu_{0.63}Sn_{0.38}Mo_6S_{6.93}$. The full
lines are Brillouin functions with $J = 7/2$.

of the hyperfine spectra do not correspond to a homogeneously mag-
netized sample. Qualitatively, the observations can be understood
as due to spin glass type ordering without long range order. The
static hyperfine field at T < 0.5 K is then arising from Eu ions
situated in magnetically ordered clusters randomly oriented in the
sample. This random orientation is possible because of the weak
magnetic interaction, and results in zero bulk magnetization at
zero external field (9). In the intermediate temperature range

0.1 K < T < 0.5 K, the ordered clusters seem to be imbedded in a paramagnetic surrounding. The Eu ions in this surrounding show slow electronic relaxation which is the origin of the line broadening. Lowering the temperature decreases the paramagnetic regions, and increases the number and/or size of the ordered clusters. The ordered clusters may be small enough to show superparamagnetic behavior with an abrupt decrease of the internal field at about 0.5 K, and a constant field at low T. Substituting Eu by Sn "dilutes" the already weak magnetic interaction, and enhances the break-off of magnetic clusters.

The inhomogeneous magnetic behavior below 0.5 K could result from statistical occupancy of the appropriate sites by Eu and Sn ions as well as from the deviation from stoichiometry. The influence of the statistical distribution of Eu and Sn is shown by the much wider temperature range over which we observe two spectra in $Eu_{0.63}Sn_{0.38}Mo_6S_{6.93}$ compared to $Eu_{0.94}Mo_6S_{6.61}$. From the independence of the internal field from the Eu concentration, and from measurements of the Mössbauer effect of $^{119}Sn$ in these compounds, it can be concluded that transferred hyperfine fields are less than 1 T and 0.1 T at the Eu and Sn sites, respectively (9).

## VALENCE INSTABILITY OF Eu IONS IN $(\underline{Sc}Eu)Al_2$

Some of the RE ions are stable in more than one valence configuration. Valence changes of these ions can be brought about by changing the lattice parameter because the ions need less space in the higher valence state. The change can occur by applying pressure or by placing the RE ion in a compound with the appropriate unit cell volume. In some RE compounds the two valence states are separated by energies only of order kT. In these cases, both valence states are populated, and intermediate valences can be observed. Mössbauer and x-ray photoelectron spectroscopy are the two microscopic techniques employed to study mixed valence systems. Whether an unstable electronic configuration is seen as static, fluctuating, or intermediate is a matter of time scale of the detection method. The XPS technique due to its short characteristic time scale ($10^{-17}$ sec) shows the instantaneous situation; therefore two subspectra corresponding to the two different valences appear. But this method can not distinguish between a homogeneous mixing (mixed valence at each ion) or an inhomogeneous mixing (different ions with different integer valences). Distinguishing between these possibilities is possible using the Mössbauer technique. Its characteristic time scale ($10^{-8}$ sec in case of $^{151}Eu$) is usually long compared to charge fluctuations in metallic compounds. For many compounds a mixed valence has been observed by Mössbauer spectroscopy hence deciding in favor of homogeneous mixing (14).

Europium mixed valence systems are particularly suited for Mössbauer investigations because of the large isomer shift between $Eu^{2+}$ and $Eu^{3+}$. Typical values are -9 mm/sec to -15 mm/sec for $Eu^{2+}$, and -1 mm/sec to +4 mm/sec for $Eu^{3+}$ (all values relative to $Eu_2O_3$ or $EuF_3$). (13) The variation arises from different covalences. The difference of the shift for $Eu^{3+}$ and for $Eu^{2+}$ is appreciably larger than typical Mössbauer linewidth observed for $^{151}Eu$. A discussion of results obtained for concentrated mixed valent Europium compounds has been given in Ref. 14. The results, strongly temperature depen- dent isomer shifts, were interpreted in terms of fast fluctuations ($\tau < 4 \cdot 10^{-11}$ sec) of an electron between a localized 4f level and the conduction band.

I will report here on the first Mössbauer investigation of isolated mixed valent ions which have been placed as impurities in $(\underline{Sc}Eu)Al_2$. Transport properties of this substance have been investigated at the University of Cologne and will be reported at this conference. (15) The unit cell volume of $ScAl_2$ (a=7.58 Å) is intermediate between those Laves phases in which Eu is divalent, like $EuAl_2$ (a=8.11 Å) or $EuPt_2$ (a=7.73 Å), and those in which Eu is trivalent, like $SmFe_2$ (a=7.40 Å). We may therefore expect a mixed valent state for this compound, as was already indicated by its transport properties. (15)

The investigated compound contains 0.75 at % Eu ions. It has been prepared by induction melting, and was annealed at $1000^\circ C$ for 24 hours. For comparison, Mössbauer spectra of $EuAl_2$ were also measured. In this compound, Eu is divalent, and has a single Mössbauer line at v=(-9.4 ± 0.1)mm/sec. Some of the spectra of $(\underline{Sc}Eu)Al_2$ are shown in Fig. 6. We see a weak line at (-9.2 ± 0.2) mm/sec at all temperatures above 100 K. At lower temperature this line shifts slightly to about -8.8 mm/sec (see Fig. 7). Below about 30 K, it splits into a typical $Eu^{2+}$ hyperfine spectrum. We attribute this component of the spectra to a contamination of $EuAl_2$ in our $(\underline{Sc}Eu)Al_2$ sample. But besides this weak $EuAl_2$-line we see a more intense line with a strongly temperature dependent isomer shift (see Figs. 6 and 7). Its width is temperature indepen- dent and equal to (2.6 ± 0.2)mm/sec. This line is attributed to Eu ions in $(\underline{Sc}Eu)Al_2$ showing an intermediate, temperature dependent valence.

The measured isomer shift v is given by

$$v = p_2 v_2 + p_3 v_3 , \qquad (1)$$

where $p_3$ is the probability of the fluctuating electron to be in the conduction band and to produce a $Eu^{3+}$ state, and $p_2 = 1-p_3$. $v_3$ ($v_2$) is the isomer shift of a $Eu^{3+}$-($Eu^{2+}$-)ion. The values for $p_3$ calculated from v dependent on the values of $v_2$ and $v_3$. For $v_2$

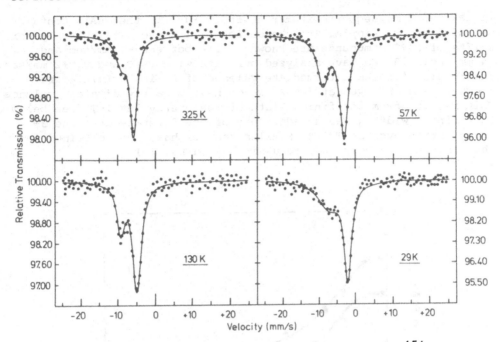

Figure 6. Mössbauer spectra of the 21.6 keV γ-line of $^{151}$Eu in
(ScEu)Al$_2$ at the indicated temperatures. The solid lines
are fits to the data points. The weak line in the spectra
is attributed to a EuAl$_2$ contamination which starts to
order magnetically at about 30 K.

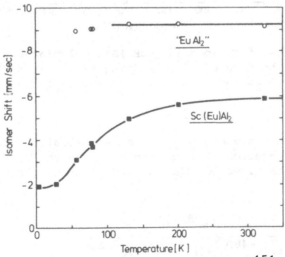

Figure 7. Isomer shift of the Mössbauer line of $^{151}$Eu in the EuAl$_2$
contamination of our sample and in (ScEu)Al$_2$.

we take -9.2 mm/sec, the isomer shift of $Eu^{2+}$ in the $EuAl_2$ contamination of our sample. The choice for $v_3$ is more difficult. Isomer
shifts of $Eu^{3+}$ compounds are known to lie between -1 mm/sec and
+4 mm/sec. (13) We have analyzed our data with $v_3=0$ and $v_3=+2$ mm/sec.
The results for the intermediate valence of Eu in (ScEu)Al$_2$ are
shown in Fig. 8. Even at T=0 K, we do have a partly divalent valence.
This results from the finite width of the energy levels, (see below).
This finite width gives an admixture of the 2+ state into the 3+
ground state even at T=0 K. Similar results have been obtained for
the concentrated compounds $EuCu_2Si_2$ and $EuRh_2$.(14)

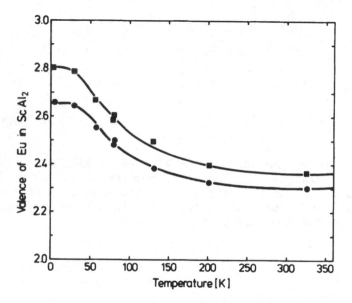

Figure 8. Valence of Eu ions in (ScEu)Al$_2$ calculated with Eq. 1
          from the data of Fig. 7, using $v_3=0$ mm/sec (■), and
          $v_3=+2.0$ mm/sec (●), respectively, (see text).

From the intermediate valence we can calculate the energy E
necessary for promotion of a 4f electron from the Fermi energy to
a localized 4f band, or for the $Eu^{3+} \rightarrow Eu^{2+}$ transition. The relevant
formula (14) is

$$\frac{P_2(T,E)}{P_3(T,E)} = \frac{8e^{-E/kT}}{1+3e^{-480/T}+5e^{-1330/T}+7e^{-2600/T}} . \tag{2}$$

This formula takes into account the multiplicity 8 of the $Eu^{2+}$ ground state at energy E above the Fermi level, and the multiplicities and energies of the $Eu^{3+}$ ground state as well as its first three excited states. Analyzing the data of Fig. 8 with Eq. 2 we get the excitation energies E as shown in Fig. 9.

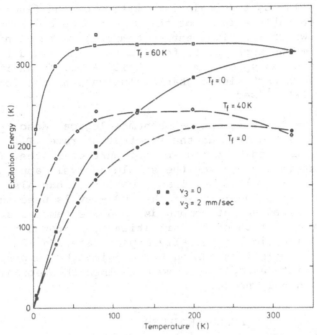

Figure 9. Excitation energy for the transition $Eu^{3+} \to Eu^{2+}$ in ($\underline{Sc}Eu$)$Al_2$ calculated from the data of Fig. 8 with Eq. 2 by replacing $T \leftrightarrow T+T_f$, and for $v_3=o$ and $v_3=2$ mm/sec, respectively, (see text).

These values for E show a very strong T-dependence, and particularly they go to zero for $T \to 0$. This behavior is unreasonable because it would mean $p_3 \to 0$ for $T \to 0$. This "defect" can be - at least partly - removed by taking into account that the energy levels have a certain width, or by introducing a characteristic temperature $T_f$ for each of them (16). Of course, we do not have anough information to analyze the results including an individual width for each of the energy states. But to go one step further in the analysis, we have replaced T in Eq. 2 by $T+T_f$ (16,17). We have made the simplifying assumption that $T_f$ is the same for all energy levels, and have chosen $T_f$ so that E ≈ constant. The results of this analysis are also shown in Fig. 9. For the analysis with $v_3=0$ mm/sec, we get a temperature independent excitation energy of 320 K or 0.03 eV at T >60 K if we choose $T_f \approx$ 60 K. If we take the values for $p_3$ of Fig. 8 which were obtained with $v_3=+2$ mm/sec, we find E=240 K for $T_f \approx$ 40 K at T > 80 K. Our results are therefore E=(280 ± 40) K and

$T_f = (50 \pm 10)$ K (17). The latter value is in reasonable agreement
with the fluctuation temperature of about 60 K, deduced from the
transport properties (15). For $T < T_f$, the above analysis is inadequate, and still results in E→0 for T→0. This has also been
observed in earlier investigations (14).

Our data allow us to obtain information on the fluctuation
rate $\tau$ of the unstable valence of the Eu ions. We observe that the
Mössbauer line width $\Gamma$ is indepent of temperature to within
0.2 mm/sec. From the appropriate formula relating $\tau$ and $\Gamma$ given
in Ref. 14 we find an upper limit $\tau \leq 5 \cdot 10^{-11}$ sec. In addition, we
can calculate a characteristic charge fluctuation time for $T_f$,
which is about $1 \cdot 10^{-13}$ sec.

The given analysis is only a phenomenological description for
the case that T is higher than the fluctuation rate $(T_f)$. Because
of their width, the ionic states are not pure 4f-states. It remains
to be shown that in these cases the simple analysis applied here
and in earlier work (14) is still applicable. It has also not yet
been shown whether the energies of the $Eu^{3+}$-levels used in Eq. 2
are correct for mixed valent compounds, and are temperature independent. A representation of the situation by a ground state of
mixed wave function might be equally appropriate. In addition, the
influence due to thermal expansion by changing the temperature
should be taken into account in a more refined theoretical treatment which is urgently needed.

## ACKNOWLEDGEMENT

The work reported here has been performed with the coauthors
of Ref. 9 and 15 whose cooperation is gratefully acknowledged. In
addition, I thank R. Rangel and Dr. R.W. McCallum for their help
with the (ScEu)Al$_2$ experiments.

## REFERENCES

1. For a recent review on superconducting spin glasses see S. Roth,
   Applied Physics 15, 1 (1978).

2. For a recent review on Chevrel phase superconductors see
   Ø. Fischer, Applied Physics 16, 1 (1978).

3. Ø. Fischer, A. Treyvaud, R. Chevrel, and M. Sergent, Solid State
   Communications 17, 721 (1975).

4. R.N. Shelton, R.W. McCallum, and H. Adrian, Physics Letters 56A,
   213 (1976).

5. W.A. Fertig, D.C. Johnston, L.E. Delong, R.W. McCallum, M.B. Maple
   and B.T. Matthias, Physical Review Letters 38, 987 (1977).

6. O.K. Anderson, W. Klose, and H. Nohl, Physical Review B17, 1209 (1978); L.F. Mattheis and C.Y. Fong, Physical Review B15, 1760 (1977).

7. For recent reviews on magnetism of superconducting RE Chevrel phases see M.B. Maple, in "The Rare Earths in Modern Science and Technology" p. 381, ed. G.J. McCarthy and J.J. Rhyne, Plenum Press 1978, and Journal de Physique C6, 1374 (1978); M. Ishikawa, Ø. Fischer, and J. Mueller, Journal de Physique C6, 1379 (1978); Ø. Fischer, M. Ishikawa, M. Pelizzone, and A. Treyvaud, CNRS Conference on "Physics of Metallic Rare Earths", St. Pierre de Chartreuse, Sept. 1978.

8. J.W. Lynn, D.E. Moncton, W. Thomlinson, G. Shirance, and R.N. Shelton, Solid State Communications 26, 493 (1978); J.W. Lynn, D.E. Moncton, G. Shirane, W. Thomlinson, J. Eckert, and R.N. Shelton, Journal Applied Physics 49, 1389 (1978); D.E. Moncton, G. Shirane, W. Thomlinson, M. Ishikawa, and Ø. Fischer, Physical Review Letters 41, 1133 (1978).

9. J. Bolz, G. Crecelius, H. Maletta, and F. Pobell, Journal Low Temperature Physics 28, 61 (1977).

10. F.Y. Fradin, G.K. Shenoy, B.D. Dunlap, A.T. Aldred, and C.W. Kimball, Physical Review Letters 38, 719 (1977).

11. C.W. Kimball, G.L. Van Landuyt, C.D. Barnet, G.K. Shenoy, B.D. Dunlap, and F.Y. Fradin, Proceedings International Conference on the Application of the Moessbauer Effect, Kyoto, 1978.

12. J.M. Baker and F.I.B. Williams, Proceedings Royal Society London, Ser. A 267, 283 (1962).

13. G.K. Shenoy and F. Wagner, "Mössbauer Isomer Shifts", North Holland Publ. 1978.

14. For reviews see E.R. Bauminger, I. Felner, D. Froindlich, D. Levron, I. Nowik, S. Ofer, and R. Yanofsky, Journal de Physique 35, C6-61 (1974); I. Nowik, in "Valence Instabilities and Related Narrow Band Phenomena", ed. R.D. Parks, Plenum Press 1976, p. 261.

15. W. Franz, F. Steglich, D. Wohlleben, W. Zell, and F. Pobell, paper B2 at this Conference.

16. B.C. Sales and D.K. Wohlleben, Physical Review Letters 35, 1240 (1975).

17. A somewhat more appropriate analysis might be performed by replacing $T$ by $\sqrt{T^2 + T_f^2}$, which gives $E = (280 \pm 20)$ K and $T_f = (80 \pm 10)$ K; (private communication by D.K. Wohlleben).

# PRESSURE TEMPERATURE BEHAVIOR OF THE ELECTRONIC TRANSITIONS IN $Sm_4Bi_3$

A. Jayaraman, R. G. Maines
Bell Laboratories, Murray Hill, New Jersey 07974
E. Bucher, Konstanz, Fed. Rep. of Germany

## ABSTRACT

The first-order valence transition boundary in $Sm_4Bi_3$ has a large positive dT/dP and terminates at a critical point above 700°K and 28 kbar. An anomalous decrease in resistance near 273°K at ambient pressure is observed, and this is attributed to a change of state from the inhomogeneously mixed to a homogeneously mixed valence state. This phase boundary has a negative slope and intersects the first-order valence transition boundary at a triple point near 225°K and 25 kbar. In the high pressure phase all the Sm ions are in the fully trivalent state.

## INTRODUCTION

Many rare-earth germanides, antimonides and bismuthides of the formula $R_4B_3$, where B is Ge, Sb or Bi, are known to crystallize in the anti-$Th_3P_4$ structure (cubic) (1-3), with the R occupying the phosphorus sites (this has been referred to as the $Gd_4Bi_3$ structure-type in reference 2). Among these, the bismuthides of Sm, Eu and Yb are reported to have anomalously large lattice parameter due to their divalent character (1-4). It is believed that in $Sm_4Bi_3$ three of the samarium ions are divalent (2,4). A pressure-induced first-order isostructural transition at $\sim$ 26 kbar hydrostatic pressure, due to a change in the valence state of the divalent Sm ions in $Sm_4Bi_3$, has recently been reported (5). The anti-$Th_3P_4$ structure is retained in the high pressure phase, but the lattice parameter decreases strikingly at the transition pressure from 9.70 to 9.4 A. In the present study we have determined the pressure-temperature behavior of this transition.

We have also found an anomalous resistance decrease near 273°K at atmospheric pressure and have followed this transition as a function of pressure. Based on the results obtained we present a P-T diagram for $Sm_4Bi_3$ and briefly discuss the implications in this paper.

EXPERIMENTAL

The method of preparation of the sample has been described elsewhere (2,4). Hydrostatic pressure was generated in a piston-cylinder device using the Teflon cell technique. For pressure-temperature studies above 400°K D.C.703 silicon fluid was used as pressure medium and isoamyl alcohol for experiments below 400°K. For resistance measurements, four indium contacts were pressed onto freshly exposed surface of the sample.

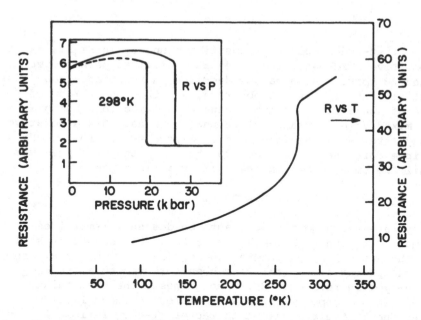

Fig. 1:   Temperature dependence of resistance of $Sm_4Bi_3$ at ambient pressure. Note the anomalous resistance decrease near 273°K. The inset shows the first-order valence change near 26 kbar at 298°K.

The results of resistance measurements are shown in Fig. 1.
At 298°K an abrupt decrease in the resistance of the sample occurs
at about 26 kbar (see inset in Fig. 1). When the sample is
pressurized at higher temperatures, the magnitude of the abrupt
decrease in resistance decreases and is preceded by a more gradual
change; as the temperature is increased the latter behavior
dominates over the abrupt decrease. This suggests that the phase
boundary has a critical point. In Fig. 1 the resistance variation
with temperature at atmospheric pressure is shown. There is a
marked and anomalous decrease in resistance near 273°K at atmospheric pressure. This anomalous resistance-decrease shifts to
lower temperatures with increasing pressure.

The results of the pressure-temperature studies, carried out
using the resistance anomaly, are presented in Fig. 2. Data
obtained on increasing pressure cycle only are plotted in the
figure. The first-order valence transition boundary designated as
I to III is almost vertical, but has a positive slope. This was
confirmed by the sign of the latent heat-change accompanying the
transition. Heat is evolved in going from the low pressure phase
(I) to the high pressure phase (III), which proves that phase I
has a higher entropy. In this respect the isostructural transition boundary I-III in $Sm_4Bi_3$ behaves like the analogous transition boundary in SmS (6). The R vs. T behavior strongly suggests
the termination of the I-III boundary at a critical point near
700°K and 28 kbar. The I-II phase boundary represents a plot of
the onset of the resistance anomaly temperature as a function of
pressure. The latter boundary intersects the valence transition
boundary I-III near 25 kbars and 225°K. If this intersection is
a triple point, another phase line has to exist and the probable
trajectory for this transition boundary designated II-III is
shown in Fig. 2 as a dotted line. We believe that this boundary
II-III has a negative slope.

DISCUSSION

The anomalous resistance decrease near 273°K at atmospheric
pressure must be due to a phase transition. Our X-ray studies
across this transition do not suggest any change in the structure,
within the precision of our measurements. We therefore believe
that it is a transition from an inhomogeneously mixed valence
state (7) to a homogeneously mixed valence situation. (The
former state represents a random mixture of $Sm^{3+}$ and $Sm^{2+}$ ions,
statically frozen in the lattice of $Sm_4Bi_3$; while in the homogeneous mixed valence situation the Sm ions are uniformly in the
same intermediate valence state.) A large configuration contribution to the entropy may be expected in the inhomogeneously mixed
valent state, which would be absent for the homogeneously mixed
state. In our view this accounts for the higher entropy of phase

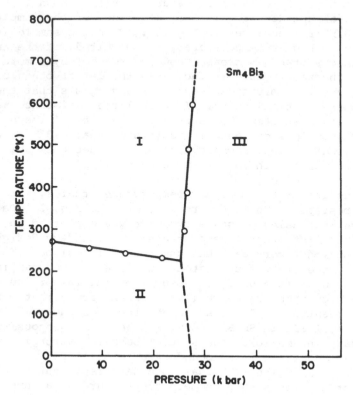

Fig. 2:  Pressure-temperature diagram for $Sm_4Bi_3$. The phase
         I-III transition boundary is isostructural accompanied
         by $\Delta V$ of $\sim$ 10%. The I-III phase boundary represents a
         plot of the anomalous resistance decrease shown in Fig. 1.

I. A transition from disordered to a more ordered state should
result in a drop in electrical resistivity, as is actually
observed. In this connection it is worth noting that no resistance anomaly is seen in $Yb_4Bi_3$. We have attributed a negative
$dT/dP$ to the II-III boundary. This would be consistent with our
view of the I-II transition.

The magnetic susceptibility behavior of $Sm_4Bi_3$ (4) is not
inconsistent with the above picture. The temperature dependence
of the susceptibility does not show any evidence of magnetic
ordering and the susceptibility is flat below 50°K, as in the
case of homogeneously mixed valence state materials (7). However,
it should be noted that there is no anomaly in the susceptibility
near 273°K. From the lattice parameter of the high pressure
phase, we believe that it is a fully collapsed phase in which all
the Sm ions are in the trivalent state.

## SUMMARY

We have shown [1] that the first-order valence transition boundary in $Sm_4Bi_3$ has a positive slope and it terminates at a critical point, [2] that the 273°K resistance anomaly at ambient pressure is due to a transition most likely from an inhomogeneously mixed valence state to a homogeneously mixed valence situation, and [3] we cannot completely rule out a subtle structure change near 273°K as the cause of the 273°K resistance anomaly.

## ACKNOWLEDGMENT

We wish to thank K. Andres for illuminating discussions.

## REFERENCES

1.  D. Hohnke and E. Parthe, "The Anti-$Th_3P_4$ Structure Type for Rare-Earth Germanides, Antimonides and Bismuthides", Acta Cryst. 21:435 (1966).

2.  R. J. Gambino, "Rare-Earth Sb and Bi Compounds With the $Gd_4Bi_3$ (Anti-$Th_3P_4$) Structure", J. Less Common Metals 12:344-52 (1967).

3.  K. Yoshihara, J. B. Taylor, L. D. Calvert and J. G. Despault, "Rare-Earth Bismuthides", J. Less Common Metals 41:329-37 (1975).

4.  E. Bucher, A. S. Cooper, D. Jaccard and J. Sierro, "Intermediate Valence Properties of a Series of $Ln_4X_3$ Compounds With Anti-$Th_3P_4$ Structure", in: <u>Valence Instabilities and Related Narrow-Band Phenomena</u>, R. D. Parks, ed., Plenum, NY (1977).

5.  A. Jayaraman, R. G. Maines and E. Bucher, "Pressure Induced Valence Instability in $Sm_4Bi_3$", Solid State Commun. 27:709-11 (1978).

6.  A. Jayaraman, P. D. Dernier and L. Longinotti, "Study of the Valence Transition in SmS Induced by Alloying, Temperature and Pressure", Phys. Rev. B 11:2783-94 (1975).

7.  C. M. Varma, "Mixed Valence Compounds", Rev. Mod. Phys. 48:219 (1976).

# MÖSSBAUER STUDIES OF $Er_{0.5}Mo_6Se_8$

James C. Glass and Frank Pobell

Department of Physics, North Dakota State Univ., Fargo,

N.D. 58105 U.S.A., and KFA Jülich, Jülich, West Germany

There is a long standing interest in the possible coexistence of magnetic order and superconductivity, and this interest has been enhanced by some recent results. Low temperature anomalies in specific heat and magnetic susceptibility in superconducting $Er_{1.0}Mo_6S_8$ and $Er_{1.0}Mo_6Se_8$ or $Er_{1.2}Mo_6Se_8$ [1] suggest the occurrence of magnetic order, while magnetic ordering is observed to destroy superconductivity in $Ho_{1.2}Mo_6S_8$ [2]. We wish to report here on the initial Mössbauer experiments on the compound $Er_xMo_6Se_8$ with x = 0.5; resistance anomalies have been observed for x = 1.0 and 1.2 at around 1.07 K [1].

Past experience with the preparation of $GdMo_6Se_8$ has shown that depending on the technique used, an excess of the rare earth can result in the formation of $Gd_2O_2Se$ [3]. Because of this, a sample of $Er_{0.5}Mo_6Se_8$ was prepared by mixing appropriate amounts of $Er_2O_3$, Mo and Se powders and treating the mixture several times at 1100°C with $H_2$ and $H_2Se$ gases. By using a relatively small amount of Er and reacting the mixture several times, it is expected that formation of $Er_2O_2Se$ is minimized. By use of the inductive method, $T_c$ was measured for a powder sample and found to be 6.41 K, with a transition width of 0.30 K. This is very similar to results reported for $ErMo_6Se_8$ [1].

The Mössbauer absorber prepared with $Er_{0.5}Mo_6Se_8$ had a [166]Er density of 9.33 mgm/cm$^2$. Until recently it has been difficult to do meaningful low temperature experiments with [166]Er because of magnetic ordering of the conventional [166]Ho sources used. It has been found, however, that dilute rare-earth dihydrides provide good single line sources down to 1.4 K. Because of this, the source used was $Ho_{0.4}Y_{0.6}H_2$, prepared by the method of Stöhr and

Cashion (4), and activated with the KFA Merlin reactor at a flux
of $10^{14}$ neutrons/$cm^2$sec. The source was checked for magnetic
ordering by using an absorber of $ErH_2$, which has been shown not
to order magnetically down to 2.4 K (5). Mössbauer spectra have
been obtained for $Er_{0.5}Mo_6Se_8$ and $ErH_2$ at source temperatures of
4.2 K and 1.7 K. The results are shown in Table 1. All spectra
showed only a single line. At 4.2 K, the line width for $ErH_2$
agrees well with that obtained by Stöhr and Cashion, while at 1.7 K
the width is only slightly higher than that obtained for the more
magnetically dilute $Ho_{0.15}Y_{0.85}H_2$ at 1.4 K. The dramatic increase
in line width for $Er_{0.5}Mo_6Se_8$ at 1.7 K can therefore not be
attributed to the source and may be due to an unresolved magnetic
hyperfine splitting of the absorber. This will be investigated
further by Mössbauer studies of $Er_{0.5}Mo_6Se_8$ at temperatures less
than 1.7 K.

TABLE 1.   Absorber Parameters Obtained with a $Ho_{0.4}Y_{0.6}H_2$ Source

| Compound | T(K) | Isomer Shift (mm/sec) | Line Width (mm/sec) |
|---|---|---|---|
| $ErH_2$ | 4.24 | −0.25±0.18 | 9.02±0.54 |
| | 2.50* | −0.10±0.19 | 9.84±0.58 |
| $Er_{0.5}Mo_6Se_8$ | 4.24 | −0.15±0.03 | 8.03±0.43 |
| | 1.70 | +0.43±0.56 | 23.10±2.07 |

*Source temperature 1.70 K

## REFERENCES

1.  R. W. McCallum, D. C. Johnston, R. N. Shelton, W. A. Fertig,
    and M. B. Maple, "Coexistence of Superconductivity and Long-
    Range Antiferromagnetic Order in the Compound $Er_xMo_6Se_8$,"
    Solid State Comm. 24, 501 (1977).

2.  M. Ishikawa and O. Fischer, "Magnetic Ordering in the Super-
    conducting State of Rare Earth Molybdenum Sulphides,
    $(RE)_{1.2}Mo_6Se$," Solid State Comm. 24, 747 (1977).

3.  H. J. Maletta, Kernforschungsanlage Jülich, private communica-
    tion, and K. P. Nerz, Ph.D. Thesis Kernforschungsanlage
    Jülich, 1979.

4.  J. Stöhr and J. D. Cashion, "Mössbauer Studies of Concentrated
    and Diluted Rare Earth Dihydrides," Phys. Rev. B12, 4805 (1975).

5.  G. K. Shenoy, B. D. Dunlap, D. G. Westlake and A. E. Dwight,
    "Crystal Field and Magnetic Properties of $ErH_2$," Phys. Rev.
    B14, 41 (1976).

# SUPERCONDUCTIVITY IN THE YTTRIUM SESQUICARBIDE SYSTEM

F.J. Cadieu[*], Queens College of CUNY, Flushing
New York 11367
and J. J. Cuomo and R. J. Gambino
IBM Research Center, Yorktown Heights, New York 10598

At the present time there are only three known crystal structures in which superconductivity with transition temperatures above
15° K have been observed.  One of these structures is the A-15
structures which contains the $Nb_3Ge$ system, the $Nb_3Ga$ system,
the $NB_3Al$ and $V_3Si$ systems.[1]  The second structure, NaCl rock salt,
occurs in the NB-N-C system having a $T_c$ up to 18°K.[2]  The third
structure is the $Pu_2C_3$ type in which, primarily, nonstoichiomatic
alloys of Y-Th-C have been observed with a $T_c$ of 17°K.[3]  In the
A-15 and the NaCl structure systems the high $T_c$ compositions are
regarded as normal high Tc systems in that they have an e/a ratio of
4.5 - 4.8 in accord with Matthias rules.[4]  The $Pu_2C_3$ systems having
a high $T_c$, however, appear anomalous in that a simple application
of the e/a rule would give an e/a ratio of 3.8 for the high $T_c$ material.[5]  The high $T_c$ material of the $Pu_2C_3$ structure type has
only been synthesized under high pressures (15-25 Kilobars) and
high temperatures (1200-1450°C).  An apparently open question is
whether these high $T_c$ $Pu_2C_3$ structure phases are really so anomalous.
A second question is whether different synthesis techniques could
further raise the $T_c$ of this system, possibly by producing phases
in a composition range more in conformity with the e/a ratio rules
of Matthias.  It should be noted that the $Nb_3Ge$ system is one in
which bulk preparation methods have only yielded $T_c$'s of about 6°K.
Special sputtering techniques have raised the $T_c$ of this system to
approximately 23°K.  In this paper we give a new analysis and interpretation of superconductivity in the $Pu_2C_3$ type structure system.  The new analysis will answer in part whether the high $T_c$'s
observed in this structure are in violation of the valence rules
of Matthias.  We also comment whether it is likely that new pre-

[*]Supported in part by NSF DMR 78-03217

247

paration techniques applied to this system would have a reasonable expectation of producing significantly higher $T_c$'s.

If the e/a ratio is computed by the conventional method of summing the total number of outer electrons and dividing by the number of atoms in the structure then an e/a value of 3.6 results of lanthanum sesquicarbide and for yttrium sesquicarbide. This was done by Giorgi who noted that 3.6 is a very anomalous value for a high $T_c$ system.[5] Instead of viewing this structure in the normal manner we have chosen to regard the structure as being of the anti--$Th_3P_4$ type with pairs of carbon atoms replacing the thorium atoms. The carbon pair then constitutes a variable valence group whose valence is a function of the carbon-carbon spacing in the particular system. The closer the carbon-carbon spacing is the more electrons are released into the structure. The expected C-C valence versus the C-C spacing has been computed by Carter et al.[6] Instead of writing the compound as $Y_2C_3$ we use the anti-$Th_3P_4$ structural formula, $Y_4(C_2)_3$. Regarding the carbon pair as a unit in the structure of valence six we can then compute the e/a ratio for seven atoms in the structure to give an e/a value of 4.3. For computing the e/a value in off stoichiomentric cases we would then write $Y_{4+x}(C_2)_{3-x}$. The composition observed to have the highest $T_c$ of 17°K then corresponds to $(Y_{0.7}Th_{0.3})_{4-0.056}(C_2)_{3+0.056}$ which is within experimental accuracy the stoichiometric compound. The e/a ratio computed on this new basis for this high $T_c$ composition is now 4.5. This new e/a ratio now corresponds to the expected value of e/a for the highest $T_c$ to be observed in a system. To further see the utility of this viewpoint in Fig. 1 we show the dependence of the $T_c$ and of the lattice parameter in the $(Y_{0.7}Th_{0.3})_{4+x}(C_2)_{3-x}$ system. Also shown is the yield of sesquicarbide phase in high pressure, high temperature synthesized bulk samples. This data is from Krupka and Giorgi.[3] It should be noted that the $T_c$ and lattice parameter vary monotonically up to the stoichiometric composition at x = o0 and beyond which the structure does not form. This graph should be contrasted to Fig. 3 of Krupka et al.[3] which plots the lattice parameter for all compositions on the same graph. The maximum $T_c$ in the system then corresponds to an intermediate value of the lattice parameter which is anomalous. The behavior shown in our Fig. 1 is typical of high $T_c$ systems such as $NB_3Ge$ which generally form over a range of compositions on the transition metal rich side of the phase boundary. That the structure forms to one side of the phase boundary reflects the nonequivalence of the atoms for occupying sites in the structure. Thus in, for example, $NB_3Ge$, the Nb atoms can occupy Ge or B atom sites for the Ge atoms cannot provide proper bonding when they occupy A atom or chain sites.[1] In the case of the sesquicarbides the highly directed bonds of Y, La, and possibly f character of Th are necessary for forming the sequicarbide phase.[6] We then have a cubic structure with highly directional bonding within the

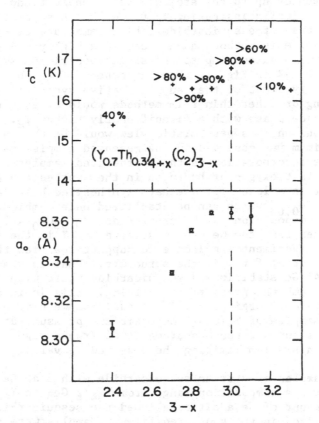

Fig. 1. The superconducting transition temperature and lattice parameter for the $(Y_{0.7}Th_{0.3})$ sesquicarbide systems versus 3 - X. The vertical dashed line denotes the stoichiometric phase boundary. The percentage figures are estimated phase yields for high temperature and pressure synthesis. This date is from Krupka et al. Ref. 3.

unit cell. The structure forms from one side up to a stoichiometric boundary but not beyond. In the case of $Nb_3Ge$[7] and $Nb_3Ga$[8]
early formation methods produced an A-15 phase which had excess
Nb atoms in the structure. The resulting $T_c$ was low in both cases.
For $Nb_3Ge$ special sputtering techniques[9,10] then allowed an A-15
phase of $Nb_3Ge$ to be formed which had a monotonically decreasing
lattice parameter up to the stoichiometric phase boundary[10] and
also a monotonically increasing $T_c$ up to the phase boundary. In
the present case of sesquicarbides, high temperature, high pressure bulk synthesis methods have produced a high yield sesquicarbide phase which extends up to the stoichiometric phase boundary as
at the line 3-x=3 in Fig. 1. For the case of $(Y_{0.7}Th_{0.3})_{4+x}(C_2)_{3-x}$
it is then very doubtful that an alternative synthesis means such
as sputtering or other thin film methods would be able to produce
a sesquicarbide phase with a signnificantly higher $T_c$. Possible
qualifications on this pessimistic view would be that a lower
$T_c$ than optimum was observed in the measured samples due to undetected gas incorporation in the synthesized samples. It would
be expected that oxygen or hydrogen in the synthesized samples
would result in a lower $T_c$. We have synthesized bulk samples of
$Y_{0.39}C_{0.58}$ $Ge_{0.03}$ which can be stabilized under ambient pressures. This composition can be written as $(Y_{0.93}Ge_{0.07})_{4.2}(C_2)_{2.9}$
which proposes that the Ge can substitute for Y in the structure.
There is no experimental evidence to support the view that the
Ge is substituting for Y in the structure. What is observed is that
0.02 to 0.04% Ge stabilizes a sesquicarbide phase with a $T_c$ up to
11.6°K under ambient pressure arc melting. This is in agreement
with the data of Giorgi et al.[11] In addition we have produced superconducting samples of $Y_4(C_2)_3$ under ambient pressure arc melting.
That our $T_c$'s support the measurements of Krupka argue against a gas
incorporation problem limiting the observed $T_c$ value.

Some samples of yttrium sesquicarbide with 3 a% Ge addition were
sputtered over a composition range from $Y_{34}C_{63}Ge_3$ to $Y_{44}C_{53}Ge_3$.
The small amount of Ge addition allowed the sesquicarbide structure
to be stabilized in the sputtered films. Samples were sputtered
in a trisputtering system so as to produce a composition gradiant
along a set of sapphire substrates. Films which were not overcoated
in place before exposure to air remained opaque and shiny for only
about 15 seconds before reacting with the air. The resulting reaction with moisture in the air produced acetylene vapors which indicates the bonding of the C-C groups in this system. Films which
were overcoated with Mo before removal from the sputtering system
remained bright and shiny when viewed through the substrate. Samples produced at a fairly high sputtering rate onto 800 C sapphire
substrates exhibited superconductivity for the higher Y compositions.
The $T_c$'s were measured inductively so that direct electrical contact
was not required. Full inductive superconducting transitions up to
6 K were observed. We do not consider the observed sputtered films
to be optimized, but they are the only examples of superconducting

yytrium sesquicarbide thin films that we know of.

Another very important aspect of this system is that the addition of appreciable amounts of a third element, in this case Th, has resulted in a substantial increase in the $T_c$ of the system. This is in agreement with the composite transition metal atom hypothesis of one of the authors.[12,13,14] In this case, the high yield of the phase up to the same phase boundary as for Y-C alone indicates that the Th is substituting partially for Y in the structure. The effective transition metal atom in this system is then a composite atom formed by averaging of the Y and Th electronic states in the system. In the case of the A-15 structure it has been very difficult to achieve an increase in any binary A-15 system by the partial substitution of a third element whether the third element be a simple metal or a transition metal. In the case of $Nb_3Al$ it was observed that the addition of Ge resulted in an increased $T_c$ from 18.5°K to 21°K, but subsequently it has been shown that $Nb_3Ge$ alone has a higher $T_c$ than any known Nb-Al-Ge combination.[1] The reason why it is easier to achieve a third element substitution resulting in improved $T_c$ in this $Pu_2C_3$ structure we believe is connected with the C-C grouping observed in this structure. Our original thought was that by changing the C-C separation in the system this group would act as a variable valence group. An important consequence would then be that various substitutions or systems which resulted in changes in this C-C separation would then be effectively releasing a variable number of electrons to the structure. By utilizing this mechanism, changes in the C-C separation would result in a change in the ratio of the number of electrons connected with bonding the C-C group as opposed to those principally involved in bonding the C-C group into the sequicarbide structure. What may be more important, however, to account for the relative ease in adding Th to $Y_4(C_2)_3$ is the flexibility of the C-C group to accomodate different size atoms into the structure. This would come about as a slight changing of the C-C separation near Th atoms or as a slight tilting of the C-C group near different sized atoms. In this particular case, the ionic radius for Y(+3) is 0.93 Å and that for Th (+4) is 0.95 Å. These radii are nearly the same so that very little if any distortion of the structure is required to substitute Th for Y. In this way the C-C groups would have a greater ability to accommodate different sized atoms into the structure than the single B-type atoms such as Ge in the A-15 structure.

The high $T_c$ sesquicarbides observed in the Y-C and (Y-Th)-C systems have been shown not to be anomalous when viewed as compounds of the anti-$Th_3P_4$ type structure with the C-C group viewed as an atom group with a variable valence which depends on the C-C separation. When viewed as a compound of the type $Y_4(C_2)_3$ the e/a ratio for yttrium sesquicarbide is 4.3 if we take the C-C group as a single group of valence 6. The higher Tc $(Y_{0.7}Th_{0.3})_4(C_2)_3$ composition then has an e/a value of 4.5.

This value is the expected value for a high $T_c$ system. Evidence for this viewpoint is provided by the asymmetry of the sesquicarbide phase field as the Y/C ratio is varied. This asymmetry is characteristic of highly hybridized systems as $Nb_3Ge$. It is then indicated that the Y or Th atoms can occupy either site in the structure but that the C-C group cannot occupy the transition metal atom sites without seriously altering the structure. Such behavior is also observed in the case of the Nb-Ge system. This is to be contrasted with a defect type structure. The $(Y_{1-y}Th_y)_{4+x}(C_2)_{3-x}$ system is also argued to be an example of a composite transition metal atom system in that the Y-Th function as a composite atom. The electronic properties of the system are determined by an averaging of the electronic states of the Y and Th atoms. This seems to be made possible by a fortuitous similarity in the metallic radius of the Y(+3) ion and Th(+4) ion which allows either one to fit into the transition metal atom sites with only a gradual change in lattice parameter and properties being observed.

REFERENCES

1. For a review of A-15 compounds sec; D. Dew-Hughes, Cryogenics, 435, August, 1975.
2. J. K. Hulm and R. D. Blaugher, A.I.P. Conf. Proc. No. 4, 1, ed. D. H. Douglass, 1972.
3. M. C. Krupka, A. L. Giorgi, N. H. Krikorian, and E. G. Szklarz, J. Less-Common Metals 19, 113 (1969).
4. B.T. Matthias, Progress in Low Temperature Physics, Vol. II, p. 138, ed. C. J. Gorter, North-Holland Printing Co., 1957.
5. A.L. Giorgi, E.G. Szklarz and M.C. Krupka, A.I.P. Conf. Proc. No. 4, 147, ed. D. H. Douglass. 1972.
6. F. L. Carter, T. L. Francavilla, and R. A. Hein, Rare Earth Conference, 1974.
7. B. T. Matthias, T. H. Geballe, R. H. Willens, E. Corenzwit, and G. W. Hull, Jr. Phys. Rev. 139, A1501 (1965).
8. G. W. Webb, A.I.P. Conf. Proc. No. 4, 139, ed. D. H. Douglass, 1972.
9. J. R. Gavaler, Appl. Phys. Letters 23, 480 (1973).
10. N. Chencinski and F. J. Cadieu, J. Low Tem. Phys. 16, 507 (1974).
11. A. L. Giorgi, H. H. Hill, E. G. Szkarz and R. W. White, in Superconductivity in d-and f-Band Metals, p. 361, ed. D. H. Douglass, 1976.
12. F. J. Cadieu, N. Chencinski, C. Z. Rosen, J. Appl. Phys. 48, 686 (1977).
13. F. J. Cadieu and N. Chencinski, Inst. Phys. Conf. Ser. No. 39, Chapter 8, 642 (1978).
14. F.J. Cadieu, J. Low Temp. Phys. 3, 393 (1970).

# SUPERCONDUCTIVITY IN La-Y-Mn ALLOYS

R. J. Stierman, O. D. McMasters and K. A.
Gschneidner, Jr.
Ames Laboratory-DOE and Department of Materials
Science and Engineering
Iowa State University, Ames, IA 50011, U.S.A.

Crystal structure changes occur along many of the lanthanide compound series, such as between adjacent lanthanide compounds, or a single compound which exhibits polymorphic forms. Since superconductivity is sometimes associated with lattice instabilities, one or both compounds at the structure change might be expected to be superconducting if they have no unpaired 4f electrons. For systems where the particular La compound does not form, or where it is several atomic numbers removed from such an instability, the proper proportions of La and Y or La and Lu can be alloyed to form the pseudo-lanthanide compound. The system investigated was $RMn_2$, which exhibits a structure change between hexagonal C14 and cubic C15 Laves phases. The C14 phase exists for Nd, Pr, Sm, Lu and Ho to Tm, the C15 phase exists for Sm and Gd to Ho, while no $RMn_2$ compounds form in the La, Ce, Eu and Yb systems. Sm and Ho are polymorphic. Samples of $La_{1-x}Y_xMn_2$, $0.2 \le x \le 1.0$ were prepared by arc melting and subsequent heat treatment. X-ray diffraction powder patterns, metallography, magnetic susceptibility and low temperature heat capacity measurements were made. All alloys contained at least two phases, with the C15 structure predominate for $x \ge 0.6$. Two superconducting transitions were detected at about 4.5 and 3.2 K in the magnetic susceptibility measurements for compositions $x \le 0.7$. Samples with composition $0.7 < x < 0.9$ showed one transition which varied from 1.5 to 3.2 K, with $T_C$ increasing as the La content increased. The lattice parameter also increased as La content increased in the range $0.6 < x < 1$. On the basis of these $T_C$ and x-ray data the 1.5 to 3.2 K transition was thought to be due to the $RMn_2$ phase. However, heat capacity measurements showed the lower (1.5 - 3.2 K) transition to be due to an impurity phase which constitutes about 10% of the sample. The impurity phase could be a La-Y

solid solution alloy which did not combine with Mn to form an
intermetallic compound.   The upper superconducting transition
temperature is thought to be due to a phase other than the
$La_{1-x}Y_xMn_2$ compound,  but the composition of this phase is
unknown.

THE ANOMALOUS PHYSICAL PROPERTIES OF RESn$_3$ AND (LaRE)Sn$_3$ COMPOUNDS

AND ALLOYS

Lance E. DeLong

Department of Physics, University of Virginia

Charlottesville, VA 22901

## INTRODUCTION

The purpose of this paper is to provide a short perspective on the available body of data concerning the physical properties of RESn$_3$ compounds and (LaRE)Sn$_3$ alloys.

A large amount of experimental and theoretical work has yielded a classification of dilute, superconducting RE alloys into three categories, depending on the value of a "characteristic temperature" $T_o$ relative to the pure host superconducting transition temperature $T_{co}$ (1). Measurements of the superconducting properties of (LaRE)Sn$_3$ alloys can provide valuable information concerning the magnitude of the RE "magnetic moment lifetime" $\tau_o \approx k_B T_o / h$.

In addition, both dilute alloys and materials containing a high concentration of magnetic RE ions may exhibit similar types of anomalies in normal state electric and magnetic properties in the presence of the Kondo effect or a related type of 4f shell instability such as intermediate valence (IV) (2). The origin of the similarities in the behavior of dilute alloys and more concentrated alloys and compounds is currently a subject of controversy (3). For example, many concentrated materials display anomalous values of the lattice parameter, thermal expansion coefficient or compressibility over some range of temperature, and it is not clear what analogies may, or may not exist between the elastic properties of representatives from the two concentration regimes. High pressure experiments may prove to be invaluable in exploring these questions.

255

The growing volume of experimental results concerning the physical properties of (LaRE)Sn$_3$ alloys and RESn$_3$ compounds has indicated that existing theories and phenomenologies are not adequate for a self-consistent explanation of the behavior of many of these materials (4). We will briefly review some recent experiments and relate these results to other existing data for several of the more interesting (LaRE)Sn$_3$ and RESn$_3$ representatives.

## (LaCe)Sn$_3$ AND CeSn$_3$

Measurements of the superconducting and normal state properties of (LaCe)Sn$_3$ alloys are consistent with a nonmagnetic state for the Ce impurities with a value $T_o \approx 10^2$K $\gg T_{co} = 6.4$K(4,5,6). Relatively large values of both the zero pressure magnitude of the initial depression of the superconducting transition temperature $-(dT_c/dn)_{n=0}$ (n=RE concentration) and its pressure derivative have been observed for (LaCe)Sn$_3$ alloys (4,5,7). These results are to be compared with the observation of a sizable reduction in the paramagnetic susceptibility with increasing pressure and an anomalously large compressibility measured for the CeSn$_3$ compound (8). The characteristic temperature appropriate to CeSn$_3$ is $T_o \approx 10^2$K, similar to the (LaCe)Sn$_3$ alloy case (2).

Although there appears to be some disagreement as to whether CeSn$_3$ is an IV compound (9) or a "Kondo lattice" (10), the superconducting state data clearly demonstrate the inappropriateness of a Kondo model for the nonmagnetic (LaCe)Sn$_3$ alloys (5,6).

## (LaPr)Sn$_3$ AND PrSn$_3$

Recent high pressure magnetization studies (11) have shown that the Neél temperature $T_N$ of PrSn$_3$ increases rapidly with applied pressure P, in analogy to the large increase of $-(dT_c/dn)_{n=0}$ of (LaPr)Sn$_3$ alloys with increasing pressure (7). These results suggest that the exchange interaction between the conduction and 4f shell electrons may be rapidly increasing with pressure (7,11). The relatively large magnitude and pressure dependence (13) of $-(dT_c/dn)_{n=0}$ (12), and the observation of a weak minimum in the temperature dependence of the electrical resistivity $\rho$ (14,15) have been cited as evidence of the presence of a Kondo effect in both (LaPr)Sn$_3$ alloys and the PrSn$_3$ compound (13,14,15). However, the superconducting state data for the jump in heat capacity at $T_c$, $\Delta C$, for (LaPr)Sn$_3$ alloys (16), and the thermal expansion (17) and compressibility (8) of the PrSn$_3$ compound show no evidence of a Kondo effect in either praseodymium concentration regime.

The CEF splitting of the Pr$^{3+}$ spin-orbit configuration

results in a nonmagnetic impurity ground state in the case of (LaPr)Sn$_3$ alloys, making any comparison of experimental data with Kondo models ambiguous. In addition, the appropriate values of $T_0$ for either (LaPr)Sn$_3$ alloys or the PrSn$_3$ compound are probably very low, if not zero. The recent data for $T_N$ vs P for PrSn$_3$ are, however, qualitatively consistent with a one-dimensional Kondo lattice model (18) which predicts that in the limit of very small $T_0$, $T_N$ should first _increase_ with an increasing exchange interaction parameter $J$, reach a maximum, then decrease as $J$ approaches a critical value for which $T_N = 0$.

### (LaNd)Sn$_3$ AND NdSn$_3$

The compound NdSn$_3$ also exhibits an appreciable increase of $T_N$ with applied pressure (11), and this result correlates with the observation of a relatively large increase in $-(dT_c/dn)_{n=0}$ with pressure in the case of (LaNd)Sn$_3$ alloys (4,7).

The superconducting state properties of (LaNd)Sn$_3$ alloys definitely reflect the influence of a CEF splitting of the Nd$^{3+}$ spin-orbit configuration and interimpurity magnetic interactions (4,19). The (LaNd)Sn$_3$ and NdSn$_3$ materials display many similarities to their praeseodymium counterparts (4); however, $\rho$ vs T data fail to show any anomalies attributable to a Kondo effect in either Nd concentration limit (14,20). In addition, the relatively large zero pressure value of $-(dT_c/dn)_{n=0}$ (21) and the $\Delta C$ vs $T_c$ data for superconducting (LaNd)Sn$_3$ alloys are difficult to reconcile with existing theoretical models, even if CEF effects are taken into account (4,7,19).

### (LaSm)Sn$_3$ AND SmSn$_3$

(LaSm)Sn$_3$ alloys have been shown (22,23) to constitute the first intrinsic example of unstable 4f shell behavior for a dilute samarium alloy with an associated value of $10K \leq T_0 \leq 10^2 K$. It is remarkable that a relatively _small_ pressure dependence of $-(dT_c/dn)_{n=0}$ is observed in this system (4,7). Kondo-like anomalies have been found in data for the electrical resistivity (23), $T_c$ vs n, $\Delta C$ vs $T_c$, and the normal state heat capacity and magnetic susceptibility (22) of (LaSm)Sn$_3$ alloys.

There has been relatively little experimental work reported for the properties of SmSn$_3$, although a very pronounced minimum in $\rho$ vs T has been observed above the Neél temperature $T_N \approx 12K$ (24). Additional data concerning the low temperature properties of SmSn$_3$ would be highly desirable in view of the unusual behavior of (LaSm)Sn$_3$ alloys.

## HIGH PRESSURE BEHAVIOR

"Isotropic exchange" models predict that both $-(dt_c/dn)_{n=0}$ and $T_N$ should be proportional to the quantity $N(E_F)J^2(g-1)^2 J(J+1)$, where $N(E_F)$ is the conduction electron density of states at the Fermi level, and g is Landé g-factor and J the total angular momentum appropriate to the RE ion under consideration (25,26). Assuming such a "deGennes scaling" relation, the pressure dependence of the quantity $\Gamma(P) \equiv N(E_F) J^2$ may be extracted from experimental data, and the results are summarized in Table 1. Note that modifications of the simple deGennes scaling due to CEF effects have been taken into account only in the cases of $(\underline{La}Pr)Sn_3$ (7) and $(\underline{La}Nd)Sn_3$ (19) alloys (CEF effects are negligible for the Eu- and Gd-based materials). Note also the relatively large size of $\Gamma^{-1}$ $(d\Gamma/dP)_{p=0}$ for $(\underline{La}Nd)Sn_3$ alloys and the $NdSn_3$ compound, as well as the "small" pressure dependence estimated for $(\underline{La}Sm)Sn_3$ alloys. "Simple" isotropic exchange models would demand that $\Gamma^{-1}$ $(d\Gamma/dP)_{p=0}$ be roughly constant within each column, independent of the particular RE considered.

Table I.  Pressure Dependence of the Parameter $\Gamma$ for
$(\underline{La}RE)Sn_3$ Alloys and $RESn_3$ Compounds

| RE | $\Gamma^{-1}(d\Gamma/dP)_{p=0}$ (alloy) (kbar$^{-1}$) | $\Gamma^{-1}(d\Gamma/dP)_{p=0}$ (compound) (kbar$^{-1}$) |
|---|---|---|
| Pr | $1.2 \times 10^{-2}$* | $2.5 \times 10^{-2}$ † |
| Nd | $1.9 \times 10^{-2}$** | $1.1 \times 10^{-2}$ † |
| Sm | "small"* | -- |
| Eu | $0.6 \times 10^{-2}$* | -- |
| Gd | $0.6 \times 10^{-2}$* | $0 \pm 4 \times 10^{-2}$ † |

*After Ref. 7   **After Ref. 19      †After Ref. 11

## SUMMARY

The $(\underline{La}RE)Sn_3$ and $RESn_3$ materials display remarkably varied and exciting physical phenomena. We would hasten to point out, however, that metallurgical difficulties, interimpurity magnetic interactions, CEF and novel exchange interaction phenomena may all complicate the acquisition and analysis of experimental data.

Although most of these materials form congruently from the melt, recent results (4,7,11) have demonstrated the importance of controlling the tin stoichiometry and reactivity of samples. A summary of arguments which support the assumption of nearly complete crystallographic ordering among the RE and tin sites has been given in ref. 4, but the potential presence and effects of small amounts of antisite disorder have never been fully addressed for these materials (19).

Evidence for the presence of interimpurity magnetic interactions has been found by various workers in nearly all of the (LaRE)Sn$_3$ alloy systems (4), particularly (LaNd)Sn$_3$ (19). We would suggest that conclusions drawn from data for RE concentrations $n > 1$ at. % RE in La must be regarded with caution.

Finally, existing data for $-(dT_c/dn)_{n=0}$ for (LaRE)Sn$_3$ alloys (21) are sufficient to show that popular "isotropic exchange" models are not adequate to describe the physical properties of these materials, even if CEF effects are taken into account (4,21, 7,19). Although it is reasonable to suspect the presence of 4f shell instabilities in a number of the (LaRE)Sn$_3$ and RESn$_3$ materials, such phenomena have proved (so far) to be rare in dilute systems based on RE ions other than Ce (2). Hirst (27) has recently summarized a large number of possibilities for more "generalized couplings" between the RE 4f and conduction electrons. The (LaRE)Sn$_3$ and RESn$_3$ materials may ultimately prove to provide a rich testing ground for such generalized coupling models.

## REFERENCES

1. M. B. Maple, Appl. Phys. 9, 179 (1976).
2. M. B. Maple, L. E. DeLong and B. C. Sales in: "Handbook on the Physics and Chemistry of Rare Earths," K. A. Gschneidner and L. Eyring, eds., North-Holland, Amsterdam (1978), Chapter 11.
3. R. D. Parks, ed., "Valence Instabilities and Related Narrow Band Phenomena," Plenum Press, New York (1977).
4. L. E. DeLong, Ph.D. Thesis, University of California, San Diego, 1977 (unpublished).
5. L. E. DeLong, M. B. Maple and M. Tovar, Solid State Commun. 26, 469 (1978).
6. J. Takeuchi and Y. Masuda, J. Phys. Soc. Japan 44, 402 (1977).
7. L. E. DeLong, M. B. Maple, R. W. McCallum, L. D. Woolf, R. N. Shelton and D. C. Johnston, J. Low Temp. Phys. 34, 445 (1979).
8. J. Beille, D. Bloch, J. Voiron and G. Parisot, Physica 86–88B, 231 (1977).
9. J. Lawrence and D. Murphy, Phys. Rev. Lett. 40, 961 (1978).

10. P. Lethuillier and C. Lacroix-Lyon-Caen, J. de Phys. 39, 1105 (1978).
11. L. E. DeLong, R. P. Guertin and S. Foner, Bull. Am. Phys. Soc. 24, 238 (1979). (Details to be published elsewhere.)
12. E. Bucher, K. Andres, J. P. Maita and G. W. Hull, Jr., Helv. Phys. Acta 41, 723 (1968).
13. L. E. DeLong, R. W. McCallum and M. B. Maple in: "Low Temperature Physics - LT14," M. Krusius and M. Vuorio, eds., North-Holland, Amsterdam (1975), Vol. 2, p. 541.
14. A. I. Abou-Aly, S. Bakanowski, N. F. Berk, J. E. Crow and T. Mihalisin, Phys. Rev. Lett. 35, 1387 (1975).
15. P. Lethuillier and P. Haen, Phys Rev. Lett. 35, 1391 (1975).
16. R. W. McCallum, et al., Phys. Rev. Lett. 34, 1620 (1975).
17. I. R. Harris and G. V. Raynor, J. Less Common Metals 9, 7 (1965).
18. S. Doniach, in ref. 3, p. 169.
19. L. E. DeLong, M. B. Maple, M. Tovar, and D. C. Johnston, Bull. Am. Phys. Soc. 23, 307 (1978); L. E. DeLong, M. Tovar, L. D. Woolf, M. B. Maple, D. C. Johnston and J. Keller, to be published; W. Schmid, E. Umlauf, C. D. Bredl and F. Steglich, J. de Phys. (Coll. C6) 39, C6-880 (1978); C. D. Bredl, F. Steglich, W. Schmid and E. Umlauf, J. de Phys. (Coll. C6) 39, C6-882 (1978).
20. B. Stalinski, Z. Kletowski and Z. Henkie, Phys. Stat. Sol. 19a, K165 (1973).
21. W. Schmid and E. Umlauf, Commun. Phys. 1, 67 (1976).
22. L. E. DeLong, R. W. McCallum, W. A. Fertig, M. B. Maple and J. G. Huber, Solid State Commun. 22, 241 (1977).
23. S. Bakanowski, J. E. Crow and T. Mihalisin, Solid State Commun. 22, 241 (1977).
24. B. Stalinski, Z. Kletowski and Z. Henkie, Bull. Acad. Sci. Poland (Chem. Ser.) XXIII, 827 (1975).
25. P. G. deGennes, J. Phys. Rad. 23, 510 (1962).
26. P. G. deGennes and G. Sarma, J. Appl. Phys. 34, 1380 (1963).
27. L. L. Hirst, Adv. Phys. 27, 231 (1978).

# 4f AND BAND CONTRIBUTION TO THE SAMARIUM FORM FACTOR IN SAMARIUM INTERMETALLICS

J.X. Boucherle[+],  D. Givord[†*], J. Laforest[*], P. Morin[*]

[*]Laboratoire Louis Néel, C.N.R.S., U.S.M.G., 166X,

38042 Grenoble-cédex, France

[†]Institut Laue-Langevin, 156X, 38042 Grenoble-cédex,

[+]D.R.F./D.N., C.E.N.-G., 85X, 38041 Grenoble-cédex

## INTRODUCTION

During the last years, crystalline electric field (C.E.F.) and exchange interactions in rare earth ferro or ferrimagnetic alloys were successfully deduced from magnetic measurements on single crystals for numerous compounds (1-3). Indeed, the magnetization is closely related to exchange and crystalline fields. However, in HoZn (3) for instance, where exchange and C.E.F. are known from other experiments, the calculated magnetization curves along the three main crystallographic directions lie systematically below the experimental ones. This small discrepancy can be attributed to the polarization of conduction electrons which has been neglected. The failure of this approximation is emphasized in the case of samarium alloys where the spin and the orbital contributions to the total magnetic moment are opposite. In this case the band polarization, parallel to the 4f spin, may be of the same order of magnitude as the total moment. The inaccuracy in estimating the band polarization on magnetization curves is then directly reflected in a large uncertainty on C.E.F. and exchange interactions.

Polarized neutron diffraction experiments appear to be a suitable method for overcoming this difficulty. Fourier analysis of the measured magnetic structure factors yields the magnetization density in the crystallographic cell. The localization in the direct space, which is very different for the conduction

electron polarization and for the 4f magnetization, allows the
separation between the two densities (4,5). In this case the
analysis of the 4f form factor may lead to the 4f wave function
(6,7).

The large absorption cross-section of natural Sm for thermal
neutrons has prevented such studies until now. This problem could
be solved by using $^{154}$Sm isotope which is not absorbing (8,9).
However, the availability of this isotope is very limited. We have
been able to perform experiments using natural samarium exploiting
the lower absorption cross-section at shorter wavelengths. Indeed
at $\lambda$ = 0.5 Å, wavelength close to the peak neutron flux of the
hot source at I.L.L., the absorption cross-section is reduced to
200 barns.

Before describing experimental results, we first present in
some details the peculiar features of the Sm form factor and the
method used for deducing the Sm wave function. From this wave-
function, Heisenberg exchange interactions and cubic crystal field
parameters may be deduced as shown in the first alloy described,
SmAl$_2$. Then, in SmCo$_5$, the necessity of experiments at different
temperatures is shown for the case of hexagonal symmetry and the
coupling between Sm and Co moments which has not been firmly
established yet, is determined. Finally, in SmZn, the role of the
conduction electrons polarized by the 4f ones is emphasized, and
shown to be important on magnetization measurement results. In all
cases, the comparisons with previous experiments (10-14) show the
advantage of the polarized neutron technique.

## Sm 4f MOMENT AND FORM FACTOR

The ground multiplet of the Sm$^{3+}$ ion is characterized by the
quantum numbers L = 5, S = 5/2, J = L - S = 5/2. Under an applied
or exchange field, the degeneracy of the multiplet is lifted. The
magnetic moment of the ground state $L_z + 2S_z = g_J(L - S)$ is
0.714 $\mu_B$. It results from the difference between an orbital contribution and a slightly smaller spin contribution. Since orbital
magnetization density is more localized than that due to spin,
the resultant magnetic density consists of regions of space with
a positive magnetization density (near the atomic sites) and
regions with a negative one (away from the sites). The Sm form
factor associated with such a magnetic density is maximum at a
non zero value of $\sin\theta/\lambda$. It is drawn in Fig. 1 within the dipole
approximation :

$$\mu f = 2 <j_o>S_z + (<j_o> + <j_2>)L_z \qquad (1)$$

where $<j_o>$ and $<j_2>$ are the Sm$^{3+}$ radial integrals (15) shown in
the inset.

The cubic C.E.F. reduces the magnetization defined by exchange or applied magnetic field. For the ground multiplet alone $S_z$ and $L_z$, proportional to $J_z$, are reduced by a similar ratio and also the form factor (Fig. 1,b).

The excited multiplets in Sm lie only 1400 K and 3200 K above the ground multiplet. It is well known, from the Van Vleck papers, that exchange mixes states between different multiplets. $J_z$ being a good quantum number for the free ion states ($<JM|J_z|J'M'> = 0$ for $J \neq J'$) the mixing leads to changes nearly opposed in $L_z$ and $S_z$ ($J_z = L_z + S_z$). The spin exchange interactions induce a net increase $\Delta S_z$ in $|S_z|$. In the dipole approximation the associated change in the form factor (Fig. 1,c) is :

$$\Delta \mu f = - (<j_o> - <j_2>)\Delta S_z \qquad (2)$$

This change, large mainly at small scattering angles increases the peak shape of the form factor and is very different from the one induced by C.E.F. inside the ground multiplet. Finally considering excited multiplets leads to C.E.F. 6th order terms absent for $J = 5/2$. The C.E.F. mixing between states from different multiplets is described similarly as the exchange one. It appears to be weaker.

## EXPERIMENTAL AND METHOD OF ANALYSIS OF RESULTS

The experiments have been performed on the polarized neutron spectrometer D5 at I.L.L. The classical flipping ratio R was measured up to $\sin\theta/\lambda \simeq 1.5$ Å$^{-1}$. To a very good approximation R may be written as :

$$R = \frac{1 + 2\gamma \sin^2\theta + \gamma^2 \sin^2\theta}{1 - 2\gamma \sin^2\theta + \gamma^2 \sin^2\theta} \qquad (3)$$

where $\theta$ is the angle between the magnetic field and the scattering vector, and $\gamma$ is the ratio $F_M/F_N$ between the magnetic and nuclear structure factors.

In order to perform the necessary extinction corrections two neutron wavelengths were used, $\lambda = 0.50$ Å and $\lambda = 0.42$ Å for which the Sm absorption cross-sections are respectively 200 barns and 130 barns. Due to the proximity of a resonant absorption peak, the Sm scattering length is wavelength dependent. We measured it (16) to be 0.532(5) barns and 0.480(5) barns at $\lambda = 0.50$ Å and 0.42 Å respectively.

The magnetic structure factors $F_M$ experimentally determined lead to the Sm form factor $\mu f(\vec{h})$ which, in the tensor operator formalism (17), may be written as :

$$\mu f(\vec{h}) \propto \sum_{K'',Q''} Y_{Q''}^{K''}(\hat{h}) \sum_{J',Q'} \left[ \sum_{\substack{J'M' \\ J\ M}} a_{J'M'}^{*}\ a_{JM} \left\{ A(K'',K') + B(K'',K') \right\} \right.$$

$$\left. \langle K'Q'J'M' | JM \rangle \right] \langle K''Q''K'Q' | 10 \rangle \qquad\qquad (4)$$

where $Y_{Q''}^{K''}$ are the spherical harmonics, the $A(K'',K')$ and $B(K'',K')$ terms represent respectively the orbital and spin part of the magnetic interaction with neutrons ; $\langle K'Q'J'M' | JM \rangle$ and $\langle K''Q''K'Q' | 10 \rangle$ are Clebsch-Gordan coefficients. The radial integrals $\langle j_i \rangle$ (i = 0, 2, 4, 6) (15) are included in the A's and B's. They were taken into account in the exact calculation of the form factor (18), the dipole approximation not being sufficient. The $a_{JM}$'s are the coefficients of the wave function $\Psi = \sum_J \sum_M a_{JM} | JM \rangle$ ; they are only parameters involved in the form factor fit.

The measurements were performed at low temperatures in order to determine the wave function of the ground state since only it is then populated.

In a following step, assuming Heisenberg and C.E.F. interactions the relevant parameters were deduced for each compound. Since the Sm wave function involves admixed terms with $J \neq J'$ it has been possible to determine both exchange and C.E.F. separately. It is worth noticing that for other rare earths where the ground multiplet is involved alone, only the ratio between exchange and C.E.F. may be attained.

DETERMINATION OF THE GROUND STATE WAVE-FUNCTION IN SmAl$_2$ (19,20)

SmAl$_2$ crystallizes in the cubic Laves phase structure of MgCu$_2$ type. According to magnetic measurements, it orders ferromagnetically at 120 K, the easy magnetization direction being [111]. Along this direction, the measured magnetization at 4.2 K in 16.5 kOe is 0.23(1) $\mu_B$/SmAl$_2$. The polarized neutron study was performed at 4.2 K, under a vertical magnetic field of 16.5 kOe, parallel to the |111| direction of the single crystalline sample. The flipping ratio R of Bragg reflections in the zero layer were measured up to $\sin\theta/\lambda = 1.64$ Å$^{-1}$.

The deduced magnetic amplitudes per Sm atom appear in Fig. 2 as a function of $\sin\theta/\lambda$. Comparing to the Sm |5/2, 5/2> state (dotted line), the form factor at any scattering angle is reduced. This shows mixing between states of the 5/2 multiplet by C.E.F. effects. Mixing with other multiplets, due mainly to exchange effects, is deduced from the large positive slope at low scatte-

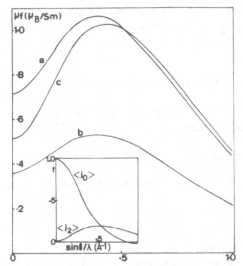

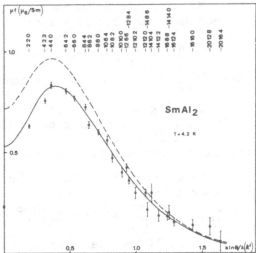

Fig. 1 : Sm$^{3+}$ form factor in the
dipole approximation ; a) free
ion ; b) in a cubic C.E.F. on
the ground multiplet ; c) in $H_{ex}$
mixing states between multiplets.

Fig. 2 : Form factor of Sm$^{3+}$ in
SmAl$_2$. The dotted line corres-
ponds to the $|5/2, 5/2>$ state,
the solid line one to the
fitted wave function.

ring angles of the Sm form factor. Choosing the easy direction
[111] as z-axis, the ground wave function may be written as :

$$\Psi = \alpha|5/2, 5/2> + \beta|5/2, -1/2> + \gamma|7/2, 5/2> + \delta|7/2, -1/2>+..$$
$$(5)$$

Fitting the form factor leads to the 4 only significant coefficients

$$\alpha = 0.952(10) \; ; \; \beta = -0.299(20) \; ; \; \gamma = -0.038(30) \; ;$$
$$\delta = 0.044(30).$$

the calculated anisotropy of the form factor is weak enough to be
neglected. The experimental values are in agreement with the calculated solid line, except for the first reflections
($\sin\theta/\lambda < 0.44$ Å-1). Especially, the calculated 4f magnetic moment
0.53(5) $\mu_B$ is larger than the measured bulk magnetization of
0.23 $\mu_B$. Such a discrepancy is usually explained in terms of
conduction electron polarization which reaches here
$0.53 - 0.23 = 0.30(5)$ $\mu_B$.

In the formalism already mentioned, the total Hamiltonian
defining the Sm ground state may be written as :

$$\mathcal{H} = \lambda\vec{L}\vec{S} + \mu_B \, H_{app}(L_z + 2S_z) + 2 \, \mu_B \, H_{ex}S_z + \mathcal{H}_{cc} \qquad (6)$$

The first term represents the spin-orbit coupling where $\lambda/k_B$ was
taken as 410 K. $H_{app}$ is the applied magnetic field. The exchange

field $H_{ex}$ represents the molecular field due to spin interactions between Sm atoms. $H_{ex}$ is related to $S_z$ by the relation :

$$2\mu_B \, H_{ex} = -\mathcal{J}_{ff} \, <S_z> \tag{7}$$

where $\mathcal{J}_{ff}$ is the exchange constant between spins.

According to de Wijn et al (11), the value of the exchange field which yields divergence of the paramagnetic susceptibility at $T_c$ = 120 K is $\mu_B \, H_{ex}/k_B$ = 50 K. This value has been kept in the calculation although $S_z$ here estimated is slightly different.

The crystal field Hamiltonian $\mathcal{H}_{cc}$ at a rare earth site may be expanded as :

$$\mathcal{H}_{cc} \quad \sum_{k,q} A_k^q \, <r^k> \, Y_k^q(\theta,\varphi) \tag{8}$$

where $Y_k^q(\theta,\varphi)$ are the standard spherical harmonics, $<r^k>$ is the mean value of the $k^{th}$ power of the 4f electron radius and the $A_k^q$ are parameters related to the strength of the crystal field.

Since we are concerned with mixing between different J states, the method of the Stevens operator equivalents is not appropriate. It is easier to use the more direct approach of the tensor-operator techniques (21). For the cubic symmetry of the Sm site, the factors $A_4^0$ and $A_4^3$ on the one hand, $A_6^0$, $A_6^3$ and $A_6^6$ on the other hand are proportional to one another. The crystal field may be expressed as a function of two parameters $A_4$ and $A_6$ which describe the strength of the C.E.F. irrespective of the direction of the z-axis.

The total Hamiltonian has been diagonalized taking into account both the J = 7/2 and 9/2 multiplets together with J = 5/2. Within this model, the following C.E.F. parameters $A_4<r^4>/k_B$ = 110(30) K ; $A_6<r^6>/k_B$ = 0(30) K lead to a satisfactory description of the Sm wave-function, the Sm magnetic moment being calculated to be 0.45 $\mu_B$ in comparison with the fitted value of 0.53 $\mu_B$ (Fig. 2). The C.E.F. parameter values are lower than those proposed by de Wijn et al (10,11). This discrepancy originates from the difference in the values taken for the conduction electron polarization, i) estimated (11) to be 0.1 $\mu_B$ ; ii) here measured as 0.30(5) $\mu_B$. That proves the necessity of the polarized neutron experiment in this case.

DIFFERENT TEMPERATURE DEPENDENCES OF THE SPIN AND<br>ORBITAL CONTRIBUTIONS TO THE FORM FACTOR IN SmCo$_5$ (16)

The hexagonal compound SmCo$_5$ orders magnetically at 1040 K. The easy magnetization direction is the $\vec{c}$-axis. The crystallographic structure was refined at 300 K and shown to be similar to that of YCo$_5$ (22). The polarized neutron study was performed at

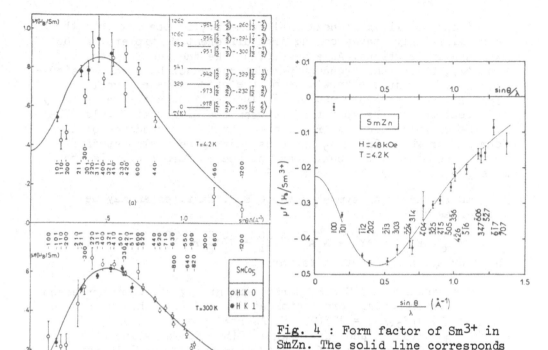

Fig. 4 : Form factor of Sm³⁺ in SmZn. The solid line corresponds to the fitted wave function.

Fig. 3 : Experimental Sm magnetic amplitude and calculated form factor at 4.2 and 300 K. Inset : level scheme under $H_{ex}$ and C.E.F.

4.2 K and 300 K, the magnetic field being applied along the $\vec{c}$-axis. In SmCo₅ the magnetic structure factors are essentially the sum of a cobalt and a samarium contribution. The individual form factors on both cobalt crystallographic sites (2c and 3g) were assumed to be the same as in YCo₅ (22). In the CaCu₅ structure, there are 8 types of structure factors in which the cobalt contribution differs. After subtracting this contribution to each magnetic structure factor, the Sm magnetic amplitudes lie on a smooth curve as expected for a uniaxial symmetry (Fig. 3). This procedure allowed us to determine the cobalt moments of both sites. The same values have been obtained at 4.2 K and 300 K for the cobalt moments $\mu_{CoI}$ (2c site) = 1.86 $\mu_B$ and $\mu_{CoII}$ (3g site) = 1.75 $\mu_B$. These values are close to those obtained in YCo₅, $\mu_{CoI}$ = 1.77 $\mu_B$ and $\mu_{CoII}$ = 1.72 $\mu_B$. Analyzing the form factor at 4.2 K led to the Sm ground state wave-function :

$$\Psi_G = .978(5)|5/2, 5/2> -.205(20)|7/2, 5/2> +.038(20)|9/2, 5/2> \tag{9}$$

where the c-axis of the hexagonal structure was taken as z-axis. The calculated form factor curve at 4.2 K is compared to the experimental points in Fig. 3.

A crystal field of hexagonal symmetry, if one neglects the $A_6^6$ terms mixes only states coming from different multiplets, as do exchange interactions. As a result, unlike the case of cubic symmetry, C.E.F. and exchange parameters cannot be obtained simultaneously from just a knowledge of the ground state wave function. However, exchange and C.E.F. interactions act in a very ry different way on the splitting between the different levels. They can therefore be deduced from the temperature dependence of the form factor which in SmCo$_5$ has consequently been also studied at room temperature : the peak shape of the Sm form factor is then more pronounced as shown in Fig. 3.

In the uniaxial symmetry the C.E.F. Hamiltonian may be written as :

$$H_{cc} = N_2^0 A_2^0 <r^2> Y_2^0 + N_4^0 A_4^0 <r^4> Y_4^0 + N_6^0 A_6^0 <r^6> Y_6^0 \qquad (10)$$

where $N_k^c$ are normalization factors (21).

The exchange and C.E.F. parameters giving the best agreement with the experimental form factor at 300 and 4.2 K has been determined to be $\mu_B H_{ex}/k_B = 175(25)$ K, $A_2^0 <r^2>/k_B = -200(50)$ K, $A_4^0 <r^4>/k_B = 0(50)$ K and $A_6^0 <r^6>/k_B = 50(50)$ K. The value of $H_{ex}$ is the same at 300 K and 4.2 K. Indeed, the molecular field acting on Sm ions is created by very high Co 3d interactions. The fitted parameters are in agreement with previous determinations from anisotropy measurements (12,13). At 4.2 K the Sm moment is 0.38 $\mu_B$, at 300 K it has decreased to 0.04 $\mu_B$. For the excited states, the inset of Fig. 3 shows that the mixing with J = 7/2, 9/2 states is more important than for the ground state. This explains why, at 300 K when they are populated, the unusual peak shape of the Sm form factor is more pronounced.

In conclusion, at 4.2 K the Sm moment has a dominant orbital character. At 300 K, the spin character of the moments has increased but the resultant moment is still orbital. A "crossover" temperature at which the spin moment equals the orbital one has been calculated to occur at 350 K. In 4f-3d alloys an antiparallel coupling between spins is always observed (23), as a consequence, below 350 K SmCo$_5$ is ferromagnetic and above 350 K it is ferrimagnetic.

BAND POLARIZATION EXCEEDING THE 4f MOMENT IN SmZn (24)

According to the magnetic measurements, SmZn orders ferromagnetically at 127 K, the direction [111] being of easy magnetization (14). The spontaneous magnetization is extremely weak, 0.055 $\mu_B$. This may indicate strong C.E.F. effects, and/or large band contribution, or even raise some doubt on the exact nature of the magnetic structure.

The experiments were performed at 4.2 K. A non polarized neutron study did not reveal any superstructure reflections corresponding to simple antiferromagnetic structure. Polarized neutron experiments were performed under a 48 kOe magnetic field applied along the [111] direction in order to eliminate any technical magnetization process (14). A flipping ratio different from 1 was observed in all the nuclear lines proving the existence of a ferromagnetic contribution. The measured magnetic amplitudes appear in Fig. 4. They are always negative, opposite in sign to the magnetic contribution determined in macroscopic measurements. Forgetting the negative sign, the results show the characteristic peak shape of the Sm form factor.

The data were analysed in terms of 4f Sm form factor. After changing sign, the curve shown in fig. 4 corresponds to a Sm ground wave function :

$$\Psi = .945(10)|5/2, 3/2> + .325(20)|5/2, -3/2> \tag{11}$$
$$- .040(30)|7/2, 3/2> - .040(30)|7/2, -3/2>$$

the associated magnetic moment reaches 0.16 $\mu_B$. This calculated form factor exhibits a dominant orbital character at all scattering angles. Comparing with experimental data, the agreement is good except for the magnetic amplitudes of the first reflections, especially for the origin of value which corresponds to the magnetic measurements of 0.055 $\mu_B$/SmZn. Such a disagreement is reminiscent of the one observed in SmAl$_2$. In SmZn, the band polarization is deduced to be 0.055 + 0.22 ≐ 0.28 $\mu_B$ (Fig. 4) : it is opposed to the 4f total moment, since Sm is a J = L-S ion, and larger than it in value. In order for the resultant magnetization of the alloy to be parallel to an applied magnetic field, the band polarization is parallel to the field but the 4f moment opposed to it. This is the case in the polarized neutron experiment, performed under H = 48 kOe. But for sinθ/λ values different from zero, the localized 4f moment dominates the magnetic scattering, the unusual negative sign of the measured magnetic amplitudes results from this unusual behaviour of SmZn.

The large band polarization measured in SmZn, as in other rare earth intermetallics, cannot be explained assuming a pure $\vec{S}s$ 4f-shell band coupling. Indeed the spin band contribution may be expressed as a function of the state density at the Fermi level and of the spin coupling constant between the 4f shell and the band :

$$M_s = \Gamma_{Ss} \, n(E_F)S_z \tag{12}$$

The $\Gamma_{Ss} \, n(E_F)$ exchange constant is estimated from the total magnetization measured in GdZn at liquid helium temperature (14) (7.3 $\mu_B$ instead of 7.0 $\mu_B$). In SmZn where $S_z = 1.2$, we deduce then a spin band contribution of  0.09 $\mu_B$, much smaller than the one observed.

Two explanations have been proposed for such a discrepancy (25) which occurs also in RAl$_2$ alloys for instance : i)  the spin exchange constant may be larger for light rare earth compounds ; ii)  an orbital contribution to the band polarization may be present, originating from the strong d character of the conduction electrons. Measurements on a series of isomorphous compounds would be needed in order to clarify this phenomenon.

CONCLUSION

In conclusion, polarized neutron diffraction gives a precise way of determining rare earth ground state in intermetallics. The influence of excited multiplets in the case of Sm make the analysis more difficult but more fruitful. The determination of C.E.F. and exchange interactions on the one hand, and the separation between localized 4f magnetic density and diffuse band polarization density on the other hand were shown to be important in order to pin-point the mechanism of exchange interactions in rare earth magnetism. Peculiar properties were observed which are related to the smallness of the total Sm 4f moment and to the influence of excited multiplets : i)"cross-over" of the Sm magnetic moment at a temperature of 350 K in SmCo$_5$, ii)  antiparallel coupling of the band exceeding the Sm 4f moment in SmZn.

ACKNOWLEDGEMENTS

We gratefully acknowledge J. Brown, R. Lemaire and J. Schweizer for very helpful discussions.

REFERENCES

1. B. Barbara, J.X. Boucherle, M.F. Rossignol, Phys. Stat. Sol., 25, 165 (1974).
2. D. Gignoux, F. Givord, R. Lemaire, Phys. Rev., B 12, 3878(1975).
3. P. Morin, D. Schmitt, J. Phys. F, 8, 951 (1978).
6. G.H. Lander, T.O. Brun, O. Vogt, Phys. Rev., B 7, 1988 (1973).
7. B. Barbara, J.X. Boucherle, J.P. Desclaux, M.F. Rossignol, J. Schweizer, Crystal field effects in metals and alloys, 168 (1977), Ed. by A. Furrer, Plenum Press, New York.
4. R.M. Moon, W.C. Koehler, Phys. Rev. Lett., 27, 407 (1971).
5. J.X. Boucherle, J. Schweizer, Physica, 86-88B, 174 (1977).
8. W.C. Koehler, R.M. Moon, Phys. Rev. Lett., 29, 1468 (1972).
9. R.M. Moon, W.C. Koehler, D.B. McWhan, F. Holtzberg, J. Appl. Phys., 49, 2107 (1978).
10. H.W. de Wijn, A.M. Van Diepen, K.H.J. Buschow, Phys. Rev., B 7, 524 (1973).

11. K.H.J. Buschow, A.M. Van Diepen, H.W. de Wijn, Phys. Rev., B 8, 5134 (1973).
12. K.H.J. Buschow, A.M. Van Diepen, H.W. de Wijn, Sol. Stat. Commun., 15, 903 (1974).
13. S.G. Sankar  V.U.S. Rao, E. Segal, W.E. Wallace, W.G.D. Frederick, H.J. Garrett, Phys. Rev., B 11, 435 (1975).
14. P. Morin, Thèse d'Etat (A.O. CNRS 9323), University of Grenoble (1975).
15. A.J. Freeman, J.P. Desclaux, J. Mag. and Mag. Mat., 12, 11 (1979).
16. D. Givord, J. Laforest, J. Schweizer, F. Tasset, J. Appl. Phys. (to appear).
17. S.W. Lovesey, R.E. Rimmer, Rep. Prog. Phys., 32, 333 (1969).
18. E. Balcar, S.W. Lovesey, R.E. Rimmer, SCAMAG Program (ILL) (1975).
19. J.X. Boucherle, D. Givord, J. Laforest, J. Schweizer, F. Tasset, J. de Physique (to appear).
20. J.X. Boucherle, D. Givord, J. Schweizer, to be published.
21. M.J. Weber, R.W. Bierig, Phys. Rev., A 134, 1492 (1964).
22. F. Tasset, J. Schweizer, to be published.
23. B. Barbara, D. Gignoux, D. Givord, F. Givord, R. Lemaire, Int. J. Magnetism, 4, 77 (1973).
24. D. Givord, P. Morin, D. Schmitt, to be published.
25. Y. Berthier, R.A.B. Devine, E. Belorizky, Phys. Rev., B 17, 4137 (1978).

# ON VARIOUS ANOMALIES IN THE TEMPERATURE DERIVATIVE OF TRANSPORT

# PROPERTIES IN RE METALLIC COMPOUNDS

M. Ausloos, Institut de Physique, Univ. de Liège, 4000
Sart Tilman/Liège 1, Belgium; and
J. B. Sousa, M. M. Amado, R. F. Pinto, J. M. Moreira,
M. E. Bragal and M. F. Pinheiro, Laboratorio de Fisica,
Univ. de Porto, Porto, Portugal

## INTRODUCTION

Some renewed interest in the anomalies of transport coefficients of metallic magnets undergoing a phase transition has recently taken place. This has led to new experimental results. In
the following, we examine the need for further investigations leading to a more complete experimental proof of the so-called "universality hypothesis" (section I). We also consider the "orientational"
magnetic transition in TbZn single crystals (1), in the light of the
understanding of a similar effect in high purity gadolinium (2)
(section II).

Finally we examine the behavior of the electrical resistivity
as a function of several anisotropy parameters (section III).


## STATUS OF UNIVERSALITY HYPOTHESIS

For second order ferro-paramagnetic phase transitions, it is
now accepted that in the vicinity of the Curie temperature $T_c$, the
temperature derivative of the electrical resistivity $(d\rho/dT)$ is proportional to the specific heat (C),

$$\rho' \equiv d\rho/dT \simeq C \qquad [1]$$

i.e., the magnetic contributions to both properties diverge with
the same critical exponents.

This is certainly a restricted aspect of "universitality," involving a specific transport coefficient $(d\rho/dT)$ and a purely

273

thermodynamic quantity (C). In particular, it is necessary to check universality relations between the transport coefficients themselves, e.g., between electrical resistivity and thermoelectric power (S). For ferromagnetic systems, recent theoretical calculations show that both for elastic and inelastic scattering contributions (3)

$$d\rho/dT \simeq dS/dT. \tag{2}$$

However, no experimental check has been done on the validity of such an expression in ferromagnetic materials.

For antiferromagnetic systems the theory does not seem definitive, but recent experimental work (4) has likely indicated the validity of relation [2]. Unfortunately, the system studied does not display a second order phase transition. It is therefore suggested that relation [2] be re-investigated using rare earth systems with well localized magnetic moments, and presenting a simple ferro- or antiferro-magnetic transition.

Among other unsolved problems, we mention the need to investigate the critical behavior of the thermal conductivity (K) and to extend relations of type [2] to the temperature derivative dk/dT.

## SPIN REORIENTATION TRANSITION (SRT)

One such example is found in Gd, which has the advantage of displaying a 2nd-order SRT ($T_R \simeq$ 220K). In particular, it has been theoretically shown that in Gd the "orientational" fluctuations should play a virtually negligible role practically restricted to the immediate vicinity of $T_R$ ($\sim 10^{-6}$ in terms of reduced temperature). Therefore, Landau mean-field theory is expected to be applicable up to the vicinity of the critical point. Such features have been confirmed by very recent measurements of $\rho$ on single crystals of high purity gadolinium (Sousa, et al., unpublished). Indeed, below $T_R$, $\rho'$ can be described by the classical Landau critical exponent (1/2); the experiment also suggests the virtual absence of fluctuations above $T_R$: $\rho'$ shows a very steep decrease at $T_R^*$, followed by a sharp kink preceeding a linear (mean-field-like) temperature dependence at higher temperatures. From the experimental data, an upper limit $(T-T_C)T_C \sim 5 \times 10^{-4}$ can be put on the temperature $T_C$ range where fluctuations might play a role.

Whence, it is of interest to re-examine the SRT in TbZn ($T_R \simeq$ 62K), where the magnetization rotates from a binary (T < $T_R$) to a quaternary axis and the lattice symmetry changes from cubic (T > $T_R$) to tetragonal. Notice that the absence of important critical fluctuations at $T_R$ is expected to be a fairly general feature, the more so in TbZn where the SRT is known to be of first order (5). Therefore, one should expect in TbZn a very sharp (and narrow) peak in $\rho'$

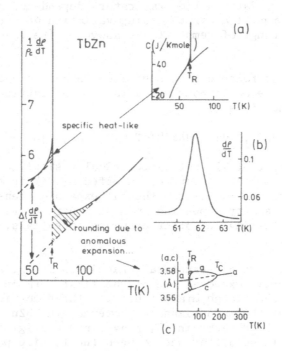

Figure 1.  Behavior of $\rho'/\rho_C$ along a quaternary axis in TbZn near the
spin reorientation temperature $T_R$.  Different contribu-
tions correspond to insets:  (a) specific heat jump with
(b) rounding of the singularity in $(d\rho_o/dT)$ near $T_R$ and
much above $T_R$ due to (c) anomalous expansion, metastable
states, ...  Notice the large discontinuity in slope
$\Delta(d\rho/dT)$ at $T_R$ due to the structural transition (see text
and inset [c]).  $t = T/T_C-1$.

at $T_R$ (ideally infinitely high, due to the discontinuity $\Delta\rho$ in $\rho$),
without any appreciable rounding (or precursor) effects on both immediate sides of $T_R$.

     Indeed, the sharp peak at $T_R$ (Fig. 1) corresponds by eqn. [1]
to that observed in the specific heat (inset a); its small but finite
width indicating that experimental data were taken with a finite rate
of change of T ($\sim$mK·min$^{-1}$).  An expanded-scale plot of $\rho'$ in the very
vicinity of $T_R$ ($\Delta T \sim 3K$; inset b) confirms that fluctuations do not
extend far on either side of the critical point.  Therefore, the

unexpected and broad dip of $\rho'$ observed over a much wider range of temperatures <u>above</u> $T_R$ (50K; see Fig. 1) requires another interpretation. In fact, data on the temperature dependence of the lattice parameters (5) reveal a corresponding variation of such parameters over a similar range of temperatures above $T_R$ (tetragonal phase) (inset <u>c</u>).

A simple approach can be envisaged to obtain a description of $\rho'$ in terms of the continuous variation of the lattice parameter deformations. One finds (6)

$$d\rho_i/dT = d\rho(^\circ_i)/dT + d\Sigma ijk/dT \ \underline{u}jK + \Sigma ijk\alpha jk \qquad [3]$$

where $i,j$ = x,y,z, $\rho(^\circ_i)$ is the electrical resistivity in the absence of thermal deformation and expansion effects, $\underline{u}jk$ is the intrinsic deformation tensor, and $\alpha jk$ the thermal expansion tensor. $\Sigma ijk$ is a complicated quantity depending on system parameters like the Fermi energy, the scattering and other exchange potentials, the phonon and magnon spectra.

In the absence of structural transition at $T_R$, a finite (but small) $\Delta\rho'$ is still expected due to the first term in [3], and is reminiscent of the change in angle between the electrical current and the direction of the magnetization; however, in TbZn, it seems that the second and third terms in [3] play a non-negligible role. Hence, the relatively large difference between the lattice parameters in the tetragonal phase may give a significant contribution to the second r.h.s. term in [3]. Such a contribution, usually neglected at 2nd order transitions, is a necessary ingredient for the interpretation of 1st order phase transitions.

## ANISOTROPY EFFECTS

In uniaxial ferromagnets, band structure and spin fluctuation asymmetry are expected to influence the overall shape of $d\rho/dT$ near a critical temperature. It has been predicted that the dip in $d\rho/dT$ above $T_C$ depends on whether $k_F^{(i)}a_i \gtrless 1$ where $k_F^{(i)}$ is the radius of the Fermi surface and $a_i$ the i axis lattice parameter. Assuming that the long range part of the spin correlation function $\Gamma$ dominates the electron critical scattering, it is easy to generalize the Ornstein-Zernicke form ($\Gamma oz$) to take into account the fluctuation uniaxial symmetry such that $\Gamma$ becomes a tensor of components

$$\Gamma_{zz} = \Gamma_o(1 + q_z^2 \xi_{zz,z}^2 + q^2 \xi_{zz,\perp}^2)^{-1},$$

$$\Gamma_{\perp\perp} = \Gamma_o(1 + q_z^2 \xi_{\perp\perp,z}^2 + q^2 \xi_{\perp\perp,\perp}^2)^{-1},$$

with $\xi_{ii,j} = \xi_o \eta_j (\delta_{jz} + \lambda\delta_{j\perp})$ \qquad [4]

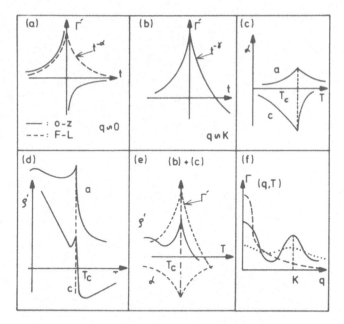

Figure 2.  Behavior of $\Gamma \equiv d /dT$ near $T_c$ as predicted from contributions as (a) Ornstein-Zernicke (O-Z) correlation function and Fisher-Langer (F-L) scaling law; (b) scaling law behavior for short range correlation function; (c) typical anisotropic anomalous expansion coefficient; (d) typical anisotropic $\rho'$ behavior (Tb-Gd); (e) combination of $\Gamma'$ and $\alpha$ contributions needed to describe $\rho'$; (f) sketch of generalized anisotropic $\Gamma:-,\cdots$; $\Gamma_{oz}:--\cdot$.

where $\lambda$ describes the fluctuation anisotropy and $\eta_j$ is a dimensionless function of $T-T_C$. Then, it can be shown that another parameter determines the behavior of $\rho'$ above $T_C$, i.e. $\xi_o a_i^{-1}$ even in the totally isotropic case. Furthermore, it can be argued that the short range part of the spin correlation function plays a role as important as $\Gamma_{oz}$ near $T_C$, due e.g. to mean-path-free effect considerations. Several generalizations of $\Gamma_{oz}$ including short range effects, however, conserve the overall Lorentzian shape of $\Gamma$.

Considerations based on the enhancing effect of Umklapp processes in uniaxial metals due to the anisotropy of the Fermi surface lead us to propose a new generalization of the correlation function modifying appreciably its shape, e.g.

$$\Gamma(q,T) = A\Gamma_{oz}(q \sim o,T) + \Sigma_i B_i \Gamma_{oz}(q + \kappa_i,T) \qquad [5]$$

where $\kappa_i$ is a set of characteristic wave vectors for which Umklapp

# MARTENSITIC TRANSFORMATION AND MAGNETIC STRUCTURES BETWEEN THE

# Sm-TYPE AND THE HCP STRUCTURES IN Tb-LIGHT RARE EARTH ALLOYS

Norio Achiwa and Shinji Kawano

Research Reactor Institute Kyoto University

Kumatori-cho, Sennan-gun, Osaka, 590-04 Japan

## INTRODUCTION

An interesting sequence of close packed structures such as hcp→Sm-type→double hcp→fcc, occurs in the heavy and the light intra--rare earth binary alloys with increasing the composition of lighter elements(1,2). Some alloys located at the boundary of two crystal phases in the intra-rare earth phase diagram, exhibit a temperature induced martensitic transformation between the two close packed structures (3,4). In the previous paper (4), we have reported the preliminary neutron diffraction studies of crystal and magnetic structures of $Tb_{0.8}-La_{0.2}$, $Tb_{0.8}-Pr_{0.2}$ and $Tb_{0.75}-Nd_{0.25}$ using single crystals. Below 30 K, such a martensitic transformation from the Sm-type to the hcp structure was observed for the latter two crystals. The Sm-type crystals order in a sinusoidal magnetic structure (o++o--o++..) along the c-axis and zeroes correspond to the cubic site, while the hcp crystals have ferromagnetic structure. In the present paper, neutron diffraction studies on $Tb_{0.803}-Pr_{0.197}$ and $Tb_{0.798}-Pr_{0.202}$ single crystals were carried out extensively in both phases of Sm-type and hcp structures at the various stages of cooling and heating cycles that were accompanied by the martensitic transformations between the two phases. The martensitic crystals of hcp and Sm-type structures have stacking faults that increase with increasing number of martensitic cycles. The diffusive patterns from the faulted crystals were interpreted by using Kakinoki and Komura theory (5) for one dimensionally disordered crystals. It is important to study the modulations of magnetic structures by the stacking faults which randomly mix the two close packed structures, because the RKKY exchange interaction (6) that rules the magnetism in these alloys, is very sensitive to the modulations of crystal structures.

## EXPERIMENTAL

Tb-light rare earth alloys were prepared in an arc furnace under purified argon. The weight loss after melting several times was not more than 0.1%. The purity of the rare earth metals is 99.9% and they contain less than 500 ppm of other rare earth metals. Single crystals of rare earth alloys were prepared by a strain anneal method. Single crystals of 4mm and 3mm side length cubes for $Tb_{0.803}-Pr_{0.197}$ and $Tb_{0.798}-Pr_{0.202}$ respectively, were used for the neutron diffraction studies. The neutron diffraction data were collected by using the double axis neutron diffractomator at Kyoto University Research Reactor over the temperature range 4.2 K~300 K. The neutron wavelength used in the experiments is 1.007 A and $\lambda/2$ contamination of the monochromatic neutrons is not more than 0.2 %. Measurements were carried out in the a*-c* reciprocal planes of Sm-type and hcp structures by step wise linear scanning parallel to the c*-axis.

## EXPERIMENTAL RESULTS

Figure 1 shows the first cycle of Sm-type→hcp martensitic transformation in $Tb_{0.803}-Pr_{0.197}$ crystal. $T_N$(130.5 K), $T_C$(163.5 K) indicate the Néel and the Curie temperatures in the Sm-type and hcp phases and $T_M$(30 K) the martensite temperature from the Sm-type to the hcp phase. The reverse transformation starts just above the Curie tempetature. Across the $T_M$, the directions of a* and c* principal axes are unchanged.

Figure 2 shows (10L) scanning patterns in the Sm-type $Tb_{0.798}-$ $Pr_{0.202}$ at 4.2 K and 300 K. This specimen also exhibits the martensitic transformation at low temperature at least on its first cooling cycle. But, after keeping the crystal at room temperature for a year, it has no martensitic transformation down to 4.2 K. The diffraction patterns at 300 K correspond to the nuclear reflections from a submicroscopically twinned Sm-type crystal half of which is combined with the other by a rotation of 180° about the c-axis (7). The additional diffraction patterns at 4.2 K come from the sinusoidal magnetic structure with the average magnetic moment μ parallel to the c-plane as expressed in the equation,

$$M_n = \mu \sin(\tau r_n + \gamma_n) \qquad\qquad (1)$$

where $\tau=3c*/2$, $\gamma_{3N}=0$, and $\gamma_{3N\pm1}=\pm\pi/6$. $n=3N$ and $n=3N\pm1$ correspond to the cubic and the hexagonal sites, respectively. In the calculation of magnetic structure factors, an equal population of three magnetic domains in one of the twin component of Sm-type crystal owing to the three equivalent spin direction in the c-plane were assumed and the magnetic form factors calculated by Blume, Freeman and Watson (8) were used. The calculated and the experimental structure factors were compared and the obtained reliability factors

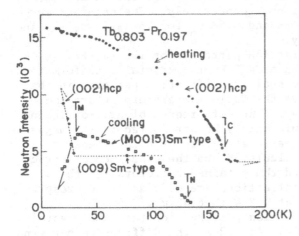

Fig. 1. The first cycle
of Sm-type→hcp martensitic
transformation in $Tb_{0.803}$-
$Pr_{0.197}$. $T_N$(130.5 K) and
$T_C$(163.5 K) the magnetic
ordering temperatures in
the Sm-type and the hcp
phases, respectively.
$T_M$(30 K) the martensite
temperature.

Fig. 2. (10L) scanning
patterns of Sm-type
$Tb_{0.798}$-$Pr_{0.202}$ at 4.2 K
and room temperature.

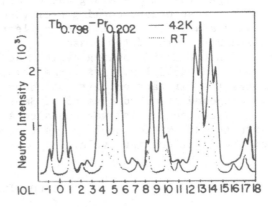

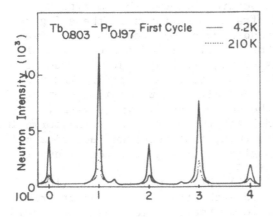

Fig. 3. (10L) scanning
patterns of hcp $Tb_{0.803}$--
$Pr_{0.197}$ on its first
martensite cycle, at 4.2 K
and 210 K.

defined as $R = \Sigma |F_c - F_o| / \Sigma |F_o|$ are 0.06 and 0.07 for the 33 and the 18 nuclear and magnetic reflections, respectively. The obtained magnetic moments at 4.2 K are 6.34 and 6.58 $\mu_B$ for $Tb_{0.803}-Pr_{0.197}$ and $Tb_{0.798}-Pr_{0.202}$, respectively.

Figure 3 shows the diffraction patterns for $Tb_{0.803}-Pr_{0.197}$ on its first martensite cycle and at 4.2 K, the crystal transformation was perfectly completed. The nuclear diffraction patterns in the hcp phase could be taken above the Curie temperature before the reverse transformation started. The differences of neutron intensities at 4.2 K and 210 K could mainly be attributed to the magnetic reflections from the ferromagnetic structure with the magnetic moment in the c-plane. The calculated and the experimental structure factors were compared and the obtained R factors for the 9 nuclear and the 14 magnetic reflections are 0.05 and 0.06, respectively. The magnetic moment at 4.2 K is 6.42 $\mu_B$.

Figure 4 shows the diffraction patterns for the same crystal on its second martensite cycle. At 4.2 K, the diffraction patterns seem to belong to the hcp type reflections but anomalous broadening of Bragg reflections along the c*-axis was observed. On warming after the secondary cooling cycle, the reverse transformation partially started at 190 K. The nuclear patterns at 190 K show the mixed structure of two phases.

On the third cooling cycle, the martensite transformation proceeded partially. After the crystal was held at 78 K for two weeks the diffraction patterns shown in Fig. 5 was obtained. It is composed of mixed crystal of Sm-type and hcp phases.

Once the crystal was annealed at 800° C for a day, the martensite transformation could be restarted from the diffraction patterns of the first cooling cycle.


DISCUSSION


The experimental diffraction patterns in Fig. 2~5 are the various stages of martensite structures between the Sm-type and the hcp phases. The nuclear diffraction patterns in these figures could be explained by Kakinoki and Komura (5) theory of one dimensionally disordered crystal. Komura (9) derived the equation of diffuse scatterning intensities having stacking faults between the hcp and the Sm-type structures. That is, three close packed ABA layers may be followed by BAB or BCB with the probability 1-α or α and α=0 or α=1 corresponds to the hcp or the Sm-type crystal. Koch (3) proposed a similar model for the martensitic transformation from the hcp to the Sm-type structure, that requires succesive glide on three adjacent planes and no glide on the next three planes in the stacking sequence as shown in Fig. 6.

The paired peak positions (10$\bar{1}$)(101), (104)(105) and (108) (109)'s in Fig. 2 shift closely to each other about 0.05 c* from the normal positions. The profile of these nuclear patterns could be fitted by the theoretical calculation using the Kakinoki and

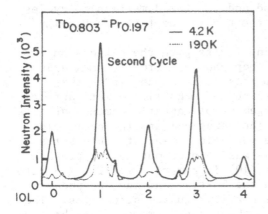

Fig. 4. (10L) scanning
patterns of hcp $Tb_{0.803}$--
$Pr_{0.197}$ on its second
martensite cycle.

Fig. 5. (10L) scanning
patterns of $Tb_{0.803}$--
$Pr_{0.197}$ after the third
martensite cycle and
holding the crystal
at 78 K for two weeks.

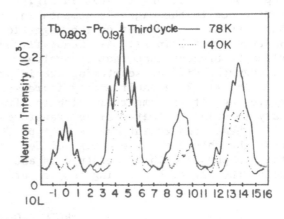

```
         (hcp)

A B A B A B A B A B A B A
 ∆ ∇ ∆ ∇ ∆ ∇ ∆ ∇ ∆ ∇ ∆ ∇
      + + +       + + +
       v v v       v v v
 ∆ ∇ ∆ ∆ ∇ ∆ ∆ ∇ ∆ ∆ ∇ ∆
A B A B C B C A C A B A B

       (Sm-type)
```

Fig. 6. A model for the transformation hcp→Sm-type structure proposed
by Koch et al. (3). ∆ and ∇ are
Frank's stacking operators and v
denotes a glide movement.

Komura theory and the obtained α parameter is 0.85. The nuclear diffraction patterns at the second and the third martensite cycles are composed of reflections from the two phases that are heavily faulted.

Then consider how the stacking faults give the effects on the magnetic structure of Sm-type and hcp phases. From the magnetic structures in the Sm-type and hcp phases, it is concluded that $hh$ packing favors parallel arrangement of magnetic moments and $hch$ packing favors antiparallel arrangement of magnetic moments at hexagonal sites. Here $h$ and $c$ denote the hexagonal and the cubic close packed layers, respectively. The magnetic moment at cubic sites behave as if it is paramagnetic as was pointed out in the previous paper (4) and by Koehler and Moon (10) in samarium metal. The rule of magnetic moment arrangement in the Sm-type and the hcp phases seems to hold in the intermediate heavily faulted structures. The magnetic diffuse scattering due to the magnetic moments on the faulted $hhh$ packing in the Sm-type crystal is distributed on the hcp type reflections as seen in Fig. 2. The magnetic diffuse scattering due to the magnetic moments on the faulted $hch$ packing in the hcp crystal spreads out on the magnetic superlattice position of Sm-type crystal in Fig. 4. Generally half width of magnetic reflections is broader than that of nuclrear reflections. Calculations of magnetic diffuse scattering are in progress.

The stacking fault energy increases with increasing martensite cycles and it is comparable with the free energy difference necessary to raise the martensitic transformation. Recently, our specific heat measurement on the Sm-type $Tb_{0.78}-La_{0.22}$ indicates that the change of free energy difference between the Sm-type and the hcp phases with temperature is ascribed to the magnetic origin below the magnetic ordering temperatures.

## REFERENCES

1.  C. C. Koch, J. Less-Common Metals, $\underline{22}$, 149 (1970).
2.  B. Johansson, Phys. Rev. B $\underline{11}$, 2836 (1975).
3.  C. C. Koch, P. G. Mardon and C. J. McHargue, Met. Transactions, $\underline{2}$, 1095 (1971).
4.  N. Achiwa and S. Kawano, J. Phys. Soc. Japan, $\underline{35}$, 303 (1973).
5.  J. Kakinoki and Y. Komura, Acta Cryst. $\underline{19}$, 137 (1965); J. Kakinoki, Acta Cryst. $\underline{23}$, 875 (1967).
6.  M. A. Rudermann and C. Kittel, Phys. Rev. $\underline{96}$, 99 (1954); T. Kasuya, Prog. Theor. Phys. Japan, $\underline{16}$, 45 (1956); K. Yoshida, Phys. Rev. $\underline{106}$, 893 (1957).
7.  A. H. Daane, R. E. Rundle, H. G. Smith and F. H. Spedding, Acta Cryst. $\underline{7}$, 532 (1954).
8.  M. Blume, A. J. Freeman and R. E. Watson, J. Chem. Phys. $\underline{37}$, 1245 (1962).
9.  Y. Komura, Acta Cryst. $\underline{15}$, 770 (1962).
10.   W. C. Koehler and R. M. Moon, Phys. Rev. Lett. $\underline{29}$, 1468 (1972).

γ-RAY EMISSION FROM ORIENTED NUCLEI IN A MULTIAXIS

NUCLEAR SPIN SYSTEM: $^{166m}$Ho$^{165}$Ho

H. Marshak
National Bureau of Standards, Washington, D.C.   20234

B.G. Turrell
Dept. of Physics, Univ. of B.C., Vancouver, B.C.   V6T 1W5

A new method(1) has been investigated to measure multiaxes nuclear spin structures using γ-ray emission from oriented radioactive nuclei. If the nuclear magnetization is produced by hyperfine interaction, then the atomic magnetic structure can be inferred. Thus, this technique, which depends only upon angular momentum theory, could be useful in certain cases in determining the distribution of atomic as well as nuclear moments. To demonstrate this method we report measurements of the spatial distribution of γ-rays emitted in the decay of $^{166m}$Ho in a single crystal of holmium metal.

The $^{166m}$Ho was produced *in situ* in a $^{165}$Ho single crystal sample by neutron irradiation. The sample was in the form of a disk which was soldered to the cold finger of our dilution refrigerator. Angular distribution measurement, $W(\theta,\phi)$, were made on 25 of the more intense γ-rays, where $\theta$ and $\phi$ are angles to the cone axis (c-axis) and in the basal plane respectively.

Theoretical expressions for $W(\theta,\phi)$ have been obtained(1) for conical spin structures like holmium, and we show that such measurements can yield the semi-cone angle, $\beta$, and the turn angle, $\tau$. However, to determine $\tau=180°$ we need $L>1$, L being the multipolarity of the observed γ-ray, for $\tau=90°$, we need $L>2$, and for holmium, if indeed $\tau=30°$, we would need $L>6$! (2). Although γ-ray intensity measurements along $\theta=0°$ and in the basal plane ($\theta=\pi/2$) enable one to deduce $\beta$ and $\tau$ respectively (if the latter can be determined); measurements for $0°<\theta<90°$ allow other conjected structures (e.g. double cones) to be examined.

Since the multipolarity of many of the 25 γ-rays used in our angular distribution measurements had not been accurately determined, a preliminary experiment was performed to measure as many of these as possible. These measurements showed, as might be expected, that there was no substantial contribution of L>2 multipole characters

in any of the 25 transitions studied. Thus, on the basis of the
theory outlined above we will not be able to measure a turn angle
of <90°. The angular distribution measurements taken in the basal
plane, $W(\theta=\pi/2,\phi)$, were all independent of $\phi$ within the accuracy of
our measurements. The angular distribution measurements perpendicular to the basal plane did show large asymmetries. Results for the
810.3(E2)keV transition are shown in Fig. 1. As one can see, we
determine the semi-cone angle, $\beta$, quite accurately and rule out a
double cone spin structure. These results for $\beta$ are in excellent
agreement with the neutron diffraction results.(3).

REFERENCES

1. H. Marshak and B.G. Turrell, Solid State Commun. <u>30</u>, 677 (1979).

2. This method does not determine the interlayer turn-angle as is
   done in neutron diffraction studies. The turn-angle is obtained
   from the number of quantization axes, thus it cannot distinguish
   between commensurate and incommensurate structures.

3. W.C. Koehler, J.W. Cable, M.K. Wilkinson and E.D. Wollan, Phys.
   Rev. <u>151</u>, 414 (1966).

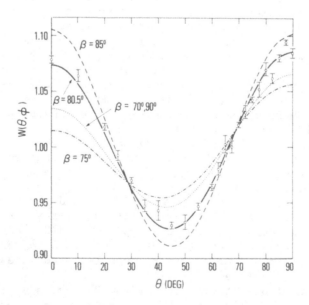

Fig. 1. $W(\theta,\phi)$ verses $\theta$ for the 810.3(E2)keV $\gamma$-ray. The solid curve
$\beta=80.5°$ is a least square fit of the data to the theory. For comparison we show the theoretical predictions for $\beta=75$, $\beta=85°$, and for
a double cone spin structure, $\beta=70°$, $90°$.

# MIRROR PROPERTIES OF RARE EARTHS BY ELECTRONS AND POSITRONS

P. S. Takhar, Dept. of Physics, Punjab Agricultural University, Ludhiana-141004, India; and

T. S. Gill, Aligarh Muslim University, Aligarh, India

## ABSTRACT

There have been a number of investigations into the physical properties of the rare earth metals. However, very little previous work deals with their mirror properties. These are the first such measurements. The present paper describes the mirror properties of yttrium, neodymium, holmium and ytterbium by penetration of electrons and positrons. A comparison of theoretical and experimental work is discussed.

## INTRODUCTION

Somewhat surprisingly there have been very little previous investigations (1-2) into the mirror properties of rare earths and their comparison with the other solids. This is the first such report to predict clearly that rare earths such as yttrium, neodymium, holmium and ytterbium do indicate mirror behavior (3-4) by penetration of electrons and positrons.

The aim of this paper is to report the mirror properties of rare earths and other solids. An attempt is made to relate the shell structure of rare earths and their mirror behavior. Also current experiments in this field are discussed.

### Mirror Properties of Rare Earths

Recent (4) experimental data on the relative penetration of electrons (1.77 MeV) and positrons (1.89 MeV) for yttrium, neodymium,

287

Table 1

(a)  Absorption coefficient in $gm^1 cm^2$ and ratio of ranges of posi-
     trons and electrons for rare earths.

|              | E    | Y39   | Nd60  | Ho67  | Yb70  |
|--------------|------|-------|-------|-------|-------|
| $\mu(e^+)$   | 1.88 | 7.55  | 8.16  | 7.26  | 8.12  |
| $\mu(e^-)$   | 1.77 | 10.91 | 13.59 | 13.44 | 11.98 |
| r+/r-        | -    | 1.44  | 1.67  | 1.85  | 1.47  |
| Rp+/Rp-      | 2.00 | 1.16  | 1.16  | 1.23  | 1.25  |

(b)  Physical properties of rare earths.

|                          |      |      |      |
|--------------------------|------|------|------|
| Unpaired electrons       | 3    | 4    | 1    |
| Magmetic moment $\mu_M$  | 1.30 | 3.81 | 1.65 |

holmium and ytterbium indicate that the positron transmission curve
lies above the electron curve (see Fig. 2 of ref. [4]). The absorption coefficients for positrons $\mu(e^+)$ and electrons $\mu(e^-)$ were calculated by a least squares fit to the linear portion of the transmission curve. The errors in the calculations varied from 4 to 7 percent.

The data (see Table 1[a]) indicate that the absorption coefficients for positrons and electrons vary inversely (i.e. as $\mu[e^+]$
goes down, $\mu[e^-]$ goes up). This clearly depicts the mirror behavior
of positrons and electrons in rare earths. This has been confirmed
by the current experiments in our laboratory with single crystals.
The number of unpaired electrons in 4f subshell in rare earth elements such as neodymium, holmium and ytterbium are 3, 4 and 1 respectively. The electron absorption coefficients (e$^-$) (see Table 1[a]
for these elements) decrease in accordance with the number of unpaired
electrons (see Table 1). However, positron absorption appears to be
independent of 4f electrons in these rare earth elements. Also,
electrons in the outer shell appear to play an important role in the
absorption of positrons. A brief comparison of various properties
of rare earth metals for neodymium, holmium and ytterbium is presented in Table 1(b). The similar trend as shown by the ratio of
ranges r+/r- (see Table 1). Table 1 also lists the theoretical ratio
of projected ranges Rp+/Rp-.

In some materials anomalies in the ratio r+/r- were discovered
in the values of ratios less than unity.

ACKNOWLEDGEMENT

The authors would like to thank Dr. Batra for his assistance in preparation of this manuscript.

Also we would like to express our sincere thanks to Professor Sehgal for providing research facilities in A. M. University.

REFERENCES

1.   K. A. Gschmeider, Jr., "Rare Earths, The Fraternal Fifteen," published by United States Atomic Energy Commission, Division of Technical Information (1966).
2.   F. H. Spedding and A. H. Danne, "The Rare Earths," John Wiley and Sons, Inc., NY (1961).
3.   P. S Takhar, Phys. Lett. 28A:423 (1968).
4.   P. S. Takhar, T. S. Gill and M. L. Sehgal, "Proceedings 11th Rare Earth Research Conference," Vol. I, 82 (1974).

# THE VICKERS MICROHARDNESS AND PENDULUM HARDNESS OF POLYCRYSTALLINE

# LANTHANIDE DICARBIDES

D.R. Bourne and I.J. McColm

School of Industrial Technology,  The University of

Bradford, Bradford BD7 1DP, England

Almost no data on the mechanical properties of lanthanide dicarbides exist, the available data being limited to a brief report by Krikorian on the Knoop microhardness(1).  One  aim of this work was to extend the Krikorian data.

A complicating factor in work involving lanthanide dicarbides is the ease with which they are hydrolysed by water vapour, hence oil films are needed to give the surface some protection.  Thin films were found to give only a limited protection and so a pyramid diamond sclerometer was built within which the dicarbides could be contained beneath 10mm of oil whilst being tested.  The pendulum hardness sclerometer has recently been used to study the chemomechanical properties of surfaces(2).  Hardness values are given by equation 1 as an arbitary measure of the rate of pendulum damping.

$$H = \frac{t}{2.303 \ (\log A_o - \log A)} \qquad (1)$$

Here H is hardness, t is time taken for initial amplitude $A_o$ to decrease to final value A.  Hardness determined in this manner is found to be independent of initial amplitude $A_o$ within the value 25–50mm.

## EXPERIMENTAL

The dicarbides were made by arc melting the metals (Rare Earth Products Ltd) with specpure graphite (Johnson Mathey Ltd) under a stream of purified argon gas.  In all cases 2-5 at.% excess of carbon was used to ensure removal of oxide and prevent any $Ln_2C_3$ formation.  Phase purity was determined by microscopic

and X-ray examination.  An 11.4 cm diameter Debye-Scherrer camera
was used with nickel filtered copper Kα radiation.  For metallographic examination samples were ground on emery papers with grits
down to 600, using CCl₄ as a lubricant.  A final polish was
obtained with 1 μm diamond dust using a non-aqueous silicone
suspension, after which the surface was immediately covered with
liquid paraffin.

For Vickers microhardness determinations a Leitz Miniload
hardness tester with a 136° angle Vickers diamond was used.  Indents
were made through the oil layer.  A specimen contact time of 10
seconds with loads in the range 50-300 g was used and a series of
4-10 indents with each load were made.  Because of the failure of
the oil layer to prevent hydrolysis surfaces were repolished
between each series of 4 indents.

A pendulum sclerometer was constructed following a design by
Westwood(2).  The total weight of the assembly was 800g with a
period of oscillation on edge rocking of one second.  Samples were
covered to a depth of 1cm with dry paraffin.  A fixed initial
amplitude of 2cm was used and the amplitude was measured again
after 1 minute.

Table 1. Vickers and Pendulum Hardness of Dicarbides

| $LnC_2$ | Vickers Microhardness Kg mm$^{-2}$ | Pendulum Hardness Arbitrary Units |
|---------|-----------------------------------|-----------------------------------|
| La | 245 ± 30 | 150 ± 10 |
| Ce | 100 ± 20 | 72 ± 15 |
| Pr | 140 ± 20 | 93 ± 5 |
| Nd | 240 ± 30 | 140 ± 25 |
| Sm | 650 ± 50 | 322 ± 5 |
| Gd | 472 ± 50 | 278 ± 5 |
| Dy | 582 ± 50 | 300 ± 5 |
| Ho | 572 ± 50 | 335 ± 5 |
| Er | 456 ± 55 | 376 ± 5 |
| Lu | 600 ± 60 | 330 ± 5 |
| Sc | 595 ± 20 | 265 ± 15 |
| Y | 510 ± 30 | 241 ± 15 |

RESULTS

Only the dicarbides in Table 1 could be prepared satisfactorily.  In the Vickers examination, despite the oil films, the
surfaces did hydrolyse which accounts for the relatively large un-
certainties in the data.  A micrograph showing the relative size
of indents and grains is shown as Fig.1.  Immediately the effectiveness of the pendulum method is demonstrated by the greater

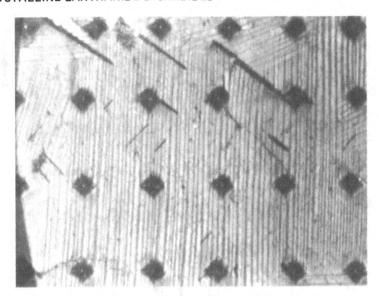

Fig.1. Vickers Indents in $LaC_2$.  Magnification x 400

precision of the data obtained.  Nevertheless, the same trends are
obvious in the sets of data for both show that the light lanthanides,
$LaC_2$ - $NdC_2$, form a relatively soft group while hardness increases
sharply after $NdC_2$ to form a hard group from $SmC_2$ to $LuC_2$.  Within
the soft group a minimum hardness occurs at $CeC_2$ while $GdC_2$ and
$ErC_2$ are the softer members of the hard group.  Both types of
measurement place the non-lanthanides $YC_2$ and $ScC_2$ in the range of
hardness characteristic of the heavy lanthanide group.  All these
data are displayed in Fig.2. along with the Knoop data obtained by
Krikorian(1).

## DISCUSSION

The Vickers and the pendulum hardness results are consistent
in clearly denoting a hard group, the heavy lanthanides $SmC_2$ - $LuC_2$,
and a soft group $LaC_2$ - $NdC_2$.  An interesting result in relation
to this division is the inclusion of $YC_2$ and $ScC_2$ in the hard group.

Considering the dicarbides on the basis of the model proposed
by Atoji(3) they possess mixed ionic-metallic bonding from the
$Ln^{3+}C_2^{2-}$ + band $e^-$ description.  The delocalised electrons are
accommodated in a band consisting of metal s, p, d and f orbitals
with $C_2^{2-}$ anion $2p^*_\pi$ antibonding orbitals.  Increased f character
in the metal band at the expense of s and d character could lead
to decreased hardness.  The ionic component of the bonding since
there is a 14% lanthanide contraction from $La^{3+}$ to $Lu^{3+}$ leads one
to expect a gradual increase in hardness and not the observed sharp
change.  Thus it seems that we might look for an intrinsic reason

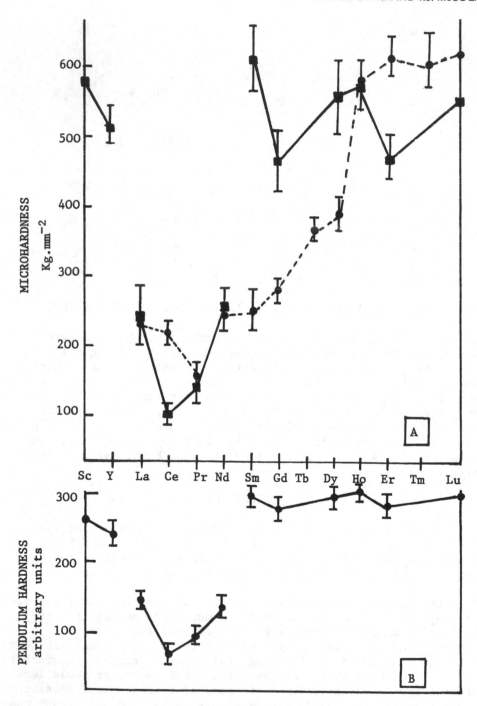

FIG.2. [A] ----●---- knoop microhardness from (1). ——■——
Vickers hardness results for LnC$_2$ phases. [B] Pendulum hardness
results for LnC$_2$ phases.

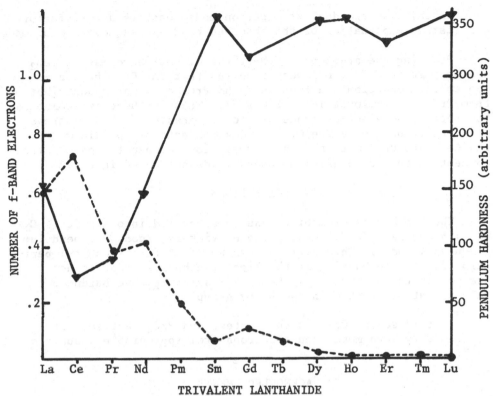

Figure 3. Correlation between f-band electron concentration for
$Ln^{3+}$(4) and pendulum hardness.  ----●---- number of f-band
electrons.  ——▼—— Pendulum hardness for $LnC_2$

for the change in hardness in an f-electron participation in bonding. Gschneidner(4) has provided an analysis of lanthanide metal
melting points and heats of sublimation that leads to a model for
the electronic configuration of the trivalent lanthanides which
predicts two kinds of 4f electron. There are the 'atomic' 4f
electrons which are commonly associated with the ions and whose
number varies from zero for $La^{3+}$ to 14 for $Lu^{3+}$ and account for the
observed magnetic susceptibilities. Then there are the 4f band
electrons which occur because the empty 4f energy levels, that is
14 for $La^{3+}$ to zero for $Lu^{3+}$, lie close to the Fermi level and are
thus partially occupied by the conduction electrons. That is, the
conduction electrons involved in the metal $-2p^{*}_{\pi}$ band have varying
f character. Gschneidner has calculated that the 4f band electron
varies from 0.7 of an electron for the light lanthanides and drops
sharply to 0.1 at samarium and zero for lutetium. In $YC_2$ and $ScC_2$
where the 4f levels are too far above the Fermi surface to give
rise to any 4f band electrons we see that the hardness is close to
$LuC_2$.

In this way the dual 4f electron model enables a satisfactory qualitative explanation of the observed hardness sequence to be made.

Pursuing the argument further, the minimum hardness at $CeC_2$ can be predicted. It has been reported that the $CeC_2$ has an appreciable contribution from $Ce^{4+}$ and $PrC_2$ also has a contribution from $Pr^{4+}$ but somewhat less at 40%(5). Thus in these two compounds the total band electron concentration is greater with an increase of 4f electrons above the 0.7 and 0.4 respectively predicted for the $Ce^{3+}$ and $Pr^{3+}$ metals. The correlation between f band electron concentration and pendulum hardness is demonstrated in Fig.3.

## CONCLUSIONS

The lanthanide dicarbides can be separated into a soft ($LaC_2$ - $NdC_2$) and a hard ($SmC_2$-$LuC_2$) group by Vickers, Knoop and pendulum hardness methods. This division can be related to f-orbital participation in the bonding of the light lanthanides that is not present in the heavy group. Hence $YC_2$ and $ScC_2$ have hardness values that place them in the heavy group.

In this series $CeC_2$ is the softest and $PrC_2$ next softest because they have more band electrons from appreciable amounts of $Ln^{4+}$.

## ACKNOWLEDGEMENTS

1.    N.H. Krikorian, T.C. Wallace and M.G. Bowman, Report of Int. Colloq. on Semi Metal Derivatives, Paris (1967) 489.

2.    A.R.C. Westwood, N.H. Macmillan and R.C. Kalyoncu, "Environment-Sensitive Hardness and Machinability of $Al_2O_3$", J.Am. Ceram. Soc., 56 (1973) 258.

3.    M. Atoji and R.C. Medrud, "Structures of Calcium Dicarbide and Uranium Dicarbide by Neutron Diffraction", J.Chem.Phys., 31 (1959) 332.

4.    K.A. Gschneider Jr., "On the Nature of 4f Bonding in the Lanthanide Elements and their Compounds", J.Less Common Metals, 25 (1971) 405.

5.    M. Atoji, "Neutron Diffraction Studies of Higher Carbides of Heavy Metals", J.Phys.Soc., Japan, 17, Supplement B-II (1962) 395.

# CRYSTAL CHEMISTRY OF RARE EARTH - TRANSITION METAL COMPOUNDS

Erwin Parthé
Laboratoire de Cristallographie aux Rayons X, Université
de Genève, quai Ernest Amsermet 24, CH-1211 Genève 4,
Switzerland

Compounds formed from rare earth elements and transition metals exist only with the latter from groups 7, 8, 9 and 10. An unexpectedly large number of different structure types have been found experimentally, some of them with unusual stoichiometry and large unit cells. The structure types change not only with an interchange of the transition metal partner, but often also within the series of rare earth alloys. The structure types can conveniently be grouped into those in which trigonal prisms can be recognized and those having other coordination polyhedra (1).

Some 20 structure types are exclusively built up of transition-metal-centered trigonal prisms of rare earth atoms. Partially successful efforts have been made to find geometrical relationships between these structure types. Certain of these of the fixed stoichiometry have been recognized as different stacking variations of a common structural slab ($Y_3Co_2$-$Dy_3Ni_2$; CrB-TbNi-FeB; $AlB_2$-$ThSi_2$). Others of varying stoichiometry can be explained as stacking variations of two different base slabs (FeB-$Y_3Ni_2$-$Y_8Co_3$) (2). The concept of periodic micro-twinning of close-packed base structures, originally proposed by Andersson and Hyde in 1974, has also been found useful for the proper classification of these structures with different stackings, 12 structure types can thus be interpreted (4). The concept can be used to predict new ternary FeB-CrB stacking variants which occur in the pseudobinary $R_{1-x}R'_xNi$ systems (R, R' = rare earth element) (5).

For the structure types not containing trigonal prisms it is more difficult to find geometrical relationships. Eight structure types can be grouped in the $R_{n+2}T_{5n+4}$ structure series, which is built up of one $R_2T_4$ slab (as in the Laves phases) and n $RT_5$ slabs (as in $CaCu_5$) (6). A second structure series includes the structure types of $W_5Si_3$, $Pu_{31}Pt_{20}$ and $Y_3Rh_2$. Moreover, there remain a few structure types like, for example, $Ce_{23}Co_{11}$ and $Y_{64}Rh_{37}$ (202 atoms per unit cell) which have so far resisted all efforts at classifying their atom arrangement. It appears that at least some of these structure types like $Sc_{11}Ir_4$ or $Sm_{11}Cd_{45}$ (448 atoms in unit cell) can be well described using the cluster concept (7).

## REFERENCES

1.   E. Parthé and J. M. Moreau, J. Less-Common Metals $\underline{53}$:1-24 (1977).
2.   J. LeRoy, J. M. Moreau, D. Paccard and E. Parthé, Acta Cryst. $\underline{B33}$:3406-3409 (1977).
3.   E. Parthé, Acta Cryst $\underline{B32}$:2813-2819 (1976).
4.   B. Chabot and E. Parthé, Acta Cryst $\underline{B34}$:3173-3177 (1978).
5.   K. Klepp and E. Parthé, Acta Cryst (submitted).
6.   E. Parthé and R. Lemaire, Acta Cryst. $\underline{B31}$, 1879-1889 (1975).
7.   M. L. Fornasini, B. Chabot and E. Parthé, Acta Cryst. $\underline{B34}$:2093-2099 (1978).

# ATOMIC VOLUME CONTRACTION IN RARE EARTH NICKEL INTERMETALLICS AS A FUNCTION OF PARTIAL COORDINATION NUMBER COEFFICIENT

Forrest L. Carter

Chemistry Division, Naval Research Laboratory

Washington, D. C.   20375

## INTRODUCTION

Molecular volume contraction upon formation of rare earth inter-metallic compounds from the elements is a very well-known phenomenon. While a method for calculating intermetallic atomic volumes was intro-duced by the author in 1971 as a partial argument for predicting the unstability of $SmCo_5$ (1) and again in 1974 in the discussion of the $Pu_2C_3$ structure (2), systematic studies have only recently included rare earth semimetals and intermetallics (3,4). As might be expected from their large size it is the rare earth atoms that are primarily responsible for the volume contraction. The rare earth atom volume is found herein to decrease approximately linearly with respect to its coefficient of partial coordination number with nickel. This latter concept is a measure of the relative importance and number of rare earth-nickel bonds and arises from the generalization of the concept of coordination number (3,5). The amount of the rare earth volume contraction is surprisingly large, i.e., greater than 30%. Nickel on the other hand shows a smaller volume expansion upon alloy-ing. These volume changes are thought to be due primarily to size difference effects upon bonding density distribution.

## APPROACH

Polyhedral Atomic Volumes (PAVs) are the volumes of cells con-structed as per G. Voronoi (6) but with the inclusion of the effects of finite atomic size. PAV cells differ from Voronoi (or Wigner-Seitz) cells for atoms of different sizes by placing the bisecting plane equally distant from the atoms' outer spherical surfaces (even if they overlap).

For the rare earth nickel compounds considered here the radii employed are the single bond metallic radii $R_1(i)$ of Pauling (7). The beauty of Pauling's approach is that number of electrons involved in bonding, i.e., the valence Vc, can be calculated from the bond distances via a simple calculation for bond orders $n_{ij}$, then $Vc(i) = \Sigma \, n_{ij}$. By requiring that the calculated valence Vc is equal to an estimated valence Ve based on the number of electrons available to the rare earth then Pauling's metallic radii can be applied self consistently. In the "neutral cell" approximation (1) charge is transferred from Ni to the rare earth R in an amount equal to that lost back from R to Ni by polarization of its bonds because of the R-Ni electronegativity difference ($\Delta$x). For the normal trivalent rare earths Ve is about 3 + 1/3. The author's experience with both the rare earth $CaCu_5$ and Laves phase compounds (1,8) indicate that while the transition metals are normal the rare earth radii, Rl, are much too large. Self-consistency and reasonable valences are achievable only by increasing the rare earth bonding d character and hence decreasing the rare earth radii, Rl(i).

By generalizing (3,5) the concept of coordination number CN, one is led to the concept of the partial CN (R-Ni), that is of the CN of the rare earth R with only Ni neighbors, and of the concept of the partial CN coefficient f(R-Ni) which is a measure of the relative total importance of all the R-Ni bonds. While many measures of an interaction between two atoms can be used, like bond energy or bond order, $n_{ij}$, for the purposes of these calculations we use a pyramidal volume defined by the common polygon face of two PAV cells as a pyramid base and the atom of interest as its apex.

RESULTS

Twenty-two R-Ni compounds were selected for study to give both a wide range of rare earths and of R-Ni composition with single crystal results preferred. The results are summarized in Table 1 and its continuation. The first column gives the Voronoi cell volume in $\mathring{A}^3$ and mathematically these fill all space. The atomic radii here are assumed to be either zero or equal. The shape of the cell in terms of the number of polygon faces and vertexes is the same as for the PAV cells (fourth column) except for DyNi, where the Voronoi data is Dy: 18/32 and Ni: 12/20, and for Ni in $Er_3Ni_2$, Ni: 13/22. The second column gives the single bond radii Rl, for the rare earths which are used in the PAV cell calculations. The nickel Rl is constant at 1.155 A. The program used to calculate the metallic radii data employed the "neutral cell" approach and adjusted the rare earth d-character to achieve self consistency. The rare earths were generally trivalent except for Ce in which the valence was taken to be that of the ambient element, i.e., 3.2. The high d-character for cerium in the nickel rich compositions suggest that they may be primarily tetravalent, certainly Ce2 in $CeNi_3$ and Ce1 in $Ce_2Ni_7$. The metallic radii results given in Table 1 columns two and three

## Table 1. Selected Rare Earth Nickel Compounds

| Cpd.* | | Vor. Vol. | R1 | Calc. Val. | Faces/ Vert. | PAV Vol. | CN$_V$ | Ref. |
|---|---|---|---|---|---|---|---|---|
| La$_7$Ni$_3$ | La1 | 30.74 | 1.637 | 3.19 | 15/26 | 34.65 | 14.58 | (9) |
| | La2 | 31.51 | 1.689 | 2.95 | 15/24 | 37.03 | 14.45 | |
| | La3 | 30.30 | 1.612 | 3.19 | 15/26 | 33.19 | 14.00 | |
| | Ni | 23.65 | 1.155 | 3.78 | 9/14 | 13.92 | 8.38 | |
| La$_2$Ni$_3$ | La | 23.11 | 1.533 | 3.26 | 17/30 | 28.68 | 15.35 | (10) |
| | Ni1 | 17.04 | 1.155 | 3.45 | 10/16 | 13.35 | 9.93 | |
| | Ni2 | 18.08 | 1.155 | 3.21 | 12/20 | 14.30 | 11.72 | |
| La$_2$Ni$_7$ | La1 | 19.17 | 1.427 | 3.31 | 16/28 | 23.28 | 15.65 | (11) |
| | La2 | 19.24 | 1.474 | 3.37 | 20/36 | 25.12 | 19.00 | |
| | Ni1 | 15.86 | 1.155 | 2.83 | 12/20 | 13.79 | 11.99 | |
| | Ni2 | 13.51 | 1.155 | 4.41 | 12/20 | 12.24 | 11.34 | |
| | Ni3 | 13.49 | 1.155 | 4.46 | 12/20 | 12.21 | 11.34 | |
| | Ni4 | 13.82 | 1.155 | 4.10 | 12/20 | 12.54 | 11.99 | |
| | Ni5 | 14.27 | 1.155 | 4.17 | 12/20 | 12.77 | 11.71 | |
| Ce$_7$Ni$_3$ s.c. | Ce1 | 28.99 | 1.633 | 3.40 | 15/26 | 33.37 | 14.47 | (12) |
| | Ce2 | 29.48 | 1.633 | 3.41 | 15/24 | 34.05 | 14.45 | |
| | Ce3 | 28.40 | 1.583 | 3.36 | 15/26 | 31.07 | 14.01 | |
| | Ni | 22.37 | 1.155 | 3.97 | 9/14 | 13.64 | 8.42 | |
| CeNi | Ce | 23.08 | 1.574 | 3.56 | 17/30 | 29.06 | 15.26 | (13) |
| | Ni | 18.96 | 1.155 | 3.83 | 11/18 | 12.98 | 8.70 | |
| CeNi$_3$ s.c. | Ce1 | 18.55 | 1.456 | 3.58 | 20/36 | 24.00 | 18.96 | (14) |
| | Ce2 | 17.79 | 1.396 | 3.52 | 16/28 | 21.15 | 15.79 | |
| | Ni1 | 14.40 | 1.155 | 3.68 | 12/20 | 12.71 | 12.00 | |
| | Ni2 | 13.33 | 1.155 | 4.46 | 12/20 | 12.11 | 11.44 | |
| | Ni3 | 13.34 | 1.155 | 4.45 | 12/20 | 12.11 | 11.44 | |
| | Ni4 | 13.74 | 1.155 | 4.32 | 12/20 | 12.40 | 11.86 | |
| Ce$_2$Ni$_7$ s.c. | Ce1 | 17.91 | 1.396 | 3.53 | 16/28 | 21.35 | 15.75 | (15) |
| | Ce2 | 18.70 | 1.462 | 3.60 | 20/36 | 24.33 | 19.05 | |
| | Ni1 | 14.42 | 1.155 | 3.71 | 12/20 | 12.74 | 12.00 | |
| | Ni2 | 13.22 | 1.155 | 4.65 | 12/20 | 11.99 | 11.42 | |
| | Ni3 | 13.30 | 1.155 | 4.53 | 12/20 | 12.07 | 11.41 | |
| | Ni4 | 13.34 | 1.155 | 4.61 | 12/20 | 12.13 | 11.99 | |
| | Ni5 | 13.77 | 1.155 | 4.45 | 12/20 | 12.45 | 11.77 | |
| NdNi$_2$ | Nd | 18.29 | 1.526 | 3.36 | 16/28 | 23.44 | 15.55 | (13) |
| | Ni | 14.87 | 1.155 | 4.78 | 12/20 | 12.30 | 12.00 | |
| GdNi | Gd | 22.56 | 1.542 | 3.30 | 17/30 | 28.05 | 15.11 | (16) |
| | Ni | 18.67 | 1.155 | 3.64 | 11/18 | 13.18 | 8.73 | |
| GdNi$_2$ | Gd | 17.81 | 1.494 | 3.33 | 16/28 | 22.42 | 15.61 | (17) |
| | Ni | 14.48 | 1.155 | 5.01 | 12/20 | 12.17 | 12.00 | |
| TbNi (HT) s.c. | Tb1 | 21.86 | 1.501 | 3.18 | 17/30 | 26.89 | 14.71 | (18) |
| | Tb2 | 22.13 | 1.501 | 3.13 | 17/30 | 27.03 | 14.96 | |
| | Tb3 | 21.98 | 1.501 | 3.45 | 17/30 | 26.82 | 15.06 | |
| | Ni1 | 18.60 | 1.156 | 3.27 | 11/18 | 13.61 | 8.72 | |
| | Ni2 | 18.26 | 1.156 | 3.47 | 11/18 | 13.38 | 8.78 | |
| | Ni3 | 18.43 | 1.156 | 3.38 | 11/18 | 13.51 | 8.76 | |
| TbNi (LT) s.c. | Tb1 | 22.37 | 1.490 | 3.29 | 17/30 | 27.06 | 14.93 | (18) |
| | Tb2 | 22.19 | 1.490 | 3.56 | 17/30 | 26.82 | 15.11 | |
| | Tb3 | 21.90 | 1.490 | 3.60 | 17/30 | 26.58 | 15.04 | |
| | Tb4 | 22.37 | 1.490 | 3.27 | 17/30 | 27.04 | 15.24 | |
| | Tb5 | 22.07 | 1.490 | 3.53 | 17/30 | 26.70 | 15.01 | |
| | Tb6 | 22.14 | 1.490 | 3.54 | 17/30 | 26.90 | 14.77 | |
| | Ni1 | 18.16 | 1.156 | 3.70 | 11/18 | 13.44 | 8.72 | |
| | Ni2 | 18.00 | 1.156 | 3.80 | 11/18 | 13.39 | 8.85 | |
| | Ni3 | 18.15 | 1.156 | 3.50 | 11/18 | 13.47 | 8.83 | |
| | Ni4 | 18.06 | 1.156 | 3.65 | 11/18 | 13.42 | 8.80 | |
| | Ni5 | 18.11 | 1.156 | 3.65 | 11/18 | 13.43 | 8.80 | |
| | Ni6 | 18.31 | 1.156 | 3.46 | 11/18 | 13.58 | 8.76 | |
| Dy$_3$Ni$_2$ | Dy1 | 21.87 | 1.469 | 3.23 | 17/30 | 25.88 | 15.00 | (19) |
| | Dy2 | 26.46 | 1.523 | 3.20 | 14/24 | 29.46 | 12.99 | |
| | Dy3 | 25.85 | 1.528 | 3.20 | 17/30 | 29.38 | 14.13 | |
| | Ni1 | 18.05 | 1.155 | 3.67 | 11/18 | 13.26 | 8.88 | |
| | Ni2 | 19.43 | 1.155 | 3.68 | 11/18 | 13.68 | 8.54 | |
| DyNi | Dy | 21.49 | 1.477 | 3.27 | 17/30 | 26.18 | 14.71 | (16) |
| | Ni | 18.47 | 1.155 | 3.16 | 11/18 | 13.78 | 8.71 | |
| DyNi$_2$ | Dy | 17.34 | 1.483 | 3.32 | 16/28 | 21.73 | 15.63 | (17) |
| | Ni | 14.10 | 1.155 | 5.41 | 12/20 | 11.91 | 12.00 | |

*s.c. indicates refined single crystal data.

## Table 1 - Continued

| Cpd.* | | Vor. Vol. | R1 | Calc. Val. | Faces/ Vert. | PAV Vol. | CN$_V$ | Ref. |
|---|---|---|---|---|---|---|---|---|
| Er$_3$Ni$_2$ | Er1 | 20.35 | 1.413 | 3.19 | 14/24 | 22.32 | 13.99 | (20) |
| s.c. | Er2 | 21.76 | 1.416 | 3.20 | 16/28 | 23.83 | 14.80 | |
| | Er3 | 24.79 | 1.532 | 3.17 | 16/27 | 29.12 | 14.59 | |
| | Ni | 18.72 | 1.155 | 3.54 | 9/14 | 13.36 | 8.79 | |
| ErNi$_3$ | Er1 | 18.18 | 1.417 | 3.29 | 20/36 | 22.87 | 19.08 | (21) |
| | Er2 | 16.96 | 1.343 | 3.23 | 16/28 | 19.46 | 15.89 | |
| | Ni1 | 13.53 | 1.155 | 4.24 | 12/20 | 12.32 | 11.99 | |
| | Ni2 | 13.03 | 1.155 | 4.66 | 12/20 | 11.99 | 11.47 | |
| | Ni3 | 13.31 | 1.155 | 4.58 | 12/20 | 12.24 | 11.88 | |
| ErNi$_4$ | Er1 | 17.69 | 1.380 | 3.29 | 20/36 | 21.66 | 19.23 | (22) |
| | Er2 | 16.47 | 1.328 | 3.17 | 16/27 | 19.27 | 15.81 | |
| | Ni1 | 12.61 | 1.155 | 4.98 | 12/20 | 11.73 | 11.54 | |
| | Ni2 | 12.68 | 1.155 | 5.07 | 12/20 | 11.94 | 11.93 | |
| | Ni3 | 12.58 | 1.155 | 5.19 | 12/20 | 11.86 | 11.93 | |
| | Ni4 | 13.81 | 1.155 | 3.77 | 13/22 | 12.96 | 11.82 | |
| YbNi | Yb | 20.54 | 1.428 | 3.24 | 17/30 | 24.42 | 14.46 | (23) |
| | Ni | 17.63 | 1.155 | 3.19 | 11/18 | 13.74 | 8.78 | |
| YbNi$_2$ | Yb | 16.99 | 1.462 | 3.29 | 16/28 | 21.04 | 15.66 | (23) |
| | Ni | 13.82 | 1.155 | 5.63 | 12/20 | 11.79 | 12.00 | |
| YbNi$_3$ | Yb1 | 17.89 | 1.399 | 3.27 | 20/36 | 22.19 | 19.10 | (23) |
| | Yb2 | 16.69 | 1.323 | 3.23 | 16/28 | 19.00 | 15.89 | |
| | Ni1 | 13.32 | 1.155 | 4.40 | 12/20 | 12.20 | 11.99 | |
| | Ni2 | 12.82 | 1.155 | 4.83 | 12/20 | 11.87 | 11.49 | |
| | Ni3 | 13.10 | 1.155 | 4.78 | 12/20 | 12.12 | 11.87 | |
| LuNi | Lu | 20.32 | 1.436 | 3.50 | 17/30 | 24.26 | 14.50 | (16) |
| | Ni | 17.44 | 1.155 | 3.42 | 11/18 | 13.50 | 8.77 | |

* s.c. indicates refined single crystal data.

also indicate that further stoichiometric and positional refinement is desirable in several of the compounds. In particular we note that La$_2$ in La$_7$Ni$_3$, Ni1 in La$_2$Ni$_7$, and Ni4 in ErNi$_4$ have somewhat suspiciously low valences and correspondingly high PAV values.

In three columns of Table 1 are tabulated the data for the PAV cells including the polyhedron Face and Vertex numbers, the PAV volume and the coordination numbers CN$_y$ based on pyramidal volumes. The Face Vertex numbers are characteristic of well-known coordinations: 20/36 corresponds to that of La in LaNi$_5$, 16/28 to the Friauf polyhedron, and 12/20 to the transition metal icosahedron. The CN$_y$ are usually smaller than the number of PAV faces and can be equal to the number of faces when all the interactions are equivalent. The PAV volumes are also illustrated graphically in Figure 1. The trivalent rare earth PAV have however been normalized to that of Gd in this figure.

## DISCUSSION AND CONCLUSION

The data contained in Table 1 and illustrated in Figure 1 demonstrates a remarkable volume contraction, as much as 35-40% for the rare earths as the partial coefficient f(R-Ni) approaches 1.0. To emphasize that this effect is real and not an artifact of the method of selecting the rare earth R1 we note: 1: That the rare earth PAV decrease is much larger than the Ni PAV decrease for increasing nickel content; that is, the volume lost by the rare earth is several times larger than that which can be accounted for by Ni PAV gain. 2: That the ratio of the smallest to largest PAV for each rare earth element is essentially the same ratio as for the corresponding

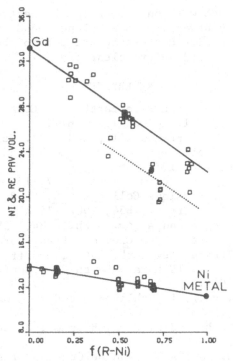

Fig. 1.   Rare earth PAV's, normalized to Gd, show a sharp decline
          with the partial coordination number coefficient f(R-Ni).
          The solid lines indicate normal trends while the dotted
          line suggests unusual or anisotropic bonding hybridization.

Voronoi cells.  Thus, the crystal structures have large Voronoi cells
for the rare earth positions and smaller cells for the Ni position
with the results that the rare earth volume percentage contractions
are virtually independent of the rare earth radius.

     The usual picture for rare earth transition metal intermetallics
would transfer the three R valence electrons to the Ni d-band, however,
the author believes this picture to be incorrect for the following
reasons:  1:  The electronegativity difference (0.6) for R and Ni is
small and would account for only 0.5 electron transfer at most.  2:  A
very similar contraction occurs in R-Al systems as well (4) and here
the electronegative difference is even smaller (0.3).  3:  In the R-Al
intermetallics the Al has no d-bands to be filled.  Finally, we note
that the magnetic data supporting the transfer of three R valence
electrons to the Ni d-band can be readily interpreted in terms of the

valence bond approach with only fractional electron transfers.  We
suggest that this contraction is a valence bond effect roughly proportional to atomic size difference and due to the contraction of
bonding electrons toward internuclear axes.

## REFERENCES

1.  F. L. Carter, Proc. 9th Rare Earth Res. Conf., ed., D. E. Field,
    Blacksburg, Va., Vol. 2 (Oct 1971) 617.
2.  F. L. Carter, T. L. Francavilla, and R. A. Hein, Proc. 11th Rare
    Earth Research Conf., eds., J. M. Haschke and H. A. Eick, Traverse
    City, Michigan, (Oct 1974) 36.
3.  F. L. Carter, J. Less-Common Metals, $\underline{47}$, 157 (1976).
4.  F. L. Carter, J. Physique Colloque, $\underline{40}$, C5-216 (1979).
5.  F. L. Carter, Acta Cryst., B$\underline{34}$, 2962 (1978).
6.  G. Voronoi, J Reine und Angew. Math., $\underline{134}$, 198 (1908).
7.  L. Pauling, "Nature of the Chemical Bond," 3rd ed. (Cornell
    Univ. Press, Ithaca, New York, 1960), 398 ff.
8.  F. L. Carter, Proc. 10th Rare Earth Res. Conf., Carefree, Ariz.
    (April 1973) 1044.
9.  P. Fischer, W. Halg, L. Schlapbach and K. Yvon, J. Less-Common
    Metals, $\underline{60}$, 1 (1978).
10. J. H. N. Van Vucht and K. H. J. Buschow, J. Less-Common Metals,
    $\underline{46}$, 133 (1976).
11. K. H. J. Buschow and A. S. Van DerGoot, J. Less-Common Metals,
    $\underline{22}$, 419 (1970).
12. R. B. Roof, Jr., A. C. Larson, and D. T. Cromer, Acta Cryst.,
    $\underline{14}$, 1084 (1961).
13. W. B. Pearson, "A Handbook of Lattice Spacings and Structures of
    Metals and Alloys," Vol. 2, (Pergamon Press Ltd., Oxford, London,
    1967).
14. D. T. Cromer and C. E. Olsen, Acta Cryst., $\underline{12}$, 689 (1959).
15. D. T. Cromer and A. C. Larson, Acta Cryst. $\underline{12}$, 855 (1959).
16. A. E. Dwight, R. A. Conner, Jr., and J. W. Downey, Acta Cryst.,
    $\underline{18}$, 837 (1965).
17. N. C. Baenziger and J. L. Moriarty, Jr., Acta Cryst., $\underline{14}$, 946
    (1961).
18. R. Lemaire and D. Paccard, Bull. Soc. Fr. Miner. Crist., $\underline{90}$,
    311-315 (1967).
19. J. M. Moreau, D. Paccard, and E. Parthe, Acta Cryst., B$\underline{30}$, 2583
    (1974).
20. J. M. Moreau, D. Paccard, and D. Gignoux, Acta Cryst., B$\underline{30}$,
    2122 (1974).
21. A. E. Dwight, Acta Cryst., B$\underline{24}$, 1395 (1968).
22. J. C. Barrick and W. J. James, abstract, Acta Cryst., A$\underline{31}$, S96,
    Part 5 (1975).
23. K. H. J. Buschow, J. Less-Common Metals, $\underline{26}$, 329 (1972).

# COMPARISON BETWEEN THE MAGNETIC PROPERTIES OF CRYSTALLINE AND AMORPHOUS RARE EARTH TRANSITION METAL ALLOYS

D. Gignoux, D. Givord, R. Lemaire, A. Liénard[+] and

J.P. Rebouillat

Laboratoire Louis Néel, C.N.R.S., 166X,

38042 Grenoble-cédex, France

[+]Commissariat à l'Energie Atomique, LETI, NCE, 85X,

38041 Grenoble-cédex, France

The atomic arrangement of the rare earth (R)-transition metal (M) alloys in both crystalline and amorphous states may be described as a packing of spheres of different radii. Their magnetic properties result mainly from the large difference in electronegativity between the two constituents (1). However the loss of rotational and translational symmetries in the amorphous state may influence strongly some of these properties.

A transfer of 5d electrons into the 3d band occurs in the alloys. This gives account for the decrease of the 3d moment in rich Co alloys which are strong ferromagnets. Furthermore, an hybridization of 3d and 5d states gives rise to the formation of a tail at the top of the 3d band (2). It is then possible to explain the weak ferromagnetism in Y-Ni alloys and in Y-Co ones when the Y content is larger than in $YCo_{3.5}$ as due to the subsequent decrease of the state density at the Fermi level. In the vicinity of the critical conditions for the onset of 3d magnetism, amorphous and crystalline alloys behave much differently. The states in the tail at the top of the 3d band favor the occurrence of an itinerant electron metamagnetism observed in $RCo_2$, $ThCo_5$ and $Y_2Ni_{17}$ crystalline alloys (3). In such a case the 3d moment is strongly dependent on interatomic exchange interactions, i.e. on the surroundings which in amorphous alloys are very diverse. Indeed the weaker decrease of Co and Fe moments in the amorphous alloys has been

described (5) within the Jaccarino-Walker model. The tail of 3d
states also allows one to explain the resurgence of Ni magnetism
in both crystalline and amorphous alloys.

In Co and Ni alloys, the 3d band being nearly filled, positive
interactions are predominent and the 3d moments are parallel in
amorphous as in crystalline state. The 3d band of Fe being slightly
more than half-filled, interactions between Fe atoms may be either
positive or negative, strongly depending on interatomic distances.
These two types of interactions are competing : i) $R_2Fe_{17}$ may
exhibit helimagnetic structures ; ii) because of the diversity of
atomic environments amorphous Y-Fe show spin-glass behavior (5).
When the 3d and 4f alloyed atoms are both magnetic the 4f spin is
always coupled antiparallelly to the 3d spin. This property may be
formulated as follows (1) : in the considered alloys, the predominant magnetic interactions originate from the 3d spins. The 3d-5d
(4d) hybridization leads to a negative polarization of the 5d (4d)
electrons as observed for s electrons in 3d metals. The positive
contact interaction between conduction electrons and 4f spins leads
then to the observed 3d-4f antiparallel coupling.

The orbital contribution of the rare earth magnetic moment is
strongly sensitive to the surroundings giving rise to a strong
local anisotropy. In high symmetry compounds the local easy magnetization directions are parallel and the magnetic structures are
collinear. In low symmetry compounds as well as in amorphous materials these directions may differ from one R atom to another. At
low temperature when the local anisotropy is strong enough the
magnetic arrangement is non collinear. In the crystalline $Tb_3Co$,
six local easy directions are observed. In the amorphous $RNi_3$,
where the Ni is not magnetic each direction of the space can be a
local easy direction. The magnetic arrangement is called speroma-
gnetic (4). When the 3d metal is magnetic sperimagnetic arrangements are observed, i.e. the directions of the R moments are
distributed inside a cone around the 3d moment.

## REFERENCES

1. B. Barbara, D. Gignoux, D. Givord, F. Givord and R. Lemaire,
   Int. J. Magnetism, 4 (1973) 77.
2. M. Cyrot and M. Lavagna, J. Appl. Phys., 50 (1979) 2333.
3. D. Gignoux, F. Givord, J. Phys. F, 1979 (to appear) ;
   D. Givord, J. Laforest, R. Lemaire, J. Appl. Phys. (to appear).
4. J.M.D. Coey, J. Appl. Phys., 49 (1978) 1646.
5. K.H.J. Buschow, M. Brouha, J.W.M. Biesterbos, A.G. Dirks,
   Physica, 91B (1977) 261.

# ELECTRIC FIELD GRADIENTS AND EXCHANGE INTERACTIONS IN AMORPHOUS RE$_4$Au ALLOYS

A. Berrada, J. Durand, N. Hassanain and B. Loegel

L.M.S.E.S. (LA 306) Institut Le Bel – U.L.P.

4, rue Blaise Pascal 67000 Strasbourg, France

## INTRODUCTION

Amorphous magnetic rare-earth alloys have been the subject of considerable interest in the recent past (1). Particular emphasis has been placed in the literature on amorphous systems made of rare earths with a large amount of transition metals. While great progress has been made toward understanding of the respective roles of local magnetic anisotropy and exchange interactions in heavy-rare-earth based amorphous compounds (2), amorphous alloys with light lanthanides have received far less attention (3-4). We have undertaken a systematic study of the magnetic properties of the amorphous RE$_4$Au alloys. Such amorphous alloys obtained by splat-cooling seemed to be attractive for magnetic studies for the two following reasons. First, the gold constituent is not likely to have a substantial magnetic moment. Second, owing to the large RE content (80 at. %), we lie in a situation relatively close to that of pure amorphous rare earths.

The sample foils were prepared by the splat-cooling technique as described elsewhere (5). The amorphous structure of each foil was checked by X-ray scanning with a Philips diffractometer (Mo K$\alpha$ radiation). Zero-field susceptibility measurements were performed down to 1.4°K with a standard ac inductance bridge with variable frequencies (35 to 35.10$^3$ Hz). Magnetization was measured using an induction technique between 2.2 and 300°K in applied fields up to 50 kG.

In this paper, we summarize the results obtained for the whole RE$_4$Au series. A detailed analysis of the data for the Gd (6)

Table 1.  Magnetic Properties of Amorphous RE$_4$Au Alloys.

| RE | $\mu_{eff}$ ($\mu_B$/RE at) | $\theta$ (°K) | $T_M$ (°K) | |
|----|----|----|----|----|
| Ce | 2.5 | − 43 | 1.9 | |
| Pr | 3.8 | − 28 | 5 | |
| Nd | 3.7 | − 10 | 12 | |
| Sm | − | | (∿150) | |
| Eu | 8.4 | 115 | 87 | (120) |
| Gd | 8.9 | 164 | 149 | |
| Tb | 8.1 | 215 | 69 | (120, 220) |
| Dy | 10.2 | 103 | 39 | (85, 180) |
| Ho | 10.8 | 50 | 21 | (7, 14) |
| Er | 9.5 | 31 | 12 | |
| Tm | 7.2 | 5 | 4 | |

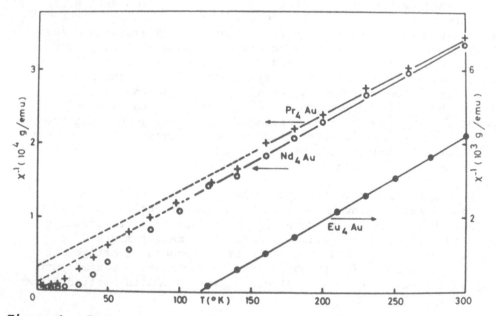

Figure 1.   Inverse susceptibility versus temperature for amorphous Pr$_4$Au, Nd$_4$Au and Eu$_4$Au.

and Ce (7) based alloys was already reported. On the other hand, the results for the alloys with heavy lanthanides (from Tb to Tm) will be presented somewhere else (8). In the following, we will focus on amorphous $RE_4Au$ alloys with RE = Pr, Nd, Sm and Eu.

## DISCUSSION

The magnetic properties (effective moments $\mu_{eff}$, paramagnetic Curie temperature $\theta$ and ac susceptibility maxima $T_M$) of the amorphous $RE_4Au$ alloys are summarized in Table 1. For most of the alloys, the effective moments as determined from the high-temperature Curie-Weiss behaviour compare rather well with the $3^+$ ionic values. No Curie-Weiss law was observed between 1.8 and 300°K for the initial susceptibility of amorphous $Sm_4Au$. A slight anomaly occurs in the susceptibility around 150°K, the magnitude of which strongly varies from foil to foil. Such an anomaly together with the small remanence occurring at temperatures close to 150°K leads us to describe this amorphous $Sm_4Au$ as an assembly of fine magnetic grains (whose concentration varies from 0.1 to 1 at. % depending upon the quality of the sample)

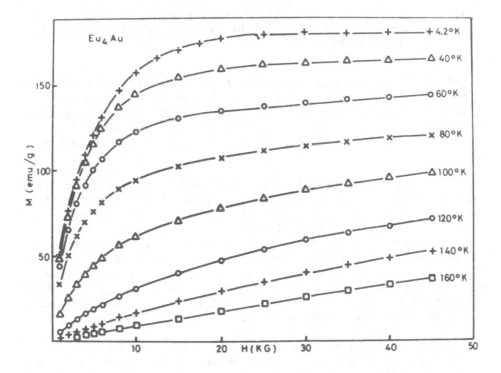

Figure 2.   Magnetization versus applied field at different temperature for amorphous $Eu_4Au$ (no correction was made for demagnetizing field).

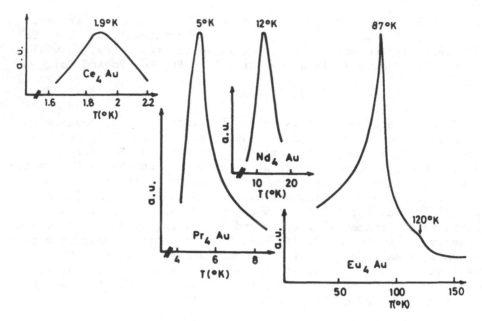

Figure 3.   Ac susceptibility (in arbitrary unit) at 3500 Hz ver-
            sus temperature for light rare-earth based amorphous
            RE$_4$Au alloys.

in solution within a non-magnetic matrix. Thus, the Sm ions in a-
Sm$_4$Au would be basically divalent, while the fine grains would be
trivalent. Photoemission and EXAFS measurements are in progress.
They could allow us to decide whether the presence of Sm$^{3+}$ in
addition to the Sm$^{2+}$ ions is intrinsic or due to sample contamination.

The divalent nature of Eu in amorphous Eu$_4$Au can be inferred
from the effective moment (Table 1) together with the saturation
moment of 6.4 $\mu_B$/Eu at. obtained at 4.2°K (Fig. 2). The sharp peak
in the ac susceptibility at 87°K (Fig. 3) (± 3°K according to
different sample foils) does not correspond to a paramagnetic to
ferromagnetic conventional transition for the bulk sample. According to dc field measurements (Fig. 2) such a transition seems to
occur around 120°K (see anomaly on the ac curve on Fig. 3), which
temperature is close to that obtained for $\theta$. The real significance
of the sharp peak at 87°K is not clear to us at present.

The high temperature Curie-Weiss law for amorphous Tb$_4$Au
yields for $\mu_{eff}$ a value which is anomalously low for a Tb$^{3+}$ ion
and for $\theta$ a value which is considerably higher than that obtained
for amorphous Gd$_4$Au. Similar anomalies were reported for crystalline Tb$_2$Au and Gd$_2$Au compounds (10). On the other hand, the ac

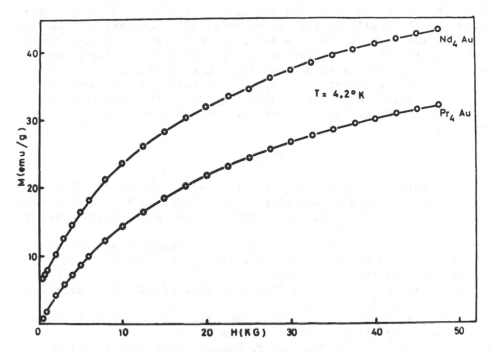

Figure 4.   Magnetization versus applied field at 4.2°K for amor-
            phous Pr$_4$Au and Nd$_4$Au alloys.


susceptibility measurements exhibited a slight anomaly at about
120°K and a steep decrease at around 220°K, in addition to a sharp
peak at 69°K. Similar features were obtained for the temperature
dependence of the ac susceptibility for amorphous Dy$_4$Au and Ho$_4$Au.
This will be discussed elsewhere (8) together with coercive field
and after-effect studies.

     The variation of θ through the RE$_4$Au series reflects in some
way the variation of the de Gennes's factor (F$_{DG}$). But, even after
taking apart the anomalous cases of Eu and Tb, no analytic dependence of θ on F$_{DG}$ is obtained, which might suggest that a RKKY
treatment of the magnetic interactions is not adequate for these
amorphous systems with very short electronic mean free  path.

     Finally, let us note that negative values of θ for Ce$_4$Au,
Pr$_4$Au and Nd$_4$Au are too large to be due only to antiferromagnetic
interactions (Fig. 4), but they reflect at least partly the effects of electric field gradients on the temperature dependence of
the susceptibility. The susceptibility maxima seem to be characteristic of concentrated spin glasses or mictomagnets (11). The de-
partures from the high-temperature Curie-Weiss law observed for
both a- Pr$_4$Au and Nd$_4$Au alloys at temperatures between 30 and 120°K

(see Fig. 1) are thought to arise mainly from "crystal field" effects. For both alloys, the ionic Curie-Weiss constant is thus reduced at low temperature by a factor of 0.65, which is more important that the effect (factor of 0.8) predicted by a simple uniaxial model of random anisotropy (2). This implies that nonaxial
electric field gradients (12) have to be taken into account. A
detailed analysis of these results together with those obtained
on dilute amorphous $La_4Au$ alloys with Pr and Nd will be given separately.

## REFERENCES

1.  J.J. Rhyne, Amorphous magnetic rare-earth alloys, in "Handbook
    on the Physics and Chemistry of Rare Earths", K.A. Gschneidner,
    Jr and L. Eyring, ed., North-Holland, Amsterdam (1979), vol.2,
    p. 259.
2.  R.W. Cochrane, R. Harris and M.J. Zuckermann, "The Role of
    Structure in the Magnetic Properties of Amorphous Alloys",
    Physics Reports 48 : 1 (1978).
3.  N.Heimann and N. Kazama, "Magnetization of $RE_{1-x}Cu_x$ amorphous
    alloys", J. Appl. Phys. 49 : 1686 (1978).
4.  K. Pappa, "Crystal fields and magnetic interactions in amor-
    phous RE-Ag alloys", Thesis, Paris (1979).
5.  W.L. Johnson, S.J. Poon and P. Duwez, "Amorphous superconduc-
    ting lanthanum-gold alloys obtained by rapid-quenching",
    Phys. Rev. B 11 : 150 (1975).
6.  S.J. Poon and J. Durand, "Critical phenomena and magnetic pro-
    perties of an amorphous ferromagnet : gadolinium-gold", Phys.
    Rev. B 16 : 317 (1977).
7.  J. Durand and S.J. Poon, "Crystal field effects and magnetic
    interactions in amorphous $Ce_4Au$", J. Appl. Phys. 49 : 702
    (1978) and to be published.
8.  A. Berrada, J. Durand, N. Hassanain and B. Loegel, "Anisotro-
    py versus exchange in amorphous heavy-rare-earth based alloys",
    Intermag - MMM Conference, New-York (July 1979).
9.  K.A. Mc Ewen, "Magnetic and transport properties of the rare
    earths", in reference 1, vol. 1, p. 411.
10. J.K. Yakinthos, P.F. Ikonomou and T. Anagnostopoulos, "Magne-
    tic properties of $RE_2Au$ compounds", Journal of Magnetism and
    Magnetic Materials $8^2$: 308 (1978).
11. W. Felsch, private communication, and Internat. Conf. on
    Magnetism, München (September 1979).
12. A. Fert and I.A. Campbell, "Nonaxial electric field gradients
    in amorphous rare-earth alloys", J. Phys. F 8 : L 57 (1978).

# MAGNETIC EXCITATIONS IN TbFe$_2$

J. J. Rhyne
National Bureau of Standards, Washington, DC  20234

N. C. Koon
Naval Research Laboratory, Washington, DC  20375

H. A. Alperin
Naval Surface Weapons Center, White Oak, MD  20910

Magnetic inelastic scattering studies have been performed as a function of temperature on ErFe$_2$ (1), HoFe$_2$ (2) and TbFe$_2$. All the heavy rare earth iron compounds of composition RFe$_2$ crystallize in the C15 Laves phase structure with lattice constants a $\approx$ 7.3 Å. They have Curie temperatures in the range 575 K to 700°K and exhibit a ferrimagnetic alignment of iron and rare earth spins. The iron moment is typically 1.5 $\mu_B$, and the rare earth has the full free ion moment at 0 K.

The six atoms in the primitive unit cell give rise to six ground-state spin wave modes, only three of which are at energies low enough to investigate by thermal neutron scattering. A linear spin wave model has been utilized (1) for these compounds which accurately represents these three modes as shown in the previous work on HoFe$_2$ and ErFe$_2$. Figure 1 shows the results of this model calculation compared to the observed excitation groups in TbFe$_2$. The lower "acoustic" mode corresponds to an in-phase precession of all spins with a bandwidth determined principally by the rare earth-iron exchange. The flat mode which, in contrast to ErFe$_2$ and HoFe$_2$, has not been observed in this study of TbFe$_2$, represents an out-of-phase precession of the rare earth spins. The highest steeply-dispersive mode is an in-phase precession of the iron-spins. This mode, which was observed in the previous studies of HoFe$_2$ and ErFe$_2$, has a dispersion ($\omega = Dq^2$) nearly identical to that in iron metal. The absence of scattering in TbFe$_2$ corresponding to the two higher modes is not understood, although it would be expected to be weak due to the small size of the crystal and the relatively higher energy of the modes in TbFe$_2$. The values shown in the figure for the Fe and Tb angular momenta have been determined in a separate magnetic diffraction experiment. The terbium-iron exchange parameter was determined

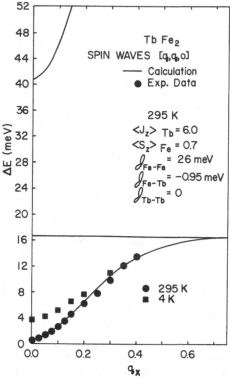

Figure 1.   Inelastic magnetic scattering data and model calculation
for TbFe$_2$.   Acoustic mode results are shown for both room
temperature and at 4 K (experimental points only) where
the q = 0 gap is increased to 3.75 meV due to the crystal
field anisotropy.

from fitting the model to the observed acoustic-mode data, while the
remaining two exchange constants were fixed at values found from the
previous HoFe$_2$ and ErFe$_2$ studies.

   An anomalous broadening and decrease in intensity of the spin
wave groups at all temperatures studied was observed for q $\gtrsim$ 0.15 A$^{-1}$.
For q values larger than those shown, the spin waves were not resolv-
able.   This effect may result from lifetime broadening or by spurious
scattering from additional crystallites which effectively make q an
invalid quantum number except for q $\approx$ 0.

REFERENCES

1.   J. J. Rhyne, N. C. Koon, J. B. Milstein and H. A. Alperin, "Spin
Waves in ErFe$_2$," Physica 86–88B, 149 (1977) and N. C. Koon and
J. J. Rhyne, "Excited State Spin Waves in ErFe$_2$," Solid State
Commun. 26, 537 (1978).

2.   J. J. Rhyne and N. C. Koon, "Magnetic Excitation in HoFe$_2$," J.
Appl. Phys. 19, 2133 (1978).

# MAGNETIC STRUCTURES OF THE Mn-RICH $Y_6(Fe_{1-x}Mn_x)_{23}$ COMPOUNDS

Kay Hardman, W. J. James, and Gary J. Long

Department of Chemistry and the Graduate Center for
Materials Research, University of Missouri-Rolla,
Rolla, MO 65401, and

W. B. Yelon

University of Missouri Research Reactor, Columbia, MO
65201, and

B. Kebe
C.N.R.S., Laboratoire Louis Néel, Grenoble, France

## ABSTRACT

The crystallographic and magnetic structures of $Y_6(Fe_{0.1}Mn_{0.9})_{23}$ and $Y_6(Fe_{0.2}Mn_{0.8})_{23}$ have been determined at 473 and 77 K. Atomic ordering effects show Fe atoms prefer the $f_1$ site and Mn atoms prefer the b and $f_2$ sites. Mossbauer effect studies show the Fe atoms to have no spontaneous moment. Therefore, long range magnetic ordering depends solely on Mn-Mn interactions. The Mn moments in these ternary compounds are greatly reduced from those found in $Y_6Mn_{23}$; however, the ferrimagnetic coupling remains the same.

## INTRODUCTION

The $Y_6(Fe_{1-x}Mn_x)_{23}$ compounds exhibit unusual magnetic behavior in that both the Curie temperature and bulk magnetization values are reduced considerably from those of the end members. In the intermediate composition region, x = 0.5 to 0.75, there exists no

315

long range magnetic ordering.  Both $Y_6Mn_{23}$ and $Y_6Fe_{23}$ are magnetically ordered with different magnetic moments on the four transition
metal crystallographic sites. (1)  The Fe moments in $Y_6Fe_{23}$ are
ferromagnetically coupled while the Mn moments on the b and d sites
are coupled antiparallel to those on the two f sites in $Y_6Mn_{23}$.

    Several experiments have been undertaken in an effort to under-
stand the magnetic interactions of the transition metal atoms in
these compounds.  The temperature dependence of the electrical
resistivity of $Y_6(Fe_{1-x}Mn_x)_{23}$ compounds indicates that the binaries
are good conductors. (2)  There is a pronounced increase of the
residual resistivity in the ternaries.  A striking feature of these
ternaries is the decrease in resistivity with increasing temperatures
in the compositional range x = 0.2 to 0.95 from 4 K to 300 K.

    Heat capacity results by Bechman et al. (3) showed that the
electronic contribution ($\gamma$) values reach a maximum at x = 0.4 and
are large throughout the composition range where the $T_c$ values are
zero.  These authors concluded that the disappearance of the magnetization in the intermediate concentration range can be explained
by an increasing antiferromagnetic coupling between the 3d moments.
When Fe is introduced in the ferrimagnetic compound $Y_6Mn_{23}$, the
lattice contracts.  The reduced nearest-neighbor distance enhances
the antiferromagnetic Mn-Mn exchange, which in turn decreases the
magnetization and the Curie temperature.  The lattice contraction
could also affect the valence of the Mn atoms and thus the magnitude of the Mn moments.

    The pronounced decrease in magnetization in the $Y_6(Fe_{1-x}Mn_x)_{23}$
system is not well understood.  Because the decrease is particularly
striking on the Mn-rich side of the system, Mossbauer effect and
neutron diffraction studies of $Y_6(Fe_{0.1}Mn_{0.9})_{23}$ and $Y_6(Fe_{0.2}Mn_{0.8})_{23}$
have been undertaken.  The magnetic structure of the ternary com-
pounds are complex in that both Mn and Fe moments might exist on
each of the four transition metal sites.  Because the nuclear
scattering lengths of Fe and Mn are quite different, $b_{Mn}$ = -0.37 and
$b_{Fe}$ = 0.95 x $10^{-12}$ cm, the number of Mn and Fe atoms on each site
can be determined by neutron diffraction techniques.  The
Mn and Fe magnetic form factors are of about the same value and
hence it is difficult to distinguish the magnetic contributions of
each atom on a particular site; however, the total magnetic moment
on each site is well known.  Thus Mossbauer effect studies will
determine the magnetic properties of the Fe atoms on each of the
four crystallographic sites.


                              EXPERIMENTAL

    Ingots of $Y_6(Fe_{0.1}Mn_{0.9})_{23}$ and $Y_6(Fe_{0.2}Mn_{0.8})_{23}$ were prepared
by induction melting of the elements (99.9% purity) in a water-cooled
Cu boat under an Ar atmosphere.  X-ray diffraction photographs of

both annealed and unannealed samples showed no evidence of other
phases.  The neutron diffraction data for the two powder samples
were collected on a two-axis neutron diffractometer at the
University of Missouri Research Reactor by using $\lambda = 1.103$ Å (80%
Mn) and $\lambda = 1.293$ Å (90% Mn).

Refinement of the data was carried out by using the Rietveld
profile method. (4)  The magnetic form factors for Mn and Fe were
obtained from the literature. (5)  No constraint was used for the
total Mn and Fe content but a constraint was imposed such that each
transition metal site was completely occupied by Fe and Mn atoms.
This permitted a refinement of the stoichiometry of the ternary
compound.  The magnetization per formula unit for each compound
was taken from the literature (3) and imposed as a constraint on
the summation of the magnetic moments for all possible sites.

The Mossbauer effect spectra were obtained on a Ranger Elec-
tonics Inc. constant-acceleration spectrometer which utilizes a
room-temperature rhodium-matrix source and was calibrated with
natural $\alpha$-iron foil.  The Mossbauer spectra were evaluated by using
least-squares minimization computer programs.  The Mossbauer effect
parameters are estimated to be accurate to at least $\pm 0.005$ mm/s.
The fit to a quadrupole doublet was constrained to have equal areas
and line widths for the two components of the doublets.

## RESULTS AND DISCUSSION

The neutron diffraction results are summarized in Table I.  The
atomic ordering in $Y_6(Fe_{0.2}Mn_{0.8})_{23}$ and $Y_6(Fe_{0.1}Mn_{0.9})_{23}$ was determined above the Curie temperature (473 K).  It is evident that the
Mn and Fe atoms exhibit strong site preference.  If it is assumed
that Fe and Mn distribute themselves stoichiometrically on each of
the four crystallographic sites, a high correlation factor ($R_{stoic}$)
results.  However, if the number of Mn and Fe atoms on each of the
sites is allowed to vary, the R-factor ($R_{order}$) is lowered considerably.  The Fe atoms prefer the $f_1$ site and the Mn atoms prefer
the $f_2$ and b sites.  The Mn and Fe atoms are distributed stoichiometrically on the d site.  Annealing appears to have little effect
on the ordering of the Mn and Fe atoms and the magnetic properties
of these compounds.

Generally, the magnitude of the Fe moments is not dependent on
composition.  One would expect the Fe saturation moment to be about
2 $\mu_B$.  However, if the d band were to be filled through electron
transfer, a moment of zero would result.  Neutron diffraction techniques were unable to differentiate between the two models based on
these moments.  Accordingly Mossbauer spectra were obtained for both
compounds at 295 and 78 K.  The spectra of $Y_6(Fe_{0.2}Mn_{0.8})_{23}$ above
and below the Curie temperature are shown in Figure 1.  The results
for both compounds are given in Table II.  The striking result is

Table I.  Neutron Diffraction Results of $Y_6(Fe_{0.2}Mn_{0.8})_{23}$ and $Y_6(Fe_{0.1}Mn_{0.9})_{23}$

| | NUCLEAR (473 K)* | | MAGNETIC (77 K)** | |
| --- | --- | --- | --- | --- |
| | $Y_6(Fe_{0.2}Mn_{0.8})_{23}$ | $Y_6(Fe_{0.1}Mn_{0.9})_{23}$ | $Y_6(Fe_{0.2}Mn_{0.8})_{23}$ | $Y_6(Fe_{0.1}Mn_{0.9})_{23}$ |
| Y | 6.0(100) | 6.0(100) | -- | --   -0.10(0.04) |
| Mn1 | 0.896(90) | 0.990(99.0) | -3.42(0.3) | -3.71   -3.53(0.14) |
| Fe1 | 0.102(10) | 0.010(1.0) | -- | --   -- |
| Mn2 | 4.793(79.9) | 5.426(90.0) | -0.49(0.15) | -1.78   -1.79(0.06) |
| Fe2 | 1.208(20.1) | 0.575(10.0) | -- | --   -- |
| Mn3 | 5.406(67.6) | 6.602(82.5) | 0.69(0.13) | 1.56   1.58(0.05) |
| Fe3 | 2.597(32.4) | 1.401(17.5) | -- | --   -- |
| Mn4 | 7.632(95.4) | 7.890(98.6) | 0.87(0.06) | 1.23   1.32(0.03) |
| Fe4 | 0.371(4.6) | 0.113(1.4) | -- | --   -- |
| Mntotal | 18.727(81.4) | 20.908(90.9) | | |
| Fetotal | 4.278(18.6) | 2.986(9.1) | | |
| $\mu_B$/f.u. | | | 5.0 | 7.0   7.0 |
| $R_{order}$ | 5.1 | 4.7 | 4.7 | 4.9   4.5 |
| $R_{stoic}$ | 45.3 | 25.2 | | |
| $R_{mag}$ | | | 7.8 | 6.0   5.0 |

*Values are given in the number of atoms per site with the percentage of Mn or Fe for that site given in parentheses.

**Values are given in $\mu_B$/Mn.  The standard deviation of these values is given in parentheses. The negative sign indicates the direction of the Mn moment.

Table II.  Mossbauer Effect Spectra Parameters*

| COMPOUND | T(K) | $\delta$ | $\Delta E_Q$ | $\Gamma$ | $\chi^2$ | MODEL |
|---|---|---|---|---|---|---|
| $Y_6(Fe_{0.1}Mn_{0.9})_{23}$ | 295 | −0.12 | −− | 0.86 | 1.00 | Singlet |
| | | −0.12 | 0.20 | 0.81 | 1.00 | Doublet |
| | 78 | −0.01 | −− | 0.96 | 1.00 | Singlet |
| | | −0.01 | 0.01 | 0.96 | 1.00 | Doublet |
| $Y_6(Fe_{0.2}Mn_{0.8})_{23}$ | 293 | −0.13 | −− | 0.49 | 1.02 | Singlet |
| | | −0.13 | 0.01 | 0.46 | 1.02 | Doublet |
| | 78 | −0.03 | −− | 0.53 | 1.21 | Singlet |
| | | −0.03 | 0.01 | 0.53 | 0.98 | Doublet |

*All data in mm/s relative to natural α-iron foil.

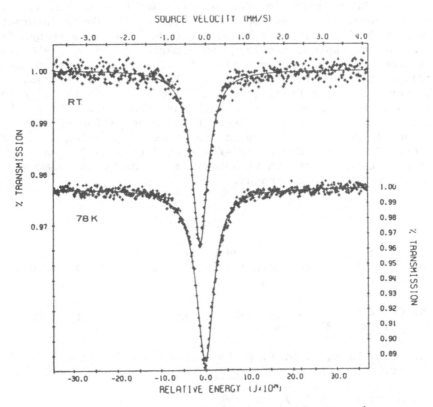

Figure 1. Mossbauer Spectra of $Y_6(Fe_{0.2}Mn_{0.8})_{23}$

that Fe has no spontaneous moment.  Therefore, the long range
magnetic ordering arises solely from Mn-Mn interactions.  The chemical
shift of Fe in these compounds is very similar to that of $\alpha$-Fe foil.
The fit to a doublet could arise either from a reduced symmetry at
the iron sites or from slightly different chemical environments.
The $\Delta E_Q$ value given in Table II would represent the maximum possible
quadrupole interaction.  The line width of the $Y_6(Fe_{0.1}Mn_{0.9})_{23}$
spectra is greater than that of $Y_6(Fe_{0.2}Mn_{0.8})_{23}$ suggesting a lower
degree of atomic order in the former compound.

   The magnetic properties of the compounds derived through neutron
diffraction techniques are given in Table I.  The coupling of the Mn
moments is the same as that in $Y_6Mn_{23}$ with the b and d sites coupled
antiparallel to the two f sites.  However, the magnitudes of the Mn
moments, except for site b, are greatly reduced from those in $Y_6Mn_{23}$.
The Mn moments on the b site probably remain constant in that Mn
atoms are surrounded by Mn atoms on the $f_2$ site.  The $R_{mag}$ factor
for the $Y_6(Fe_{0.2}Mn_{0.8})_{23}$ compound is reasonable in that there is
very little coherent magnetic scattering present.

   The Mn-Mn interactions are weaker than those in $Y_6Mn_{23}$.  The
substitution of Fe for Mn appears to disrupt the Mn-Mn interactions
by increasing the distances between the Mn atoms.  As enough interactions are disrupted, the long range magnetic ordering disappears.
It is known that the concentration of delocalized electrons has
decreased in the Mn-rich region. (2)  Therefore, it is possible
that there are not sufficient delocalized electrons to align the
Mn moments for long range ordering.  The atomic ordering in these
compounds and, to a lesser extent, the concentration of delocalized
electrons leads to a decrease in the magnetic ordering of the Mn
moments.  Presently, Mossbauer effect studies at 4.2 K in a magnetic
field have been undertaken to determine if the Fe is paramagnetic or
nonmagnetic in these compounds.


REFERENCES

1.  K. Hardman, W. J. James, and W. Yelon, "The Rare Earths in
    Modern Science and Technology," Plenum Publishing Corporation,
    New York (1978), p. 403.

2.  E. Gratz and H. R. Kirchmayr, J. of Magnetism and Magnetic
    Materials 2:187 (1976).

3.  C. A. Bechman, K. S. V. L. Narasimhan, W. E. Wallace, R. S.
    Craig, and R. A. Butera, J. Phys. Chem. Solids 37:245 (1976).

4.  H. M. Rietveld, J. Appl. Cryst. 2:65 (1969).

5.  G. E. Bacon, "Neutron Diffraction," Clarendon Press, Oxford,
    England (1975), p. 227.

# MAGNETOSTRICTION IN SOME $R_6Fe_{23}$ RARE EARTH INTERMETALLIC COMPOUNDS[*]

F. Pourarian, W. E. Wallace, S. G. Sankar, Department of
Chemistry, University of Pittsburgh, Pittsburgh, PA 15260

and R. Obermyer, Department of Physics, Pennsylvania
State University, McKeesport, PA 15632

## ABSTRACT

Anisotropic magnetostriction measurements of $R_6Fe_{23}$, R = (Tb,
Dy, Ho and Er) have been carried out from 77 K to room temperature.
Magnetic fields up to 2.1 Tesla were applied.  All the compounds
exhibited large magnetostrictions at 77 K, the largest effect being
obtained for $Tb_6Fe_{23}$.  Saturation magnetostriction values $\lambda_s$ for the
compounds were determined at 77.4 K and room temperature.  Results
of the temperature dependence of magnetostriction for $Er_6Fe_{23}$ are
in good agreement with Callen and Callen's single ion theory.  Therefore, the main source of magnetostriction in this compound is the
Er ion.

## INTRODUCTION

The magnetic properties of rare earth (R) and transition (T)
elements have been the subject of great interest [1].  The compounds
of formula $R_6Fe_{23}$ crystallize with cubic $Th_6Mn_{23}$ structure type [2].
The iron atoms occupy four nonequivalent lattice sites.  From bulk
magnetization measurements of Hilscher and Rais [3] on $Er_6Fe_{23}$,
Van der Goot and Buschow [4] on $Dy_6Fe_{23}$, and Buschow [1] on $Ho_6Fe_{23}$,
it is evidenced that in these compounds the rare earth moments are
coupled ferrimagnetically to the resultant of the iron moments below
the magnetic ordering temperature.  In all cases the magnetic ordering temperatures determined by Salmons [5] were found to be well
above room temperature.

---

[*]Work assisted by a grant from the National Aeronautics and Space
Administration.

In magnetic materials the magnetoelastic coupling arises from the strain dependence of the exchange interaction and the orbital interaction between a magnetic ion and the local crystalline electric field.  Large magnetostriction values have been observed for the cubic Laves $RFe_2$ (6) and $RCo_2$ (7) compounds.  In $RFe_2$, the source of this huge magnetostrictive effect was established to arise from the single-ion anisotropy of R sublattice moments (6,8).  In this paper we present the first experimental results of magnetostrictive strains in the polycrystalline samples of $Tb_6Fe_{23}$, $Dy_6Fe_{23}$, $Ho_6Fe_{23}$ and $Er_6Fe_{23}$ over the temperature range 77-300 K.  Some of these compounds, due to their large magnetostriction values, may have a considerable importance for technical applications.

## EXPERIMENTAL PROCEDURE

The bulk samples were prepared by melting the constituents in an induction furnace in a water-cooled copper boat under an argon atmosphere.  The purity of the constituents was 99.9% for rare earth metals and 99.99% for iron.  The resultant buttons were then annealed under high vacuum at 1100°C for periods between two to six weeks. X-ray analysis of the powder of the annealed samples obtained revealed the presence of a small amount of second phase.  For the measurements, the samples were cut into cylinders 5 mm in diameter and 2 mm long.

Measurements of linear magnetostriction were made using electrical resistance strain gauges (Micromeasurements type WK-05-031DE-350) in an applied magnetic field of 2.1 Tesla.  The strains were measured with the field applied in parallel and then perpendicular to the gauge.  These two strains are denoted by $\lambda_\parallel$ and $\lambda_\perp$ .

## RESULTS AND DISCUSSION

The linear magnetostriction for cubic crystals may be described in terms of the conventional magnetostriction,

$$\frac{\delta \ell}{\ell} = 3/2 \; \lambda_{100}(C_1 - 1/3) + 3 \; \lambda_{111} C_2,$$

where $\lambda_{111}$ and $\lambda_{100}$ are magnetostriction constants and $C_1$, $C_2$ are constants related to the average directions of magnetization and measurements.  In the case of polycrystalline materials in which the preferred domain or grain orientation is absent, the magnetostrictive strain at the saturation can be shown to take the form: $\lambda_s = 2/5 \; \lambda_{100} + 3/5 \; \lambda_{111}$.  Our results of magnetostriction are presented in terms of $\lambda_t = \lambda_\parallel - \lambda_\perp$ and for true saturation this equals $3/2 \; \lambda_s$.  The isotropic distribution of the domains in our polycrystalline sample below the magnetic ordering temperature was verified by performing the measurements in the direction parallel and perpendicular (cooling direction) to the plane of the disc of the samples. The strains observed for the two cases differed by approximately 2 to 3%, which is within the experimental errors.

The magnetostriction isotherms for the compounds studied well
below the Curie point are shown in Fig. 1a,b,c. Saturation values,
$\lambda_s$ determined at room temperature and 77.4 K were obtained from
extrapolation of the high field region (> 1.4 Tesla) of the strain-
field curves to H = ∞, except for Ho$_6$Fe$_{23}$ and Er$_6$Fe$_{23}$ at 77.4 K,
where $\lambda_s$ values were defined at 2.1 Tesla. The results are given
in Table 1.

Table 1.   Saturation Magnetostriction ($\lambda_s$) of Some R$_6$Fe$_{23}$ Compounds

| Compound | $10^6 \lambda_s (=2/3 \lambda_t)$ | |
|---|---|---|
| | 295 K | 77.4 K |
| Tb$_6$Fe$_{23}$ | 633 | 2130 |
| Dy$_6$Fe$_{23}$ | 293 | 240 |
| Ho$_6$Fe$_{23}$ | 60[a] | 137[a] |
| Er$_6$Fe$_{23}$ | - 43[a] | -230 |

a.   Values determined at 2.1 Tesla. Other values obtained by
extrapolation to 1/H = 0.

Among all the compounds studied, the highest magnetostriction
was observed for Tb$_6$Fe$_{23}$. The lack of saturation of strains observed at maximum field of 2.1 Tesla (Fig. 1a) indicates that this compound is highly anisotropic. The strain value resulting from extrapolation to H = ∞ at 77.4 K is much larger than the value determined at 2.1 Tesla, at the same temperature. This implied that
fields of the order of tens of Tesla will be needed to produce saturation at 77.4 K (or at 4.2 K). The room temperature value of $\lambda_t$
is very close to that expected for a Tb content of 43% by wt. from
the work of Clark (6).

Similar to the case of Tb$_6$Fe$_{23}$, the $\lambda_t$-H curves (not shown)
for Dy$_6$Fe$_{23}$ did not show saturation up to maximum applied fields.
A remarkable decrease in the magnitude of the strain at 2.1 Tesla
below about 235 K was observed for this material. Magnetization
measurements which were reported previously (4) indicated that the
compound exhibited a compensation point at about 265 K, where the
magnetization shows a minima. The reduction in strain suggests a
rather strong increase in the anisotropy energy of the compound
below the compensation point.

In the case of Ho$_6$Fe$_{23}$, the strain-field curves all appear to
saturate at about 0.8 Tesla (Fig. 1b). The temperature dependence
of $\lambda_t$ at saturation exhibited a normal behavior. Extrapolation of

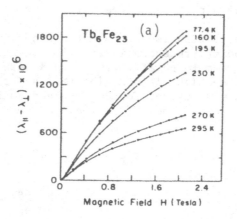

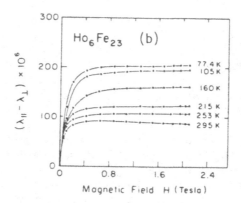

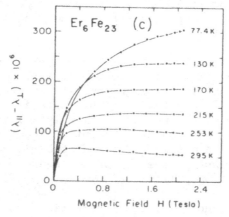

Fig. 1. Magnetostriction isotherms for $Tb_6Fe_{23}$, $Ho_6Fe_{23}$ and $Er_6Fe_{23}$.

this curve below 77.4 K yielded a value of 237 x $10^{-6}$ at 0 K.

For $Er_6Fe_{23}$, the magnetostriction isotherm curves exhibited saturation above 130 K (Fig. 2). At 77.4 K saturation of strains was not observed up to an applied field of 2.1 Tesla. Results of temperature dependence of magnetization reported by Hilscher and Rais (3) showed a compensation point at 105 K. These authors predicated that at this temperature a noncollinear arrangement of the rare earth with iron moments may occur, by applying a strong magnetic field (> 3 Tesla). Our results of lack of saturation of strain at 77 K confirm their conclusion that the compound below 105 K becomes highly anisotropic, in which the anisotropy constants increase drastically.

For all the compounds studied the form of $\lambda_t$-H curves suggest that magnetostriction constants $\lambda_{100}$ and $\lambda_{111}$ for each compound are of the same sign.

The temperature dependence of magnetostriction of $Er_6Fe_{23}$ was analyzed and found to follow the single-ion magnetoelastic theory, as treated by Callen and Callen (9):

$$\lambda_t(T) = \lambda_t(o) \, I_{5/2}[L^{-1}(m_R)]$$

where $I_{5/2}$ is the reduced hyperbolic Bessel function, $L^{-1}$ is the inverse Langevin function and $m_R$, reduced sublattice magnetization of the rare earth ions. Values of sublattice magnetization of Er as a function of temperature were taken from the recent magnetization work of Hilscher and Rais (3)   Using $\lambda_t(o) = -400$ x $10^{-6}$, the

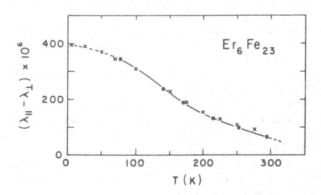

Fig. 2.   Temperature dependence of magnetostriction of $Er_6Fe_{23}$. The crosses are values calculated using the single-ion theory. The solid curve is best fit through the experimental data.

temperature dependence is deduced from the above equation and is shown in Fig. 2. Excellent agreement of the results with the single-ion model is evident. This indicates that the source of the magneto-strictive effects in $Er_6Fe_{23}$ is mainly the strain-dependent anisotropy of the Er sublattice.

Assuming the observed magnetostriction originates primarily with the interaction of the rare earth ions with the crystalline electric field, we can make use of the theory of Tsuya et al. (10) to estimate the saturation magnetostriction of the several $R_6Fe_{23}$ compounds at 0 K, based upon the extrapolated value of $\lambda_t$ at 0 K ($-400 \times 10^{-6}$) for $Er_6Fe_{23}$; $\lambda_t(o) = C \, \alpha_J \, <r_f^2> \, <<O_2^0(J_z)>>$ where $\alpha_J$ is the Steven's coefficient (11), $<r_f^2>$ is the mean square of the 4f orbital radius and $<O_2^0>$ is the expectation value associated with crystal field operator $O_2^0$. Using the approximation form of the latter in terms of the angular momentum J, $<O_2^0>_{J=J_z} \simeq 2J(J-1/2)$, then the total strain $\lambda_t(o) = C \, \alpha_J \, <r_f^2> \, J \, (2J-1)$. The calculated results using the above method are: $+412 \times 10^{-6}$, $+1040 \times 10^{-6}$ and $+1080 \times 10^{-3}$ for $Ho_6Fe_{23}$, $Dy_6Fe_{23}$ and $Tb_6Fe_{23}$, respectively. These were compared with the values deduced experimentally. In all cases the signs of $\lambda_t(o)$ deduced from the measurements are in agreement with those predicted by the theory for the corresponding R ions. In the case of $Dy_6Fe_{23}$ and $Tb_6Fe_{23}$ the calculated values of $\lambda_t(o)$ are the same, whereas the observed saturation results show a large difference. This may be partly due to uncertainty in the extrapolation of $\lambda_t$ to $H = \infty$ for $Tb_6Fe_{23}$.

## REFERENCES

(1)  W. E. Wallace, Rare Earth Intermetallics, Academic Press Inc., 1973, Chap. 11. See also K. H. J. Buschow, Rep. Progr. Phys. 40, 1179 (1977).

(2)  O. S. Zarechnyuk and P. I. Kripyakevich, Dopov. Akad. Nauk. Ukr. R. S. R., 1593 (1964).

(3)  G. Hilscher and H. Rais, J. Phys. F. 8-3, 511 (1978).

(4)  A. S. Van der Goot and K. H. J. Buschow, J. Less-Common Metals 21, 151 (1970).

(5)  R. L. Salmans, Tech. Rep. for Wright-Patterson Air Force Base, AFML-TR-68-159 (1968).

(6)  A. E. Clark, AIP Conf. Proc. 18, 1015 (1974).

(7)  E. W. Lee and F. Pourarian, phys. stat. sol.(a) 34, 383 (1976).

(8)  A. E. Clark, R. Abbundi, H. T. Savage and O. D. McMasters, physica 86-88B, 73 (1977).

(9)  E. R. Callen and H. B. Callen, Phys. Rev. 129, 578 (1963).

(10) N. Tsuya, A. E. Clark and R. M. Bozorth, Proc. Int. Conf. Magn., Nottingham, 250 (1964).

(11) K. W. N. Stevens, Proc. Phys. Soc.(London) 65, 209 (1952).

# TRANSPORT AND MAGNETIC PROPERTIES OF THE $Gd(Co_xNi_{1-x})$ SERIES

# $(0 \leq x \leq 0.3)$

E. Gratz, G. Hilscher, H. Kirchmayr and H. Sassik

Institut für Experimentalphysik, Technical University

Karlsplatz 13, A-1040 Vienna / Austria

## INTRODUCTION

It is known that of the 3d-metals only Ni forms with RE metals compounds with the most simple stoichiometry AB, i.e. 1:1. These compounds can crystallize in the FeB or CrB structure (1). The primary candidate for R was Gd, because $Gd^{3+}$ is a definite S-state ion. Since GdNi possesses the CrB-structure, one possible candidate for a comparison where the magnetic moment at the A site is 0 is La. In order to study the influence of a partial substitution of Ni by Co the series $Gd(Co_xNi_{1-x})$ was investigated in detail, because one can expect that by substituting Co for Ni the Fermi-level decreases in energy and a partially filled d-band is formed. Possible consequences of this fact are the appearence of a 3d-moment, which changes the magnetic and transport properties, especially the ordering temperature, the electrical resistivity, the thermopower, etc. Irrespective of these properties a change of the Fermi level may induce a change of structure (2).

## RESULTS AND DISCUSSION

The samples were prepared by high frequency induction melting in a cold boat. The purity of the materials was 99.9 %. The phase purity was checked by x-ray diffraction and metallographic methods. From the change of the lattice constants with x a solid solubility limit of approximately x = 0.25 (i.e. 25 % Co) has been established. This critical concentration is in agreement with the change of the Curie temperature $T_c$ and of the magnetic moment M versus x (fig. 1). In both cases near $\bar{x}$ = 0.25 a more or less constant value for $T_c$ and M is reached. The fluctuations of M can be explained by unavoidable inhomogenities due to the presence of the second phase, probably $Gd_4(Co_xNi_{1-x})_3$. This pseudobinary system $(0.75 \equiv x \equiv 1)$ is at present being investigated along similar lines.

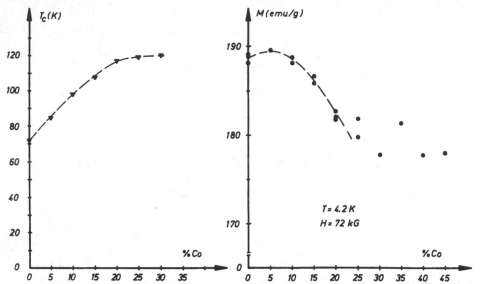

Fig. 1: Curie temperatures and magnetizations vs. Co-concentration.

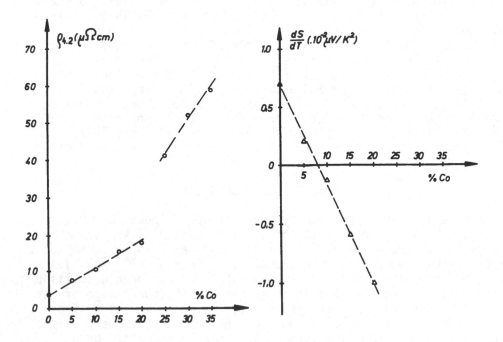

Fig. 2: $\varrho_{4.2}$ and dS/dT vs. Co-concentration.

The residual resistivity $\varrho_{4.2}$, i.e. the resistivity at 4.2 K, plotted in fig. 2, shows the same discontinuity at x = 0.25. The following experimental results are concerned with the homogeneous region, i.e. for x = 0 to x = 0.25. In fig. 3 the results of resistivity measurements versus temperature and composition are given.

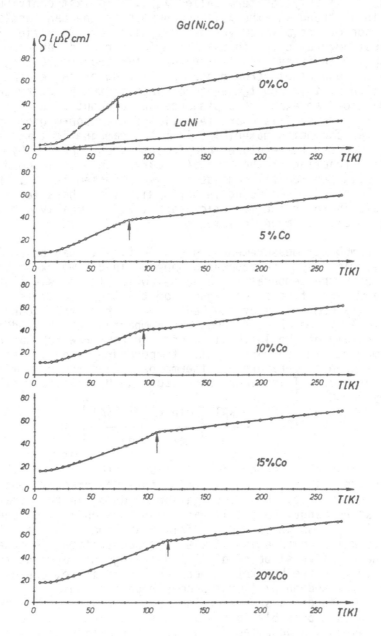

Fig. 3: $\varrho$(T) for various Co-concentrations ( ⬆ indicates T$_c$).

All the $\varrho$(T) curves are characterized by a sharp bend at the ordering temperature. As described in (3) the total resistivity is the sum of three contributions, namely $\varrho_0$ corresponding to the temperature independent scattering process of conduction electrons by lattice imperfections, $\varrho_{ph}$ arising from electron-phonon scattering processes and finally the term called $\varrho_{mag}$. This last contribution dictates in a strong way the $\varrho$(T) dependence as one can conclude by comparison of the $\varrho$(T) curves of GdNi with LaNi (see fig. 3). In the isostructural compound LaNi the $\varrho$(T) curves depend only on $\varrho_0$ and $\varrho_{ph}$. With increasing temperature starting from 4.2 K, $\varrho_{mag}$ increases because of electron-magnon scattering processes. An investigation of this low temperature range in the boundary compound GdNi shows an exact $T^2$ dependence up to about 17 K (4). Such a behaviour of $\varrho$ vs. T is characteristic of electron-magnon scattering in ferromagnetic materials. In the paramagnetic temperature range $\varrho_{mag}$ becomes temperature independent as one can see by a suitable substraction of the LaNi $\varrho$(T) curve. This is in good agreement with theoretical considerations discussed in (3). The shape of the $\varrho$(T) curves found in Gd(Co$_x$Ni$_{1-x}$) is characteristic of materials in which the magnetic properties are mainly dictated by strong localized magnetic moments.

The thermopower measurements, given in fig. 4, can be analyzed in the following way. Thermopower S depends in a complex manner on temperature in the temperature range below $T_c$. There is a clear evidence of a strong influence of magnons on the thermopower as can be concluded by a comparison of the LaNi behaviour to that found in the Gd(Ni,Co) system. Fig. 4 shows, by comparing LaNi and GdNi, that the anomaly of the latter is absent in former which has no magnetic moment. In the paramagnetic temperature range a linear relation between temperature and thermopower is detected similar to the $\varrho$(T) behaviour. This linear behaviour can be satisfactorily explained by formula (1)

$$S(T) = \frac{\pi^2 k_B^2 T}{3e} \left[ \frac{d\ln\Lambda}{d\varepsilon} + \frac{d\ln S(\varepsilon)}{d\varepsilon} \right]_{\varepsilon=\varepsilon_F} \tag{1}$$

$\Lambda$, mean free path of conduction electrons; S($\varepsilon$), area of the energy surface with energy $\varepsilon$, assuming that the bracket does not depend on temperature. The investigation of dS/dT in the paramagnetic temperature range shows a linear dependence on x (i.e. the Co concentration) as shown in fig. 2. This means that the bracket in equ. (1) is a linear function of the Co concentration. In order to explain the negative slope of dS/dT and taking into account that the first term in the bracket is necessarily positive, one has to assume that the second part must become negative, which is only compatible for specific shapes of the Fermi surface. It is therefore in principle possible to explain the constant negative slope of dS/dT; however, a detailed discussion must await further experimental or theoretical information concerning the density of states function and the shape of the Fermi surface in the alloys under discussion.

For GdNi a rather large high field susceptibility ($\chi_{HF}$ = 1.75. 10$^{-4}$ cm$^3$/g) is observed and values of the magnetic moment at 4.2K and 72 kG are found to vary from 7.25 - 7.3 $\mu_B$ for different samples. From the plot of the magnetic moment versus (R:Co)-ratio Poldy and Taylor (2) deduced that Ni has a zero transition metal moment which

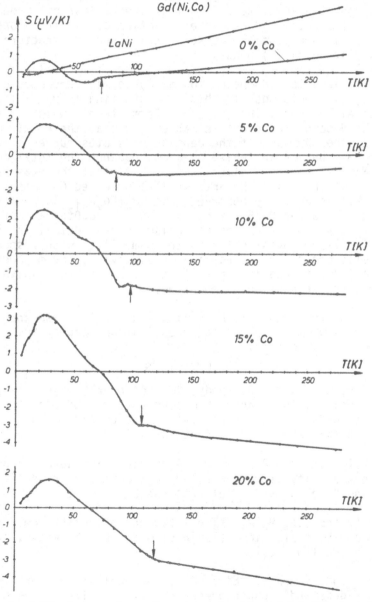

Fig. 4: S(T) for various Co-concentration ( ↟ indicated T$_c$).

is in accordance with the small difference between the saturation
moment and the paramagnetic moment of GdNi. With increasing Co-concentration the total magnetization is lowered approximately 5% up
to the border of the phase stability. Therefore we have assumed
that in the Ni-Co band a magnetic moment appears, which is coupled
antiferromagnetically to the Gd moment. This fact supports the assumption that by reducing the electron concentration the Fermi level
may reach at x = 0.25 a high density of states range. The same fact
can be described in a slightly different way: The reduction of the
magnetic moment exactly to 7.0 $\mu_B$ indicates a loss of conduction
electron polarisation, due to the reduction of electron concentration, which is one possible cause of the observed structural instability of the Gd(Ni,Co) phase.

The complementary compound to the Gd(Co,Ni) compounds are the
$Gd_4(Co_xNi_{1-x})_3$ compounds. In this case starting from x = 1 the addition of Ni obviously increases the Fermi level, which is situated
within the d-band. Therefore we can expect that the Fermi level
reaches a steep decrease of the density of states curves at which the
structure may become instable. As a consequence the total moment of
Co plus Ni must become smaller, which means that the measured magnetic moment including the antiparallel oriented Gd moment becomes
larger. This is actually observed. The $Gd_4(Co_xNi_{1-x})_3$ serie which
is stable for values $0.75 \leq x \leq 1$ is there a counterpart of the
$Gd(Co_xNi_{1-x})$ serie which has been discussed in detail in this
paper and will be investigated on the same lines with respect to
the magnetic properties, electrical resistivity, thermopower, etc.
These measurements and interpretations have not been finished and
will therefore be published separately.

We are grateful to D.I. R. Pfliegl and to P. Schwarzhans for
supporting us by metallographic investigations and by thermopower
measurements respectively.

## REFERENCES

1. H.R. Kirchmayr and C.A. Poldy, "Magnetic properties of rare
   earth metals", in: "Handbook on the physics and chemistry of
   rare earths", Vol. 2, Chapt. 14, K.A. Gschneidner, jr., ed.,
   North Holland Publ. Comp., Amsterdam (1978).

2. C.A. Poldy and K.N.R. Taylor, "A possible influence of 3d-states
   on the stability of rare earth-rich rare earth-transition metal
   compounds", phys. stat. sol. (a) 18:123 (1973).

3. E. Gratz and C.A. Poldy, "The influence of antiferromagnetic
   order on the evaluation of spin disorder resistivity", phys.stat.
   sol. (b) 82:159 (1977).

4. G. Adam and E. Gratz, "Einfluß von Spinwellen auf den elektri-
   schen Widerstand", paper presented at the fall-meeting of the
   Austrian Physical Society, Innsbruck (1978), to be published.

# NEUTRON DIFFRACTION STUDY OF $Th_2(Co,Fe)_{17}$

W. J. James and P. E. Johnson

Departments of Chemistry and Metallurgical Engineering

The University of Missouri-Rolla, Rolla, MO 65401

## INTRODUCTION

The $R_2M_{17}$ structures are related to the $RM_5$ structure by the ordered substitution of two transition metal atoms for every third rare earth atom. If the substitution repeats along $\underline{c}$ in every second $RM_5$ subcell, the hexagonal $P6_3/mmc$ structure shown in Figure 1a results. This may be written

$$6RM_5 - 2R + 4M \rightarrow 2R_2M_{17}$$

which is not reduced to the lowest denominator so as to indicate the actual cell content of 4R atoms and 34M atoms.

If the substitution repeats along $\underline{c}$ in every third $RM_5$ subcell then the rhombohedral $R\bar{3}m$ structure shown in Figure 1b results. This may be written

$$9RM_5 - 3R + 6M \rightarrow 3R_2M_{17}$$

where again the equation is not reduced to the lowest denominator. Since, for either the hexagonal or for the rhombohedral structure, one rare earth in three $RM_5$ subcells is substituted, the perturbation of the $RM_5$ 2c site reported by Deportes et al (1) will occur for all of the 2c sites resulting in the formation of either the rhombohedral $R_2M_{17}$ 18f site shown in Figure 1b, or the hexagonal 12j site shown in Figure 1a. Deportes et al (2) have also shown that this perturbation of the $RM_5$ 2c site is the major cause of the low anisotropy of the $R_2M_{17}$ compounds.

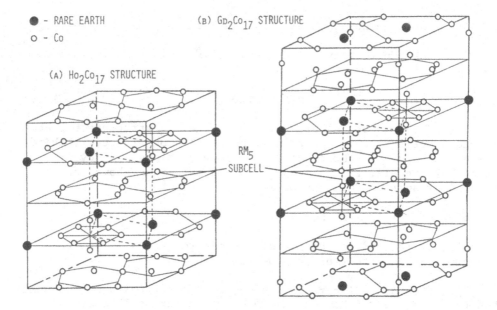

Figure 1.   Crystal structures of $Ho_2Co_{17}$ and $Gd_2Co_{17}$
as reported by Bouchet et al (7).

Since most of the $R_2M_{17}$ compounds exhibit basal plane magnetization, they are not important as permanent magnets.  However with
the substitution of some Fe for Co, many of the $R_2Co_{17}$ compounds
show a change of magnetization from the easy basal plane to an easy
c axis, indicating an increase in the anisotropy and in their possible
usefulness as permanent magnets.

Neutron diffraction studies of the $Th_2(Fe_{1-x}Co_x)_{17}$ system were
performed to determine possible ordering of the Fe and Co atoms on
nonequivalent crystallographic sites, which might explain the change
in the magnetic easy direction with the addition of Fe.  The Fe-Co
ratio for the transition metal 6c site is of particular interest
because this site is comparable to the transition metal pair substitutional site of the $RM_5$ structure.

## EXPERIMENTAL

### Preparation of the Samples

The compounds were prepared using Fe (99.95%), Co (99.9%) and
Th (99.9%) mixed in stoichiometric ratios to obtain various
$Th(Co_{1-x}Fe_x)_8$ compositions.  Although the hexagonal form for the
Th compounds does not exist, this stoichiometry was chosen to match
that used in studies of the $Y_2(CoFe)_{17}$ series where the hexagonal
form does occur.  Five to ten gram samples were melted by induction
heating in a cold copper crucible under a purified argon atmosphere.
The sample ingot was reversed several times to insure homogeneity.
The ingots obtained were annealed at 1000°C for 3 to 4 days under
vacuum.  Room temperature x-ray spectra of each sample indicated a
single phase $R\bar{3}m$ structure to be present.  The ingots were then
ground to about 100 mesh size.

### Collection of Neutron Data

The data were collected on the double axis, graphite monochro-
mated, Mitsubishi diffractometer located at the University of
Missouri-Columbia Research Reactor.

The ordering temperature, $T_c$, was determined by measuring the
intensity of several of the peaks as a function of temperature.
Neutron scans ($\lambda = 1.1113$ Å) from 8 to 55° in $2\theta$ were taken at 50°
above $T_c$, at ambient, and at liquid nitrogen temperatures.  Empty
furnace and empty cryostat scans had been obtained during previous
studies.

## Refinement

A modified Rietveld profile refinement program (3) was used initially to refine the atomic structure of $Th_2Co_{7.5}Fe_{8.5}$ using the data taken above $T_c$. This composition was chosen for refinement since the change to the easy $\underline{c}$ axis of magnetization had occurred. Only the data above $T_c$ were refined since any ordering of Fe and Co could be obtained from the nuclear data without the added complication of the magnetic structure contribution.

The rhombohedral $R_2M_{17}$ structure was used as the starting model. Coherent nuclear scattering factors of 1.03, 0.96 and $0.25 \times 10^{-12}$ cm were used for Th, Fe and Co respectively (4). The ratio of Fe to Co was allowed to vary on each transition metal site. However, when Co was placed on the 6c site, its population parameter went negative. Accordingly, this site was restricted to Fe atoms only. The best nuclear R factor obtained was 11.3%.

It was decided at this time to prepare Fourier difference maps

$$\Delta F_{hk\ell} = (|F_o| - |F_c|)e^{i\alpha_c}$$

where $F_o$ and $F_c$ are the observed and calculated structure factors respectively and $\alpha_c$ is the phase of the calculated structure factor.

Since the profile refinement program did not have provisions for obtaining Fourier maps, the refinement procedure was continued using the X-RAY 72 system (5) of crystallographic programs. Accordingly, 15 observed and integrated neutron intensities were obtained from the output of the profile program, which separated overlapping peaks based upon the ratio of the calculated intensities. The 15 reflections were chosen based upon the degree of resolution of the scan data. These reflections were introduced into the X-RAY 72 system where difference maps indicated some substitution at the position (0,0,0). These data were refined using the CRYLSQ program of the X-RAY 72 system. Again the ratios of Fe to Co were allowed to vary on each site including the 6c. In addition, some of the transition metal atoms of the 6c site were replaced by Th.

## RESULTS AND DISCUSSION

A final R factor of 3.5% was obtained for the parameters and structure factors given in Table 1. The final model resulted in a hyperstoichiometric ratio of $R_2M_{15}$ or, based upon the $R_2M_{17}$ unit cell, $Th_{2.22}Co_{8.52}Fe_{8.04}$, where the results are written this way to indicate the extent of loss of Fe-Co and the gain of Th.

There is a very strong preference for Fe on the 6c site as seen by the negative population parameter in the profile refinements

Table 1. Structure Parameters and Observed and Calculated Structure Factors.

Space Group $R\bar{3}m$
$\underline{a} = 8.521\text{Å}$  $\underline{c} = 12.441\text{Å}$ (Hexagonal Description)  $B = 1.8(6)\text{Å}^2$

| | Positions | Atoms | Occupancy |
|---|---|---|---|
| 3a | $(0,0,0)$ | Th | 0.22(18) |
| 6c | $\pm(0,0,z)$ $z = 0.350(8)$ | Th | 1.0 |
| 6c | $\pm(0,0,z)$ $z = 0.103(7)$ | Fe | 0.8(1) |
| | | Co | 0.011(2) |
| 9d | $(1/2,0,1/2;0,1/2,1/2;$ | Fe | 0.43(3) |
| | $1/2,1/2,1/2)$ | Co | 0.57(4) |
| 18f | $\pm(x,0,0;0,x,0;x,x,0)$ | Fe | 0.46(2) |
| | $x = 0.286(4)$ | Co | 0.54(2) |
| 18h | $\pm(x,\bar{x},z;x,2x,z;2\bar{x},x,z)$ | Fe | 0.41(1) |
| | $x = 0.166(9)$ $z = 0.488(9)$ | Co | 0.59(2) |

with translations of $(0,0,0;2/3,1/3,1/3;1/3,2/3,2/3)$

| hkl | $F_{cal.}$ | $F_{obs.}$ | hkl | $F_{cal.}$ | $F_{obs.}$ |
|---|---|---|---|---|---|
| 101 | 2.44 | 2.39 | 033 | 15.87 | 15.92 |
| 012 | 1.27 | 1.44 | 233 | -12.08 | -12.48 |
| 110 | 1.67 | 2.32 | 413 | 6.48 | 6.10 |
| 003 | 0.79 | 1.36 | 143 | 6.63 | 6.31 |
| 021 | -1.68 | -1.77 | 217 | -7.24 | -7.30 |
| 300 | 13.21 | 12.43 | 324 | -9.77 | -9.46 |
| 220 | 19.87 | 20.11 | 226 | 10.17 | 10.24 |
| 303 | 16.04 | 16.19 | | | |

$R_F = 3.5\%$

and in the final ratio of 72.7 iron atoms to cobalt atoms on the 6c site resulting from the CRYLSQ refinement. The other transition metal sites appear to have a slight preference for Co over Fe.

Perkins and Fischer (6) in a similar study of the rhombohedral $Y_2(CoFe)_{17}$ system also found a strong preference of Fe for the 6c site. Their work indicated a stoichiometry of $Y_2(Co_{0.31}Fe_{0.69})_{19.2}$, as obtained by microprobe analysis. In their model, the excess transition metal was found on pair substitution sites similar to those reported for hexagonal $R_2M_{17}$ compounds (7). This type of substitution is unlikely in the rhombohedral phase. If it did occur, the distance between the disordered pair substitution site and the 6c site would be too short (1.9Å). Perkins and Fischer (6) did state that an alternate model involving vacancies on the Y site would give equally good agreement with the data. In our study, a population parameter of about 22% occupancy for Th is found at the 3a site.

An interesting result of this study is that the refinement shows the Th 6c site not at 0,0,1/3 as has been reported elsewhere (7) but displaced to the position (0,0,0.35). Results similar to these were observed by Ray (8) for the compound $PrFe_7$, where he indicated a statistical distribution of one Pr atom and four Fe pairs in the rhombohedral $R_2M_{17}$ cell, with a shift of the atoms from their $Th_2Zn_{17}$ positions to those found by Zalkin, Sands and Krikorian (9) for $Nb_2Be_{17}$.

## ACKNOWLEDGEMENT

The authors acknowledge the support of the International Division of the National Science Foundation, Grant No. NSF INT76 02665.

## REFERENCES

1.  J. Deportes, D. Givord, J. Schweizer, and F. Tasset, Joint MMM Intermag. Conf., Paper 8E-3 (1976).

2.  J. Deportes, D. Givord, R. Lemaire, H. Nagai, and Y. T. Yung, J. Less-Common Metals 44:273 (1976).

3.  H. M. Rietveld, J. Appl. Cryst. 2:65 (1969).

4.  G. E. Bacon, Acta Cryst. A28:357 (1972).

5.  J. M. Stewart, G. J. Kruger, H. L. Ammon, C. Dickinson, and S. R. Hall, "The X-Ray System," Computer Science Center, University of Maryland, Maryland (1972).

6.  R. S. Perkins and P. Fischer, Solid State Communications 20:1013 (1976).

7.  G. Bouchet, J. Laforest, R. Lemaire, and J. Schweizer, C.R. Acad. Sc. Paris t. 262:1227 (1966).

8.  A. E. Ray, Acta Cryst. 21:426 (1966).

9.  A. Zalkin, D. E. Sands, and O. H. Kirkorian, Acta Cryst. 12:1713 (1959).

# MAGNETIC BEHAVIOUR OF TmAl$_2$ AT LOW TEMPERATURES

M. Loewenhaupt[+], S. Horn[+], H. Scheuer[*],
W. Schäfer[o] and F. Steglich[**]
[+]Institut für Festkörperforschung, KFA Jülich
[*]Institut Laue-Langevin Grenoble
[o]Mineralogisches Institut der Univ. Bonn
[**]Technische Hochschule Darmstadt

Most of the REAl$_2$ compounds (RE rare earth) are reported to order ferromagnetically /1/. Exceptions are the
intermediate valence system YbAl$_2$ /2/, the Kondo system
CeAl$_2$ /3/ and EuAl$_2$ /1/, which is the only member of
this family containing divalent RE ions. The REAl$_2$ compounds crystallize in the MgCu$_2$ structure, where the
RE ions build up a diamond lattice. The point symmetry
of these sites is cubic. In the case of TmAl$_2$, the
$(2J+1) = 13$ fold degeneracy of the Tm$^{3+}$ is partially
lifted by the crystal field, leaving a $\Gamma_5$ triplet as
ground state /4/.

In this note we report neutron scattering experiments
on TmAl$_2$,which were performed at the HFR of the ILL in
Grenoble. For inelastic scattering we used the time of
flight (TOF) instrument D7 with incident neutrons of
energy 3.55 meV and a resolution in energy $\Delta E/E \approx$ o.1.
The elastic measurements were performed for several
temperatures between 2 K and 1oo K at the D1B instrument. Here a multidetector allows for simultaneous recording of a range of 8o° of the neutron diffraction
pattern. The incident wave length of the neutrons, used
at D1B was 2.54 Å. Absolute intensities were obtained
from the nuclear Bragg reflections with a nuclear scattering length $b_{Tm} = $ o.72 1o$^{-12}$cm for Tm and $b_{Al} = $ o.35
1o$^{-12}$cm for Al.
The TOF spectra show a rapid decrease of diffuse quasi-
elastic magnetic scattering below 4 K, indicating the
onset of long range magnetic order below this temperature. In addition, there is a change in the shape of
the inelastic part of the spectrum.

The neutron diffraction pattern of $TmAl_2$ at 100 K shows
only nuclear Bragg reflections and paramagnetic background scattering. At 2 K we observe superlattice reflections, but there is also additional magnetic intensity at the positions of the nuclear reflections. The
positions of the superlattice reflections can be described by a propagation vector $Q = (1/2, 1/2, 1/2)$.
Taking also into account the additional intensity at
the nuclear Bragg reflections, we obtain a canted ferromagnetic structure with a ferromagnetic component of
$(4.8 \pm 0.3)\mu_B$ and an antiferromagnetic component of
$(2.6 \pm 0.3)\mu_B$ at 2 K. The magnetic moments of the Tm-
ions within one $\langle 111 \rangle$ -plane are aligned parallel to
each other, with an angle of $\approx 56°$ between the moments
of adjacent $\langle 111 \rangle$ -planes, if we assume, that the size
of magnetic moment is the same for all Tm-ions.

Replacing 10% $TmAl_2$ by $LaAl_2$ we found only ferromagnetic long range order (for temperatures down to 2 K) with
almost the same transition temperature as in the case
of $TmAl_2$. Apparently, introduction of such a disorder
is sufficient to destroy the long range antiferromagnetic component, at least above 2 K. Instead, antiferromagnetic correlations of short range are present between 2 and 9 K. A more detailed description of this work
will be published elsewhere.

REFERENCES

/1/ W.E. Wallace, in "Rare Earth Intermetallics"
    Academic Press New York and London (1973)
/2/ J.C.P. Klaasse, Ph. D. Thesis, "Natuurkundig
    Laboratorium" of the university of Amsterdam (1977)
/3/ B. Barbara, J.X. Boucherle, J.L. Buevoz,
    M.F. Rossignol and F. Schweizer, Solid State
    Commun. 24, 481 (1977)
/4/ H. Happel, P. v. Blanckenhagen, K. Knorr, A. Murani,
    Proc. 2nd Int. Conf. on Crystal Field Effects in
    Metals and Alloys, Zürich 1976, ed. A. Furrer,
    p. 273 Plenum Press New York and London (1977).

# CRITICAL BEHAVIOUR OF TRANSPORT PROPERTIES IN GdCd COMPOUNDS

J.B. Sousa, R.P. Pinto, M.M. Amado, J.M. Moreira, M.E. Braga
Centro de Fisica Universidad do Porto, Porto, Portugal

P. Morin, R. Aleonard
C.N.R.S., Laboratoire Louis Néel, Grenoble, France

We initiated a detailed investigation of the critical behavior of tranport properties in Cs-Cl structure intermetallic compounds[1,2]. This work is now extended to ferromagnetic GdCd single crystals. Accurate data have been obtained on the electrical resistivity ($\rho$), temperature derivative $d\rho/dT$, thermoelectric power (Q) and thermal conductivity (k) along <100>, <110> and <111> directions. The crystals were prepared by direct fusion of the components in a sealed Ta crucible, cooling from the melt in a Bridgman furnace ($\overline{\nabla}T=20^\circ C.cm^{-1}$). Samples with appropriate geometry (15x1x1 mm$^3$) were obtained using a spark machine.

## ELECTRICAL RESISTIVITY ($\rho,d\rho/dT$)

Fig. 1 shows the temperature dependence of $\rho$ and $d\rho/dT$ from 100-300K, with the current along <100>. Measurements on other GdCd samples along <110> and <111> directions revealed an isotropic resistivity both in the paramagnetic and ferromagnetic states. The magnetic transition is marked by a sharp reduction in $d\rho/dT$ at $T_c$ =256K, in close agreement with the Curie temperature derived from the magnetization M (Fig. 2).

An interesting feature in $d\rho/dT$ is the persistence of fairly high values in the ferromagnetic phase, down to considerably low temperatures. Such a behavior has also been observed in the compound TbZn[2] and Tb-Gd alloys[3]. However, in these cases $d\rho/dT$ exhibited some structure below $T_c$, whereas in GdCd this derivative hardly changes in the ferromagnetic phase (except for a faint depression around 220K; see Fig. 1).

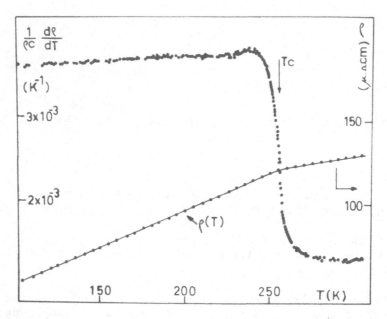

Figure 1.   Temperature dependence of ($\rho$) and ($d\rho/dT$) in GdCd.

In order to understand the anomalously high and almost constant $d\rho/dT$ values below $T_c$, we present a simple analysis based on the electron spin-disorder scattering (4):

$$\rho(T) = \rho\infty \left[ 1 - (M/M_o)^2 \right] \qquad (1)$$

where $\rho\infty$=paramagnetic spin-disorder resistivity, $M_o$=M(0) and the ionic spin S was assumed S>>1.   For GdCd, and over a wide range of temperatures (40-220 K), experiment gives:

$$M(T)/M_o \simeq 1 - B \cdot T^{3/2} \qquad (2)$$

as shown in Fig. 2 (B=1.65x $10^{-4} K^{-3/2}$).   Taking the temperature derivative of (1) and using (2) we obtain:

$$= (1/\rho\infty)(d\rho/dT) = 3B \cdot (T^{1/2} - B \cdot T^2) \qquad (3)$$

Inserting the numerical value of B for GdCd we obtain, in $10^{-3} K^{-1}$ units:   $\beta$=3.24(%=100K),3.34(150K),3.25(170K) and 3.06(200K).   These values are fairly close to the experimental ones (Fig. 1), after subtraction of the phonon background.

The vicinity of the Curie point has been analyzed in detail. The rapid decrease of $d\rho/dT$ above $T_c$ indicates that fluctuations do not extend far into the paramagnetic phase, i.e. the Ginsburg reduced temperature $\varepsilon_G$ is small. One can in fact show that (5):

$$\varepsilon_G = \text{const.} (KT_c/J).S^{-6} \tag{4}$$

and from this one concludes that $\varepsilon_G \propto Z^3$, where $Z$ = coordination number. Since GdCd has a simple cubic structure (Z=6), the fluctuations are severely restricted to the vicinity of $T_c$. For more compact structures, of course, one expects $\varepsilon_G$ to increase.

The critical behavior of $d\rho/dT$ above $T_c$ can be satisfactorily described by a log-type expression:

$$1/\rho_c (d\rho/dT) = A \ln_\varepsilon + C \tag{5}$$

with $\rho_c = \rho (T_c), \varepsilon = T/T_c - 1$. The fit is valid in the range $10^{-3} \leq \varepsilon \leq 4 \times 10^{-2}$, and $A = -0.24$, $C = 0.77$ $(10^{-3}K^{-1})$. This behavior resembles that in the specific heat, suggesting the dominance of short-range spin correlations. According to Fisher-Langer (6) this should occur in systems with $(K_F \cdot a)$ large, where $K_F$ = Fermi vector and a = lattice vector.

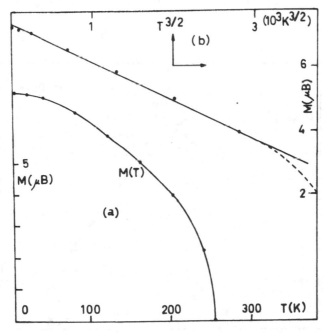

Figure 2.   (a) T-dependence of magnetization M(T); (b) Plot M vs $T^{3/2}$ shows that $M = M_o (1 - BT^{3/2})$ for T between 40K and 220K.

## THERMOELECTRIC POWER (Q)

Fig. 3 shows Q(T) near the Curie point, with $\vec{\nabla}T$ along <100>; similar behavior was found along <110> and <111>. The magnetic transition produces a clear break in Q. Above $T_c$, Q becomes rapidly linear in T, and we ascribe it to normal diffusion thermopower ($Q_n$):

$$Q_n(T) = (\pi^2/3)(K/e) \cdot (KT/E_F) \tag{6}$$

where $E_F$=Fermi energy, and $\underline{e}$ is the algebric charge of carriers. Our results indicate that e<0 (electrons).

Below $T_c$, Q is clearly depressed with respect to the extrapolation from the paramagnetic states, i.e. Q gets more negative. A similar behavior was found in TbZn compound (2). This can be qualitatively understood using Kasuya's model (7) for the magnetic contribution to the thermopower ($Q_m$):

$$Q_m(T) = 2(K/e) \cdot (H_c/E_F) \cdot \frac{1- \exp(-x)}{1+ \exp(+x)} \tag{7}$$

where $H_c$=(g-1)S(G/E_F)(M/M_0)$, $x=3/(S+1) \cdot (T/T_c)(M/M_0)$, g=Landé factor, G=(s-d) exchange integral. Then relation (7) predicts $Q_m$<0 for GdCd, when $H_c$>0.

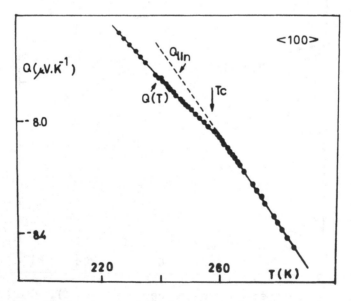

Figure 3. Temperature dependence of thermopower Q near $T_c$.

A more detailed picture is obtained if we expand $Q_m$ in the vicinity of the Curie point (x<<1) and using $Q=Q_m+Q_n$:

$$Q(T) \simeq Q_n(T) - D \cdot (M/M_o)^2 \qquad (8)$$

where D=constant.   Comparing with expression (1) for $\rho(T)$, we expect a close resemblance between Q and $\rho$ near $T_c$.   This is confirmed in Fig. 4, with a plot of $(Q-Q_n)$ vs T (inset).

In addition to the immediate vicinity of $T_c$, the measurements have been extended down to 80K.   For all crystallographic directions, Q shows a broad maximum around 170K, as shown in Fig. 4.   This is different from TbZn where Q reaches a minimum (2).   The GdCd results are inconsistent with Kasuya's model, which predicts a minimum in Q when e<0.   Since phonon and magnon drag contributions usually present large bumps at intermediate temperature (8), a definite comparison between experiment and theory cannot be done without further extension of Kasuya's model, to include these effects.

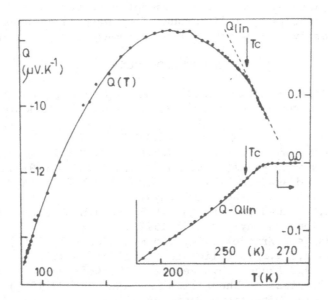

Figure 4.   Temperature dependence of thermopower Q.
            Inset:  expanded scale of the plot $(Q-Q_{lin})$ vs T.   Observ
            striking similarity to the resistivity curve in Fig. 1.

## THERMAL CONDUCTIVITY (k)

This coefficient (k) was measured along <100>,<110>,<111> directions. The qualitative behavior is remarkably similar near $T_c$: the slope dk/dt increases sharply as the sample enters the paramagnetic state, k becoming rapidly linear in T. This can be understood in terms of a simple kinetic model for the electronic component $k_e$:

$$k_e = (1/3).C_e.V_F.l_e \qquad (9)$$

where $C_e$=electronic specific heat ($\propto$T), $V_F$=Fermi velocity, $l_e$= electron mean free path. In the paramagnetic phase, spin-disorder scattering dominates, imposing a small and practically constant mean free path $l_e$. Therefore, $k_e$= const.T.

Below $T_c$, magnetic order sets in, and $l_e$ increases with decreasing T, i.e. $dl_e/dT<0$. Therefore, from (9) one can show:

$$k_e^{-1} (dk_e/dT)_{ferro} < k_e^{-1}(dk_e/dT)_{para} \qquad (10)$$

The k-measurements have been extended down to 80K. A new feature is the appearance of a noticeable anomaly in k around $T^*$=210K, where it shows a marked peak. Careful re-analysis of the S(T) data around $T^*$ confirms the existence of such an anomaly, but with much smaller amplitude. The anomaly is not detected in $\rho$, but a faint depression appears in $d\rho/dT$ around $T^*$. Work is in progress to further clarify this point in the ferromagnetic phase of GdCd.

## ACKNOWLEDGEMENTS

The financial support of I.N.I.C. (Portugal; +),C.N.R.S.(France;*) and NATO research grant no.1481(+) is greatfully acknowledged.

## REFERENCES

1.J.B.Sousa,M.M.Amado,R.P.Pinto,J.M.Moreira,M.E.Braga,M.Ausloos,P.
   Morin,J.Physique(Paris),supl.no.5,tome 40,C5-42(1979).
2.J.B.Sousa,M.M.Amado,R.P.Pinto,J.M.Moreira,M.E.Braga,P.Morin,M.
   Ausloos; to be published in J.Phys.F.
3.J.B.Sousa,M.M.Amado,M.E.Braga,R.P.Pinto,J.M.Moreira,D.Hukin,
   Communications on Physics 2:95(1977).
4.P.G.De Gennes,J.Friedel,J.Phys.Chem.Solids 4:71(1958).
5.V.L.Ginsburg,Sov.Phys.Solid State 2:1824(1961).
6.M.E.Fisher,J.S.Langer,Phys.Rev.Lett. 20:665(1968).
7.T.Kasuya,Progress on Theoretical Physics 22:227(1959).
8.S.Legvold, in "Magnetic properties of rare earth metals",ch.VII,
   ed.R.J.Elliott,Plenum,N.Y. (1972).

MAGNETIC BEHAVIOUR OF $RCu_4Al$ (R = RARE EARTH) INTERMETALLIC COMPOUNDS II[*]

S. K. Malik,[†] W. E. Wallace and T. Takeshita[**]

Department of Chemistry, University of Pittsburgh

Pittsburgh, PA  15260

## ABSTRACT

Intermetallic compounds of the general formula $RCu_4Al$, where R = La,Ce,Pr,Nd,Sm,Gd,Tb,Dy,Ho,Er and Tm, have been prepared and are found to crystallize in the hexagonal $CaCu_5$ type structure. Magnetization measurements on these compounds have been carried out in the temperature range 4.2 to 300 K.  The compounds with Gd, Tb,Dy and Ho have been reported earlier (1) to order antiferromagnetically at low temperatures.  There is no conclusive evidence about magnetic ordering in the remaining $RCu_4Al$ compounds from magnetization measurements in 21 kOe applied field.  Deviation from Curie-Weiss behaviour, at low temperatures, is observed in Er and Tm compounds and is attributed to crystal field effects.  The effective paramagnetic moments are close to free ion value except for the Ce and Pr compounds.  The paramagnetic Curie temperatures are negative for Ce,Pr and Gd compounds and positive for the remaining compounds.  The results are discussed and compared with those on $RCu_5$ and $RCu_4Ag$ compounds.

## INTRODUCTION

The $RCu_5$ (R = rare earth) intermetallic compounds exhibit some interesting crystallographic features (2,3).  The compounds with

---

[*]Work supported by the Army Research Office.
[†]Present address:  Tata Institute of Fundamental Research, Homi Bhabha Road, Bombay 400 005, India.
[**]Present address:  Ames Laboratory-DOE and Department of Materials Science and Engineering, Iowa State University, Ames, Iowa  50011.

light rare earths, namely, with La,Ce,Pr,Nd and Sm, crystallize in
the hexagonal CaCu$_5$ type structure which is also the structure type
most commonly adopted by compounds of AB$_5$ stoichiometry such as
RNi$_5$ and RCo$_5$ compounds. The compounds with Ho, Er and Tm crystallize in the less common cubic AuBe$_5$ type of structure. However,
the RCu$_5$ compounds with R = Gd, Tb and Dy coexist in the two structure types mentioned above – the CaCu$_5$ type structure being more
prominent at high temperatures. By suitably annealing, the Tb and
Dy compounds can be obtained in the cubic AuBe$_5$ type of structure,
free of the hexagonal CaCu$_5$ type phase. In earlier communications
(4,5) from this laboratory it has been shown that it is possible to
form intermetallic compounds of the type RCu$_4$Al and RCu$_4$Ag with
most of the rare earths. Further, it has been found that all of
the RCu$_4$Al compounds crystallize in CaCu$_5$ type structures while
RCu$_4$Ag (R = Nd to Tm) form in the AuBe$_5$ type structure. Thus replacement of one Cu by either Al or Ag stabilizes, respectively,
the hexagonal or the cubic type of structure. Magnetization measurements have been carried out on RCu$_4$Al compounds in the temperature range 4.2 to 300 K. The compounds with R = Gd,Tb,Dy and Ho
have been reported earlier (1) (hereafter referred to as I) to
order antiferromagnetically at low temperatures. In this communication we report the results of magnetization studies on remaining
RCu$_4$Al compounds.

## EXPERIMENTAL

The experimental procedure used for alloy preparation, for
characterisation by x-ray diffraction studies and for magnetization
measurements is the same as given in I. It may be added that a
continuous record of magnetization versus temperature is obtained
on a x-y recorder by recording the force while slowly varying the
sample temperature from 4.2 K to 300 K with the magnetic field held
constant.

## RESULTS AND DISCUSSION

The results of magnetization studies on RCu$_4$Al compounds with
R = Ce,Pr,Nd,Er and Tm are shown in Figs. 1 and 2 where inverse
molar susceptibility ($X_M^{-1}$) is plotted as a function of temperature.
There is no conclusive evidence regarding magnetic ordering in
these compounds down to 4.2 K. The magnetization at 4.2 K varies
linearly with the applied field for CeCu$_4$Al, PrCu$_4$Al, SmCu$_4$Al and
TmCu$_4$Al. For NdCu$_4$Al and ErCu$_4$Al slight deviation from the linear
dependence is observed at high fields. For the latter two compounds there is also a slight rounding in magnetization versus
temperature curves just around 4.2 K when magnetization is measured
in an applied field of 5 kOe. These results are not adequate to
draw any conclusion about the magnetic ordering in these compounds
down to 4.2 K.

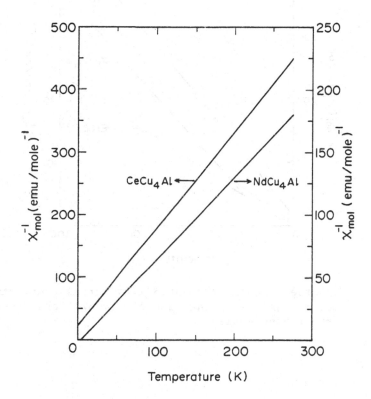

Fig. 1 Inverse magnetic susceptibility versus temperature
for CeCu₄Al (left-hand scale) and NdCu₄Al (right-
hand scale.

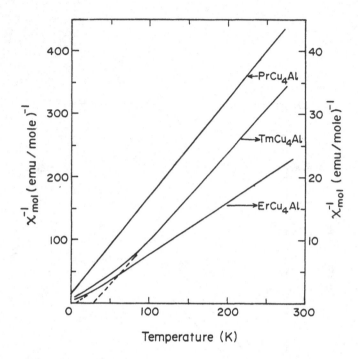

Fig. 2   Inverse magnetic susceptibility versus temperature
         for PrCu$_4$Al (left-hand scale), TmCu$_4$Al and ErCu$_4$Al
         (right-hand scale).


     In the paramagnetic state the susceptibility follows the Curie-
Weiss behaviour for CeCu$_4$Al, PrCu$_4$Al and NdCu$_4$Al throughout the temperature range investigated, while for ErCu$_4$Al and TmCu$_4$Al, deviation from Curie-Weiss behaviour is observed below 100 K and 30 K,
respectively.   The deviation presumably arises due to crystal field
effects which introduce an additional temperature dependence in the
susceptibility due to the lifting of the degeneracy of the ground
J-manifold.   In the case of SmCu$_4$Al also, deviation from Curie-Weiss
behaviour is observed and attributed to the narrow multiplet width
in Sm$^{3+}$ ion.   However, the forces involved in the measurement of
susceptibility of SmCu$_4$Al were small and the absolute values are
not very reliable - and have not been plotted for this reason.   The

results of crystallographic and magnetic measurements on $RCu_4Al$
compounds are given in Table I. For the sake of completeness the
results on Gd, Tb, Dy and Ho compounds reported in I are also
included.

Table I.   Lattice Constant, Effective Magnetic Moment, Paramagnetic
Curie Temperature and Néel Temperature of $RCu_4Al$ Compounds

| Compound | a (Å) | c (Å) | $\mu_{eff}(\mu_B)$ | $\theta_p$(K) | $T_N$(K) |
|----------|-------|-------|--------------------|---------------|----------|
| $LaCu_4Al$ | 5.238 | 4.189 | - | - | - |
| $CeCu_4Al$ | 5.195 | 4.188 | 2.26 | -14.0 | - |
| $PrCu_4Al$ | 5.186 | 4.185 | 2.27 | - 9.3 | - |
| $NdCu_4Al$ | 5.150 | 4.181 | 3.49 | 3.2 | ? |
| $SmCu_4Al$ | 5.126 | 4.168 | $\chi_{mol}^{-1}$ vs T nonlinear | | |
| $GdCu_4Al$ | 5.096 | 4.152 | 8.18 | - 1.3 | 42.0 |
| $TbCu_4Al$ | 5.073 | 4.149 | 9.03 | 13.0 | 37.0 |
| $DyCu_4Al$ | 5.064 | 4.152 | 10.11 | 20.5 | 20.0 |
| $HoCu_4Al$ | 5.047 | 4.148 | 10.36 | 12.0 | 8.5 |
| $ErCu_4Al$ | 5.039 | 4.139 | 9.84 | 4.6 | ? |
| $TmCu_4Al$ | 5.026 | 4.138 | 7.57 | 27.5 | - |

The effective paramagnetic moments ($\mu_{eff}$) are close to the respective free ion values of the rare earths involved except for Ce
and Pr compounds. No anomaly in the lattice parameters of $CeCu_4Al$
is observed, indicating that the valence of cerium in this compound
is close to 3. The $\mu_{eff}$ for the Pr compound is much smaller than
the free ion value. This may be due to crystal field effects. However, its susceptibility does not show any deviation from Curie-
Weiss behaviour. The paramagnetic Curie temperatures are distinctly
negative for Ce and Pr compounds, small positive and negative, respectively, for Nd and Gd compounds and positive for the rest of
the $RCu_4Al$ compounds. A similar trend is observed in $RCu_4Ag$ compounds (4). The dominant interaction between the rare earth ions
in these compounds is the so-called RKKY interaction mediated by
the conduction electrons. This mechanism does not usually predict
a sign change in $\theta_p$ in going from one compound to another in a given
series unless conduction electron concentration is changing and one
is close to a crossover region. It seems unlikely that such a situation occurs both in hexagonal $RCu_4Al$ compounds and in cubic $RCu_4Ag$
compounds and therefore these results may be taken to imply failure
of the RKKY theory applied to these systems.

The magnetic interactions in these compounds change appreciably in going from $RCu_5$ to $RCu_4Al$ and to $RCu_4Ag$. In those $RCu_4Al$ compounds where magnetic ordering has been observed, it is of the antiferromagnetic type, while in $RCu_5$ and $RCu_4Ag$ compounds both ferromagnetic and antiferromagnetic orderings are observed, depending on the rare earth ion involved. This shows that the magnetic interactions in these compounds are indeed dependent on the concentration of conduction electrons - Ag and Al contributing, respectively, one and three electrons to the conduction band. However, trends are not predictable and even within a series the ordering changes from ferromagnetic to antiferromagnetic type in going from one compound to another.

## REFERENCES

1.  S. K. Malik, T. Takeshita and W. E. Wallace, in "Rare Earths in Modern Science and Technology," edited by G. J. McCarthy and J. J. Rhyne, Plenum Press, New York & London (1978), pp. 429-434.
2.  K. H. J. Buschow, A. S. van der Goot and J. Birkhan, J. Less-Common Metals 19:433 (1969).
3.  K. H. J. Buschow and A. S. van der Goot, Acta Cryst. 27B:1085 (1971).
4.  T. Takeshita, S. K. Malik, A. A. Elattar and W. E. Wallace, AIP Conf. Proc. 34:230 (1976).
5.  T. Takeshita, S. K. Malik and W. E. Wallace, J. Solid State Chem. 23:225 (1978).

# THE CURIE-WEISS LAW AND THE CHEMICAL BOND

X. Oudet

Laboratoire de Magnétisme, C.N.R.S.

1, Place Aristide Briand - 92190 Meudon (France)

In our work on the Curie constant of the rare earths (1) we got poor results in contrast with the case of 3d elements. Using a characteristic bond of the 4f elements we obtain now good results. So let this work be the continuation of the first work. If $D(Ed,U)$ is expressed in full we get :

$$\chi(U) = 1.17 \, N \, \sigma^2 \, \alpha \, U^{-1}[1 + \exp \alpha(Ed. \, U^{-1} - 1)]^{-1} \quad \text{with } \alpha = 1.5049 \quad (1)$$

$\chi(U)$ exhibits a peak, a property of the paramagnetic compounds with or without magnetic order as in the case of the spin glasses.

Experimentaly it is $\chi(T)$ that is known, so it is necessary to determine U (T). This is generaly difficult but fortunatly in the temperature region of the Dulong and Petit law it is possible to express U as follows :

$$U = \Omega \, (T - T_h + b \, T_h^2 \, T^{-1}) \quad (2)$$

For this law $\Omega = 3k$ where k is Boltzmann constant. $T_h$ expresses that in low temperature region, the heat capacity is largly different from $\Omega$. The term $b \, Th^2 T^{-1}$ gives the asymptotic character of the law.

Let us write $Ed = \Omega Td$ and expand $\chi^{-1}(U)$ in power of $\alpha Ed \, U^{-1}$ and replace U by the relation (2). We obtain :

$$\chi^{-1}(T) = C^{-1} [T + \Theta + \gamma(T)] \quad (3) \quad ; \quad C = 2N \, \sigma^2 \, D \, \Omega^{-1} \quad (4)$$

$$\Theta = D \, (fTd - Th) \quad (5) \quad ; \quad \gamma(T) = D \, g \, Td^2 \, (T-Th)^{-1} + DbTh^2T^{-1} \quad (6)$$

$$D = 0.722 \quad ; \quad f = 0.379 \quad ; \quad g = 0.285$$

These different expressions give the essential characteristics of

353

the paramagnetism in the high temperature region. C is the Curie
constant well verified for the 3d elements with $\Omega$ = 3k, $\gamma$(T) shows
its asymtotic character (2). The sign of $\Theta$ gives rise to the
different types of magnetism.

In our first work on Ln we thought that the relation (4) must
be verified with $\Omega$ = 3k. But let us consider the compound $Ln_2O_2S$
(9in!) in the case of heavy Rare Earths. These compounds begin to
follow the Curie-Weiss law from temperature in the vicinity of 10K
except for Yb. Now let us have a look at their heat capacity curves.
In the vicinity of 10K the value of their heat capacity is much
less than 3k, of the order of k. So to verify the law of the paramagnetism for the rare earth elements we must take in the expression of the mean thermal energy (2) : $\Omega$ = k in place of 3k. Now in
the same way as before but making the assumption that states with
m = j are always quenched (4) we get :

| La | Ce | Pr | Nd | Sm | Eu | Tb | Dy | Ho | Er | Tm | Yb |
|------|------|------|------|------|------|------|------|------|------|------|------|
| 0.00 | 0.89 | 1.59 | 1.59 | 0.00 | 8.03 | 11.3 | 14.3 | 14.3 | 11.3 | 4.41 | 0.00 |

in good agreement with experiments. The cases of Sm, Eu, Tm and Yb
need further discution that will be given elsewhere.

Now let us consider some deduction from the relation $\Omega$ = k.
The value 3k is interpreted as a manifestation of the two degrees
of freedom, position and speed, in each of the three directions of
the space. So the value k must be interpreted as a manifestation of
the fact that the atom core of the rare earth element is bound to
the crystal, just in one direction of the space, the one by which
atom core exchanges energy.

Now the 4f electrons are deep in the atom core, all the Ln
atoms have the valency III and most of them have their Curie
constant almost independant of the compound. This leads to the
hypothesis that the bond of atom core with its neighbours is
obtained by a 4f valency electron the third valency electron and
eventually by the other similar 4f electrons which gravitate in
the same direction. We call this bond covalent as in the case of
LnIG. We have already predicted this kind of bond and the existence
of trivalent $Ln^{2+}$ written as $Ln_{III}^{2+}$ in the interpretation of the
crystal structure of $La_2O_3$ which is the same as that of $Ln_2O_2S$.

## REFERENCES

1.   X. Oudet Previous conf. Vol. page 453.

2.   L. Neel, Ann. de Phys., 3, 137, (1948).

3.   J. Rossat-Mignod et al, Phys. Stat. Sol. b 49, 147, (1972).

4.   With Y. Charreire E. R 211, and O. Gorochov Laboratoire de
     Physique des Solides the both at the same address than us,
     $La_2O_2Te$ have been found diamagnetic.

# MAGNETIC PROPERTIES OF SOME $Sm_{2-x}R_xCo_{17-y}Mn_y$ INTERMETALLIC COMPOUNDS

F. Rothwarf, H.A. Leupold, A. Tauber, J.T. Breslin,
R.L. Bergner and J.J. Winter

US Army Electronics Technology and Devices Laboratory
(ERADCOM)

Fort Monmouth, New Jersey 07703

## INTRODUCTION

Current military requirements are leading to the development
of a new generation of microwave and millimeter wave tubes whose
magnetic circuits need permanent magnets with the following properties: energy products in excess of 30 MGOe, low reversible temperature coefficients of magnetization, $\alpha$, and linear demagnetization curves. Rothwarf, Leupold and Jasper (1) have discussed
approaches to the design of such magnetic circuits and the synthesis of suitable permanent magnet materials. Some of our preliminary efforts to attain compounds with such properties were also recently reported (2,3) and reviewed (4). In the present paper we
review our earlier work and report new results on the use of heavy
rare earth substituents for some samarium in an attempt to achieve
intrinsically temperature compensated 2-17 magnets. Such an approach was first successfully applied to SmCo$_5$ by Benz, Laforce and
Martin (5). The systems being reported on are $Sm_{2-x}R_xCo_{17-y}Mn_y$,
where R=Gd, Er and Dy. The manganese is being used since we had
previously found (6) that its presence in the quaternary system
$Sm_2Mn(Co,Fe)_{16}$ significantly enhanced the anisotropy fields $H_A$ over
those measured for the corresponding compounds in the ternary system $Sm_2(Co,Fe)_{17}$. Also Nagel (7) found that using manganese or
chromium substituents for cobalt in the $Sm_2Co_{17}$ and $Sm_2(Co,Fe)_{17}$
compounds significantly enhances their coercivities and energy
products. Thus, we have been investigating the magnetic properties
of these systems with the expectation that the heavy rare earth
substituents would yield low $\alpha$'s in compounds having their
anisotropy fields enhanced by the manganese substituent.

355

## EXPERIMENTAL

The experimental procedures used to determine the magnetization $\sigma_s(T)$, the temperature coefficient of magnetization, $\alpha$, and the anisotropy field $H_A$ have been described in detail elsewhere (2-4,8). We reiterate the experimental reproducibility of the various measurements. For the magnetization $\sigma_s$ it was about ±1.5% in a maximum field of 15 kOe. The resultant $\alpha$'s have an uncertainty of ±15%/°C. The reproducibility for $H_A$ when the material is saturated ($H_A$ < 100 kOe) is about ±1.0%. The values for the extrapolated $H_A$'s are ±2.5% for 100 < $H_A$ < 125 kOe, ±5.0% for 125 < $H_A$ < 150 kOe and ±8.0% for $H_A$ > 150 kOe.

The $\alpha$'s for all the materials investigated are shown in Table 1 together with room temperature (300 K) saturation magnetization values $\sigma_s$ and the anisotropy fields $H_A$ at 273 K for the various compounds. Approximate $4\pi M_s$ values derived from $\sigma$'s are also tabulated. The densities necessary to make this conversion were obtained by assuming that the volumes per unit cell of all compounds were the same as that for $Sm_2Co_{17}$ and scaling the mass per unit cell to take into account the atomic masses of the various substituents. This was done because our preliminary x-ray diffraction results indicate that unit cell volumes of all the compounds are the same to within about one percent. The densities calculated in this manner varied from 8.61 to 8.72 g/cm$^3$.

## RESULTS AND DISCUSSION

It was intended that the $Sm_{2-x}R_xCo_{17-y}Mn_y$ systems would be investigated for the compositions x=0, 0.2, 0.4 and 0.6 and y=0, 1 and 2. To date, the complete set of data has only been obtained for the gadolinium-substituted system. The results for the various compounds studied so far are listed in Table 1 and their various correlations are shown in Figs. 1-5. For purposes of clarity the results for the gadolinium compounds will be discussed first in section A. Section B will then compare the three sets of compounds for the case y=0 where all the data are complete.

### A.    The $Sm_{2-x}Gd_xCo_{17-y}Mn_y$ Systems

The magnetization results for the gadolinium compounds are shown in Fig. 1 for the x=0, 0.2, 0.4 and 0.6 samples. From such curves an average reversible temperature coefficient of magnetization was obtained, where $\alpha$ is defined as

Table 1.  Saturation Magnetizations, Anisotropy Fields and Average
Temperature Coefficients for the Systems
$Sm_{2-x}R_xCo_{17-y}Mn_y$ (R=Gd, Er, Dy).

| Composition | | $\sigma_s$ | $4\pi M_s$ | $H_A$ | $\alpha_L$ | $\alpha_H$ |
|---|---|---|---|---|---|---|
| | | 300 K | 300 K | 273 K | 225–300 K | 300–425 K |
| y | x | (emu/g) | (kG) | (kOe) | (%/K) | (%/K) |
| | | | $Sm_{2-x}Gd_xCo_{17-y}Mn_y$ | | | |
| 0, | 0.0 | 117 | 12.7 | 85.3 | −0.010 | −0.011 |
| | 0.2 | 107 | 11.6 | 82.1 | −0.010 | −0.010 |
| | 0.4 | 96.2 | 10.5 | 78.7 | 0.000 | +0.015 |
| | 0.6 | 97.3 | 10.6 | 66.5 | 0.000 | +0.015 |
| | 2.0 | 67.0* | 7.50* | -- | +0.069* | +0.074* |
| 1, | 0.0 | 118 | 12.7 | 109 | −0.021 | −0.027 |
| | 0.2 | 110 | 11.9 | 98.5 | −0.019 | −0.017 |
| | 0.4 | 105 | 11.4 | 88.2 | 0.000 | −0.006 |
| | 0.6 | 101 | 10.9 | 85.0 | −0.002 | −0.019 |
| 2, | 0.0 | 114 | 12.3 | 90.0 | −0.032 | −0.038 |
| | 0.2 | 106 | 11.4 | 92.7 | −0.022 | −0.021 |
| | 0.4 | 91.0 | 9.82 | 86.7 | −0.020 | −0.019 |
| | 0.6 | 96.3 | 10.4 | 109 | −0.029 | −0.029 |
| | | | $Sm_{2-x}Er_xCo_{17-y}Mn_y$ | | | |
| 0, | 0.0 | 117 | 12.7 | 85.3 | −0.010 | −0.011 |
| | 0.2 | 99.8 | 10.9 | 81.4 | 0.000 | 0.000 |
| | 0.4 | 105 | 11.5 | 67.5 | +0.007 | +0.002 |
| | 0.6 | 105 | 11.5 | 64.0 | +0.015 | +0.001 |
| | 2.0 | 90.2† | 10.1† | -- | +0.099† | +0.045† |
| 1, | 0.0 | 117 | 12.7 | 109 | −0.021 | −0.027 |
| | 0.2 | 103 | 11.2 | 116 | −0.006 | −0.019 |
| 2, | 0.0 | 114 | 12.3 | 90.0 | −0.032 | −0.038 |
| | 0.2 | 102 | 11.1 | 106 | −0.031 | −0.038 |
| | 0.4 | 107 | 11.6 | 105 | −0.120 | −0.018 |
| | | | $Sm_{2-x}Dy_xCo_{17}$ | | | |
| 0, | 0.0 | 117 | 12.7 | 85.3 | −0.010 | −0.011 |
| | 0.2 | 114 | 12.5 | 69.8 | 0.000 | 0.000 |
| | 0.4 | 105 | 11.5 | 49.2 | +0.016 | +0.016 |
| | 0.6 | 95.1 | 11.4 | 60.0 | +0.014 | +0.014 |
| | 2.0 | 64.4* | 7.14* | -- | +0.188* | +0.171* |

* R. Lemaire, Cobalt 33, 301 (1966)
† A. E. Miller, T. D'Silva and H. Rodrigues, IEEE Trans.
Magnetics, MAG-12, 1066 (1976).

$$\alpha = \frac{\Delta M}{M_1 \Delta T} \times 100\% = \frac{M(T_2) - M(T_1)}{M(T_1)\ (T_2-T_1)} \times 100\ \% \tag{1}$$

We chose to list $\alpha$'s for a high temperature region, $\alpha_H$, where
300 K < T < 425 K and for a low temperature region, $\alpha_L$, where
225 K < T < 300 K.  In each case $T_1$ was taken to be 300 K.  $\Delta M$
was always chosen as the maximum magnetization change over the
temperature interval $\Delta T$.

In Fig. 1 the set of $4\pi M_s(T)$ curves for the $Sm_{2-x}Gd_xCo_{17-y}Mn_y$
compounds are shown.  It is clear that the substitution of gadolinium dramatically reduces the magnitude of the slopes of the curves
for the cases where y=0 and 1, especially for the substitution of
$Gd_{0.4}$ and $Gd_{0.6}$.  In some cases the slopes are changed from negative
to positive as more gadolinium is substituted.  These results demonstrate the validity of using heavy lanthanide substituents for
lowering the magnitude of $\alpha$ in the 2-17-based compounds.

The $\alpha_L$ and $\alpha_H$ values are plotted in Fig. 2 as a function of
gadolinium concentration for the various fixed values of the manganese substituent.  Making use of magnetization data for $Gd_2Co_{17}$
(9) and our data for $Sm_2Co_{17}$, we have plotted the linear variation
of $\alpha$ with Gd concentration expected from a simple additivity approximation (1) for the y=0 case.  Within experimental error this approximation seems to have some validity.  However, the shapes of
the y≠0 curves, which are similar to the y=0 case, lead one to
believe that the functional dependence of $\alpha$ with gadolinium concentration may be more complex.  From a practical standpoint it seems
that the compound $Sm_{1.6}Gd_{0.4}Co_{16}Mn$ would furnish a good intrinsically temperature compensated magnet material for the temperature
range of military interest, 225 K < T < 425 K.

The final choice of an optimal alloy to achieve a zero temperature coefficient (ZTC) material is influenced by the variation
of magnetization and anisotropy, as well as ultimately the coercivity and maximum energy product with manganese and gadolinium
concentration.  In Fig. 3 we plot the variation with temperature of
$4\pi M_s$, and $H_A$ with manganese concentration for the different gadolinium concentrations at temperatures of 2.3 and 273 K.  From Fig.
3A it is clear that there is a monotonic decrease in $4\pi M_s$ as the
gadolinium substituent is increased.  This behavior is what one
expects from the antiferromagnetic coupling of gadolinium to the
other moments in the system.  For a given x, however, a shallow
maximum occurs in the 273 K or 300 K curves for the substitution
of one manganese atom for cobalt.  A similar monotonic decrease
is apparent from Fig. 3B for the variation of $H_A$ with gadolinium
concentration and with the exception of x=0.6 a maximum also occurs
for the case of one manganese atom substituted for cobalt.  The
microscopic rationale for these trends is presently not well

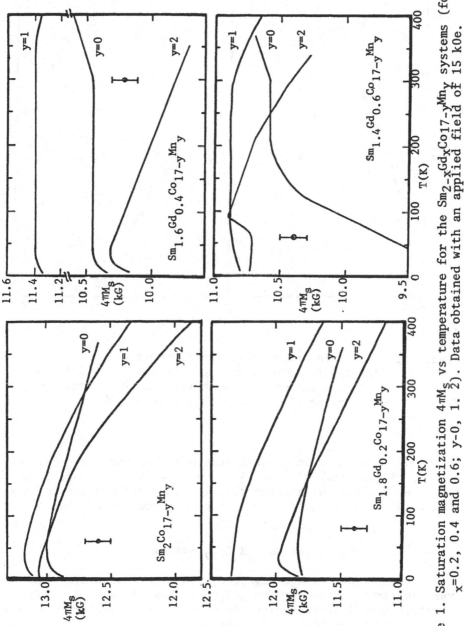

Figure 1.  Saturation magnetization $4\pi M_s$ vs temperature for the Sm$_{2-x}$Gd$_x$Co$_{17-y}$Mn$_y$ systems (for x=0.2, 0.4 and 0.6; y-0, 1. 2). Data obtained with an applied field of 15 kOe.

F. ROTHWARF ET AL.

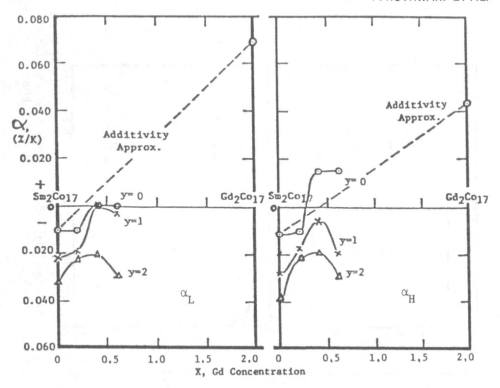

Figure 2. Reversible temperature coefficients $\alpha_L$
        (225 K < T < 300 K) and $\alpha_H$ (300 K < T < 425 K)
        vs Gd concentration x with various concentrations
        y of Mn for the systems $Sm_{2-x}Gd_xCo_{17-y}Mn_y$.

understood. However, these considerations show that the achieve-
ment of a ZTC material with x=0.4 and y=1, will most probably be
attained at the cost of a lower remanence, energy product and coercivity than would be possible with the substitution of less Gd.


B.   The $Sm_{2-x}R_xCo_{17}$ Systems

The $Sm_{2-x}Er_xCo_{17}$ and $Sm_{2-x}Dy_xCo_{17}$ compounds displayed $4\pi M_s(T)$
curves similar to those for $Sm_{2-x}Gd_xCo_{17}$ shown in Fig. 1.   The
temperature coefficients derived from such curves are given in Fig.
4.   The dotted lines result from the additivity approximation discussed above based on data for the $R_2Co_{17}$ compounds from the work
of Lemaire et al (9).   The erbium and dysprosium compounds yield
ZTC's for x=0.2.   However, in Fig. 5 where $4\pi M_s$ and $H_A$ data are
shown, it may be seen that certain trade-offs exist.   The $4\pi M_s$

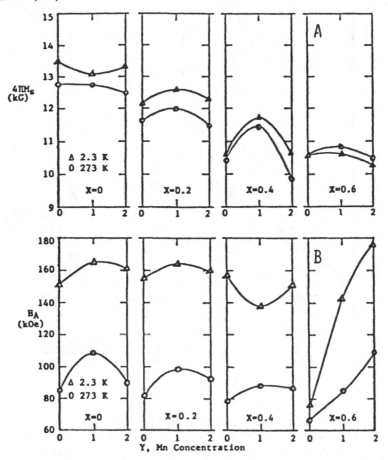

Figure 3.   (A) Saturation magnetization 4πM$_s$ and (B) anisotropy
            field H$_A$ vs Mn concentration y for the sets of compounds
            Sm$_{2-x}$Gd$_x$Co$_{17-y}$Mn$_y$ (where x=0, 0.2, 0.4, 0.6) at temper-
            atures 2.3 K (Δ) and 273 K (0).   4πM$_s$ and H$_A$ data were
            taken in applied fields up to 100 kOe.

for the dysprosium compound is not reduced as much as it is for
the erbium compound.   On the other hand, the dysprosium compound
shows a greater decline in H$_A$ than does the erbium.   A preliminary
result for the erbium system indicates that the substitution of
one manganese atom for cobalt significantly enhances H$_A$ from 85
kOe to 109 kOe (see Table 1) while maintaining relatively low α's.
This result makes the Sm$_{1.8}$Er$_{0.2}$Co$_{16}$Mn compound an attractive candidate for magnet development.   Whether the addition of manganese
to the dysprosium compound will produce similar results remains to
be determined.

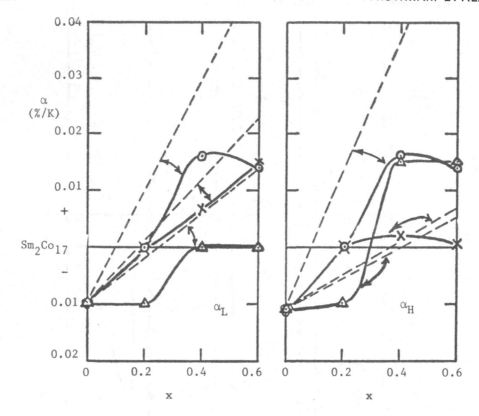

Figure 4. Reversible temperature coefficients $\alpha_L$
(225 K < T < 300 K) and $\alpha_H$ (300 K > T > 425 K) vs Gd
concentration x for systems $Sm_{2-x}R_xCo_{17}$, where
R = Gd ($\triangle$), Er (x), and Dy (⊚).

        The dashed lines in Fig. 5A represent the $4\pi M_s$ values obtained
by suitably summing the moments of the constituent atoms.  In so
doing we assumed the rare earth free atom moments (Sm, 0.714 $\mu_B$;
Gd, 7.0 $\mu_B$; Er, 9.0 $\mu_B$; and Dy, 10.0 $\mu_B$) given by Kirchmayer (10)
and the cobalt moment (1.65 $\mu_B$) given by Streever (11) for cobalt
in the $Nd_2Co_{17}$ compound.  It was further assumed that the heavy
lanthanides coupled antiferromagnetically with the samarium and
cobalt.  The trends seem to be correct.  For the dysprosium system
the theoretical curve parallels the low temperature results and
agrees with them to within experimental error.  The deviations for
the other systems probably reflect crystal field and/or band
structure effects which require further theoretical effort for
their clarification.  The same can be said for the various systematic trends seen in the figure.

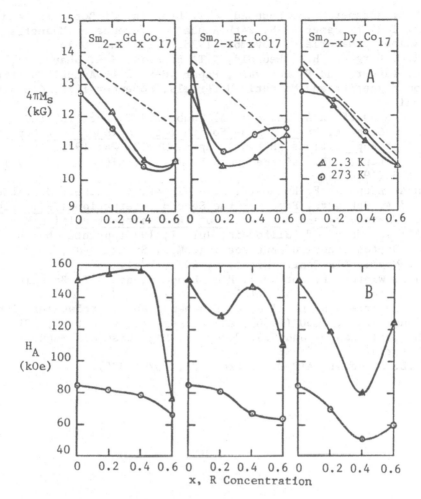

Figure 5. (A) Saturation magnetization $4\pi M_S$ and (B) anisotropy field $H_A$ vs R concentration x for the compounds $Sm_{2-x}R_xCo_{17}$ at temperatures 2.3 K ($\triangle$) and 273 K (O). $4\pi M_S$ and $H_A$ data were taken in applied fields up to 100 kOe.

## REFERENCES

1. F. Rothwarf, H.A. Leupold and L.J. Jasper, Jr., Proc. of the Third International Workshop on Rare Earth-Cobalt Magnets and Their Applications, Univ. of California, June 27-30, 1978; Printed by the Univ. of Dayton School of Engineering, K.J. Strnat, ed., p. 255.

2.  R.L. Bergner, H.A. Leupold, J.T. Breslin, F. Rothwarf, and
    A. Tauber, Paper 4C-8, 24th Annual Conference on Magnetism and
    Magnetic Materials, November 1978, Cleveland, Ohio.
3.  R.L. Bergner, H.A. Leupold, J.T. Breslin, J.R. Shappirio,
    A. Tauber, and F. Rothwarf, Paper 4C-9, 24th Annual Conference
    on Magnetism and Magnetic Materials, November 1978, Cleveland,
    Ohio.
4.  F. Rothwarf, H.A. Leupold, A. Tauber, J.T. Breslin, and
    R.L. Bergner, Proc. of the Fourth International Workshop and
    Their Applications, Hakone, Japan, May 22-24, 1979.
5.  M.G. Benz, R.P. Laforce and D.L. Martin, AIP Conf. Proc. 18,
    1173 (1974).
6.  H.A. Leupold, F. Rothwarf, J.J. Winter, A. Tauber, J.T. Breslin,
    and A. Schwartz, Proc. of the Second International Symposium
    on Coercivity and Anisotropy of Rare Earth-Transition Metal
    Alloys, Univ. of California, July 1, 1978; printed by the Univ.
    of Dayton School of Engineering, K.J. Strnat, ed., p. 87.
7.  H. Nagel, AIP Conf. Proc. 29, 603 (1976).
8.  J.J. Winter, F. Rothwarf, H.A. Leupold, and J.T. Breslin, Rev.
    Sci. Instrum. 49, 845 (1978).
9.  J. LaForest, R. Lemaire, R. Paughenet and J. Schweizer, Comptes
    Rendus 262B, 1260 (1966); and R. Lemaire, Cobalt 33, 301 (1966).
10. H.R. Kirchmayer and C.A. Poldy, J. Magnetism and Magnetic
    Materials 8, 1 (1978).
11. R.L. Streever, AIP Conf. Proc. 24, 462 (1975).

# THE EFFECT OF AN YTTRIUM-OXIDE DISPERSION ON THE YIELD AND

# 'MEMORY' CHARACTERISTICS OF NITINOL

R. Vincent Milligan
Benet Weapons Laboratory, ARRADCOM
Watervliet Arsenal, Watervliet, NY  12189

## INTRODUCTION

A report in the early 1950's (1) on the gold-cadmium system was one of the first demonstrating shape memory in a metal.  About a decade later, interest in shape memory alloys was revived by the work of Buehler and his associates (2) at the Naval Ordinance Laboratory where they studied the binary nickel-titanium alloy called Nitinol.  Since then about a dozen and a half binary or ternary alloy systems have been reported to show this effect (3).  The shape memory effect is the ability of a specimen to return to its original shape after being strained in what appears to be an elastic-plastic deformation.  After the specimen is strained, it is unloaded and then given a low temperature heat treatment to induce recovery thus enabling it to return to its original undeformed shape.  It is now generally agreed that the memory phemomenon is due to a stress-assisted martensitic phase transformation.  The transformed phase can revert to the parent phase when the material is heated to temperatures somewhat higher than the test temperature providing certain conditions are satisfied.  Some of the critical parameters are alloy composition, the amount of strain imposed, and the testing temperature.

Reported work concerning the effects of alloying additions to Nitinol is rather limited.  Wang (4) stated that small additions of iron or cobalt depress the transition temperature.  Eckelmeyer (5) reported that 1-2 percent additions of zirconium or gold increase the transition temperature while 1-2 percent of aluminum or manganese decrease it.

The purpose of this study was to determine the effects of adding a small amount of yttrium oxide powder to the melt of a 55 wt.% nickel-titanium alloy. The objective was to increase the apparent yield strength and determine whether the memory characteristics would be degraded relative to the non-yttriated base material.

## MATERIAL

The material for both the non-yttriated and yttriated heats was supplied by Reactive Metals, Inc. The purity of the nickel and titanium melting stock was 99.9 percent pure. The blending and melting of the materials for both heats was similar with the exception that 0.03 wt.% $Y_2O_3$ powder was added to obtain the yttriated heat. The material was vacuum melted twice under argon pressure. For brevity the details of the processing will not be given here but can be obtained from the author (6). A chemical analysis for both heats is given in Table 1.

Table 1. Chemical Analysis*

| Heat** | Ni | Ti | C | N | Fe | Mo | O | H |
|--------|------|-----|------|-------|------|------|-------|----------|
| A | 55.0 | Bal | 0.01 | 0.006 | 0.01 | 0.01 | 0.039 | 0.0039 |
| B | 54.6 | Bal | 0.01 | 0.006 | 0.01 | 0.01 | 0.048 | Not Det. |

*All tabular values are in wt. percent.
**Heat A is the non-yttriated or base heat.
  Heat B is the yttriated heat containing 0.03 wt. percent $Y_2O_3$.

The ASTM grain size was 3.5 for heat A and 4.0 for heat B.

Test specimens were machined to a modified ASTM configuration of 0.90 cm diameter. The modification consisted of lenthening the specimen shoulders to facilitate fastening to the loading rods of the testing machine.

## TESTING PROCEDURE

Specimens were tested at a constant strain rate of $10^{-4}$ cm/cm/ sec using a servo-controlled electrohydraulic testing machine. The strain for recording and controlling the testing machine was obtained using an LVDT type extensometer. The load was measured using a precision load cell. Temperatures were measured using chromel-alumel

thermocouples and nickel foil sensors. Load versus strain was
recorded using a standard X/Y recorder. In addition, the testing
parameters of load, strain, temperature, and time were recorded using
a multi-channel analog-to-digital recorder. A small environmental
chamber surrounded the specimen for heating and cooling. For cooling helium gas was flowed through a coil of copper tubing that was
inserted into a dewar containing liquid nitrogen. Heating the specimen was accomplished by placing coils of Nichrome wire inside the
chamber. These were powered by a DC supply. This chamber and
ancillary equipment provided a means of accurately controlling the
temperature of the specimen during straining and provided heat to
effect recovery. Since the thermal coefficient of expansion for
Nitinol is about the same as for steel, the magnitude of the thermal
strains were small for the temperature variations in this experiment.
Consequently, the thermal strains were very small compared with the
mechanical strains, and therefore neglected.

The sequential procedure for testing is as follows. The specimen is cooled to the desired test temperature. It is then strained
through the elastic region and the apparent plastic region to a maximum strain. It is then unloaded to zero load and the coolant gas
turned off. One of the nuts of the loading rod is then loosened to
permit unrestrained contraction during thermal recovery. The DC
power supply is then turned on causing the Nichrome wires to heat
up thus heating the specimen by convection. When the specimen
reaches the transformation temperature, recovery starts, and the
specimen returns to its original length thus completing the shape
memory cycle.


## RESULTS AND DISCUSSION

Figure 1 shows stress-strain curves for the yttriated and non-
yttriated material that were tested at approximately the same temperature.

It is obvious that the apparent yield strength (denoted by $\sigma_A$
and determined by a 0.1% offset) is significantly different. Another
interesting difference is the amount of strain elastically recovered
on unloading. For example, roughly 70-90 percent of the maximum
strain is recovered for heat B while only 10-35 percent is recovered
for heat A. The variation in elastic recovery strain depends on the
maximum load which in turn depends on the amount of strain imposed on
the specimen and the test temperature. The strain remaining after
unloading was completely recovered by heating to effect the reverse
transformation except for two tests where the testing temperature was
21°C. However, the thermal recovery for heat B appeared to be more
sluggish than that for heat A.

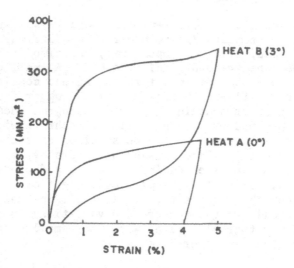

Figure 1.    Stress vs. Strain.    Curves for heats B (yttriated) and A
             (base) materials - numbers indicate testing temperature
             in °C.


        Figure 2 shows a composite of the early portion of the stress-
strain curves for heats A and B.    The sensitivity of the apparent
yield stress to test temperature is readily seen.

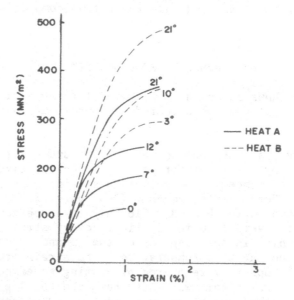

Figure 2.    Stress vs. Strain.    Initial portion of stress-strain
             curves for heats A and B tested at different temper-
             atures - degrees C.

Figure 3 is a plot of apparent yield stress vs. temperature for both heats. As the temperature decreases, the difference in yield strength between the two heats becomes significantly larger. If both curves are extrapolated to zero stress we can get an approximation for the athermal martensitic transformation temperature ($M_s$) determined by cooling under stress free conditions. The extrapolated $M_s$ values are -7°C and -40°C for heats A and B respectively.

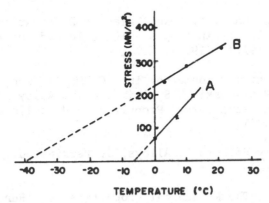

Figure 3.   Apparent yield stress vs. temperature for heats A and B.

As a check to determine whether the $M_s$ temperature had been suppressed by the $Y_2O_3$ addition, transformation temperatures were obtained using a Dupont 990 Thermoanalyzer in a differential thermal analysis (DTA) mode. $M_s$ values of -23°C and -89°C were obtained for heats A and B respectively based on the temperature where a peak occurred in the exothermal curve on cooling. This additional DTA test confirmed that the $M_s$ temperature was suppressed and that the influence of stress on the phase transformation temperature is significant for the base material and even stronger for the yttriated material. A paper by Waugh (7) dealing with the feasibility of using rare earth oxides for dispersion strengthening titanium suggests a 30% possibility for a reaction occurring between $Y_2O_3$ and titanium. Any reaction occurring between $Y_2O_3$ and titanium would result in a lowering of the $M_s$ temperature (8). Although $Y_2O_3$ particles were added to the base material efforts using scanning electron microscopy and microanalysis to identify them in the yttriated heat have not been successful. Metallography studies have not yet confirmed whether a reaction between the $Y_2O_3$ and the base material occurred.

For dispersion hardening involving such small amounts of $Y_2O_3$, one would not expect small variations in the test temperature to have such a strong effect on the yield strength. Since the base material shows the same test temperature sensitivity, the effect of the particle addition on this property is not known. In view of these

results, it is recommended that additional experiments be conducted
to separate out the effects of transformation versus dispersion
strengthening which will hopefully shed additional light on the
possible benefits to be gained from rare earth oxide additions to
Nitinol.


## REFERENCES

1.  L. C. Chang and T. A. Read, "Plastic Deformation and Diffusion-
    less Phase Change in Metals - The Gold Cadmium Beta Phase,"
    Trans AIME 191:47 (1951).

2.  c.f. Bibliography contained in:  C. M. Jackson, H. J. Wagner,
    and R. J. Wasilewski, "55 Nitinol - The Alloy with a Memory:
    Its Physical Metallurgy, Properties and Applications," NASA
    SP 5110 (1972).

3.  Shape Memory Effects in Alloys, J. Perkins ed., Plenum Press,
    New York (1975).

4.  F. E. Wang, "The Mechanical Properties as a Function of Temper-
    ature and Free Electron Concentration in Stoichiometric TiNi,
    TiCo, and TiFe Alloys," Proceedings of the First International
    Conference on Fracture, Sendai, Japan, p. 899 (1965).

5.  K. H. Eckelmeyer, "The Effect of Alloying on the Shape Memory
    Phenomenon in Nitinol," Script Metallurgica 10:667, 1976.

6.  R&D Dev. Rpt #610 RMI Co. to Watervliet Arsenal, May 1976.

7.  R. C. Waugh, "Suitable Oxides for Dispersion Strengthening of
    Titanium Alloys," Powder Metallurgy 2:85, 1976.

8.  R. J. Wasilweski, S. R. Butler, and J. E. Hanlon, "On the
    Martensic Transformation in TiNi," Journ. of Met. Sci. 1:104,
    1967.

# HEAT CAPACITIES OF LANTHANIDE COMPOUNDS, THEIR MORPHOLOGY

# SCHOTTKY, AND LATTICE CONTRIBUTION*

Edgar F. Westrum, Jr.

Department of Chemistry, University of Michigan

Ann Arbor, Michigan 48109

## ABSTRACT

With gradual accumulation of precise cryogenic calorimetric
data on the sesquioxides, chalcogenides (notably the sesquisulfides), hexaborides, hydrides, pnictides, halides, and especially the tri-hydroxides, an appreciation of the important contribution made by Schottky heat capacity contributions has arisen.  In
an attempt to understand and analyze -- and ultimately to correlate
and to predict the morphology and thermophysics of heat capacity
of transition element binary compounds -- increasing attention is
focussed on the lattice heat capacity contribution.  Accurate
knowledge of the lattice contribution is essential in the resolution of electronic, magnetic, Schottky and other contributions to
the heat capacity.  The importance of correlating the lattice contribution with molal volumes of the compounds as well as with
cation masses (of lanthanide cations) will be discussed -- in the
light of recent developments -- particularly in lanthanide halides
and trihydroxides.  The relevance of these developments to other
compounds is emphasized.

---

*This research was supported in part by the Structural Chemistry
and Chemical Thermodynamics Program of the Chemistry Section of
the National Science Foundation under contract GP-4252X at the
University of Michigan.

## INTRODUCTION

The importance and relevance of the Schottky contribution to the heat capacities and thermophysical functions -- in cryogenic regions and at super-ambient temperatures -- has been one of the new factors which chemical thermodynamicists have had to take into account during the past two decades in transition element as well as in lanthanide and actinide compounds.

## LANTHANIDE COMPOUNDS

Lanthanide Sesquioxides. Recognition of the importance of this thermophysical contribution came with a series of papers by Justice and Westrum (1-5) who studied the lanthanide sesquioxides and obtained an unusually rich yield of data concerning the energetics of the trivalent ions in these compounds. Although the Schottky contributions may also be studied spectroscopically, the general unavailability of single crystal samples for absorption spectroscopy or for paramagnetic resonance experiments had tended to favor the calorimetric approach. The initiatory measurements (1-5) on neodymium sesquioxide yielded (cf. Table 1) results in levels more than an order of magnitude smaller than those estimated by Penny and his collaborators (6) from crystal field splittings. But any discredit was of short duration, since spectroscopy (7) confirmed, in this instance, the values that had been obtained by calorimetry

Table 1.   Stark levels for $Nd_2O_3$ $(cm^{-1})$

| Level | $g_i$ | Calc. (6) | Calorimetry (1) | Spectroscopy (7) |
|-------|-------|-----------|-----------------|------------------|
| 0 | 2 | 0 | 0 | 0 |
| 1 | 2 | 492 | 21 | 22 |
| 2 | 2 | 1476 | 81 | 83 |
| 3 | 2 | 2952 | 400 | 390 |
| 4 | 2 | 4920 | --- | --- |

The method of approach involved measurement of the total heat capacity of $Nd_2O_3$, of a diamagnetic analog, $La_2O_3$, and the resolution of the difference in heat capacity of the two compounds in terms of a sequence of Schottky levels of the degeneracies predicted by crystal-field theory. The power of the cryogenic calorimetric approach thus demonstrated was later extended by the same authors to include most the lanthanides, even those containing $C_2$ and $C_{3i}$ sets of levels. It was demonstrated that the levels were valid not only for the cryogenic heat capacity contribution, but as well for temperatures in excess of 1000 K. Unfortunately, the sesquioxides crystallized in A-, B-, and C- forms thus making it harder to recognize the underlying trends.

Lanthanide Chlorides. The desirability of extending such studies to other systems led Sommers and Westrum to examine the lighter lanthanide trichlorides (8,9). Their study further heightened the understanding of the trend and regularities involved and showed the importance of the Schottky contribution to the thermophysical functions. The latter were in excellent accord with those predicted by the scheme of Westrum (10) based upon the treatment of Grønvold and Westrum (11). The excellence of the accord can be seen in Table 2.

Table 2. Comparison of some trihalide entropy estimation schemes

| Compound | $S^o(298.15K) - S^o(0)$ | | | | |
|---|---|---|---|---|---|
| | Latimer | Latimer augmented[a] | Westrum[a,b] augmented | "Spectro-scopic" | Reference No. 5 |
| $LaCl_3$ | 34.5 | 34.5 | 33.1 | 32.47 | 32.88 |
| $CeCl_3$ | 34.5 | 38.1 | 36.1 | 35.99 | (36.0)[c] |
| $PrCl_3$ | 34.5 | 38.9 | 36.8 | 36.74 | 36.64[d] |
| $NdCl_3$ | 34.6 | 39.2 | 36.8 | 36.84 | 36.67 |
| $PmCl_3$ | 34.7 | 39.5 | 36.8 | 36.82 | (37.0)[c] |
| $SmCl_3$ | 34 6 | 38.4 | 35.7 | 36.10 | 35.88 |
| $EuCl_3$ | 34.8 | 37.3 | 34.5 | 34.70 | 34.43 |
| $GdCl_3$ | 35.0 | 39.1 | 36.0 | 36.60 | 36.19 |

[a] By $R \ln(2J+1)$; the $(Cl_3)^{3-}$ ion contribution is taken as 20.7 $cal_{th}$ $K^{-1}$ $mol^{-1}$.

[b] By $R \ln(2J+1)$; the $(Cl_3)^{3-}$ ion contribution is taken as 17.9 $cal_{th}$ $K^{-1}$ $mol^{-1}$.

[c] Parentheses denote values involving interpolated lattice and calculated Schottky contributions.

[d] Based on 0.294 K.

Lanthanide Trifluorides. Apart from some heat-capacity measurements on hydrated halides (12,13) and an isolated one on a tribromide (14), the only other systematic study initiated is that of Lyon et al. on the anhydrous lanthanide trifluorides (15-17). $CeF_3$ (18) is found to give Schottky results in accord with expectation.

Although a change in crystal structure between $NdF_3$ and $SmF_3$ precludes application of the interpolation technique employed to estimate the $Ln(OH)_3$ and lighter $LnCl_3$ lattice heat capacities, the available heat capacity data for the La, Ce, Pr, and Nd homologs can be shown to be consistent with this approach. A detailed analysis must await the availability of spectroscopic data for concentrated PrF$_3$ and NdF$_3$. As shown by Chirico, Westrum, and Gruber (25), spectroscopic data determined for doped crystals may only be used to calculate a first estimate of the Schottky contribution for the corresponding concentrated salts.

Lanthanide Pnictides. The heat capacities of a set of lanthanide mononitrides has been achieved by Stuttius (19) on materials less well characterized than desirable, but has not involved a level of sophistication dealing with the trends in the lattice heat capacities as had been done subsequently. We are presently undertaking a reinterpretation of these data. His work is interesting in that he did utilize the linear crystal parameters in interpolating lattice heat capacities across the lanthanide series.

Lanthanide Hexaborides. Several of these compounds have been studied over the cryogenic range by Westrum et al. (20) and by others as well and clearly show Schottky functions. Unfortunately, the data have not yet been critically analyzed.

Lanthanide Hydrides. Bohdan Stalinski and Zygmunt Bignanski in Wroclaw have studied a number of the di- and trihydrides of the lanthanides and have summarized their findings elsewhere (21). These studies are interesting and will be examined critically as soon as access to the relevant data are available to the speaker.

THE LATTICE CONTRIBUTIONS

The principal problem of the calorimetric approach to the resolution of Schottky, magnetic, structural, and electronic contributions is the problem of how best to represent the lattice contribution to the heat capacity of the particular compounds studied. The literature abounds with various endeavors, including an attempt to resolve the Debye theta of the substance in question from normal heat-capacity regions curve itself and to use this as a means of interpolating across the region in which the heat capacity is anomalous. Several authors have proposed corresponding states theories of various types. A common approach is to use the measured heat capacity of an isostructural diamagnetic analog to represent the heat capacity of the compound. Yet one is confronted with the question of how does the stand-in compound differ from that used to represent the lattice contribution in the magnitude of this contribution? For nearly half a century the Latimer scheme (22,23) has been the favorite way of taking into account the differences between compounds in analogous series. This time-honored scheme is not without its flaws, despite the several times it has been adjusted by Latimer himself.

In two contributed papers (24,25) presented at this meeting, a significant breakthrough has occurred in the provision of the volumetric scheme to be utilized in the interpolation of heat-capacity contributions for lattices in analogous series. Here the measurements of Robert Chirico on the lanthanide trihydroxides have pointed an approach which is particularly helpful over the temperature region in which the development of entropy is significant and important, and over which many large Schottky functions are to be found.

## LANTHANIDE TRIHYDROXIDES

These compounds have the great advantage of being isostructural across the entire lanthanide series. Moreover, they may be prepared by hydrothermal synthesis provided care is taken to preclude formation of the oxyhydroxide. The change is the lattice contributions across the series is significantly larger than in either the sesquioxides or the trihalides and this series, therefore, provides a convenient testing ground for the evaluation of the lattice heat capacity and the trends with increasing atomic number across the lanthanide series. As a matter of fact, even yttrium hydroxide and uranium trihydroxide are outlying members which are also isostructural and can serve to provide extreme delimitation of the effects of volume and mass on the theories. The four definitive papers (26-30) in which these works to date have been summarized provide an appreciation of the validity of the approach.

Perhaps the best way of testing the validity of a lattice contribution is in the calculation of the calorimetric Schottky contribution and the comparison of this excess heat capacity with that calculated from spectroscopic data on the samples itself. As has been emphasized in the two contributed papers to this meeting (24,25), this comparison can only be made when one utilizes the Stark levels of the concentrated compounds. Measurements made on doped lanthanide halides, for example, need to be extrapolated by some technique (discussed elsewhere) (30) or by calculations based on crystal-field parameters.

The scheme developed in these papers is based on an interpolation on the basis of the molal volumes of the compounds in question. In particular, the formula by which the lattice heat capacity of the praseodymium trihydroxide may be calculated is indicated below:

$$C_p[Pr(OH)_3 \text{ lattice}] = xC_p[La(OH)_3] + (1-x) \, C_p[Gd(OH)_3]$$

and in which is the fractional molal volume increment

$$x = \{V[Pr(OH)_3] - V[La(OH)_3]\}/\{V[gd(OH)_3 - V[La(OH)_3$$

The importance of volume was appreciated by Westrum (31) who recognized that for the lanthanide chalcogenides the lattice contribution decreased with increasing atomic number and was, therefore, diametrically opposed to the trend in the Latimer scheme based on mass. The lanthanide contraction, however, provided the clue. Other authors (32,33) have been engaged in a polemic as to the relevance of volume versus mass in providing interpolation schemes for lattice contributions. Moreover, Kieffer (34-36) has undertaken a theoretical and experimental correlation of the lattice vibrations of minerals. This takes into account the many factors involved and discusses particularly the analysis of the vibrational contribution. This aspect also has been discussed by Sommers and Westrum (9).

## LANTHANIDE SESQUISULFIDES

Exceedingly interesting spectroscopic studies, together with heat capacity measurements at very low temperatures, are beginning to probe very interesting lanthanide sesquisulfides which are being prepared in stoichiometric single crystals, as well as in the hyper- and hypo-stoichiometric forms.

Since much of this work is as yet unpublished in definitive form, one can only herald the endeavors of Professors Gschneidner and Gruber and their collaborators for these pioneering studies.

It is the plan that the Schottky functions at the higher cryogenic temperatures will be explored by this author and his coworkers in a collaborative endeavor.

## SUMMARY AND CONCLUSIONS

It is already clearly evident that the excellence of the agreement between the Schottky functions for the condensed trichlorides and trihydroxides have verified the success of the volumentric approach to the development of lattice heat-capacity contributions. We are in the process of extending this approach to other lanthanide systems and hope to test it also as a possibly more general approach to the estimation of lattice and compound entropy contributions in a Latimer-like scheme and to assist in the estimation, evaluation and correlation of entropies of mineral systems and chemical compounds beyond the lanthanide series.

## REFERENCES

1.  B. H. Justice and E. F. Westrum, Jr., "Thermophysical properties of the lanthanide oxides. I. Heat capacities, thermodynamic properties, and some energy levels of Lanthanum (III) and Neodymium(III) oxides from 5 to $350\frac{1}{4}$K," J. Phys. Chem. 67: 339-45 (1963).

2.  B. H. Justice and E. F. Westrum, Jr., "Thermophysical properties of the lanthanide oxides. II. Heat capacities, thermodynamic properties, and some energy levels of Samarium(III), Gadolinium (III), and Ytterbium(III) oxides from 10 to 350°K", J. Phys. Chem. 67:345-51 (1963).

3.  E. F. Westrum, Jr. and B. H. Justice, "Thermophysical properties of the lanthanide oxides. III. Heat capacities, thermodynamic properties, and some energy levels of Dysprosium(III), Holmium (III), and Erbium(III) oxides", J. Phys. Chem. 67:659-65 (1963).

4.  B. H. Justice, E. F. Westrum, Jr., E. Chang, and R. Radebaugh, "Thermophysical properties of the lanthanide oxides. IV. Heat capacities, thermodynamic properties, and energy levels of Thulium(III) and Luterium(III) oxides", J. Phys. Chem. 73:333-340 (1969).

5.  B. H. Justice and E. F. Westrum, Jr., "Thermophysical properties of the lanthanide oxides. V. Heat capacity, thermodynamic properties, and energy levels of Cerium(III) oxide," J. Phys. Chem. 73:1959-1962 (1969).

6.  W. G. Penney, "Crystalline fields of Pr, Nd, and Yb from paramagnetic susceptibilities", Phys. Rev. 43:485 (1933).

7.  J. R. Henderson, M. Muramoto, and J. B. Gruber, "Spectrum of $Nd^{3+}$ in Lanthanide Oxide Crystals", J. Chem. Phys. 46:2515 (1967).

8.  J. A. Sommers and E. F. Westrum, Jr., "Thermodynamics of the lanthanide halides. I. Heat capacities and Schottky anomalies of $LaCl_3$, $PrCl_3$, and $NdCl_3$, from 5 to 350 K", J. Chem. Thermodynamics 8:1115-1137 (1976).

9.  J. A. Sommers and E. F. Westrum, Jr., "Thermodynamics of the lanthanide halides. II. Heat capacities and Schottky anomalies of $SmCl_3$, $EuCl_3$, and $GdCl_3$ from 5 to 350 K", J. Chem. Thermodynamics 9:1-26 (1977).

10. E. F. Westrum, Jr., Developments in chemical thermodynamics of the lanthanides, in "Lanthanide/actinide chemistry", R. F. Gould, ed., American Chemical Society, Washington, D. C. (1967).

11. F. Grønvold and E. F. Westrum, Jr., "Heat capacities and thermodynamic functions of $FeS_2$ (pyrite), $FeSe_2$, and $NiSe_2$ from 5 to 350°K.  The estimation of standard entropies of transition metal chalcogenides", Inorg. Chem. 1:36-48 (1962).

12.  W. Pfeffer, "Specific heats of $DyCl_3$, $NdCl_3$, $LuCl_3$, $HoCl_3$, and $ErCl_3$ hexahydrates", $\underline{Z}$. $\underline{Physik}$ 162:413-420 (1961).

13.  K. H. Hellwege, "Spezifische Warme von $PrCl_3$, $LaCl_3$, und $GdCl_3$ hydraten", $\underline{Z}$. $\underline{Physik}$ 154:301-309 (1959); 162:34-45 (1961).

14.  T. M. Deline, E. F. Westrum, Jr., J. M. Haschke, "Entropy and heat capacity of europium(III) bromide from 5 to 340 K", $\underline{J}$. $\underline{Chem.}$ $\underline{Thermodynamics}$ 7:671-676 (1975).

15.  W. G. Lyon, D. W. Osborne, H. E. Flotow, F. Grandjean, W. N. Hubbard and G. K. Johnson, "Thermodynamics of the lanthanide trifluorides. I. The heat capacity of lanthanum trifluoride, $LaF_3$ from 5 to 350°K and enthalpies from 298 to 1477°K", $\underline{J}$. $\underline{Chem}$. $\underline{Phys}$. 69(1):167-73 (1978).

16.  W. G. Lyon, D. W. Osborne, and H. E. Flotow, "Thermodynamics of the lanthanide trifluorides. II. The heat capacity of praseodymium trifluoride $PrF_3$ from 5 to 350°K", $\underline{J}$. $\underline{Chem}$. $\underline{Phys}$. 70:675-680 (1979).

17.  W. G. Lyon, H. E. Flotow (unpublished data, 1979).

18.  E. F. Westrum, Jr. and A. F. Beale, Jr., "Heat capacities of cerium(III) fluoride and of cerium(IV) oxide from 5 to 350°K", $\underline{J}$. $\underline{Phys}$. $\underline{Chem}$. 65: 353-55 (1961).

19.  W. G. Stuttius (personal communication, 1975).

20.  E. F. Westrum, Jr., H. L. Clever, J. T. S. Andrews, and G. Feick, "Thermodynamics of lanthanum and neodymium hexaborides $\underline{in}$ "Rare Earth Research III", L. Eyring, Ed., Gordon and Breach, New York (1966) and unpublished work.

21.  Z. Bieganski and B. Stalinski, "Specific heats of solids at low temperatures", $\underline{Fiz}$. $\underline{Chem}$. $\underline{Ciala}$ $\underline{Stalego}$: 295-318 (1977).

22.  W. M. Latimer, "The mass effect in the entropy of solids and gases", J. Am. Chem. Soc. 43:818 (1921).

23.  W. M. Latimer, "Methods of estimating the entropies of solid compounds", J. Am. Chem. Soc. 73:1480 (1951).

24.  R. D. Chirico and E. F. Westrum, Jr., "Thermodynamics of the lanthanide trihydroxides -- resolution of the lattice and Schottky contributions". Proceedings of this Conference.

25. E. F. Westrum, Jr., R. D. Chirico, and J. B. Gruber, "Thermophysics of the lanthanide trichlorides -- reanalysis of the Schottky heat capacity contributions". Proceedings of this Conference.

26. R. D. Chirico, E. F. Westrum, Jr., J. B. Gruber and J. Warmkessel, "Low-temperature heat capacities, thermophysical properties, optical spectra, and analysis of Schottky contributions to Pr(OH)$_3$", J. Chem. Thermodynamics 11:000-000 (1979).

27. F. Gronvold, J. Drowart, and E. F. Westrum, Jr., "The actinide chalcogenides" in The Chemical Thermodynamics of Actinide Elements and Compounds, Part 4, International Atomic Energy Agency, Vienna, Austria (1979).

28. E. F. Westrum, Jr., "Thermophysical aspects of orientational disorder in crystals", in Proceedings of 75 Aneversario de la Real Sociedad Espanola de Fisica y Quimica (1979).

29. R. D. Chirico, E. F. Westrum, Jr., J. B. Gruber, and J. Warmkessel, "Low temperature heat capacities, thermophysical properties, optical spectra, and analysis of Schottky contributions to Pr(OH)$_3$", J. Chem. Thermodynamics 11:000 ooo (1979).

30. R. D. Chirico, E. F. Westrum, Jr., and J. B. Gruber, "Thermodynamics of some lanthanide trihalides. III. Reinterpretation of LnCl$_3$ Schottky anomalies", J. Chem. Thermodynamics 11:000-000 (1979).

31. E. F. Westrum, Jr., "Exciting developments in the thermophysics of the pnictides and chalcogenides of the transition elements", Uspekhi Khimii XX:00-00 (1979).

32. S. K. Saxena, "Entropy estimate for some silicates at 298°K from molar volumes", Science 193:1241-1242 (1976).

33. S. Cantor, "Entropy estimates of garnets and other silicates", Science 198:206-207 (1977).

34. S. W. Kieffer, "Thermodynamics and lattice vibrations of minerals: 1. Mineral heat capacities and their relationships to simple lattice vibrational models", Reviews of Geophysics and Space Physics 17:1-19 (Feb. 1979).

35. S. W. Kieffer, "Thermodynamics and lattice vibrations of minerals: 2. Vibrational characteristics of silicates", Reviews of Geophysics and Space Physics 17:20-34 (Feb. 1979).

36.   S. W. Kieffer, "Thermodynamics and lattice vibrations of
      minerals: 3.  Lattice dynamics and an approximation for
      minerals with application to simple substances and framework
      silicates", Reviews of Geophysics and Space Physics 17:35-39
      (Feb. 1979).

THERMOPHYSICS OF THE LANTHANIDE TRIHYDROXIDES -- RESOLUTION

OF THE LATTICE AND SCHOTTKY CONTRIBUTIONS[a]

Robert D. Chirico and Edgar F. Westrum, Jr.

Department of Chemistry, University of Michigan

Ann Arbor, MI  48109

## ABSTRACT

Lattice heat capacities and Schottky contributions have been resolved and analyzed for a series of lanthanide trihydroxides through heat-capacity measurements from 5 to 350 K.  These compounds all crystallize with the $UCl_3$-type structure and, therefore, are a particularly suitable series for which to attempt this separation.

The lattice heat capacities of $Pr(OH)_3$, $Nd(OH)_3$, $Eu(OH)_3$, and $Tb(OH)_3$ have been approximated by means of a volume-weighted interpolation between the $La(OH)_3$ and $Gd(OH)_3$ lattice heat capacities. The resulting calorimetrically resolved Schottky contributions of $Eu(OH)_3$, $Pr(OH)_3$, and $Tb(OH)_3$ are in excellent accord with those calculated from energy levels and degeneracies spectroscopically determined for the concentrated compounds.  No spectroscopic data exist for $Nd(OH)_3$.  In its absence the Stark level energies of the $^4I_{9/2}$ state were approximated through use of the new lattice approximation and extrapolation of the energy levels of $Nd(III)$ doped $LaCl_3$ to stronger crystal fields.  The $^4I_{9/2}$ Stark levels were also approximated by means of crystal-field calculations (including J-mixing with the $^4I_{11/2}$ state) using estimated crystal-field parameters. The results of the approximation methods are compared.

---

[a]This research was supported in part by the Structural Chemistry and Chemical Thermodynamics Program of the Chemistry Section of the National Science Foundation under contract GP-4252X at the University of Michigan.  (The present address of RDC is Department of Chemistry, University of Illinois, Chicago Circle, Chicago, IL.)

Heat-capacity measurements in the 5 to 350 range are in progress for Y(OH)$_3$ and Ho(OH)$_3$. Results of these experiments will be presented together with a brief discussion of projected studies including Dy(OH)$_3$ and Eu(OH)$_3$.

## INTRODUCTION

The resolution of Schottky contributions from the generally much larger vibrational ("lattice") heat capacities of lanthanide compounds has been limited by the uncertainty in the magnitude of the lattice contribution. Consequently, such subtle effects as dependence of the Stark levels on temperature and host lattice have been heretofore undetected calorimetrically. Since the lanthanide trihydroxides are a homologous series having relatively small lattice contributions and their lower-lying Stark levels have been spectroscopically deduced for many of the concentrated compounds, the lanthanide trihydroxides are the most nearly ideal system yet studied in an attempt to resolve Schottky contributions in the 5 to 350 K temperature range.

We here present a lattice-approximation scheme based upon interpolation between the lattice heat capacities of the La and Gd homologs weighted by the fractional molal volume variation along the series. As this is an empirical approach its defense lies in the consistent interpretations afforded by its application.

## THE SCHOTTKY CONTRIBUTIONS TO THE Ln(OH)$_3$ HEAT CAPACITIES

### Eu(OH)$_3$

The Schottky contribution to the heat capacity of the Eu(III) analog is unique in that it arises entirely from thermal population of excited [SL]J-manifolds. This invariably results in the lowest excited Stark levels being much higher in energy for the Eu(III) analog than for any other series member. The calculated Schottky heat capacity is consequently relatively insensitive to small shifts in the Stark level energies and, therefore, is expected to be the most accurate approximation to the true Schottky heat capacity within any lanthanide series.

The energy levels of concentrated Eu(OH)$_3$ were determined by Cone and Faulhaber (1) from absorption and fluorescence spectra at 4.2 and 7.7 K. Stark levels arising from the $^7F_0$, $^7F_1$, $^7F_2$, and $^7F_3$ manifolds all contribute to the Schottky heat capacity below 350 K.

The lattice heat capacity of Eu(OH)$_3$ was approximated using the following formula:

$$Cp[Eu(OH)_3, \text{lattice}] = 0.13 \, Cp[La(OH)_3] \quad (1)$$
$$+ 0.87 \, Cp[Gd(OH)_3, \text{lattice}].$$

The derived <u>calorimetric</u> Schottky contribution shown in figure 1a is seen to be in excellent accord with that calculated from the spectral data.

$$Pr(OH)_3$$

The crystal-field splitting of the $^3H_4$ manifold of $Pr(OH)_3$ has been determined from the absorption spectra of mulls at 95 K (2). The observed spectra were not as highly resolved as one might obtain from measurements on single crystals. This lack of resolution is reflected in a ± 3 cm$^{-1}$ uncertainty in the Stark level energies.

The lattice heat capacity of $Pr(OH)_3$ was approximated as:

$$Cp[Pr(OH)_3, \text{lattice}] = 0.65 \, Cp[La(OH)_3 \quad (2)$$
$$+ 0.35 \, Cp[Gd(OH)_3, \text{lattice}].$$

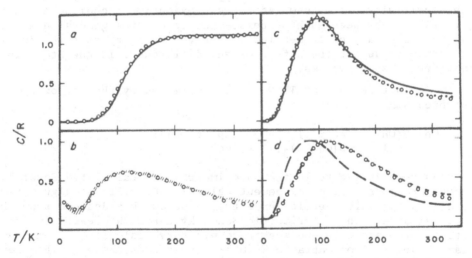

Figure 1. The open circles represent the calorimetrically deduced Schottky contributions. The uninterrupted curves represent the Schottky contributions calculated from spectroscopically derived Stark levels of the concentrated compounds. The dotted curve in 1c represents the Schottky contribution calculated from the energy levels of Tb$^{+3}$ doped Y(OH)$_3$. The long and short dashed curves of 1d represent the Schottky heat capacity contribution calculated from the energy levels of Nd$^{+3}$ doped LaCl$_3$ and those estimated for concentrated Nd(OH)$_3$ (see text). 1a, Eu(OH)$_3$; 1b, Pr(OH)$_3$; c, Tb(OH)$_3$; and 1d, Nd(OH)$_3$.

As seen in figure 1b the calorimetric and calculated Schottky curves are in very good agreement between 15 and 230 K. Below 25 K a cooperative magnetic contribution of unknown magnitude plus the uncertainty in the energy of the lowest excited Stark level preclude any attempt to accurately determine the Schottky contribution in this temperature region. Above 260 K the calorimetric curve trends below the calculated band. Such a decrease in the high temperature calorimetric Schottky curve could be due to a gradually decreasing crystal field intensity within the $Pr(OH)_3$ crystals as the lattice expands with temperature. A gradual shift to lower energies of 5 to 10 $cm^{-1}$ by the four highest Stark components of the $^3H_4$-manifold between 100 and 350 K would account for the observed deviation.

$$Tb(OH)_3$$

The energy levels of the lowest four manifolds of concentrated $Tb(OH)_3$ and $Tb^{+3}$ doped $Y(OH)_3$ were determined by Scott, Meissner, and Crosswhite (3). The observed Schottky below 350 K is due almost entirely to population of the $^7F_6$ manifold. The availability of spectroscopically determined energy levels for both the $Tb(OH)_3$ and $Y(OH)_3$ host lattices provides an opportunity to directly observe the sensitivity of the new lattice approximation technique in differentiating between such systems. Heretofore the general assumption has been that any calorimetrically derived Schottky contribution is too crude to detect the effect of any differences in the Stark level energies of such systems.

The lattice heat of $Tb(OH)_3$ was estimated by the following weighted sum:

$$Cp[Tb(OH)_3, lattice] = -0.12\ Cp[La(OH)_3] \qquad (3)$$
$$+ 1.12\ Cp[Gd(OH)_3, lattice].$$

As seen in figure 1c the calorimetric and calculated Schottky curves are in excellent agreement below 160 K, while at higher temperatures the calorimetric curve trends below that deduced from the spectral data. The calorimetric Schottky curve is clearly in for better agreement with the spectroscopic curve calculated from the Stark levels of concentrated $Tb(OH)_3$ rather than those of $Tb^{+3}$ doped $Y(OH)_3$. Above 160 K the difference between the calorimetric and calculated curves may be accounted for if the Stark levels are assumed to undergo approximately a six per cent shift to lower energies between 77 and 350 K. Such a shift may be postulated as being due to the decrease of the crystal-field intensity as the lattice expands with increasing temperature.

Nd(OH)$_3$

Although spectral data are yet unavailable for Nd(OH)$_3$ several interesting comparisons with isomorphous NdCl$_3$ can be made. The lattice heat capacity of Nd(OH)$_3$ was approximated as:

$$Cp[Nd(OH)_3, lattice] = 0.44 \ Cp[La(OH)_3] \qquad (4)$$
$$+ \ 0.56 \ Cp[Gd(OH)_3, lattice].$$

The derived calorimetric Schottky heat capacity is shown in figure 1d along with the spectroscopic Schottky curve calculated from the Stark level energies of Nd$^{+3}$ doped LaCl$_3$. As can be seen, the maximum heights of the Schottky contributions (C/R) are within 0.015, while the stronger crystal field within the trihydroxide shifts the Nd(OH)$_3$ curve to higher temperatures. The relative splitting of the $^4I_{9/2}$ levels of Nd(OH)$_3$ is expected to be related to the fractional shift in temperature necessary to bring the Nd$^{+3}$ doped LaCl$_3$ Schottky peak into coincidence with that of the Nd(OH)$_3$ calorimetric curve. The energy levels of the $^4I_{9/2}$ manifold of Nd(OH)$_3$ estimated in this manner are 0, 165, 175, 355 and 365 cm$^{-1}$. The Schottky contribution calculated from these estimated energies is shown in figure 1d. The uncertainty in the energies is estimated to be about four per cent of the separation from the ground state.

Crystal field calculations, which included J-J mixing with the $^4I_{11/2}$ [SL]J-state, using the spectroscopically derived crystal-field parameters of Eu(OH)$_3$ (1) and Pr(OH)$_3$ (2) yielded Stark level energies only within $\sim$30 cm$^{-1}$ of those deduced calorimetrically. Corroboration of the energy levels must await a full spectral analysis of Nd(OH)$_3$.

## REFERENCES

1.  R. L. Cone and R. Faulhaber, "Optical studies of energy levels in Eu(OH)$_3$," J. Chem. Phys. 55:5198 (1971).

2.  R. D. Chirico, E. F. Westrum, Jr., J. B. Gruber, and J. Warmkessel, "Low-temperature heat capacities, thermophysical properties, optical spectra, and analysis of Schottky contributions to Pr(OH)$_3$," J. Chem. Thermodynamics (In press).

3.  P. D. Scott, H. E. Meissner, and H. M. Crosswhite, "Crystal fields for terbium(III) in hydroxides," Phys. Lett. 28A:489 (1969).

# THERMOPHYSICS OF THE LANTHANIDE TRICHLORIDES--REANALYSIS

# OF THE SCHOTTKY HEAT CAPACITY CONTRIBUTIONS*

Edgar F. Westrum, Jr., Robert D. Chirico
Department of Chemistry, University of Michigan,
Ann Arbor, MI 48109
John B. Gruber, Department of Physics,
North Dakota State University, Fargo, ND 58102

## ABSTRACT

The heat capacity data on the anhydrous lanthanide trichlorides from 5 to 350 K of Sommers and Westrum provide an excellent opportunity to further test the volumetric lattice heat-capacity approximation method. Schottky contributions in $PrCl_3$, $SmCl_3$, and $EuCl_3$ were calorimetrically derived using the volume-weighted interpolation between the lattice heat capacities of the La and Gd homologs. Previously qualitative agreement had been observed upon comparison of these calorimetrically derived Schottky contributions with those calculated from spectroscopic data obtained for Ln(III) doped $LaCl_3$ crystals. Excellent accord between "spectroscopic" and "calorimetric" Schottky contributions is achieved by adjusting the Stark-level energies to represent those of the concentrated salts through extrapolation of the Ln(III) doped $LaCl_3$ energies to stronger crystal fields either empirically or by means of estimated crystal-field parameters. These methods are described and the resulting energy levels are compared with the spectroscopic data that do exist for the concentrated trichlorides.

---

*This research was supported in part by the Structural Chemistry and Chemical Thermodynamics Program of the Chemistry Section of the National Science Foundation under contract GP-4252X at the University of Michigan. (The present address of RDC is Department of Chemistry, University of Illinois, Chicago Circle, Chicago, IL.

## INTRODUCTION

Excellent resolutions of the Schottky contributions to the heat capacities of Pr(OH)$_3$ (1), Eu(OH)$_3$ (2), and Tb(OH)$_3$ (3) have recently been achieved by means of a vibrational ("lattice") heat-capacity estimation-scheme based upon interpolation between the lattice heat capacities of the lanthanum and gadolinium analogs weighted by the fractional molal volume variation along the series. The heat-capacity determinations of the anhydrous lanthanide trichlorides from 5 to 350 K by adiabatic calorimetry of Sommers and Westrum (4,5) plus the wealth of spectroscopic data -- which provided primary impetus for the original investigation -- provide a further validation of the volumetric lattice-heat-capacity estimation scheme.

## THE SCHOTTKY CONTRIBUTIONS TO THE LnCl$_3$ HEAT CAPACITIES

### EuCl$_3$

The Schottky heat-capacity contribution of EuCl$_3$ is unique in that it arises entirely from thermal population of excited [SL]J-manifolds with the first excited states near 355 and 405 cm$^{-1}$. The Schottky contributions was calculated from energy levels of (1 and 4 percent) Eu$^{+3}$ doped LaCl$_3$ determined from the absorption spectrum at 4 K and the fluorescence spectrum at 4 and 77 K studied by Deshazer and Dieke (6). The energy levels of concentrated EuCl$_3$ are not expected to be identical to those of Eu$^{+3}$ doped LaCl$_3$. The stronger crystal field in concentrated EuCl$_3$ -- compared to that in the LaCl$_3$ host -- is expected to increase the Stark splitting and simultaneously to lower the center of gravity of the $^7$F$_1$-manifold, i.e., to lower the energy of the $\mu$ = 1 doublet and to leave the $\mu$ = 0 level essentially unchanged. The effect of the stronger crystal field will be countered to some extent by expansion of the EuCl$_3$ lattice at higher temperatures (i.e., in the region of the Schottky maximum); however, this is anticipated to be insufficient to fully nullify the effect. Because the energies of the Stark levels contributing to the Schottky heat capacity are relatively high, a shift of the $\mu$ = 1 doublet by as much as 10 to 15 cm$^{-1}$ will have but a small effect on the calculated Schottky contribution.

The lattice heat capacity of EuCl$_3$ was approximated by the following weighted sum:

$$C_p(\text{EuCl}_3, \text{lattice}) = 0.12 \, C_p(\text{LaCl}_3) + 0.88 \, C_p(\text{GdCl}_3, \text{lattice}).$$

The weighting factors were determined on molal volumes. The derived calorimetric Schottky heat capacity shown in figure 1a is seen to be in excellent accord with that derived from the spectroscopic data.

$$NdCl_3$$

The Schottky contribution below 350 K to this compound is due almost entirely to thermal population of excited Stark levels within the $^4I_{9/2}$ ground manifold.  The center of gravity of the first excited manifold ($^4I_{11/2}$) is approximately 2100 $cm^{-1}$; consequently, these levels make only a small contribution to the Schottky heat capacity below 350 K.

Carlson and Dieke (7) have determined the energy levels of (0.2 and 2 per cent) $Nd^{+3}$ doped $LaCl_3$ from fluorescence and absorption spectra at 4 and 77 K.  The relative splitting of the $^4I_{9/2}$ levels of concentrated $NdCl_3$ are expected to be related to the fractional shift in temperature necessary to bring the $Nd^{+3}$ doped $LaCl_3$ Schottky peak into coincidence with that of the calorimetric Schottky peak.  The Stark levels of concentrated $NdCl_3$ derived in this manner yield the agreement between the calculated and calorimetric Schottky curves shown in figure 1b.  The uncertainty in these energies is estimated to be approximately three per cent of the separation from the ground state.  Above 230 K, some of the observed deviation may be occasioned by compression of the ground J-manifold due to expansion of the lattice.  Further corroboration of the $^4I_{9/2}$ Stark level energies estimated calorimetrically is provided by the absorption spectral data obtained for concentrated $NdCl_3$ by Prinz (8).  The agreement between the two independently derived sets of energy levels is generally well within three per cent.

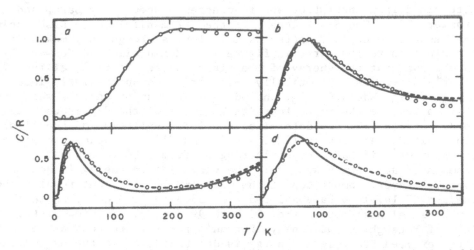

Figure 1.  The calorimetric Schottky curves are shown by dashed lines and the spectroscopic Schottky curves for doped $LaCl_3$ crystals by solid lines; those deduced from estimated energy levels for the concentrated halides by O.  1a, $EuCl_3$.  1b, $NdCl_3$.  1c, $SmCl_3$.  1d, $PrCl_3$.

## SmCl$_3$

The Schottky contribution below 200 K is occasioned by thermal population of the excited Stark levels of the $^6H_{5/2}$ ground [SL]J-manifold. Near 150 K, population of the first excited manifold, centered near 1000 cm$^{-1}$, begins to make an appreciable contribution to the Schottky heat capacity and ultimately causes the Schottky curve to rise again after the first maximum (see figure 1c).

The energy levels of (0.2 and 2 per cent) Sm$^{+3}$ doped LaCl$_3$ were determined by Magno and Dieke (9) from absorption and fluorescence spectra at 4 and 77 K. The Stark levels of the $^6H_{5/2}$, $^6H_{7/2}$, and $^6H_{9/2}$ [SL]J states were used in the Schottky calculation; however, the contribution of the $^6H_{9/2}$ manifold, centered near 2300 cm$^{-1}$, is very small below 350 K. The magnitudes of the spectroscopic level adjustments were estimated by the same technique used for the $^4I_{9/2}$ levels in NdCl$_3$. Excellent agreement is obtained between the calcu-lated and spectroscopic Schottky heat capacities below 230 K when the first two excited Stark levels are adjusted by 10 and 22 cm$^{-1}$.

## PrCl$_3$

The analysis of the Schottky contribution to the heat capacity of PrCl$_3$ is complicated by unusual shifts in the Stark level energies as the intensity of the crystalline field is varied. However, the basic arguments remain essentially unchanged from those applied to the preceding compounds. The energy levels of Pr$^{+3}$ doped LaCl$_3$ were determined from absorption and fluorescence spectra by Sarup and Crozier (10). The Stark levels of the $^3H_5$ manifold were also included in the spectroscopic Schottky calculation, although they make little contribution to the curve in figure 1d. McLaughlin and Conway (11) have shown that the energy of the first excited Stark level ($\mu$ = 3) of Pr$^{+3}$ decreases monotonically for Pr$^{+3}$ doped successively into LaCl$_3$, CeCl$_3$, NdCl$_3$, SmCl$_3$, and GdCl$_3$. They unfortunately were un-able to observe the upper four Stark levels of the $^3H_4$ ground mani-fold. Dorman (12) has determined the Stark splitting of the first two excited Stark levels ($^3H_4$, $\mu$ = 3$^-$ and 2') for dopant concentra-tions of Pr$^{+3}$ in LaCl$_3$ crystals ranging from 0.26 to 100 per cent from their absorption spectra. The variation in energy is not mono-tonic; however, comparison of the 0.26 and 100 per cent results shows the $\mu$ = 3$^-$ level to have shifted from 33.3 to 31.8 cm$^{-1}$, while the. $\mu$ = 2' level has shifted from 95.9 to 99.5 cm$^{-1}$. In view of the find-ings of McLaughlin and Conway, Dorman's results are reasonable. The first excited Stark level energy is decreased, while the second ex-cited level increases in energy for the stronger crystal line field of PrCl$_3$. Overlap with the vibronic spectrum prevented Dorman from observing the upper three Stark levels of the $^3H_4$ manifold. If these upper levels shifted to higher energies in the concentrated salt, the (observed) lowering of the maximum heat-capacity peak height would ensue.  Energies of the upper three Stark levels of the $^3H_4$ manifold for concentrated $PrCl_3$ were evaluated by maintaining the energies of the first two excited Stark levels determined by Dorman, while the remaining Stark-level energies were shifted approximately 25 to 35 $cm^{-1}$.  Good agreement between the Schottky contributions is noted (figure 1d).

## SPECTRAL DATA AND A TEST

Adsorption spectral values and new electronic Raman scattering data on the four compounds (1) are in excellent accord with the values extrapolated by the empirical scheme described as well as with levels derived from estimated crystal-field parameters

## $CeCl_3$

An interesting test of the validity of the deduced crystal-field parameters for the concentrated lanthanide trichlorides and the new lattice approximation technique could be performed by determining the heat capacity of $CeCl_3$ from 5 to 350 K.  Because the $Ce^{+3}$ ion has only one f-electron only two [SL]J-states occur within the [Xe] $4f^1$ configuration and these are separated by approximately 2200 $cm^{-1}$. If the observed trends in the crystal-field parameters remain valid, the energies of the $^2F_{5/2}$-state should be calculable to a high degree of accuracy and then compared with those deduced calorimetrically. The simplicity of the $Ce^{+3}$ ground state should enable the Stark splitting to be readily determinable from calorimetric data and lend insight into the Stark splitting of the comparably simple ground manifold of concentrated $SmCl_3$.

## REFERENCES

1.  R. D. Chirico, E. F. Westrum, Jr., J. B. Gruber, and J. Warmkessel, "Low -temperature heat capacities, thermophysical properties, optical spectra, and analysis of Schottky contributions to $Pr(OH)_3$," J. Chem. Thermodynamics. [In press]

2.  R. D. Chirico and E. F. Westrum, Jr., "Thermodynamics of the lanthanide hydroxides.  I.  Heat capacities of $La(OH)_3$, $Gd(OH)_3$, and $Eu(OH)_3$ from near 5 to 350 K.  Lattice and Schottky contributions," and "II. Heat capacities from 10 to 350 K of $Nd(OH)_3$ and $Tb(OH)_3$.  Lattice and Schottky contributions." [In press.]

3.  R. D. Chirico and E. F. Westrum, Jr., Thermophysics of the lanthanide hydroxides. I. Heat capacities of $La(OH)_3$, $Gd(OH)_3$, and $Eu(OH)_3$ from near 5 to 350 K.  Lattice and Schottky contributions."

4.  J. A. Sommers and E. F. Westrum, Jr., "Thermodynamics of the lan-
    thanide halides.  I. Heat-capacities and Schottky anomalies of
    $LaCl_3$, $PrCl_3$, and $NdCl_3$ from 5 to 350 K," J. Chem. Thermodynamis
    8:1115 (1976).

5.  J. A. Sommers and E. F. Westrum, Jr., "Thermodynamics of the lan-
    thanide halides.  II. Heat-capacities and Schottky anomalies of
    $SmCl_3$, $EuCl_3$, and $GdCl_3$," J. Chem. Thermodynamics 9:1 (1977).

6.  L. C. DeShazer and G. H. Dieke, "Spectra and energy levels of
    $Eu^{+3}$ in $LaCl_3$," J. Chem. Phys. 38:2190 (1963).

7.  E. H. Carlson and G. H. Dieke, "Fluorescence spectrum and low
    levels of $NdCl_3$," J. Chem. Phys. 34:1602 (1961).

8.  G. A. Prinz, Doctoral Dissertation, "Adsorption spectra of $NdCl_3$
    and $NdBr_3$."  Johns Hopkins University, Baltimore, Md. (1966).

9.  M. S. Magno and G. H. Dieke, "Absorption and fluorescence spec-
    tra of hexagonal $SmCl_3$ and their Zeeman effects," J. Chem. Phys.
    37:2354 (1962).

10. R. Sarup and M. H. Crozier, "Analysis of the eigenstates of $Pr^{3+}$
    in $LaCl_3$ using the Zeeman effect in high fields," J. Chem. Phys
    42:371 (1965).

11. R. D. McLaughlin and J. G. Conway, "Anion effects on the $Pr^{3+}$
    spectrum," J. Chem. Phys. 38:1037 (1963).

12. E. Dorman, "Concentration broadening and oscillator strengths
    in $Pr:LaCl_3$," J. Chem. Phys. 44:2910 (1966).

# MAGNETIC PROPERTIES OF THE PYROCHLORES $(RE)_2V_2O_7$; RE = Tm, Yb, Lu.
# NEW FERRO- AND FERRIMAGNETIC SEMICONDUCTORS

Lynne Soderholm and J.E. Greedan

Institute for Materials Research, McMaster University

Hamilton, Ontario, Canada  L8S 4M1

There are few examples in the literature of materials which are simultaneously ferromagnetic and semiconducting. Two recent reports (1,2) indicate that $Lu_2V_2O_7$ falls into this category. Bazuev (1) also interprets susceptibility data for $Yb_2V_2O_7$ and $Tm_2V_2O_7$ as indicative of a ferromagnet with an inflection in the $X_m^{-1} = f(T)$ curve at T = 190°K. No critical temperatures or saturation moments were reported for any of the compounds. Therefore work was undertaken to augment existing results and clarify the confusion in interpretation of the $Yb_2V_2O_7$ and $Tm_2V_2O_7$ data. Some results are shown in Table 1.

## Table 1

Summary of magnetic data for $Tm_2V_2O_7$, $Yb_2V_2O_7$ and $Lu_2V_2O_7$.

| | $T_c$ (K) | | | $\mu_{SAT}$ (4.2K, B.M.) | | | $C_m$ (cm³ mole⁻¹ K⁻¹) | | |
|---|---|---|---|---|---|---|---|---|---|
| | (1) | (2) | This Work (±1 K) | (1) | (2) | This Work | (1) | (2) | This Work |
| $Lu_2V_2O_7$ | 80 | 77-90 | 72.5 | 1.86 | - | 1.92 | 0.85 | 1.0 | 0.92 |
| $Yb_2V_2O_7$ | 85-90 | - | 73 | - | - | 5.36 | 4.6 / 2.9 | 5.2 | 6.0 |
| $Tm_2V_2O_7$ | 85-90 | - | 74 | - | - | -* | 12.7 | 14.8 | 14.8 |

*Does not saturate to 1.5T.

Critical temperatures were determined to be the same, 72 ±
1°K, within experimental error, for all three compounds. Measurements of magnetization as a function of applied field (0-1.5 T) and
susceptibility as a function of temperature (4.2°K-300°K) for
$Lu_2V_2O_7$, $Yb_2V_2O_7$ and $Tm_2V_2O_7$ were obtained (see Figure 1). The
results confirmed that $Lu_2V_2O_7$ is ferromagnetic with $\mu_{eff}/v$ = 1.90
$\mu_B$ and $\mu_{SAT}$ (4.2°K)/v = .93 $\mu_B$. This is in good agreement with the
work done in (1). Note that $\mu_{SAT}$ is smaller than the theoretical
1.0 $\mu_B$ while $\mu_{eff}$ is larger than expected for a $d^1$ system (1.73
$\mu_B$). Chemical analysis showed the large value of $\mu_{eff}$ was not due
to the presence of $V^{+3}$ (3).

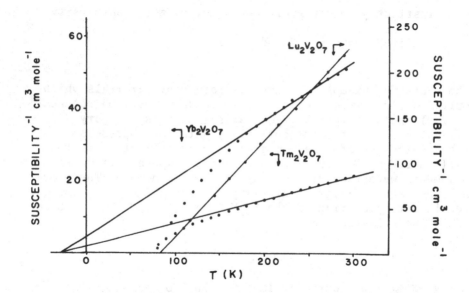

FIGURE 1:   Susceptibility data for $Tm_2V_2O_7$, $Yb_2V_2O_7$ and $Lu_2V_2O_7$.

$Tm_2V_2O_7$ and $Yb_2V_2O_7$ were found to be ferrimagnetic. For each
the $X^{-1}$ vs T curve was analyzed in terms of a classical Neel Hyper-
bola. This description clarifies the confusion regarding the interpretation of existing data for these two materials (1,2).

1.  G.V. Bazuev et al, Akad Navk SSSR, 230: 869 (1976).

2.  T. Shin-ike et al, Mat. Res. Bull., 12: 1149 (1977).

3.  J.E. Greedan, Mat. Res. Bull., 14: 13 (1979).

# Magnetic and Transport Properties of LaTiO$_3$

David A. MacLean and J.E. Greedan

Department of Chemistry, McMaster University

Hamilton, Ontario, Canada   L8S 4M1

LaTiO$_3$ has been known for some time, but neither its crystal structure nor its magnetic and transport properties have been thoroughly and accurately documented. Having determined the crystal structure of LaTiO$_3$[1], we wish to report here the magnetic and transport properties of this material.

Figure 1 shows the temperature dependence of the resistivity and magnetic susceptibility for LaTiO$_3$. The susceptibility data were collected on a Faraday balance using a polycrystalline sample, and the resistivity data were obtained using the van der Pauw technique on a single crystal wafer.

Above 125 K LaTiO$_3$ behaves like a metal. This is indicated by the low, temperature-independent susceptibility typical of a Pauli paramagnet, and by the decreasing resistivity with decreasing temperature. Below 125 K the susceptibility increases dramatically and the transport properties change to those of a semiconductor. Bazuev et al[2] obtained similar susceptibility results and from transport studies on a polycrystalline sample between 290 K and 1173 K concluded that LaTiO$_3$ is metallic in that temperature range.

The field dependence of the magnetization of a polycrystalline sample of LaTiO$_3$ at 4.2 K is shown in Figure 2. The magnetization reaches a saturation value of $\sim$ .007 $\mu_B$ per formula unit, and the material exhibits a distinct hysterysis. These data indicate that LaTiO$_3$ is a weak ferromagnet in the low temperature region, and thus account for the increased susceptibility observed below 125 K.

This behaviour for LaTiO$_3$ is not unlike that observed for SrVO$_3$[3], a similar d$^1$ compound. In that case there was an onset of weak ferromagnetism ($\sigma_0 \sim$ .007 $\mu_B$) below 85 K attributed to a ferromagnetic spin-density wave.

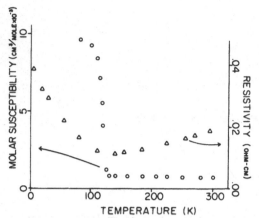

Fig. 1 Temperature dependence of the resistivity and susceptibility
       for LaTiO₃.

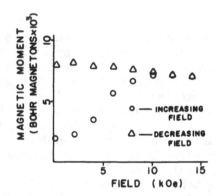

Fig. 2 Field dependence of the magnetization of LaTiO₃ at 4.2 K.

Acknowledgements

      The authors wish to thank I. Batalla (McMaster University)
and R.D. Shannon (E.I. duPont, Wilmington, DE) for their help in
obtaining the resistivity data, and A.B.P. Lever (York University,
Downsview, Ontario) for performing the susceptibility experiment.

References

1.  D.A. MacLean, H.N. Ng and J.E. Greedan, J. Solid State Chem.,
    in press.

2.  G.V. Bazuev and G.P. Shveikin, Izv. Akad. Nauk. SSSR, Neorg.
    Mater., 14, 267 (1978).

3.  P. Dougier, J.C.C. Fan and J.B. Goodenough, J. Solid State Chem.,
    14, 247 (1975).

# THE MAGNETIC STRUCTURE OF ErTiO$_3$

Carl W. Turner,[a] Malcolm F. Collins,[b]  J.E. Greedan[a]

Department of Chemistry[a] and Physics[b], McMaster University

Hamilton, Ontario, L8S 4M1  Canada

Magnetic data were collected from polycrystalline samples of ErTiO$_3$ on a PAR vibrating sample magnetometer.  The ordering temperature was estimated from magnetization as a function of temperature data at a residual field of .0045 Tesla.  As only one inflection point was observed it was concluded that the Er and Ti sublattices order and couple together at the same critical temperature of 39 ± K.[1]

Figure 1 shows a plot of inverse susceptibility as a function of temperature, the hyperbolic curve indicating that ErTiO$_3$ is a ferrimagnetic.  The Curie constant is considerably less than the theoretical value.  This may be the result of preferred orientation in the polycrystalline sample.

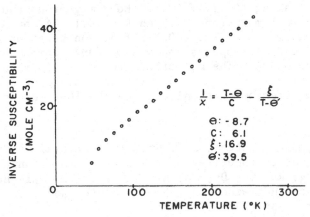

$$\frac{1}{\chi} = \frac{T-\Theta}{C} - \frac{\xi}{T-\Theta'}$$

$\Theta: -8.7$
$C: 6.1$
$\xi: 16.9$
$\Theta': 39.5$

Fig. 1 Temperature dependence of the reciprocal susceptibility for ErTiO$_3$.

397

Neutron diffraction data were collected at the McMaster Nuclear Reactor at a wavelength of 1.401 Å. All reflections at 4.2 K could be indexed on the nuclear cell so the magnetic cell is the same size as the nuclear cell. The temperature difference pattern showing diffraction effects which have developed between 298 K and 4.2 K is shown below.

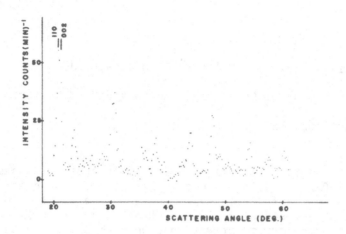

Fig. 2 Temperature difference plot showing diffraction effects which have developed between 298 K and 4.2 K for $ErTiO_3$.

The magnetic intensities were analyzed assuming the $Er^{3+}$ moment and the $Ti^{3+}$ moment to be colinear. A least squares fit gave an Er moment along [001] of 8.0±.5 Bohr magnetons and a Ti moment of -0.8.3 Bohr magnetons with an R of 8.1. A moment direction of [001] requires that the (002) reflection be absent. The unresolved peak at 20.9° appears to have no (002) component so this supports the model. The magnetic structures of $ErFeO_3$[2] and $ErCrO_3$[3] have previously been solved. In these structures the Er moments couple antiferromagnetically along the [001] direction.

References

1.  C.W. Turner, D.A. MacLean, J.E. Greedan, "Crystal Chemistry and Some Magnetic Properties of the Rare-Earth Titanium(III) Oxides, $RTiO_3$, R = Gd, Dy, Ho, Er, Lu" in: The Rare Earths in Modern Science and Technology, G.J. McCarthy and J.J. Rhyne, eds., Plenum Press (1978), p. 201.

2.  Koehler, W.C., E.O. Wollan, and M.K. Wilkinson, Phys. Rev. 118 (1960) 58.

3.  Bertaut, E.F., and J. Mareschal, Solid State Comm. 5 (1967) 93.

# MAGNETIC PROPERTIES OF SINGLE CRYSTAL NdF$_3$ IN THE LIQUID HELIUM TEMPERATURE REGION

J. Picard[+], M. Guillot[++], H. Le Gall[*], C. Leycuras[*], P. Feldmann[*]

+ University Nancy 1 - 54037 Nancy, France
++ Laboratoire Louis Néel, Laboratoire Propre du CNRS associé à l'USMG - 166 X, 38042 Grenoble-Cedex, France
* CNRS, Laboratoire de Magnétisme et d'Optique des Solides - 92190 Meudon Bellevue, France

## INTRODUCTION

Investigations of magnetic and magnetooptical properties of ionic solids containing rare earth ions have been developped during the last twenty years. Former studies have considered only compounds with high R.E. site symmetry (chlorides, bromides, ethyl sulphates). Recently much effort has been devoted to low site symmetry systems such as fluorides, double nitrates, oxides (1) and garnets (2) (3).

In this paper, magnetic properties of NdF$_3$ in magnetic fields up to 26 kOe at five temperatures 2 K, 2.5 K, 3 K, 3.5 K and 4.2 K are reported.

The specimen was a single crystal prepared by Dr. Picard using the Stockebarger method (4) ; the trifluoride was molten in an HF atmosphere at about 1650 K. The crystallisation was obtained by a slow extraction of the crucible in the temperature gradient (extraction speed of about .5 mm per hour). The crucibles are in graphite or iridium or platinium. A spherical single crystal of 2 mm diameter was used (the mass of the specimen is 21.72 mg) ; it was allowed to rotate freely in the applied magnetic field. NdF$_3$ has the structure of natural mineral tysonite (space group P $\bar{3}$ c-D$_{3d}^4$) and is water insensitive. Recently the TbF$_3$ heavy rare earth fluoride has been extensively studied ; it becomes ferromagnetic below T$_c$=3.95 K (5).

The magnetization curves, M(H), are shown in Fig. 1 for five values of the temperature T : 2, 2.5, 3, 3.5 and 4.2 K. In the whole temperature range, no spontaneous magnetization has been found and the fluoride remains paramagnetic. The low applied field susceptibility, $\chi$, can thus be deduced ; the variation of $\chi^{-1}$ as a function of T is shown in Fig. 2. It follows a Curie Weiss law : $\chi^{-1} = (T-\theta_p)/C$ with $\theta_p = (0.43 \pm 0.05)$ K, $C = (0.42 \pm 0.05)$ e.m.u. °K/mole. As the Curie constant of the $^4I_{9/2}$ free ion ground state of $Nd^{3+}$ is 1.64 e.m.u. °K/mole, it is clear that

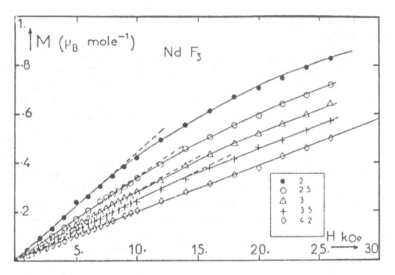

Fig. 1 - Magnetization of a sphere allowed to rotate freely in the magnetic field at different temperatures.

the crystal field is an important perturbation of the $^4I_{9/2}$ ground state. Such a conclusion is strengthened by the fact that the magnetization curve M (H/T) at 2 K is the same as those observed for other temperatures provided that the data are plotted in terms of H/T (fig. 3). The experimental curve was fitted to an equation of the form :

$$M(H/T) = Nm_0 \tanh \frac{m_0 H}{kT} \qquad (1)$$

where k is the Boltzmann constant and N the Avogadro number. $m_0$ is a constant equal to $(1.1 \pm 0.1)$ Bohr magneton mole$^{-1}$. For the free ion, the absolute saturation moment is 3.27 $\mu_B$. Such a result confirms that the lowest level is a Kramers doublet which is well separated from the other doublets. From equation (1), we

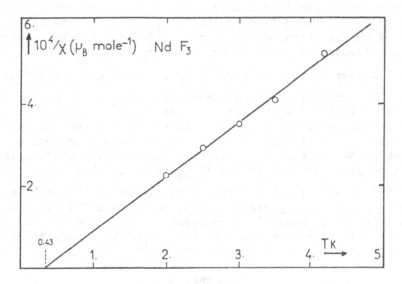

Fig. 2 - Inverse of the low field susceptibility versus temperature

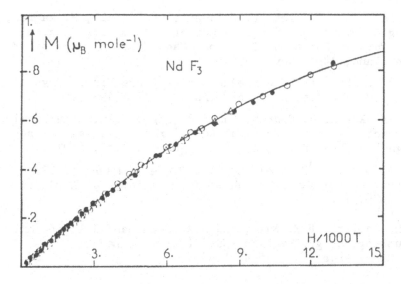

Fig. 3 - Magnetization versus the reduced variable H/1000 T
(expressed in Oersted/degree).

can calculate $C = Nm_0^2/k = 0.47$ e.m.u. $^\circ K$/mole, this value agrees
with the previous result derived from the magnetization values
measured in low field (H<6 kOe). Recently, the Zeeman splitting
of the lowest Kramers level of the $^4I_{9/2}$ ground state has been
measured by far infrared techniques (6). The principal components
of the g tensor were derived ($g_x = 1.72$, $g_y = 1.03$, $g_z = 3.02$).

If we consider that $m_0$ is given by g $S\mu_B$ with $g^2 = g_x^2 + g_y^2 + g_z^2$ and
S=1/2, the agreement between the experimental (1.1 $\mu_B$) and the
calculated (1.8 $\mu_B$) $m_0$ values is rather poor. So far as we know
the calculation of the crystal field parameters has not yet been
performed and no theoretical value of the saturation moment is
available. Nevertheless, our experimental results confirm the
important level spacing in the $^4I_{9/2}$ state as obtained by Caspers
et al (7) in $Nd^{3+}$ : $LaF_3$ crystal field levels at 0, 45, 136, 296,
500 $cm^{-1}$ giving the spacings 45, 91, 160, 204 $cm^{-1}$). To clarify
the origin of the magnetic properties further experimental investigations are being performed such as magnetic susceptibility measurements in the 4-300 K temperature range.

## REFERENCES

1. D. Bloor, G.M. Copland, Far infrared spectra of magnetic ions
   in crystals, Rep. Prog. Phys. 35, 1173, (1972).

2. E.V. Berdennikova, R.V. Pisarev, Sublattice contributions to
   the Faraday effect in rare earth iron garnets, Sov. Phys. Sol.
   State, 18, 45 (1976).

3. M. Fadly, M. Guillot, P. Feldmann, H. Le Gall, H. Makram,
   Single crystal Holmium Iron Garnet : Anomalous Behaviour of
   the Faraday Rotation. Conf. on "Magnetism and Magnetic Materials"
   Cleveland, 14-18 Nov. 1978.

4. D.C. Stockbarger, The Production of Large Single Crystals of
   Lithium Fluoride, Rev. Sci. Inst. 7, 133 (1936).

5. J. Brinkmann, R. Courths and H.J. Guggenheim, Logarithmic
   Corrections to the Critical Behavior of Uniaxial, Ferromagnetic $TbF_3$, Phys. Rev. Let., 40, 19, 1287, (1978).

6. D. Bauerle, G. Borstel and A.J. Sievers, Magnetic Field
   dependence of far-infrared spectra of $NdF_3$, J. Appl. Phys.,
   49, 676, (1978).

7. H.H. Caspers, H.E. Rast, and R.A. Buchanan, Intermediate
   coupling Energy Level for $Nd^{3+}$ (4 $f^3$) in $LaF_3$.
   J. Chem. Phys., 42, 3214, (1965).

# MAGNETIC AND ELECTRICAL PROPERTIES OF THE RARE EARTH OXYNITRIDES

$Eu_{1-x}Nd_xO_{1-x}N_x$

B. Chevalier, J. Etourneau and P. Hagenmuller, Laboratoire
de Chimie du Solide du C.N.R.S.; and R. Georges, Laboratoire de Physique du Solide, Université de Bordeaux I,
351 cours de la Libération, 33405 Talence, Cedex,
France

## INTRODUCTION

Rare earth oxynitrides with the formula $Eu_{1-x}Nd_xO_{1-x}N_x$ and a
NaCl-type structure have been prepared (1). Generally Nd exists in
the trivalent state whereas Eu can be found in both the divalent and
the trivalent states, with quite different magnetic behaviors. $Eu^{2+}$
with $4f^7$ configuration has a $^8S_{7/2}$ ground state and its susceptibility follows a Curie-Weiss law. $Eu^{3+}$ with $4f^6$ configuration has
a $^7F_0$ ground state and shows a Van Vleck type susceptibility,
temperature-independent below 100 K. Since these two types of susceptibility are quite different (magnitude and thermal variation),
the valence state of europium can be monitored by susceptibility
measurements.

The present paper reports the experimental results and their
discussion concerning the lattice parameters, and the electrical and
magnetic measurements in the $Eu_{1-x}Nd_xO_{1-x}N_x$ system for $0 \leqslant x < 1$.

## EXPERIMENTAL RESULTS

### Lattice Parameter

The lattice parameter (a) has been measured by x-ray powder
techniques at 300 K and its variation as a function of x is given
in Fig. 1. As x increases, (a) slowly decreases from x = 0 to x =
0.26 and then shows a deep minimum at about x = 0.80. Similar measurements down to 4.2 K indicate no symmetry change except for the

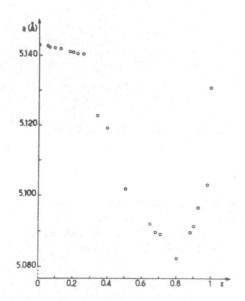

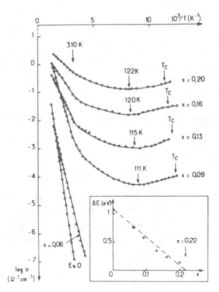

Figure 1.   Lattice paramenters vs. composition X for $Eu_{1-x}Nd_x$ $O_{1-x}N_x$ at 300K.

Figure 2.   Conductivity as a function of temperature for $Eu_{1-x}$ $Nd_xO_{1-x}N_x$ (x<0.22).  The variations of the activation energy vs x is shown in the inset.

limit phase NdN which undergoes a cubic → tetragonal distortion at the ferromagnetic order temperature (2).

## Electrical Properties

Four point probe d.c. resistivity measurements have been carried out on sintered samples for the complete series, between 4.2 K and 800 K.  For x < 0.22 a semiconducting behavior is observed and the activation energy ΔE decreases almost linearly with rising x and vanishes at about x = 0.22 (Fig. 2).  At low temperatures a semiconductor-metal transition occurs as for $EuO_{1-\delta}$ and $EuO_{1-x-\delta}N_x$ (3,4).

The compounds are metallic for x > 0.22 (Fig. 3).

## Magnetic Properties

Magnetization and susceptibility measurements have been performed between 4.2 K and 300 K.  The saturation magnetization was not generally reached up to 20 kOe applied magnetic field at 4.2 K,

except for Eu-rich compounds (x ≤ 0.50). At high temperatures the
susceptibility follows a Curie-Weiss law (Fig. 4). In the range
0 ≤ x ≤ 0.20 the experimental molar Curie constant $C_m$ (exp.) is in
good agreement with the calculated one $C_m$ (calc.) assuming that all
Eu ions are divalent. For x > 0.20 $C_m$ (exp.) is smaller than $C_m$
(calc.) (Table 1). The deviation from the Curie-Weiss law at low
temperatures is due to the crystal field splitting of the $^4I_{9/2}$
ground state of the $Nd^{3+}$ ion. With rising x, up to x = 0.26, the
paramagnetic Curie temperature $\theta_p$ remains constant and positive,
then decreases to become negative and goes through a minimum at
about x = 0.80. As x increases the magnetic ordering temperature
$T_C$ shows a maximum at about x = 0.22 (Table 1).

## DISCUSSION

The energy diagram of the oxynitrides can be deduced from those
of the RE monochalcogenides obtained on the basis of optical absorption experiments and direct calculations (5,6,7). The valence

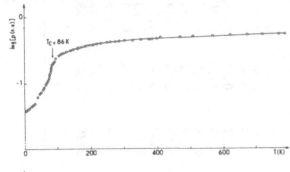

Figure 3. Resistivity as
a function of tempera-
ture for $Eu_{1-x}Nd_xO_{1-x}N_x$
(x=0.24).

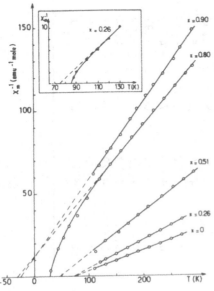

Figure 4. Temperature dependence
of reciprocal susceptibility for
five samples in the $Eu_{1-x}Nd_xO_{1-x}N_x$
system.

Table 1. Magnetic data for $Eu_{1-x}Nd_xO_{1-x}N_x$. $C_m$ has been calculated and assuming the presence of only $Eu^{2+}$ and $Nd^{3+}$. The saturation magnetization has been calculated taking into account the $Eu^{2+}$, $Eu^{3+}$ and $Nd^{3+}$ ion concentration.

| x | $\theta_p$ (K) | $T_C$ (K) | $C_m$ exp. | $C_m$ calc. | $Eu^{2+}_{1-x-y}Eu^{3+}_yNd^{3+}_xO_{1-x}N_x$ $Eu^{2+}$ | $Eu^{3+}$ | $Nd^{3+}$ | $M_{sat.}$ ($\mu_B/mole$) at 4.2 K exp. | $M_{sat.}$ calc. |
|---|---|---|---|---|---|---|---|---|---|
| 0 | + 77 | 69.5 | 7.88 | 7.90 | 1 | 0 | 0 | 6.77 ± 0.20 | 7 |
| 0.09 | + 73 | 80.3 | 7.41 | 7.34 | 0.91 | " | 0.09 | 6.29 ± 0.20 | 6.20 |
| 0.13 | + 75 | 82.5 | 7.11 | 7.08 | 0.87 | " | 0.13 | 5.74 ± 0.17 | 5.85 |
| 0.18 | + 74 | 85.5 | 6.52 | 6.77 | 0.82 | " | 0.18 | 5.15 ± 0.15 | 5.40 |
| 0.20 | + 74 | 87.5 | 6.36 | 6.64 | 0.76 | 0.04 | 0.20 | 4.70 ± 0.14 | 4.94 |
| 0.23 | + 74 | 87 | 6.08 | 6.46 | 0.71 | 0.06 | 0.23 | 4.45 ± 0.13 | 4.54 |
| 0.26 | + 74 | 85 | 5.73 | 6.26 | 0.66 | 0.08 | 0.26 | 4.18 ± 0.13 | 4.13 |
| 0.34 | + 72 | 80 | 4.96 | 5.77 | 0.54 | 0.12 | >0.34 | 3.16 ± 0.10 | 3.14 |
| 0.40 | + 70 | 76 | 4.43 | 5.38 | 0.44 | 0.16 | 0.40 | 2.55 ± 0.08 | 2.78 |
| 0.51 | + 44 | 66 | 3.97 | 4.69 | 0.36 | 0.13 | 0.51 | 1.6 ± 0.05 | 1.56 |
| 0.65 | + 26 | -- | 3.03 | 3.83 | 0.22 | 0.13 | 0.65 | | |
| 0.68 | + 13 | -- | 2.74 | 3.64 | 0.17 | 0.15 | 0.68 | | |
| 0.80 | - 29 | -- | 2.51 | 2.89 | 0.14 | 0.06 | 0.80 | | |
| 0.90 | - 25 | -- | 2.15 | 2.26 | 0.07 | 0.03 | 0.90 | | |
| 1 | + 20 | 25 | 1.61 | 1.64 | 0 | 0 | 1 | | |

band which derives from the anionic p levels is separated by an
energy gap from the 5d-6s conduction band. Whereas the $Eu^{2+}:4f^7$
levels lie in this gap for the semiconducting phases, the $Nd^{3+}:4f^3$
levels are located below the top of the valence band. The lower
part of the conduction band is mainly of $5d-^{5d}t_{2g}$ character. With
rising x the bottom of the conduction band is lowered due to both
the increase of the $Nd^{3+}$ ion content and the decrease of the lattice
parameter, thus giving $d\Delta E/dx < 0$.

The semiconductor-metal transition could be explained by the
crossing of the $Eu^{2+}:4f^7$ levels with the conduction band, involving
a change from the divalent state to the trivalent state of europium.
Comparing the experimental molar susceptibility to the calculated
one by the expression $X_m = (1-x-y)X_{Eu2+} + yX_{Eu2+} + xX_{Nd3+}$, we have
determined the $Eu^{3+}$ ion content (Table 1). The sharp shrinkage of
the unit cell is closely related to the $Eu^{3+}$ ion content
$(r_{Eu3+} < r_{Eu2+})$.

The experimental saturation magnetization is consistent with
the $Nd^{3+}$ magnetic moment antiparallel to those of $Eu^{2+}$. It is worthwhile to notice that the magnetic moment of $Nd^{3+}$ is opposite to its
spin momentum: $\vec{\mu}_{Nd3+} = [g_J/(g_J-1]\mu_B \vec{S}_{Nd3+}$, $(g_J = 8/11)$. Therefore
the spin-spin interactions are ferromagnetic in the oxynitrides and
they can be described by the mechanism involving a virtual electron
transfer between nearest neighbors (8,9,10).

$$Eu^{2+}(4f^7 5d^0) - Eu^{2+}(4f^7 5d^0) \rightarrow Eu^{3+}(4f^6 5d^0) = Eu^+(5f^7 5d^1)$$

$$Eu^{2+}(4f^7 5d^0) - Nd^{3+}(4f^3 5d^0) \rightarrow Eu^{3+}(4f^6 5d^0) - Nd^{2+}(4f^3 5d^1).$$

The strength of the ferromagnetic interactions depend mainly
on the excitation energy between the $4f^7$ and $^{5d}t_{2g}$ states. Thus the
$Eu^{2+}-Eu^{2+}$ and $Eu^{2+}-Nd^{3+}$ ferromagnetic interactions are the strongest
and can explain the increase of $T_C$ with x up to x = 0.22 despite
both the lower spin value of $Nd^{3+}$ compared to that of $Eu^{2+}$ and the
increase of the covalency with nitrogen which favor the 180°
antiferromagnetic couplings via the orbitals of the anions.

The decreases of $T_C$ with rising x in the metallic range (x >
0.22) could be the consequence of the presence of $Eu^{3+}$ ions and conduction electrons for which the influence is not yet well understood.

Finally, an elementary analysis taking into account the relative magnitude of the exchange between nearest neighbors ($Nd^{3+}-Nd^{3+}$,
$Eu^{2+}-Eu^{2+}$ and $Eu^{2+}-Nd^{3+}$) and the fact that the magnetic moments of
$Nd^{3+}$ and $Eu^{2+}$ are antiparallel, clearly explains the minimum in $\theta_p =$
$f(x)$ for negative values of $\theta_p$, in agreement with the experimental
data.

## CONCLUSIONS

The oxynitrides $Eu_{1-x}Nd_xO_{1-x}N_x$ are new magnetic compounds exhibiting semiconductor-metal transitions as a function of both x and temperature. In the semiconducting range (0 < x < 0.22) with rising x, the Curie temperature $T_C$ increases whereas the activation energy $\Delta E$ decreases. The metallic oxynitrides (x > 0.22) are mixed valence compounds in which europium has both the divalent and trivalent states.

## REFERENCES

1.  B. Chevalier, J. Etourneau and P. Hagenmuller, C. R. Acad. Sc. 828:375 (1976).
2.  C. Mourgout, B. Chevalier, J. Etourneau, J. Portier, P. Hagenmuller and R. Georges, Rev. Int. Htes. Temp. et Refract. 14:89 (1977).
3.  T. Penney, M. W. Shafer, and J. B. Torrance, Phys. Rev. B5: 3669 (1972).
4.  B. Chevalier, J. Etourneau, J. Portier, P. Hagenmuller, R. Georges, and J. B. Goodenough, J. Phys. Chem. Solids 39:539 (1978).
5.  R. Suryanarayan, Phys. Stat. Sol. 85:9 (1978).
6.  S. J. Chao, Phys. Rev.B1:4589 (1970).
7.  P. Watcher, Physics Reports 44:161 (1978).
8.  P. Watcher, CRC Crit. Rev. Solid State Sciences 3:189 (1972).
9.  J. B. Goodenough, Magnetism and the Chemical Bond, Wiley, NY, p. 149 (1963).
10. T. Kasuya, IBM Res. Develop. 14:214 (1970).

# THE MAGNETIC STRUCTURE OF $Ho_{11}Ge_{8.75}Si_{1.25}$ DETERMINED BY NEUTRON DIFFRACTION

Penelope Schobinger-Papmantellos

Institut für Kristallographie und Petrographie

ETHZ, Sonneggstrasse 5, 8006 Zürich

## ABSTRACT

The magnetic structure of the compound $Ho_{11}Ge_{8.75}Si_{1.25}$, space group I4/mmm, a = 10.79, c = 16.18Å, with 4 symmetry positions for the 44 Ho atoms, is determined by powder neutron diffraction. Below $T_N$ = 27 K and up to 5.8 K k = 0 (no cell enlargement), the main magnetic moment component is ferromagnetic, in contrast to the other near equi-atomic heavy RE-Ge alloys, and close to the 4-fold axis. The structure is described in the I4/mm'm' magnetic space group; only atoms Ho(1) and (4) and 16(n) have antiferromagnetic components along x or y. The refined magnetic moment values for Ho(1) to (4) are 3.3. 0.6, 5.5, 7.9 $\mu_B$ respectively. The distinct and partly quenched moment values reflect the complexity of the structure containing polyhedra with high coordination numbers, the strong and locally varying crystal field and the crystal anisotropy.

## INTRODUCTION

During recent investigations of the $Ho_5Ge_{4-x}Si_x$ mixed compound, single crystals of $Ho_{11}Ge_{10}(2)$ were identified by x-ray diffraction, next to $Ho_5Ge_4(1)$. This structure type, common for the heavy rare earth (RE) atoms, is described in the tetragonal, I4/mmm space group, with 4 formula units per elementary cell. The 44 RE atoms are distributed in 4 equivalent symmetry positions, while the 40 germanium atoms are in five positions. The lattice parameters for $Ho_{11}Ge_{10}$ are a = 10.79, c = 16.23Å.

All the known RE-Ge near equiatomic compounds crystallizing in the FeB, CrB and $RE_5Ge_4$ structure types characteristically contain trigonal prisms and deformed cubes as coordination polyhedra for RE atoms. In contrast, the polyhedra of $Ho_{11}Ge_{10}$ are complicated and the RE-coordination number varies between 15-17. A further characteristic of this compound is the very short interatomic distance of Ho(4)-Ho(4) (3.327Å), which indicates a tendency toward pair formation or covalent bonding. Other atoms, e.g. Ho(3)-Ho(3) are so far apart along the c-axis, that no bonding could be formed. Similarly germanium forms pairs and clusters.

For these reasons, a study of the magnetic properties and structure of thus unusual compound seemed of interest. We report here a model for the magnetic structure (k = 0) of the isomorphic $Ho_{11}Ge_{8.75}Si_{1.25}$ based on neutron powder data recorded at 5.8 K and on the dependence of the magnetic intensity on temperature.


EXPERIMENTAL

We investigated two powder samples (I and II): $Ho_5Ge_4$ (I), $Ho_5Ge_{3.5}Si_{0.5}$(II), both prepared by arc melting of the elements under argon atmosphere and annealed in a tantal crucible under vacuum for 48 hours at 1580°C. X-ray powder data with the Guinier focusing camera (CuKα), showed a few weak lines next to the $Ho_{11}Ge_{10}$, lines which are mainly $Ho_5Ge_3$ in sample I. Electron microprobe analysis shows that single crystals of samples I are $Ho_{11}Ge_{10}$ and the powder of sample II is $Ho_{11}Ge_{8.75}Si_{1.25}$.


Neutron Diffraction

The neutron powder intensities of samples I and II were collected with the double axis spectrometer in the reactor Saphir in Würenlinger (2.346Å) at the temperatures 293, 4.2 and 5.8 K. For the data evaluation we used sample II which is almost a pure phase. Silicon is assumed to be statistically distributed in the Ge sites. The observed data corrected for absorption were evaluated by means of line profile analysis (references in [1]).

The refinement of the nuclear intensities was based on the parameters given in (2). The scattering lengths, $b_{Ho}$ = 0.85 and $b_{Ge-Si}$ = 0.769 x $10^{-12}$ cm were used. Figure 1a shows the 293 neutron data. All refined parameters are summarized in Table 1. The resulting reliability factors Rn 0.07 and Rwp (weighted profile) 0.13 are satisfactory.

The magnetic structure at 5.8 K. The 5.8 K neutron diagrams of both samples (Fig. 1b) show that the magnetic lines are superimposed on the nuclear ones (k = 0), characteristic of a ferromagnetic

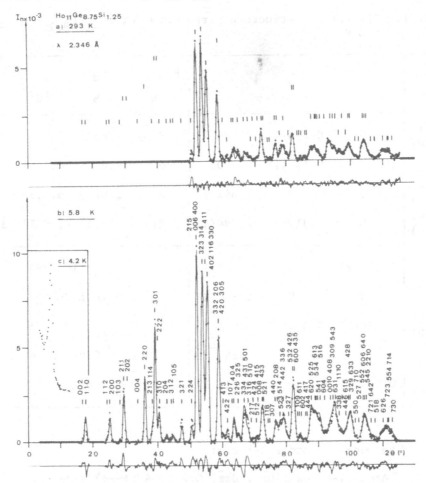

Fig. 1.    Observed and calculated neutron-diffraction profiles from
Ho$_{11}$Ge$_{8.75}$Si$_{1.25}$ at:  (a) 293, (b) 5.8 and insert (c) 4.2
K.  Dots:  observed data I$_o$; solid lines:  calculated pro-
files I$_c$.  Below is the difference diagram I$_o$-I$_c$.

component.  The absence of all (001) magnetic reflections suggest
that this component is Fz (along the 4-fold axis).

Assuming no symmetry lowering we find that the only magnetic
space group associated with k = 0, and an Fz magnetic component, is
the Ir/mm'm' or Shubnikov Sh $\frac{537}{139}$ (3).  The magnetic moment direction
for the special positions is restricted by symmetry:  the 4(e) atoms
along the tetrad can have only an Fz component, the same holds for
the 8(h) atoms, at the intersections of the mirror plane m$_z$ and anti-
mirror m$_{xy}$.  Besides having an Fz component the 16(n) atoms at anti-
mirror planes m$'_x$ or m$'_y$ may also have an A$_x$(+--+) or A$_y$(+--+) magnetic

P. SCHOBINGER-PAPMANTELLOS

Table 1.  The refined structure parameters for $Ho_{11}Ge_{8.75}Si_{1.25}$ at 5.8 K.

| Atom/ site | x | y | z | $\mu_{xory}$ $(\mu_B)$ | $\mu_z$ $(\mu_B)$ | $\mu$ $(\mu_B)$ | $\theta x$ (°) |
|---|---|---|---|---|---|---|---|
| Ho(1) 16(n) | 0 | 0.253(1) | 0.186(1) | 2.1(3) | 2.6(1) | 3.3(2) | 38.3 |
| Ho(2) 8(h) | 0.178(1) | 0.178(1) | 0 | 0 | 0.6(1) | 0.6(1) | 0 |
| Ho(e) 4(e) | 0 | 0 | 0.341(1) | 0 | 5.5(1) | 5.5(1) | 0 |
| Ho(4) 16(n) | 0 | 0.319(1) | 0.394(1) | 1.8(4) | 7.7(1) | 7.9(1) | 13.5 |
| Ge(1) 8(i) | 0.357(2) | 0 | 0 | $a = b = 10.779Å$ | | | |
| Ge(2) 4(e) | 0 | 0 | 0.120(3) | $c = 16.187Å$ | | | |
| Ge(3) 8(h) | 0.376(2) | 0.376(2) | 0 | $B_{overall} = 0.75Å$ | | | |
| Ge(4) 4(d) | 0 | 0.50 | 0.25 | $R_n = 0.07$, $R_m = 0.09$ | | | |
| Ge(5) 16(m) | 0.210(1) | 0.210(1) | 0.317(1) | $R_{wp} = 0.14$ | | | |

mode.  The modes of atoms 1-8 are:

Atoms 1-4  $0xz$, $0\bar{x}z$, $0x\bar{z}$, $0\bar{x}\bar{z}$    $A_y(+--+)Fz(++++)$

16(n) 4-8  $x0z$, $\bar{x}0z$, $x0\bar{z}$, $\bar{x}0\bar{z}$    $A_x(+--+)Fz(++++)$

The remaining 16(n) atoms obtained by the body centering operation maintain the same magnetic modes.  The refinement of the magnetic structure results in the parameters given in Table 1, and the relia- bility factors Rn 0.07, Rm 0.09, Rwp 0.14 support the chosen model. The Ho magnetic form factor is taken from (4).  In Fig. 2a we show one of the 4 nets formed by the Ho and Ge atoms at the heights z = 0.10-0.19, 0.31-0.40, 0.81-0.90, 0.60-0.69.  The structure can be easily described by noting that net two is derived through inversion of net one, over the symmetry center at 1/4 1/4 1/4, which does not affect the magnetic moment direction.  On the other hand nets three and four obtained through reflection over the mirror plane at z = 1/2 of net one and two respectively, have $\mu_x$ and $\mu_y$ inverted but not $\mu_z$. In Fig. 2b we show for clarity only the (101) $\bar{m}_y$ plane containing Ho(1), (3), and (4).  Atom (2), not belonging to that plane, has zero moment.  The distribution of Ho(1) and (4) on $m'_y$ and $m'_x$

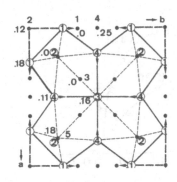 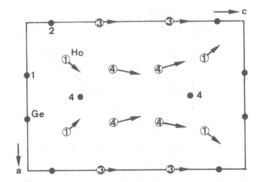

Fig. 2.    (a) The net of Ho and Ge atoms at z = 0.10–0.19. Only Ho(1) and Ho(2) at 16(n) have x or y magnetic moment components, $u_z$ has the same for all atoms. Full lines: Ho–Ho bonds. Dotted lines: Ge–Ho and Ge–Ge bonds. Non-connected atoms are at z = 0,1/4,1/2.

         (b) The (101) plane, showing the elliptic distribution of atoms Ho(1), (4), and the small $u_z$ component of (3).

describes an elongated ellipse starting from (000) or (1/2 1/2 1/2) and with the tetrad as the major axis. The moments of Ho(1) and (4) are tangential to the ellipse, apparently the largest moment value is in the middle of the orbit where the Ho(4) distance is the shortest. Ho(2) at x = 0.17 is near the origin of the ellipse and its moment is zero.

     The temperature dependence measurements of the (301)– ferro- and antiferro-magnetic intensity and of the (220)– which is purely ferro-magnetic (Fig. 3b) result in the ordering temperature $T_N$ = 27 K. The same $T_N$ was obtained from the susceptibility vs. T measurements kindly provided by Miedan-Gross and Clerc. Below 5.8 K (Fig. 1c) an additional weak line at $2\theta$ = 7.2° appears which indicates a cell doubling in all directions. Since all reflections remain unchanged we suggest that atom Ho(2) orders with a weak moment in the magnetic space group $I_p4/mm'm'$ or $P_I4/mmc$. This means a symmetry lowering of the whole structure to the P 4/mm'm' space group.

## DISCUSSION

     $Ho_{11}Ge_{8.75}Si_{1.25}$ and $Ho_{11}Ge_{10}$ are the first Ho–Ge compounds which have a ferromagnetic component. The whole structure can be considered as a weak ferrimagnet, the main component being ferromagnetic and oriented along c. This structure explains the neutron and the magnetic susceptibility data. The distinct magnetic moment values of the 4 Ho positions reflect the complexity of this structure and the locally

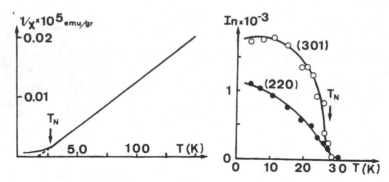

Fig. 3.   (a) Reciprocal susceptibility vs. temperature of
              $Ho_{11}Ge_{8.75}Si_{1.25}$.
          (b) The temperature dependence of neutron intensities (301),
              (220).

varying crystal field.  It seems that a correlation between nearest
neighbor distance and moment value exists, e.g. the Ho(4) atom which
has the highest moment value forms a covalent bond Ho(4)-Ho(4) and
has the shortest Ho-Ho average distance with its nearest neighbor.
The reverse occurs for Ho(2) which has zero moment at 5.8 K, orders
below with another space group and has the largest interatomic distances.

## REFERENCES

1.   P. Schobinger-Papamantellos and A. Niggli, The two magnetic
     structures of $Ho_5Ge_4$.  A neutron diffraction study.  (To appear
     in J. de Phys, 1979).
2.   G. S. Smith, Q. Johnson and A. G. Tharp, Acta Cryst. 23:640
     (1967).
3.   W. Opechowski and R. Guccione, In Treatise of Magnetism (Edited
     by H. Suhl and G. Rado), Vol. III, p. 186.  Academic Press, NY
     (1963).
4.   G. H. Lander and T. O. Brun, J. Chem. Phys. 53:1387 (1970).

MAGNETIC SUSCEPTIBILITIES OF SmP, SmAs, SmSb, SmBi, AND THEIR SOLID

SOLUTIONS WITH SmS*

R. B. Beeken, E. D. Cater, R. L. Graham, D. C. Henry,
Wm. R. Savage, J. W. Schweitzer, and K. J. Sisson

The University of Iowa

Iowa City, Iowa  52242

The study of the phenomenon of intermediate valence in rare earth compounds and alloys has become a very active field of research, and many unusual properties associated with intermediate valency have been investigated (1).  The magnetic properties are perhaps the most interesting since they appear at present to be the most difficult to explain (2).  The behavior of SmS under pressure is typical.  At a pressure of 6.5 k bar, SmS undergoes a large isostructual volume decrease, a large increase in the electrical conductivity, and a change in color from black to gold.  Various studies have shown that this transition involves a change in the electronic configuration of the Sm ion from the $4f^6$ configuration to a mixture of the $4f^5$ and $4f^6$ configurations.  However, the magnetic susceptibility in the collapsed phase shows neither a low temperature Curie law divergence nor a transition to an ordered magnetic phase.  Instead, the magnetic susceptibility saturates to a constant value as the temperature goes to zero.  This behavior is not easy to reconcile with the fact that the $4f^5$ configuration has a magnetic ground state.  The lowest level of the $4f^5$ configuration is $^6H_{5/2}$ in which the degeneracy is only partially lifted by the crystal field to give a $\Gamma_7$ doublet and a $\Gamma_8$ quartet where the doublet is lower in energy.  This absence of magnetic order and local moments at low temperatures in the intermediate valence phase of SmS appears to be a characteristic of intermediate valency when the ground state of one of the valence states being mixed is nonmagnetic.  The lowest level of the $4f^6$ configuration is the nonmagnetic $^7F_o$ state.

At the transition in SmS at 6.5 k bar, the magnetic behavior goes from Van Vleck paramagnetism to the intermediate valent

415

behavior.  Presumably, at some much greater pressure the Sm ion
would have the pure trivalent $4f^6$ configuration.  The trivalent SmS
should order magnetically at low temperatures.  Near the crossover
from intermediate valency to trivalency it is reasonable to expect
that there will be interesting effects associated with the competition between the interatomic exchange interactions which tend to
stabilize the moments and the freezing out of moments which is
characteristic of intermediate valency.  Unfortunately, the high
pressures required would make it very difficult to study this crossover.  However, the solid solutions of SmS with the samarium monopnictides (SmN, SmP, SmAs, SmSb, and SmBi) provide very convenient
systems for investigating this phenomenon without applied pressure.

Intermediate valence can be chemically induced in SmS by both
cation and anion substitutions, however, the substitutions for Sm
also introduce magnetic disorder which is an undesirable complication.  Intermediate valence produced by replacing some of the
divalent S ions by trivalent ions was first studied in the solid
solutions of SmS with SmAs (3-6).  The Sm ion is trivalent in SmAs
which is a semimetallic material having the same NaCl structure as
SmS.  With increasing As concentration, the $SmS_{1-x}As_x$ alloy
exhibits a large decrease in the lattice parameter and a change in
color from black to gold in the concentration range from 5 to 10%
As.  This transition is similar to the transition in pure SmS under
pressure at 6.5 k bar.  In the collapsed phase of $SmS_{1-x}As_x$,
$x \geq 0.1$, the Sm ions have a homogeneous intermediate valence which
is estimated to be 2.8 at $x = 0.1$ and increases with increasing x
until it reaches the pure trivalent state at approximately $x = 0.7$.
However, unlike SmS for pressures less than 6.5 k bar where the Sm
is divalent, the alloys for $x \leq 0.05$ show behavior consistent with
there being an inhomogeneous mixed valent state of $Sm^{2+}$ and $Sm^{3+}$
ions where Sm ions with one or more As as nearest neighbors are
trivalent.  Fortunately, such local environment effects do not seem
to be important in the collapsed phase.

Recently, intermediate valence has also been observed in the
solid solutions of SmS with SmP (7) and SmSb (8).  The properties of
the $SmS_{1-x}P_x$ alloys are very similar to those of the $SmS_{1-x}As_x$
alloys.  However, the $SmS_{1-x}Sb_x$ solid solutions do not undergo a
transition to a collapsed golden phase.  This is probably owing to
the much larger size of the Sb ion in comparison with the As and P
ions.  Nevertheless there is evidence for an intermediate valence
which deviates significantly from the $2 + x$ that is inherent in the
ionic charge transfer associated with the substitution of a fraction
x of trivalent Sb for the divalent S.

Since we are primarily interested in the competition between
magnetic ordering and the demagnetization associated with intermediate valency, we shall focus our attention on solid solutions of
SmS with the samarium monopnictides in the concentration range
where the Sm valence approaches the trivalent value.  However, first

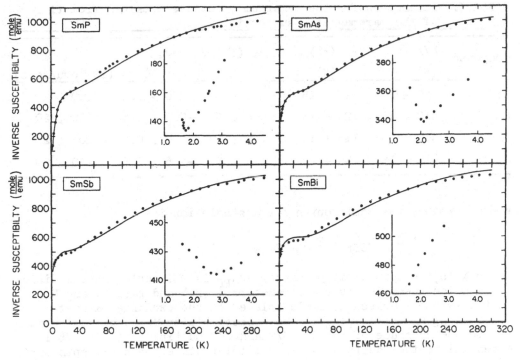

Figure 1

let us consider the samarium monopnictides where the Sm ion is pure
trivalent. We have measured the magnetic susceptibilities of SmP,
SmAs, SmSb, and SmBi as a function of temperature between 1.5 and
300 K. The measurements were made using the Faraday method in fields
of 4 and 6 kilogauss. The results of the measurements are shown in
Fig. 1 where the inverse susceptibilities are plotted versus temperature. The inserts show the data below 4 K plotted on an expanded
scale. SmP, SmAs, and SmSb exhibit antiferromagnetic transitions at
$1.7 \pm 0.1$ K, $2.0 \pm 0.1$ K, and $2.8 \pm 0.1$ K respectively, while no
transition is observed in SmBi down to 1.6 K. A recent study of the
magnetic susceptibilities of the samarium monopnictides by Hulliger,
et al. (9) obtained similar results except that they found evidence
for a magnetic transition in SmBi at 9.0 K using single crystals
oriented with the [001] axis along the magnetic field.

We have attempted to analyze the paramagnetic susceptibility of
the samarium monopnictides in terms of spin-orbit, crystal-field,
and exchange interactions. The measured susceptibility is expected
to consist of two contributions: $\chi_o$, the temperature independent
contribution due primarily to the diamagnetism of the ion cores and
filled bands; and $\chi_f(T)$, the contribution from the 4f electrons.
The 4f contribution is assumed to be given by a model consisting of
$Sm^3(4f^6)^6H$ ions in a cubic crystal field which interact by an isotropic exchange coupling of the spins. Within the molecular field

TABLE I.   Parameters for the samarium pnictides obtained from fits
           of the paramagnetic susceptibilities.

| Compound | $\lambda/k$ (K) | $A_4 \langle r^4 \rangle /k$ (K) | $A_6 \langle r^6 \rangle /k$ (K) | $J_{ff}/k$ (K) | $X_o$ ($10^{-6}$ emu/mole) |
|----------|-----------------|----------------------------------|----------------------------------|----------------|----------------------------|
| SmP      | $450 \pm 20$    | $125 \pm 15$                     | $8 \pm 5$                        | $2.0 \pm 0.5$  | $-60 \pm 20$               |
| SmAs     | $410 \pm 15$    | $120 \pm 15$                     | $4 \pm 5$                        | $-2.0 \pm 0.5$ | $-90 \pm 20$               |
| SmSb     | $410 \pm 15$    | $115 \pm 15$                     | $-2 \pm 5$                       | $-3.0 \pm 0.5$ | $-105 \pm 20$              |
| SmBi     | $415 \pm 15$    | $140 \pm 20$                     | $-4 \pm 5$                       | $-3.5 \pm 0.5$ | $-120 \pm 20$              |

approximation, the Hamiltonian for a single ion is

$$H = \lambda \, L \cdot S + H_{CEF} + 2 \, \mu_B \, S_Z \, H_{ex} \quad ,$$

where $\lambda$ is the spin-orbit parameter, $H_{CEF}$ is the cubic crystal-field
Hamiltonian specified by the parameters $A_4 \langle r^4 \rangle$ and $A_6 \langle r^6 \rangle$, and $H_{ex}$
is the exchange molecular field related to the exchange parameter
$J_{ff}$ by $2 \, \mu_B \, H_{ex} = - \, J_{ff} \langle S_Z \rangle$. The susceptibility $X_f(T)$ was calculated
within the subspace of the $J = 5/2$ and $7/2$ multiplets (10). It is
essential to include the $J = 7/2$ multiplet since the small spin-orbit
splitting gives rise to a large VanVleck contribution and appreciable admixtures of the excited multiplet by the crystal-field and
exchange interactions. Least-squares fits of the measured susceptibilities between 3 and 300 K were used to determine the parameters
$\lambda$, $A_4 \langle r^4 \rangle$, $A_6 \langle r^6 \rangle$, $J_{ff}$, and $X_o$. The results are listed in Table I.
The solid curves in Fig. 1 show the calculated inverse susceptibil-
ities using the best-fit parameters.

The spin-orbit parameter $\lambda$ agrees with the free ion value of
410 K to within the estimated error except for the case of SmP where
the fit is quite poor. The crystal-field parameter $A_4 \langle r^4 \rangle$ also does
not vary significantly between the different compounds. Our results
for $A_4 \langle r^4 \rangle$ agree well with previous estimates by Jones (11) from
Knight shift measurements, however, they are inconsistent with esti-
mates of $A_4 \langle r^4 \rangle = 103 \pm 4$ K for SmP (12) and $72 \pm 3$ K for SmSb (13)
from specific heat measurements. We can not reconcile this discrep-
ancy between the measurements. The $A_6 \langle r^6 \rangle$ parameter is not well
determined by the fits since this term effects the susceptibility
only through admixtures of the excited multiplet. The variation in
the exchange parameter between the compounds is significant, and it
explains the large differences between compounds in the susceptibility below 50 K. The temperature independent contribution $X_o$ is
diamagnetic and the diamagnetism increases from SmP to SmBi as is
expected. The $-60 \times 10^{-6}$ emu/mole found for SmP agrees with the
measured susceptibility of LaP which has no 4f contribution.

TABLE II.   Lattice parameters for the $SmS_{1-x}As_x$ and $SmS_{1-x}Sb_x$ alloys studied with the estimates of the Sm valence determined by the lattice constants.

| Composition X | $SmS_{1-x}As_x$ Lattice Parameter(Å) | Valence | $SmS_{1-x}Sb_x$ Lattice Parameter(Å) | Valence |
|---|---|---|---|---|
| 0.00 | 5.70 | 2.00 | 5.70 | 2.00 |
| 0.40 | 5.77 | 2.91 | | |
| 0.45 | | | 6.07 | 2.55 |
| 0.55 | 5.80 | 2.95 | 6.11 | 2.62 |
| 0.70 | 5.83 | 2.99 | 6.145 | 2.80 |
| 0.75 | | | 6.155 | 2.86 |
| 0.80 | 5.86 | 2.99 | 6.18 | 2.90 |
| 0.85 | | | 6.19 | 2.94 |
| 0.90 | 5.89 | 3.00 | | |
| 0.95 | 5.90 | 3.00 | 6.24 | 2.99 |
| 1.00 | 5.92 | 3.00 | 6.27 | 3.00 |

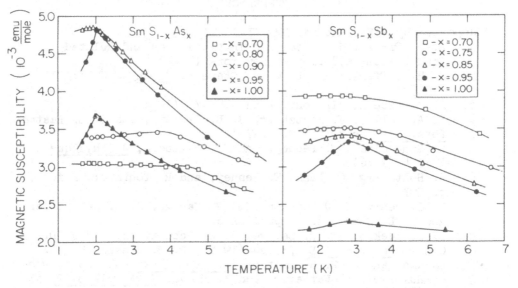

Figure 2

In order to study the competition between the antiferromagnetic ordering and the demagnetization associated with intermediate valency we have measured the magnetic susceptibilities of a series of $SmS_{1-x}As_x$ (14) and $Sm_{1-x}Sb_x$ (8) alloys.   Table II lists the Sm valence as a function of the composition.   The Sm valence was calculated from the measured lattice parameters in the usual manner. The low temperature susceptibilities of several of the alloys are shown in Fig. 2.   The disappearance of magnetic ordering with increasing S concentration can be seen in both alloy systems. The cusp associated with the antiferromagnetic transition in SmAs at 2.0 K is also seen in the 5% S alloy, but it is eliminated with only 10% S substituted for As. The saturation of the susceptibility for alloys with $x \leq 0.9$ to a near constant value as the temperature approaches zero is a behavior characteristic of intermediate valence, however, the Sm valence as estimated from the room temperature lattice constant is essentially trivalent for $x \geq 0.7$. Also, in the $SmS_{1-x}As_x$ system the temperature at which the saturation takes place increases significantly with increasing S concentration. Therefore, this system behaves as if there were a characteristic energy scale, like the spin fluctuation temperature in dilute alloys, which increases with S substitution. In the $SmS_{1-x}Sb_x$ system there is no evidence for such scaling behavior. The susceptibilities of the $x = 0.95$ alloys and the pure pnictide compounds appear qualitatively similar, however, we were unable to obtain satisfactory fits for the alloys with the calculated susceptibility used to fit the pure compounds. Clearly, the magnetic behavior shown in Fig. 2 is very intriguing and awaits a microscopic explanation.

## REFERENCES

*Supported by NSF Grant No. DMR 77-26410.

1.  R. D. Parks (Ed.), "Valence Instabilities and Related arrow-Band Phenomena," Plenum, New York (1977).
2.  J. H. Jefferson and K. W. H. Stevens, J. Phys. C 11:3919 (1978).
3.  F. Holtzberg, AIP Conf. Proc. 18:478 (1974).
4.  R. A. Pollak, F. Holtzberg, J. L. Freeouf, and D. E. Eastman, Phys. Rev. Lett. 33:820 (1974).
5.  S. Von Molnar, T. Penney, and F. Holtzberg, J. Physique 37:C-4 (1976).
6.  F. Holtzberg, O. Peña, T. Penney, and R. Tournier, Ref. 1, p. 507.
7.  D. C. Henry, K. J. Sisson, Wm. R. Savage, J. W. Schweitzer, and E. D. Cater, to be published.
8.  R. B. Beeken, Wm. R. Savage, J. W. Schweitzer, and E. D. Cater, Phys. Rev. B 17:1334 (1978); R. B. Beeken, Wm. R. Savage, and J. W. Schweitzer, J. Appl. Phys. 49:2093 (1978).
9.  F. Hulliger, B. Natterer, and K. Ruegg, Z. Physik B 32, 37 (1978).
10. H. W. deWijn, A. M. van Diepen, and K. H. J. Buschow, Phys. Rev. B 7:524 (1973).
11. E. D. Jones, Phys. Rev. 180:455 (1969).
12. R. J. Birgeneau, E. Bucher, J. P. Maita, L. Passell, and K. C. Turberfield, Phys. Rev. B 8:5345 (1973).
13. M. E. Mullen, B. Luthi, P. S. Wang, E. Bucher, L. D. Longinotti, J. P. Maita, and H. R. Ott, Phys. Rev. B 10:186 (1974).
14. R. B. Beeken and J. W. Schweitzer, Phys. Rev. B 19 (1979).

# ELECTRICAL RESISTIVITY OF $(Nd_xGd_{1-x})_{3-y}S_4$: MAGNETIC EFFECTS

S. M. A. Taher, S. Schwartz, J. C. Ho
Wichita State University, Wichita, KS 67208
J. B. Gruber
North Dakota State University, Fargo, ND 58105
K. A. Gschneidner, Jr.
Ames Laboratory-DOE, Iowa State University, Ames,
IA 50011

We wish to report the electrical resistivity $\rho$ of single crystal $(Nd_xGd_{1-x})_{3-y}S_4$ where $(0 \leq x \leq 1,\ 0 < y < 0.33)$ between 2 and 300 K. X-ray diffraction analyses reveal that all samples possess the high temperature $\gamma$-phase structure which is represented as the bcc $Th_3P_4$ defect structure. The electrical resistivity was measured by a four point DC-technique with pressure contacts at the gold-coated bars of samples with dimensions of $1 \times 2 \times 8$ mm. The measured resistivity values without and with a magnetic field of 7.7 kG as a function of $1/T$ are shown in Fig. 1. As shown in the figure, the resistivity in each sample decreases linearly with decreasing temperature and goes through a minimum as the Curie temperature is reached. The variations in $\rho_{min}$ and temperatures of $\rho_{min}$ are attributed to different rare earth concentrations as well as the total metal to sulfur ratios in the samples. Application of a magnetic field reduces the resistivity below about 60 K in all samples. The $\rho$ vs T curves for all samples strongly suggest that the increase in resistivity is directly related to the state of magnetic ordering through temperature or applied magnetic field. Since the carrier concentration is greater than $8 \times 10^{19}/cm^3$ in all samples, the model of Cutler and Mott (1) suggests that $E_F$ (Fermi energy) lies near but above $E_c$ (conduction band energy).

We have followed the model of Penny et al (2) to determine the activation energy $\Delta E$ for various samples using the equation

$$\rho(T) = \rho_m(T) \exp(\Delta E/kT)$$

where $\rho_m(T)$ is the high temperature resistivity. Peaks in $\Delta E$ occur in the neighborhood of the Curie temperature, and are reduced by an applied magnetic field. We believe the change in

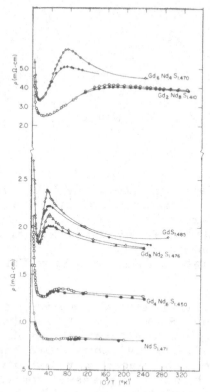

Fig. 1.  Electrical resistivity ρ(mΩ- cm) vs 1/T.  The open data
points were taken at 0 kG.  The solid data points were
taken using a magnetic field of 7. 7 kG.

$\Delta E$ may be described in terms of a localized magnetic polaron (3).  The cause for localization is due to the exchange interaction between the conduction electron and localized 4$\underline{f}$ spin. The
localized states remain stable over a wider temperature range by
the combined effect of magnetic and coulombic interactions.

## ACKNOWLEDGMENT

This work was supported by the U. S. Department of Energy,
contract No. W-7405-Eng.-82, Office of Basic Energy Sciences,
Division of Materials Sciences (AK-01-02).

## REFERENCES

1.   M. Cutler and N. F. Mott, Phys. Rev. 181:1336 (1969).
2.   T. Penney, M. W. Shafer, and J. B. Torrance, Phys. Rev.
     B 5:3669 (1972).
3.   S. Von Molnar, F. Holtzberg, T. R. McGuire, T. J. A.
     Popma, "AIP Conference Proc. 5", C. D. Graham and J. J.
     Rhyne, eds. p. 869 (1972) also S. Von Molnar and F. Holtz-
     berg, ibid, p. 1259.

# MAGNETIC AND THERMAL PROPERTIES OF $Ce_{3-x}S_4$

S. M. Taher, J. C. Ho
Wichita State University, Wichita, KS 67208
J. B. Gruber
North Dakota State University, Fargo, ND 58105
B. J. Beaudry, K. A. Gschneidner, Jr.
Ames Laboratory-DOE, Iowa State University,
Ames, IA 50011

The high temperature phase of cerium sesquisulfide is characterized by the $Th_3P_4$ bcc defect structure and usually expressed as $Ce_{3-x}V_xS_4$ ($0 \leq x \leq 0.33$) to indicate the vacancies $V_x$ in the rare earth sublattice. (1) Between the extremes of $Ce_3S_4$ ($x = 0$) and $Ce_2S_3$ ($x = 0.33$) various ratios between Ce and S are possible and when prepared and melted into ingots or single crystals all samples show the same $Th_3P_4$ defect structure. (2) This note reports the magnetic susceptibility ($X$) and the heat capacity ($C_p$) for two such samples, namely $CeS_{1.39}$ and $CeS_{1.46}$.

In Fig. 1 the reciprocal magnetic susceptibility is given as a function of temperature T(K) between 4 and 300 K. Measurements were made using a vibrating sample magnetometer. Between 60 and 300 K both samples follow Curie-Weiss behavior and display the paramagnetic characteristics of $Ce^{3+}(4f^1)$ in a crystalline environment. Over the linear portion of the curves one obtains effective magnetic moments of $2.45\mu_B$ ($CeS_{1.39}$) and $2.52 \mu_B$ ($CeS_{1.46}$). Uncertainty in these determinations is $\pm 3\%$. The theoretical value for the effective moment of $Ce^{3+}(4f^1)$ is $2.54 \mu_B$. Our values are somewhat lower than calculated since the conduction electron contribution to the susceptibility must also be considered in these degenerate n-type semiconductors. Becker and his co-workers (3) have reported similar measurements over the $Ce_2S_3$-$Ce_3S_4$ composition range. Their measurements also indicate ferromagnetic ordering at temperatures below 10 K. Our samples show similar behavior with a Curie temperature of 6.6 K ($CeS_{1.39}$) and 3.2 K ($CeS_{1.46}$).

In Fig. 2 the heat capacity is reported between 0.5 and 20 K

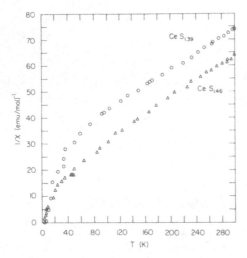

Fig. 1. Reciprocal magnetic susceptibility, $1/X(\text{emu/mol})^{-1}$ vs absolute temperature T(K) for $CeS_{1.39}$ and $CeS_{1.46}$.

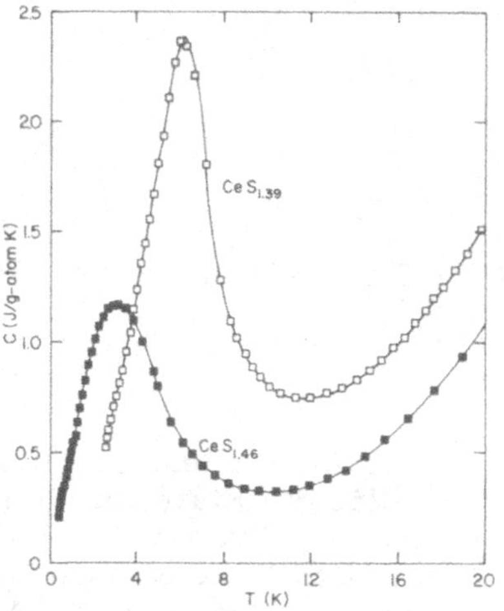

Fig. 2. Heat capacity C(J/g-atom K) vs absolute temperature T (K) for $CeS_{1.39}$ and $CeS_{1.46}$.

for $CeS_{1.46}$ and between 2 and 20 K for $CeS_{1.39}$. The measurements were made using pulse joule heating and germanium thermometry. The lattice and electronic contributions to the heat capacity were assumed to be similar to those of $LaS_{1.46}$ and $LaS_{1.41}$ for $CeS_{1.46}$ and $CeS_{1.39}$, respectively. The La-S compounds have similar crystal structures and their heat capacities were recently measured by the authors. The Schottky contribution to the heat capacity is small over the temperature range reported. The entrophy estimated from the magnetic contribution to the heat capacity is the same for both Ce samples and is 5.3 J/g-at. Ce $K^2$ which is within 10% of the theoretical value of Rlog 2.

## ACKNOWLEDGMENT

We thank Prof. S. Legvold, Ames Laboratory-DOE, Iowa State University, for use of his vibrating sample magnetometer. This work was supported by the U. S. Department of Energy, contract No. W-7405-Eng.-82, Office of Basic Energy Sciences, Division of Materials Sciences (AK-01-02).

## REFERENCES

1.  R. W. G. Wyckoff, "Crystal Structures" Wiley-Interscience, New York (1965).
2.  J. R. Henderson, M. Muramoto, E. Loh and J. B. Gruber, J. Chem. Phys. 47:3347 (1967).
3.  G. Becker, J. Feldhaus, K. Westerholt and S. Methfessel, J. Mag. and Mag. Mat. 6:14 (1977).

REFINING THE PARAMETERS OF THE REFINED SPIN-PAIRING ENERGY

DESCRIPTION

Christian K. Jørgensen

Départment de Chimie minérale, analytique et appliqúee

Université de Genève, CH 1211 Geneva 4

It is a general feature of monatomic entities (gaseous atoms or ions with charge + z) that the <u>barycenters</u> (the average energy of all the states belonging to a definite electron configuration) for a given element vary as a parabolic (quadratic) function of z (1,2) and for a given z, as a linear function of the atomic number Z. A typical example is the barycenter of [18]$3d^2$ of $Sc^+$ situated 1.24 eV (1eV=8065.73 $cm^{-1}$) above the ground state belonging to [18]3d4s having the barycenter at 0.09 eV), almost at the same distance as the unique state of [18]$4s^2$ situated at 1.45 eV. The same mechanism is involved in lanthanides, where the neutral atoms in 11 cases have the ground-state belonging to [54]$4f^q6s^2$ and of $M^+$ to [54]$4f^q6s$. In the following, M is a lanthanide.

The lowest J-level of a configuration containing <u>one</u> partly filled f shell is situated below the barycenter by an amount

$$D \left[ <S(S+1)> - S(S+1) \right] - x_q E^3 - y_q \zeta_{nf} \qquad (1)$$

where D is a definite linear combination of Slater-Condon-Shortley (or Racah) parameters of interelectronic repulsion (3); the coefficient $x_q$ indicates the additional stability of the Hund-rule (S,L) ground term, where more than one L value is compatible with the maximum S; and finally $y_q$ the spin-orbit coupling separating the lowest J-level from the barycenter of this term. By far the largest contribution to Eq.(1) is the deviation of S(S+1) from its average value $< S(S+1)> = 3q(14-q)/52$ for a partly filled $f^q$ shell, and hence, the introduction of $x_q$ and $y_q$ in Eq.(1) is referred to as the "refined" spin-pairing energy model.

There is no obvious reason why the linear variation of the

barycenter energy as a function of q(for a given z) should hold as
exactly in condensed matter, but it turned out (4,5) that electron
transfer spectra of M(III) complexes of reducing ligands agree with
Eq. (1) assuming the parameters (E-A)=0.40eV, D=0.806eV, $E^3$=0.062eV,
(this happens to be D/13) and the Landé parameters $\zeta_{4f}$ derived from
M(III) absorption spectra, where a linear q(E-A) is introduced.  In
this case, the energy change from (q+1) to q varies in a very characteristic way:

| q = 0 | 1 | 2 | 3 | 4 | 5 | 6 |
|---|---|---|---|---|---|---|
| 0.15 | 1.55 | 2.6 | 2.7 | 2.75 | 3.75 | 5.1 |

| q = 7 | 8 | 9 | 10 | 11 | 12 | 13 |
|---|---|---|---|---|---|---|
| 0.1 | 1.5 | 2.5 | 2.5 | 2.5 | 3.45 | 4.7 |

with almost identical values (in eV) for (7+q) and q.  The same
type of expression is valid for processes where the lanthanide looses
one 4f electron, such as 4f→5d transitions in the ultra-violet (or
in the visible for M(II)] and standard oxidation potentials $E^0$ (1,6).
It is perhaps more surprising that the photo-electron spectra corresponding to the ionization of the ground state (belonging to $4f^{q+1}$)
to a variety of J-levels of $4f^q$(7) show the lowest ionization energy
$I^*$ (relative to the Fermi level) varying much like the "double zig-
zag curve".  Actually, the difference between $I^*$ and the quantity
given in Eq.(2) increases smoothly from 1.7eV for metallic praseodymium (q=1) to 2.9 eV in gadolinium (q=6) and is 2.37 eV for six
elements from terbium (q=7) to lutetium (q=13) with an average deviation 0.15 eV.  The agreement could be ameliorated if D was increased
14 percent to about 0.92 eV, and this would be appealing for another
reason, that the J-level distances observed in $4f^q$ ionized states are
6 to 12 percent higher than in the absorption spectrum of the iso-
electronic aqua ion M(III) of the preceding element.  The NaCl-type
antimonides MSb have $I^*$ values which are all 1.0 ± 0.2 eV higher
than of the corresponding metallic element, and hence, the same
parameter set would suit both series equally well.  They have all
(including LaSb) $I^*$ (Sb5p) at 1.9 or 2.0 eV.

The difference 0.4 eV in Eq.(2) between q=6 and q=13 is a very
sensitive function of the linear parameter (E-A) since it is 0.07 eV
larger for each 0.01 eV increase of (E-A).  Though it is not easy to
be firmly convinced by an invariant value of D and $E^3$ for all the
trivalent lanthanides, it is also clear that the nephelauxetic ratio
(1) varies less than 3 percent in nearly all compounds of a given
M(III) and one would expect an influence of the ligands rather to
take place through (E-A).  For chemical purposes, it is particularly
interesting to compare spectroscopic (and other) observations with
the difference of Eq.(1) between (q=6) and (q=13):

$$(48D/13) - 7(E-A) + 0.2eV \qquad\qquad (3)$$

Where 0.2 eV is the differential contribution from spin-orbit coupling. In electron transfer spectra, one does not look at a removal of an electron, but an addition, and hence Eq.(3) should be compared with the wave-number difference between the first electron transfer band (4,5) of $4f^{13}$ ytterbium(III) and of $4f^6$ europium(III) with the same set of ligands. This difference

| Complex: | $MI_6^{-3}$ | $MBr^{+2}$ | $MBr_6^{-3}$ | $MCl^{+2}$ | |
|---|---|---|---|---|---|
| Solvent | $CH_3CN$ | $C_2H_5OH$ | $CH_3CN$ | $C_2H_5OH$ | |
| M = Yb($cm^{-1}$) | 17850 | 35500 | 29200 | 41000 | |
| M = Eu($cm^{-1}$) | 14800 | 31200 | 24500 | 36200 | |
| Diff.(eV) | 0.38 | 0.53 | 0.58 | 0.59 | (4) |
| (E-A)(eV) | 0.398 | 0.377 | 0.370 | 0.369 | |

can be used for evaluating (E-A) assuming invariant D and spin-orbit coupling. The first conclusion is that (E-A) only varies by 7 percent in the four pairs of complexes. It would also seem that the less covalent complexes have the slightly lower values of (E-A). This would actually agree with an increase (1,3) of (E-A) in the various transition groups 4f<5f<3d<4d. However, one has to realize that the variation of the electron transfer band energies as a function of the halide ligand $X^-$ is slightly atypical in the 4f group, in particular in the hexahalide anions, and it has been argued (3) that this, unusually strong, variation is connected with interligand anti-bonding effects in bromide and iodide much like the shift of 4d and 5d group hexahalide electron transfer spectra (8) in organic solvents (compared with water), under conditions of hydrostatic pressure, and in substituted crystal lattices favoring shortened internuclear distances. The inter-ligand anti-bonding character of the most readily excited M.O. may take exceptional proportions in the green ytterbium(III) cyclopentadienide (9) compared with purple $Eu(C_5H_5)_3$ where the distance in Eq.(4) is _negative_ and seems to be about -0.3 eV producing (E-A) close to 0.50 eV. Though the first electron transfer band of europium(III) aqua ions is detected at 53200 $cm^{-1}$, it has not been reported for Yb(III), but it is likely that (E-A) is down below 0.35 eV in aqua ions and in fluorides.

Also $E^O$ = -0.35 V for Eu(II) aqua ions and -1.15V for Yb(II) are not perfectly certain, but such a difference 0.8 eV(1,10) would produce (E-A)=0.34 eV in Eq.(4). Hence, there seems to be a general trend of (E-A) increasing around 0.06 eV from the aqua ions to hexa-iodo complexes. This range of (E-A) values is about 20 times smaller than the difference between ionization energy and _electron affinity_ of the 4f shell incondensed matter, which is known (7,11) from photoelectron spectra to 8 to 9 eV, constituting the main reason for a constant oxidation state in the lanthanides (5).

The ionization energy $I_3$ of gaseous $M^{+2}$ also varies (12) in a way similar to Eq. (2), increasing from 20.20 eV in $Ce^{+2}$ (q=1) to 24.92 eV in $Eu^{+2}$ (q=6) and from 20.33 eV (the value for [54]$4f^8$ situated 0.30 eV above the ground state belonging to [54]$4f^7$5d in $Gd^{+2}$

to 25.03 eV in $Yb^{+2}$ (q=13). Taken at their face value, the difference 4.59 eV between $Eu^{+2}$ and $Gd^{+2}$ suggests D about 10 percent smaller than in Eq.(2), and the difference 4.72 eV between $Eu^{+2}$ and $Ce^{+2}$ compared with (5.1-1.55)=3.55 eV in Eq.(2), as well as 4.70 eV between $Yb^{+2}$ and $Gd^{+2}$ compared with (4.7-0.1)=4.6eV suggest a deviation from a constant (E-A), being perhaps 0.6 eV in the first half of the 4f shell. If Eq.(3) is applied to the observed difference-0.11 eV, (E-A) is as large as 0.44 eV. However, there is no doubt that the origin of the more negative standard oxidation potential $E^o$ of Yb(II) than of Eu(II) aqua ions is related to a slightly stronger hydration energy (5) of the smaller Yb(III), inverting the difference between the marginally larger $I_3$ for gaseous $Yb^{+2}$ compared with $Eu^{+2}$. For some odd reason, the spin-pairing treatment seems to work better in condensed matter than for gaseous ions. The physical mechanism of the linear parameter (E-A) is likely to be highly complicated, the increased attraction from an additional nuclear charge (going from one element to the next) being almost completely compensated by increased repulsion between the 4f electrons. Compared with the total energies of such systems, 0.4 eV is entirely negligible. Both the 3d and 4f groups have D close to 0.8 eV, but (E-A) is 1.0 eV for M(II) in condensed matter and 1.65 eV for $M^{+2}$ of the 3d group. The situation is more extreme in the 4d group with D around 0.5 eV but (E-A) =1.85 eV for gaseous $M^{+2}$. Correspondingly, Nb, Mo, Tc and Ru in the beginning of the 4d group tend toward higher z than V, Cr, Mn, Fe, but at the end of the 4d group, silver (I) is more difficult to oxidize than copper (I). The same trend is seen in the 5f group, where Th, Pa, U, Np and Pu markedly tend to higher z, but Md and No toward lower z than the lanthanides with Z 32 units lower.

1.  C.K. Jørgensen, Oxidation Numbers and Oxidation States. Springer, Berlin, (1969).
2.  C.K. Jørgensen, Angew. Chem. (Int. Ed.) 12: 12 (1973).
3.  C.K. Jørgensen, Modern Aspects of Ligand Field Theory. North-Holland, Amsterdam, (1971).
4.  C.K. Jørgensen, Mol. Phys. 5: 271 (1962).
5.  C.K. Jørgensen, in Handbook on the Physics and Chemistry of Rare Earths (eds. : K.A. Gschneidner and LeRoy Eyring). Vol. 3, pp. 111-169. North-Holland, Amsterdam, (1979).
6.  L.J. Nugent, R.D. Baybarz, J.L. Burnett and J.L. Ryan, J. Phys. Chem. 77: 1528 (1973).
7.  C.K. Jørgensen, Structure and Bonding. 22: 49 (1975).
8.  C.K. Jørgensen, Progress Inorg. Chem., 12: 101 (1970).
9.  R. Pappalardo and C.K. Jørgensen, J. Chem. Phys. 46: 632 (1967).
10. D.A. Johnson, Adv. Inorg. Chem. Radiochem. 20: 1 (1977).
11. M. Campagna, G.K. Wertheim and E. Bucher, Structure and Bonding 30: 99 (1976).
12. J. Sugar and J. Reader, J. Chem. Phys. 59: 2083 (1973).

RESOLUTION OF INDIVIDUAL ELECTRON ORBITS FOR SOLID STATE ATOMS BY
FLUX QUANTUM ANALYSIS?
SIMPLE IIIB-COMPOUNDS AND PHASE TRANSITION OF GADOLINIUM$(A_3-A_2)$

Jules T. Muheim

Laboratory of Solid State Physics, Swiss Federal

Institute of Technology, Hönggerberg, CH-8093 Zürich

(Switzerland)

INTRODUCTION

The study of spark mass spectra has led to the discovery that
flux quantization (fq) is important both for processes connected with
chemical bonding and for the structure of atoms in solids. Magnetic
flux which is involved in any changes of state appears to be quantized with regard to well-determined planar dimensions of atoms and
molecules. From the multipositive atomic ions on the one hand (1)
and the valence effects revealed by the molecule formation from
non-molecular solids on the other hand (2) it follows that both
charge or current type processes and spin type processes involve an
identical minimum flux $\emptyset_o = h/2e$. This value corresponds to Dirac's
equation 76 in (3) (cf. also (4)). The plasma process associated
with the ion formation involves an electronic Bose-Einstein condensation for which an (electronic) quasiequilibrium may be assumed.
The individual atom behaves approximately as a harmonic oscillator
(charge oscillator, (5)). Since the flux is quantized it would in
principle be possible to obtain absolute information about microdi-
mensions. As the mass of the process-carrying quasiparticle deviates
in a bond-dependent way from the ideal boson mass $2m_e$, the information
is only relative. In the case of a compound MX information about the
shell structure of the atoms as well as the ratio of the radii
anion/cation may be obtained from the K-slopes of the normalized con-

centrations of multipositive ions as a function of the degree of
ionization. The slopes of the optical density as a function of the
logarithmic exposure contain relative information about the ionization energy of the $\nu$-positive state (corresponding to the $\nu$th
electron orbit) and thus the orbit dimension. In order to get the
absolute magnitudes of shells and orbits it is indispensable to
make use of the lattice constant which is known independently. Owing
to rather large fluctuations of the slopes a reliable orbit spectrum
ought to be based on at least four experiments in photographic ion
detection. The inaccuracy of the orbit dimension amounts to 5-10%
for four, and 20-50% for only three experiments. The existence of
electron orbits which is thereby very probable is of fundamental
importance, since quantum physics in its actual state makes no
definite statement on the existence of orbits. The orbits ought not
to be confused with the orbitals of quantum chemistry. It is very
important that from empirical orbit dimensions physical information
may be derived. In order to obtain that information, two things are
required, i.e. the empirical determination of the (geometric) orbit
dimensions and the knowledge of the interactive weight of those
orbits mainly involved in a particular physical phenomenon. In a
few cases we have succeeded in determining or guessing the approximate weights. In the case of AgBr, positron lifetimes may be quantitatively related to the orbit structure (6). A relation between
orbit structure and superconducting transition temperature of IIIB-
sulphides on the one hand, and the volume change of transforming
gadolinium ($A_3-A_2$) on the other hand has been found. For the example of IIIB-phosphides and -sulphides "complete" electron orbit
spectra are presented for the first time.


IIIB-PHOSPHIDES AND -SULPHIDES: ELECTRON ORBIT SPECTRA


      Spark mass spectra may, in favorable cases, inform about all
important chemical parameters. By means of a memory effect the atoms
preserve information about their former state in the solid. Because
of the limited space, we simply present the empirical orbit spectra.
Fig. 1 shows for both the cation and the anion electron orbit radii
of IIIB-phosphides and -sulphides as a function of the orbit order.
The order corresponds to increasing binding energy. A close correspondence between the cationic and anionic spectrum is found. The
largest change occurs between scandium and yttrium, whereby the orbits of the oxidation states are inside the atoms in the case yttrium, while the more strongly bound higher orbits become wider.
This possibly represents a spatial analog to the nephelauxetic effect (7). Dashed lines correspond to the average shell radii. These
are determined from the average slopes (K-values) of the multiposi-

tive ions. The solid lines connect the individual electron orbit radii. These cannot be derived from the K-diagram.


### ORBIT STRUCTURE AND SUPERCONDUCTIVITY OF IIIB-SULPHIDES

Because of the inherent fq-relationship between Bose-Einstein condensation and superconductivity the superconducting critical temperature $T_c$ must be directly related to the geometry of particular orbits. For a known orbit geometry and orbit weight, $T_c$-variations ought in principle to be predictable. Unfortunately, the weight is usually unknown. ScS presents an interesting case. A decrease of the lattice parameter of 0.2% only reduces $T_c$ by a factor of two. Contrary to all expectations, the first and second Sc-orbits of ScS (residuum specimen) expand by more than 12% with regard to nearly ideal ScS (sublimate). The average values are $\Delta r_{1.2}=0.18\text{Å}$, $r=1.54\text{Å}$. If now $T_c \sim \omega \sim r^{-2}$, or $\Delta T_c/T_c= -2\,\Delta r_{1.2}/r = -0.23$, then $\Delta T_c= -1.1\text{K}$, in full accord with the experimental result of H.R. Ott (private communication). In case the slightly contracting ($\Delta r_3= -0.07\text{Å}$) "3d-orbit" is also taken into account and equal weights for orbits one to three are assumed, the decrease in $T_c$ becomes

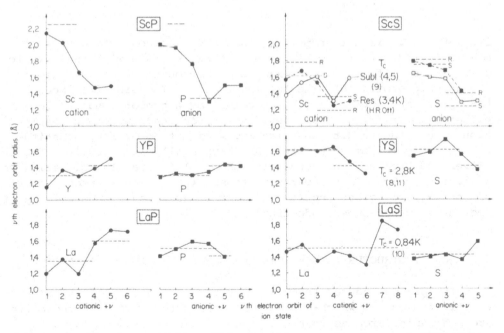

Figure 1. Electron orbit radii of cations and anions of IIIB-phosphides and -sulphides as a function of orbit order.

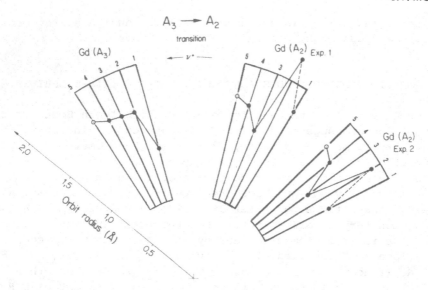

Figure 2. Electron orbit radius spectrum of transforming Gd($A_3$-$A_2$)
for initially hexagonal ($A_3$) and finally body-centered
cubic ($A_2$) states. Two different experiments designated
1,2.

$\Delta T_c$= -0.6K. It is not surprising that the superconducting properties
are connected with the cationic orbits rather than with the anionic
ones. The $e^+$-lifetime is an example of a process which is bound to
the anionic orbits (6). The falling trend of $T_c$ going from ScS to
LaS (8-11) may empirically be described by a proportionality to
either $r_3$ or $r_3/r_{1.2}$. Small deviations from the equiweight assumption affect the result considerably. As a matter of fact substitution of only 10 ppm copper in AgBr changes all matrix orbits of both
silver and bromine, and hence also the $e^+$-lifetime. The situation is
analogous to ScS. Thus imperfectness suppresses the superconductivity
of "potentially superconducting" ScS and causes superconductivity of
"potentially non-superconducting" ScN (12,13).


ORBIT STRUCTURE AND VOLUME CHANGE OF GADOLINIUM ($A_3$-$A_2$)

The plasma-induced transition of gadolinium ($A_3$-$A_2$) (5) provides
the first case for which all data necessary for physical predictions
are available. From decreases of the optical density of 152Gd+2
and 154Gd+3 associated with the phase transition, it may be concluded
that the weights of the second and third orbits are x=36% and 1-x=

64%. For the determination of the orbit structure in the hexagonal phase (A$_3$) and in the body-centered cubic phase (A$_2$) two experiments were available (Exp. 1,2 in Fig. 2). From these results the volume change may be estimated. The orbit spectrum of the hexagonal phase corresponds to the initial state, while that of A$_2$ corresponds to the final state. While the second orbit expands, the third orbit contracts: $\Delta r_2 = 0.45\text{Å}$, $\Delta r_3 = -0.2\text{Å}$; $r = 1.68\text{Å}$. Thus $\Delta V/V = 3x(\Delta r_2/r) + 3(1-x)\Delta r_3/r = -1.9\%$ in excellent agreement with the x-ray result, namely -2% (14). Thus it must be assumed that the oxidation state orbits reflect the volume of the primitive cell. Hence the empirical weight x automatically accounts for the change of the coordination number or, in other words, oxidation state orbits reflect both the volume and the chemical environment. K-analysis gives a similar result: $\Delta V/V = -(3\pm1)\%$ (1). In this case, however, the packing efficiency must explicitly be accounted for. Thus it is concluded that the higher orbits reflect the volume of the atom and not the details of the actual bonding. This observation, if correct, may lead to the assumption of an internal electronic organization in the atom.

## ACKNOWLEDGEMENT

The author wishes to thank Professors G. Busch and P. Wachter for their support of this work. He is also very indebted to R. Hauger for the specimens, Drs. F. Hulliger and H.R. Ott for helpful discussions and P. Dekumbis for his valuable assistance.

## REFERENCES

1. J.T. Muheim, "Discovery of Flux Quantization on the Atomic Scale," Helv. Phys. Acta 50: 584 (1977).
2. J.T. Muheim, "Valence Relationships in the Rf Spark Mass Spectrum from Solids," Mat. Res. Bull. 7: 1417 (1972); "Gesetze der Festkörperplasma-Valenzphänomenologie und ihre statistische Deutung," Helv. Phys. Acta 49: 739 (1976).
3. P.A.M. Dirac, "The Theory of Magnetic Poles," Phys. Rev. 74: 817 (1948).
4. C. Kittel, "Introduction to Solid State Physics," Wiley, New York (1976) p. 382.
5. J.T. Muheim, "On a New Method to Study the Polar Properties of Matrix and Impurity Cations in Dielectric Solids by Spark Source Mass Spectrography - With Particular Reference to Eu- and Gd-Chalcogenides and -Pnictides," Proc. 10th R.-E. Res. Conf. Arizona, Vol. I:208, USAEC Techn. Inf. Center, Oak Ridge, Tenn. USA (1973).

6.  J.T. Muheim, "The $e^+$-Lifetime Spectrum of Heteropolar Solids and its Relation to the Chemical Bond and Microscopic Flux Quantization," 5th Int. Conf. $e^+$-Annihil., Lake Yamanaka, Japan, 8-11 April 1979.

7.  C.K. Jörgensen, "Oxidation Numbers and Oxidation States," Springer, New York (1969) p. 79.

8.  F. Hulliger and G.W. Hull Jr., "Superconductivity of Rocksalt-Type Compounds," Solid State Comm. 8: 1379 (1970).

9.  A.R. Moodenbaugh, D.C. Johnston and R. Wiswanathan, "Superconductivity in Two NaCl Structure Compounds:$\alpha$-ZrP and ScS$_{1+x}$," Mat. Res. Bull. 9: 1671 (1974).

10. E. Bucher, A.C. Gossard, K. Andres, J.P. Maita and A.S. Cooper, "Magnetic Properties and Specific Heats of Monochalcogenides of La, Pr, and Tm," Proc. 8th R.-E. Res. Conf. Reno, Nevada, I:74, T.A. Henrie and R.E. Lindstrom, Eds. (1970).

11. A.R. Moodenbaugh, "Superconductivity of Some NaCl Structure Sulfides, Selenides, and Phosphides," thesis University of California, San Diego (1975) p. 125.

12. J.J. Veyssié, D. Brochier, A. Nemoz and J. Blanc, "Supraconductivité du nitrure de lanthane," Phys. Lett. 14: 261 (1965).

13. F. Hulliger and H.R. Ott, "Superconductivity of Lanthanum Pnictides," J. Less-Common Met. 55: 103 (1977).

14. F.H. Spedding, J.J. Hanak and A.H. Daane, "High Temperature Allotropy and Thermal Expansion of the Rare-Earth Metals," J. Less-Common Met. 3: 110 (1961).

# QUASI STATIONARY STATES AND THE Lα AND Lβ SATELLITES OF THE HEAVY RARE EARTHS

B. D. Shrivastava and P. R. Landge

School of Studies in Physics, Vikram University

Ujjain 456010, India

## INTRODUCTION

The X-ray satellites have been generally classified into two broad groups, namely the high frequency (HF) satellites and the low frequency (LF) satellites according as they occur on the high or on the low frequency side of the parent dipole lines. Although the multiple ionization theory of Wentzel (1) and Druyvesteyn (2) has been found to explain the origin of about 60% of the HF satellites (3), the theory is inherently incapable to explain the emission of LF satellites, because the frequency of a satellite, calculated on the basis of this theory, always comes out to be more than that of the parent line. An alternative theory which has been applied by many workers (4,5) to explain the HF satellites is the Hayasi's theory (6) of quasi sationary states (QSS). In the present paper we have shown that Hayasi's theory can be applied to explain satellites of both the types in the L emission spectra of the heavy rare earths.

## THEORY

Hayasi's theory (6) is based on the existence of more or less localized QSS which arise in the following manner: an electron ejected from an excited atom is reflected backward ($\theta=90^\circ$) by certain crystal planes which, in fact, prevent it from moving far from the parent atom. In this way a standing wave pattern is set up in the vicinity of the atom, which leads to the existence of a number of allowed energy states called QSS. The model suggests that when a transition between inner atomic energy levels and a simultaneous transition of an excited electron between QSS takes place, a

435

satellite is emitted. The process is diagramatically represented
in Figure 1. Though Hayasi's theory has been mostly applied to the
HF satellites, we have shown in the figure how a LF satellite can
be emitted. As the process is obviously reverse of that which gives
rise to the extended X-ray absorption fine structure (EXAFS), the
energy difference between a satellite and the parent line must be
equal to the energy difference between two maxima appearing in the
EXAFS of the absorption spectrum of the series to which the parent
line belongs. Thus,

$$E_{satellite} \quad E_{parent} \quad \begin{aligned} &=\Delta E(\text{between two QSS}) \\ &=\Delta E(\text{between two EXAFS maxima}) \end{aligned}$$

## CALCULATIONS AND RESULTS

A necessary requirement of this theory is the knowledge of the
accurate data of EXAFS of the element in metallic form. Recently,
the data for the EXAFS maxima for the metals dysprosium, holmium,
thulium, ytterbium and lutetium have been reported by Vijayvargiya
et al (7) and for gadolinium by Deshmukh et al (8) and these are
collected in Table 1. In the case of gadolinium, we have, however,
designated the maxima at 18.2 eV as A and at 24.0 eV as B* which
were originally designated as $A^+$ and B respectively by Deshmukh
et al. Similarly, in the case of dysprosium we have redesignated
the maxima at 36.0 eV to be B* instead of B. Here, we may point
out that different authors use different designations for EXAFS
maxima and for uniformity, sometimes, it becomes necessary to re-
name them. The other data for the satellites, which is required
for the calculations, has been taken from references (3,9-11).

Using these data the wavelengths of the satellites have been
calculated by us according to Hayasi's theory and are given in
Table 2. The low frequency satellite $\beta_{14}$ has been assigned by us
to electron jump between the QSS (C*-C), the jump occurring simultaneously with the transition between inner atomic energy levels
which gives rise to the parent line $L\beta_2$. Similarly, transitions
have been assigned to the high frequency satellites and are given
in Table 2.

## DISCUSSION

The occurrence of the L emission satellites of the heavy rare
earths has also been explained on the basis of multiple ionization
theory by Nigam and Mathur (3). The wavelength of the satellites
calculated by them are included in Table 2. An inspection of this
table shows that the agreement between calculated and experimental
values are better with Hayasi's theory in comparison to the multiple
ionization theory.

Table 1

Energies (in eV) of the EXAFS maxima at the
$L_3$ absorption edge of heavy rare earth metals
taken from ref. (7) and (8).

| EXAFS maxima | $^{64}Gd$ | $^{66}Dy$ | $^{67}Ho$ | $^{69}Tm$ | $^{70}Yb$ | $^{71}Lu$ |
|---|---|---|---|---|---|---|
| A  | 18.2[b]   | 13.5    | 10.9  | 14.5 | 7.3   | 6.8   |
| B  | -         | -       | 24.9  | 38.2 | 27.8  | 24.8  |
| B* | 24.0[b]   | 36.0[a] | 32.0  | -    | 34.3  | 31.8  |
| C  | 49.0      | 45.5    | 42.3  | 54.7 | 49.9  | 49.2  |
| C* | -         | 64.1    | 63.0  | 74.8 | 66.6  | 59.7  |
| D  | 76.0      | 82.6    | 79.2  | -    | 79.3  | 82.1  |
| E  | 113.0     | 117.6   | 100.1 | -    | 118.3 | 104.1 |
| F  | -         | -       | 133.7 | -    | 137.8 | 145.1 |

a) Originally it was designated as B but we have
   designated it as B*.

b) Originally they were designated as A⁺ and B but
   we have designated them as A and B* respectively

Figure 1

Energy level diagram
(not to scale) for
the emission of
satellites of $L\beta_2$.
The M levels have
not been shown.
1. parent line $L\beta_2$
2. HF satellite $\beta_2^I$
3. LF satellite $\beta_{14}$

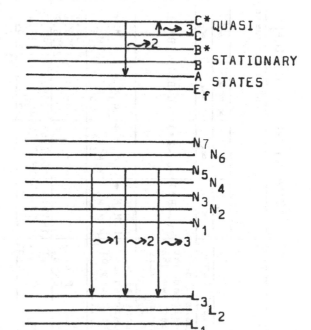

## Table 2
### Data for $L\alpha$ and $L\beta$ satellites of the heavy rare earths

| Parent line | High frequency satellites $\alpha_0$ ($\alpha_2$) | $\alpha_{III}^{Z}$ ($\alpha_2$) | $\alpha'$ or $\alpha^{IX}$ ($\alpha_1$) | $\alpha^{X}$ ($\alpha_1$) | $\alpha_5$ ($\alpha_1$) | $\beta_2^{0}$ ($\beta_2$) | $\beta_2^{I}$ ($\beta_2$) | $\beta_2^{II}$ ($\beta_2$) | Low frequency satellites $\beta_{14}$ ($\beta_2$) |
|---|---|---|---|---|---|---|---|---|---|
| Transitions assigned between two QSS on the basis of H.T. | (C–B*) | (C*–B*) | (C–B*) | (B*–A) | (E–C) | (B*–B) | (C*–A) | (D–A) | (C*–C) |
| Transitions assigned on the basis of M.I.T. | $L_3M_3M_1\rightarrow M_3M_1M_4$ | $L_3M_1M_2\rightarrow M_1M_2M_4$ | $L_3M_3\rightarrow M_3M_5$ | $L_3M_2\rightarrow M_2M_5$ | $L_3M_1\rightarrow M_1M_5$ | – | $L_3M_5\rightarrow M_5N_5$ | $L_3M_3\rightarrow M_3N_5$ | $L_3N_4\rightarrow N_4N_5$ |
| 64Gd H.T. | – | – | – | 2036.73 | – | – | – | – | 1742.87 |
| M.I.T. | | | | 2035.7 | | | | | 1748.54 |
| Expt. | | | | 2035.7 | | | | | |
| 66Dy H.T. | 1913.11 | 1907.63 | 1902.07 | 1898.28 | – | – | 1609.66b | – | 1620.29 |
| M.I.T. | 1913.58 | 1903.58 | 1903.59 | 1898.43 | | | 1609.69b | | 1620.57 |
| Expt. | 1912.3 | 1907.2 | 1901.1 | 1898.1 | | | 1609.8 | | 1625.1 |
| 67Ho H.T. | – | – | – | 1834.43b | – | – | 1553.67 | 1550.52b | 1567.93 |
| M.I.T. | | | | 1834.27b | | | 1554.0b | 1551.35b | 1563.87 |
| Expt. | | | | 1834.5a | | | 1555.1 | 1551.9a | 1567.15 |
| 69Tm H.T. | – | – | – | – | – | – | 1450.67 | – | 1464.50 |
| M.I.T. | | | | | | | 1452.28b | | – |
| Expt. | | | | | | | 1450.6 | | 1462.47 |
| 70Yb H.T. | – | – | 1664.93 | 1662.38 | 1653.18 | – | 1403.7 | – | 1415.14 |
| M.I.T. | | | 1961.19 | 1661.19 | 1661.19b | | 1407.41 | | 1413.12 |
| Expt. | | | 1665.3 | 1662.0 | 1654.1b | | 1406.1 | | {1413.8 / 1413.1} |
| 71Lu H.T. | – | – | 1612.49 | 1610.89 | – | 1366.12 | 1359.33 | – | – |
| M.I.T. | | | 1613.97 | 1610.83 | | – | 1359.26 | | |
| Expt. | | | 1613.0 | 1609.7 | | 1366.3 | 1358.7 | | |

Notes:  a Nigam et al(10).
b Mathur (11)

H.T. – Calculated values from Hayasi's theory(present calculation)
M.I.T.–Calculated values from Multiple Ionization theory from ref(3).
Expt.–Experimental values from ref. (9)except those marked.

Though Hayasi's theory has been used by several authors (4,5) to explain the origin of satellites, the main objection against it has been that the number of QSS differences greatly exceeds the number of satellites, so that some correspondence is inevitable (13). This objection has been raised, perhaps because Hayasi's theory has been applied only in a few scattered cases (4,5,13). It appears that in the past no attempt seems to have been made to correlate the result of one element with another. In the present work we have applied the theory in the heavy rare earth series and have shown that the same QSS difference gives rise to a particular satellite in all the elements. Hence, the assignment of transitions is not arbitrary in the present study.

As already pointed out in the introduction, the LF satellite cannot be explained on the basis of multiple ionization theory. Hence, various other mechanisms have been proposed, to explain the low frequency satellites (14,15), but these have been found to have limited success in a few cases of LF satellites (13). In the present work we have attempted to use Hayasi's theory to explain the LF satellites as well and it is seen from Table 2 that the agreement between the calculated and experimental results is good.

We would like to mention here that we very much wanted to correlate the intensities or the transition probabilities of the EXAFS maxima with those of the satellites so as to give our assignment of the satellites a theoretical basis. But due to lack of availability of such data, it has not been possible to attempt such a correlation at present. The present study is a humble step towards a unified theory of the low and high frequency satellites and it is clearly seen that Hayasi's theory of quasi stationary states yields good results in the case of $L\alpha$ and $L\beta$ satellites of the heavy rare earths.

## REFERENCES

1. G. Wentzel, Funkenlinien im rontgenspectrum, Ann. Phys. 66:437 (1921).

2. M. J. Druyvesteyn, Das rontgenspectrum zweiter art, Z. Phys. 43: 707 (1927).

3. \. N. Nigam and R. B. Mathur, A survey of the high frequency satellites in X-ray emission spectra. Proc. Int. Conf. Inner Shell Ionization Phenomena and its Future Applications, Atlanta, Georgia, 3:1698 (1972).

4. S. Rai and S. D. Rai, Origin of K satellites of yttrium, Acta Phys. Polon. A 50:493 (1976).

5. B. D. Shrivastava and P. R. Landge, Extended X-ray absorption fine structure and the $L\alpha$ and $L\beta$ satellites of rhenium, Nuovo Cimento B 49:118 (1979).

6.  T. Hauasi, Zue systematik der statelliten der Kα- and β- linien vom metallischen aluminium.  Sci. Rep. Tohoku University 45:221 (1961).

7.  V. P. Vijayvargiya, S. N. Gupta and B. D. Padalia, Extended X-ray absorption fine structure (EXAFS) studies of the lanthanides, Phys. Stat. Sol. (b) 80:83 (1977).

8.  P. Deshmukh, P. Deshmukh and C. Mande, $L_3$ X-ray absorption discontinuity of gadolinium in the metal and in some compounds, J. Phys. Chem. 10:3421 (1977).

9.  Y. Cauchois and C. Senemaud, Wavelengths of X-ray emission lines and absorption edges, Pergamon, Oxford (1978).

10.  A. N. Nigam, R. B. Mathur and B. G. Gokhale, New diagram and non-diagram lines in the L X-ray emission spectrum of 67Ho, Phys. Rev. A 13:1756 (1976).

11.  R. B. Mathur, Studies on the diagram and non-diagram lines in the L X-ray spectra of some rare earth elements (Thesis), University of Rajasthan, Jaipur, India (1975).

12.  J. A. Bearden, X-ray wavelengths, U.S. Atomic Energy Commission, Oak Ridge, Tennessee (1964).

13.  S. J. Edwards, X-ray satellite lines (Thesis), University of Leicester, England (1967).

14.  H. Hulubei, A new type of emission in X-ray non-diagram lines of Kα spectra.  C. R. Acad. Sci. Paris 224:770 (1947).

15.  T. Aberg, X-ray satellites and their interpretation, Proc. Int. Symp. X-ray Spectra and Electronic Structure of Matter, Munich 1:1 (1973).

CALCULATED RADIATIVE AND NONRADIATIVE TRANSITION PROBABILITIES AND

STIMULATED CROSS-SECTIONS OF SELECTED $Er^{3+}$ LINES IN OXIDE GLASSES*

R. Reisfeld, L. Boehm, E. Greenberg and N. Spector**

Department of Inorganic and Analytical Chemistry
The Hebrew University of Jerusalem, Israel
**Soreq Research Center, Yavne Israel

## ABSTRACT

Radiative transition probabilities for fluorescence bands extending from 0.523μ to 13μ of $Er^{3+}$ in phosphate, germanate and tellurite glasses were calculated using the Judd-Ofelt approach. Multiphonon relaxation rates from these levels were also calculated using the phenomenological parameters of multiphonon relaxation theory in glasses. Possible laser transitions of $Er^{3+}$ in those glasses are discussed.

Stimulated emission of several infrared lines of erbium above 1μ has been obtained in a variety of solid hosts. See Weber [1] et al and references therein. In addition to these transitions, stimulated emission arising from the $^4S_{3/2}-^4I_{13/2}$ at 0.85μ was observed in YLF by Chicklis et al [2].

Erbium can also be used as a donor in energy transfer to Ho and Tm in activating fluorescence of these ions [3].

In order to predict the possible laser lines in Er doped glasses, we have performed the calculation of the relevant spectroscopic properties using the Judd-Ofelt parameters obtained by us previously [4,5]. All the relevant formulae for the calculation of radiative transition probabilities by the Judd-Ofelt method can be found in the preceding paper [6] in this volume, and a detailed description of the method in ref. 7. In the specific case of the two levels of erbium $^4S_{3/2}$; $^2H_{11/2}$ which are separated in the glassed by an energy gap of approximately 700 $cm^{-1}$ the effect of thermalization

---

*Partially supported by U.S. Army Contract No. DAERO-76-G-006

Table I.  Spectral characteristics for the observed laser line of
          $Eu^{3+}$ in various matrices.

1.   Transition $^4S_{3/2} \to {}^4I_{9/2}$ at $\sim 1.67\mu$.

| Matrix | $fx10^6$ | A $s^{-1}$ | β | $\tau_i$ eff Sec | comments |
|---|---|---|---|---|---|
| YAℓO$_3$ | 1.21 | 107 | 0.015 | $0.14x10^{-3}$ | pulse laser at room temperature |
| Phosphate glass | 1.19 | 68 / 63* | 0.032 / 0.041* | $0.20x10^{-3}$ | |
| Germanate glass | 0.65 | 50 / 42* | 0.03 / 0.05* | $0.60x10^{-3}$ | |
| Tellurite glass | 1.55 | 188 / 172* | 0.03 / 0.04* | $0.68x10^{-4}$ | |

2.   Transition $^4S_{3/2} \to {}^4I_{11/2}$ at $1.2\mu$.

| Matrix | $fx10^6$ | A $s^{-1}$ | β | $\tau_i$ eff Sec | comments |
|---|---|---|---|---|---|
| YAℓO$_3$ | 0.38 | 51.2 | 0.007 | $0.14x10^{-3}$ | pulse laser at $77^{\circ}K$ |
| CaF$_2$ | 0.31 | 51.2 | 0.007 | $0.14x10^{-3}$ | |
| Phosphate glass | 0.33 | 37 / 32* | 0.017 / 0.021 | $0.20x10^{-3}$ | |
| Germanate glass | 0.16 | 25 / 19* | 0.216 | $0.6x10^{-3}$ | |
| Tellurite glass | 0.45 | 107 / 95* | 0.017 / 0.021 | $0.68x10^{-4}$ | |

3.   Transition $^4S_{3/2} \to {}^4I_{13/2}$ at $0.85\mu$.

| Matrix | $fx10^6$ | A $s^{-1}$ | β | $\tau_i$ eff Sec | comments |
|---|---|---|---|---|---|
| CaF$_2$ | 1.61 | 557 | 0.078 | $0.14x10^{-3}$ | operating at $77^{\circ}K$ |
| Phosphate glass | 2.13 | 421 / 440* | 0.20 / 0.283* | $0.20x10^{-3}$ | |
| Germanate glass | 0.991 | 248 / 251* | 0.149 / 0.281* | $0.60x10^{-3}$ | |
| Tellurite glass | 3.01 | 1252 / 1310* | 0.199 / 0.284* | $0.68x10^{-4}$ | |

4.   Transition $^4I_{11/2} \to {}^4I_{13/2}$ at $2.75\mu$.

| Matrix | $fx10^6$ | A $s^{-1}$ | β | $\tau_i$ eff Sec | comments |
|---|---|---|---|---|---|
| YAℓO$_3$ | 1.29 | 41.0 | 0.22 | $1.63x10^{-3}$ | |
| YAG | 1.29 | 41.0 | 0.067 | $1.63x10^{-3}$ | room temperature |
| Phosphate glass | 1.35 | 26.0 | 0.147 | $5.02x10^{-6}$ | |
| Germanate glass | 0.98 | 23.3 | 0.163 | $1.57x10^{-4}$ | |
| Tellurite glass | 1.86 | 76.0 | 0.144 | $3.16x10^{-4}$ | |

| Matrix | $f \times 10^6$ | A $s^{-1}$ | $\beta$ | $\tau_i$ eff Sec | comments |
|---|---|---|---|---|---|
| 5. Transition $^4I_{13/2} \rightarrow {}^4I_{15/2}$ at $1.54\mu$. | | | | | |
| YAℓO$_3$ | 2.04 | 211.0 | 1.00 | $4.75 \times 10^{-3}$ | |
| YAG | 2.04 | 211.0 | 1.00 | $4.75 \times 10^{-3}$ | pulse laser, room temperature |
| Phosphate glass | 2.19 | 134.0 | 1.0 | $7.42 \times 10^{-3}$ | |
| Germanate glass | 1.46 | 119.0 | 1.0 | $9.1 \times 10^{-3}$ | |
| Tellurite glass | 3.05 | 396 | 1.0 | $2.53 \times 10^{-3}$ | |

At room temperature must be taken into account. In this case effective radiative transition probabilities $A_{eff}$ and lifetimes $\tau_{eff}$ are - obtained by ere

$$A^R_{ijeff} = \frac{\sum_{ij i} q_i A^R_{ij} \exp{-(E_1-E_i)/kt}}{\sum_{ij i} g_i \exp{-(E_1-E_i)/kt}}$$

$i = {}^2H_{11/2}, \ {}^4S_{3/2}$

$j$ = terminal level

$$\tau_{eff}^{-1} = \sum_{ij} A^R_{ijeff} + \sum_{ij} W^{NR}_{ij}$$

The fromula for non-radiative transfer is given in ref. 8.

In Table I we present the oscillator strengths, transition probabilities A, branching ratios $\beta$ and effective life time $\tau_{eff}$ for transitions which have been shown to lase in crystals and in three oxide glasses: phosphate, tellurite, and germanate. (The composition of the glasses was described in ref. 5.). The calculated values are compared to those of CaF$_2$, YAℓO$_3$ and YAG.

As can be seen from the Table, the radiative transition probabilities in the glasses, in particular the tellurite glass, compare
well with those of $CaF_2$ and $YA\ell O_3$.  Also, the branching ratios are
quite favourable in the glasses.  Therefore, those glasses can be
considered as good materials for erbium lasers.


REFERENCES

1.    M.J. Weber, M. Bass & G.A. de Mars, J.Appl.Phys. 42: 301 (1971).
2.    E.P. Chicklis, C.S. Naiman & H.P. Jenssen, J.O.E. QE-13: 893
      (1977).
3.    D. Pacheco & B. Di Bartolo, J. Lumin.  16: 1 (1978).
4.    R. Reisfeld & Y. Eckstein, J.Non. Crys.Solids, 15: 125 (1974).
5.    R. Reisfeld & Y. Eckstein, J.Chem.Phys. 63: 4001 (1975).
6.    R. Reisfeld, A. Bornstein, J. Flahaut & A.M. Loireau-Lazac'h,
      This volume.
7.    R. Reisfeld & C.K. Jørgensen, Lasers & Excited States of Rare-
      Earths, Springer Verlag, 1977.
8.    R. Reisfeld, L. Boehm & N. Spector, The Rare Earths in Modern
      Science & Technology, Ed. G.J. McCarthy & J.J. Rhyne, Plenum
      Press 1977 p.513.

# Eu RICH EuO: SPECIFIC OPTICAL TRANSITIONS

C. Godart*, A. Mauger**, M. Escorne**, J.C. Achard*

*ER 209, ** Laboratoire de Physique des Solides

CNRS, 1, pl. A. Briand, Bellevue 92190 (France)

## INTRODUCTION

Eu-rich europium oxide is well known for the insulator metal transition (IMT) at T $\sim$50 K associated with oxygen vacancies which act as donors. These vacancies can be evidenced by a jump of the resistivity by an amount of several orders of magnitude for suitable concentrations (1) and by two peaks in the absorption spectra at 1.85 $\mu$ and 2.20 $\mu$, which do not exist in stoichiometric samples (2). These peaks have been imputed to an exchange resonance of the outer-electron of the oxygen vacancy in a bound magnetic polaron, split by the spin orbit interaction (2). Recently, an alternative ex-planation has been given (3), based on the existence of the two peaks in the metallic configuration, according to which the optical tran-sitions should be imputed to a $F^+$ center on the vacancy.

We report here the reflectivity spectra as a function of temp-erature for an Eu rich but nearly stoichiometric EuO single crystal. These are of interest because no reflectivity spectra have previously been reported for this material except at room temperature, and these provide some insight into the actual debate above mentioned.

## EXPERIMENTAL

1) Sample elaboration.

The EuO single crystal used in this study has been grown accor-ding to Guerci-Shafer techniques (4). A mixture of $Eu_2O_3$ and Eu is sealed in one outgased molybdenum crucible in a dry glove box under an argon atmosphere continuously purified; the ratio of the two

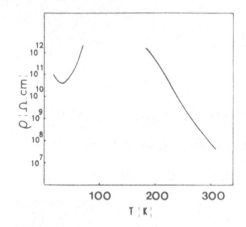

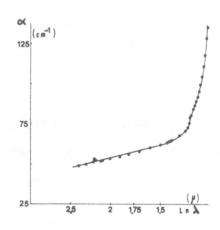

Fig. 1 :   Resistivity versus          Fig. 2 :   I.R. Absorption versus
           temperature.                           wavelength.

compounds correspond to the initial composition $Eu_{1.024}O$. A melting
bath is realized and then slowly cooled from about 2000°C at a rate
of 3.6°C/hour. The sample homogeneity is achieved owing to two
meltings with an intermediate turn of crucible. The lattice constant deduced from powder diagrams is 5.143 A°in agreement with
usual measurements (5). Metallographic, electronic and X-ray pic-
tures do not reveal the existence of more than one phase. Any $Eu^{3+}$
peak is absent from Mössbauer spectra and infrared optical spectra.

    2) Physical properties.

    The final composition of the sample has been deduced from the
measurements of transport properties (Fig. 1) and infrared absorption spectra (Fig. 2) owing to the correlation between these properties and stoichiometry established by Shafer, Torrance, Penney
(6). Measurements of resistivity versus temperature in the range
[4.2 - 300 K] show a semiconductor type behavior. At room temperature the activation energy is equal to 480 meV and the resistivity
is $\rho = 7.10^7$ $\Omega cm$. This sample does not exhibit any IMT at T < 50K.

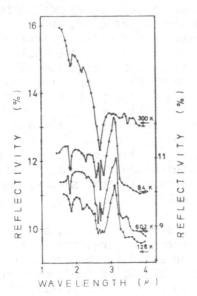 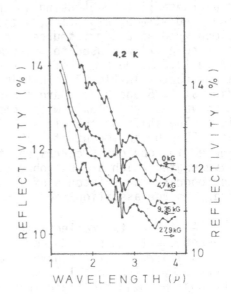

Fig. 3: Reflectivity spectra          Fig. 4: Reflectivity spectra
versus temperature.                    versus magnetic field at 4.2 K.

All these data are characteristic of nearly stoichiometric sample.
However, the IMT can be induced by a laser beam of 1mW power, 6328
Å wave length.  The resistivity which is higher than $10^{12}$ Ωcm at
T < 50 K is lowered by illumination to $5.10^{10}$ Ω cm at T = 30 K.
The occurence of the IMT reveals the existence of oxygen vacancies;
since the IMT does not appear without any laser illumination, we
can conclude that the concentration of such vacancies is low.  This
is corroborated by measurements of the absorption spectra reported
in Fig. 2, for a 200 μ thick sample, cut in the same ingot and
polished.  No $Eu^{3+}$ absorption peak is observed at 2μ.  The absorption coefficient (roughly the ratio of optical density by sample
thickness) measured at a wavelength $\lambda$ = 2 μ and at 300 K is 50 $cm^{-1}$;
this value is slightly larger than that of stochiometric samples
(about 30 $cm^{-1}$ on some of our samples).  A peak at 2.2 μ is well
observed which corresponds to one of the two peaks (at 1.85 μ and
2.20 μ) due to oxygen vacancies (2).  We can deduce from these
results that our sample is slightly Eu rich EuO but nearer to
stoichiometry than these studied by Helton and al.(3).

    3) Reflectivity spectra.

The reflectivity measurements have been achieved in the spectral range [1.5 - 4μ] using a Perkin Elmer model 112 monochromater equipped with a $CaF_2$ prism, a thermocouple as detector, and a globar rod as a light source. A magnetooptic cryostat, described elsewhere (7) was used to perform the measurements in the whole range [4.2 - 300 K]. Reflectivity spectra in the range [60- 300 K] are reported in Fig. 3. In this whole range of temperature, two dips at 1.85 and 2.20 μ are observed. The energy position of the dips varies very slowly with temperature. More accurately, there is a blue shift of the dip at 1.85 μ and a red shift of the dip at 2.20 μ by an amount of 0.10 μ when cooling from 300 K to 60 K. These dips obviously correspond to the absorption peaks at the same wavelengths which are not observed in stoichiometric samples and corroborate the existence of oxygen vacancies in this sample, contrary to the assertion made in reference (8).

At 4.2 K, the reflectivity spectra, shown in fig. 4, still present the two dips. Moreover, two extra structures appear at 1.74 μ and 2.00 μ which have also been observed in the absorption spectra by Helton et al (3). At 4.2 K in the Voigt configuration, the applied magnetic field up to 28 kG does not significantly shift the different dips but alters their intensities. Nevertheless, the too small resolution of our spectrometer makes impossible any accurate determination of the profile of the dips and no further comments will be made on this point.

The structures appearing between 2.8 and 3 μ are irreproducible results, due to absorption by $H_2O$ and $CO_2$.

## DISCUSSION

The structures at 1.85 μ and 2.20 μ were first observed by Torrance and al. (2) on absorption spectra at room temperature. These authors imputed this effect to a transitional of the outer electron of the oxygen vacancy and proposed an interpretation based on a model of a bound magnetic polaron (BMP) around the vacancy. The structures observed in optical spectra can then be imputed to a transition of this outer electron to a spin down state, reverse to the 4 f spins of the Eu atoms in the polaron. According to such a model, these structures should not be observed in the metallic configuration where BMP are no longer present. Helten and al. (3) could report absorption spectra in the whole range [4.2 - 300 K], and observed that the IMT does not affect the absorption peaks. Their interpretation based on the hypothesis of a $F^+$ center raises some questions as it has been noticed by themselves. The main problem is that the intensity of the structures correspond to a forbidden transition although it should be an allowed transition in a $F^+$ center.

We propose a different model to overcome these problems. In

the range of paramagnetic temperatures, the most loosely bound electron at the oxygen vacancy is bound on a donor level. The energy
difference $E_d$ between this donor level and the bottom of the conduction band is mainly determined by the exchange energy $\sim 0.6$ eV,
the half of which is gained in the formation of the polaron above
the Curie temperature ($E_d \sim E_{ech}/2$). Assuming the existence of
acceptors, since impurities or defects are always present, the
activation energy $E_a > 0.3$ eV, evidenced by the temperature dependence of the resistivity $\rho$, will depend on the degree of compensation.
For a high concentration of oxygen vacancies, the extrinsic resistivity satisfies the law $\rho(T) \propto \exp)-E_a/kT)$ and the activation energy
is $E_a \sim E_d \sim 0.3$ eV as excepted. For a quasi stoichiometric sample,
the resistivity is of an intrinsic type, $E_a \sim E_{gap}/2 \sim 0.6$ eV. In our
sample which contains a low concentration of oxygen vacancies, the
temperature dependence of the resistivity is intermediate between
the completely extrinsic and the intrinsic regime above described.
In other words, we can still write $\rho \propto \exp(-E_a/kT)$, but $E_a$ depends
on temperature, and ranges from $E_a \sim 0.3$ eV at $T \sim 200$ K, to
$E_a \sim 0.5$ eV at 300 K, due to the fact that the contribution from
f levels to the conductivity is no longer negligible at room temperature. We have also observed that the laser illumination of this
sample results in a lowering of its activation energy down to 435 meV
at room temperature. This can be explained by the shift of the
Fermi level induced by the enhancement of the donor concentration.

According to the model of exchange resonnance, the most loosely
bound electron can reverse its spin. The energy involved in this
transition is roughly 0.6 eV (i.e the exchange energy), and, in
fact, is split by spin orbit interaction so that two optical transitions at 550 meV and 650 meV can be observed. It is worth noticing
that such energies are not correlated to the activation energy $E_a$
measured from the resistivity curves in the paramagnetic configuration, in the sense that they do not depend on the degree of compensation. They have been actually observed at 550 meV and 650 meV
in our sample with $E_a \sim 480$ meV and in samples of Ref. 3, with
$E_a \sim 300$ meV and $E_a \sim 0$.

Below the IMT temperature, the outer electron of the vacancy is
delocalized, provided that the concentration of vacancies exceeds
that of the Mott transition. At such temperatures, however, the
materials is ferromagnetic, and when, for simplicity, we assume that
the conduction band is split as if it was a s-type band, the spin-
splitting of the conduction band is so large that the electron gas
is completely polarized in the spin up subband (9). Since the energy
required to flip the spin of these electrons is still 0.6 eV, the
same optical transitions as those observed in the insulating configuration at higher temperatures should be observed. Only the nature of the transitions has changed, since they should be interband
transitions between spin up and spin down subbands in the metallic

configuration. The intensity of the transitions for samples with either high or small concentrations of vacancies favours our model.

It should be noticed that two extra peaks are observed at very low temperatures, the existence of which cannot be explained in the oversimplified model above described. In fact the conduction band is not a s but a d-type band, and the six $t_{2g}$ states build six sub-bands instead of two. In the ferromagnetic configuration, the lower edge of the total band originates from the three spin up subbands. Notting and Schoenes (10) have shown that, at least in EuS, there is a crossing of the spectral weights of these subbands when the spin polarization $\sigma$ increases from 0.9 to 1, and the same behavior should be excepted for EuO. It is not unplausible to assume that such a phenomena results in extra optical transition between spin up and spin down subbands in the range of temperatures corresponding to $0.9 < \sigma < 1$, i.e at $T < 30$ K, ressponsible for the two extra peaks observed in the spectra.

## References

1.  M.R. Oliver, J.O. Dimmock, T.B. Reed, IBM J. Res. Develop. 14: 276 (1970)
2.  J.B. Torrance, M.W. Shafer, T.R. Mc. Guire, Abstract in : Bull. Amer. Phys. Soc. 17: 315 (1972).
3.  M. Helten, P. Grunberg, W. Zinn, Physica 89B: 63 (1977).
4.  C.F. Guerci, M.W. Shafer, J. Appl. Phys. 37: 1406 (1966).
5.  E.J. Huber, C.E. Holley, J. Chem. Thermodyn. 1: 301 (1969).
6.  M.W. Shafer, J.B. Torrance, T. Penney, J. Phys. Chem. Solids 33: 2251 (1972).
7.  M. Escorne, A. Mauger, J. Phys. E., in press.
8.  J.P. Desfours, J.P. Nadaï, M. Averous, C. Godart, Solid State Comm. 20: 691 (1976).
9.  A. Mauger, C. Godart, M. Escorne, J.C. Achard, J.P. Desfours, J. Phys. 39: 1125 (1978).
10. W. Nolting, J. Phys. C., 11: 1427 (1978) - Solid State Physics- 11: 1427 (1978). J. Schoenes and W. Nolting, J. Appl. Phys. 49: 1466 (1978).

# VIBRATIONAL SPECTRA OF TERNARY RARE EARTH SULFIDES WITH $CaFe_2O_4$, $Yb_3S_4$, $MnY_2S_4$ and $ZnGa_2S_4$ TYPE STRUCTURES

Hong Lee Park, Catherine A. Chess and William B. White
Materials Research Laboratory, The Pennsylvania State
University, University Park, PA 16802   USA

## ABSTRACT

Raman spectra of $CaFe_2O_4$ and $Yb_3S_4$ type compounds are sharp
and similar, indicating similar and ordered structures.  Spectra of
$MnY_2S_4$ type compounds exhibit three broad bands indicative of structural disorder.  $ZnGa_2S_4$ is an intermediate spectrum.  There is a
linear relationship between the highest frequency IR and Raman
bands and the average coordination numbers of the structures.

## INTRODUCTION

Ternary chalcogenides of $A_2BX_4$ stoichiometry are represented
by many structure types, some isostructural with the better-known
oxide structures and some that are unique.  The investigation
reported here deals with the vibrational spectra of ternary sulfides
of the rare earths belonging to the $CaFe_2O_4$, $Yb_3S_4$, $MnY_2S_4$ and thiogallate structure types.  Earlier reports describe the vibrational
spectra of $Th_3P_4$ type structures (1,2), their defect solid solutions
(3) and compounds of the spinel type (4,5)

The $Th_3P_4$ structure is related to fluorite.  The $CaFe_2O_4$ and
$Yb_3S_4$ structures are similar except for the 8-6 and 7-6 coordination.
The $MnY_2S_4$ structure, 6-6 coordination, is said to be disordered.
The thiogallate structure is an ordered derivative of sphalerite
with 4-4 coordination.  For further detail see the review by Flahaut
(6,7).  The number and selection rules for vibrational modes of
these structures were calculated by standard methods (Table 1).

All compounds were prepared by dry-firing oxides and carbonates
in flowing $H_2S$ at 1000 - 1100° C.  Infrared and Raman spectroscopic
measurements are described in references (2,5).

Table 1. Crystallographic and Vibrational Properties of Sulfides.

| Structure Type | Space Group | Factor Group | Z | CN | Vibrational Modes |
|---|---|---|---|---|---|
| $Th_3P_4$ | $I\bar{4}3d$ | $T_d$ | 4 | 8-8 | $A_1(R) + 2A_2(-) + 3E(R) + 5T_1(-) +$ $5T_2(IR + R)$ |
| $CaFe_2O_4$ | Pnam | $D_{2h}$ | 4 | 8-6 | $14A_g(R) + 14B_{1g}(R) + 7B_{2g}(R) +$ $7B_{3g}(R) + 7A_u(-) + 6B_{1u}(IR) +$ $13B_{2u}(IR) + 13B_{3u}(IR)$ |
| $Yb_3S_4$ | Pnam | $D_{2h}$ | 4 | 7-6 | $14A_g(R) + 7B_{1g}(R) + 14B_{2g}(R) +$ $7B_{3g}(R) + 7A_u(-) + 13B_{1u}(IR) +$ $6B_{2u}(IR) + 13B_{3u}(IR)$ |
| $MnY_2S_4$ | $Cmc2_1$ | $C_{2v}$ | 4 | 6-6 | $13A_1(IR + R) + 7A_2(R) + 6B_1(IR + R)$ $+ 13B_2(IR + R)$ |
| Spinel | Fd3m | $O_h$ | 8 | 6-4 | $A_{1g}(R) + E_g(R) + T_{1g}(-) + 3T_{2g}(R) +$ $2A_{2u}(-) + 2E_u(-) + 4T_{1u}(IR) + 2T_{2u}(-)$ |
| $ZnGa_2S_4$ | $I\bar{4}$ | $S_4$ | 2 | 4-4 | $3A(R) + 6B(IR + R) + 6E(IR + R)$ |

## RESULTS AND DISCUSSION

Raman spectra of selected compounds are given in figures 1-3; IR spectra in figure 4. Both Raman and IR spectra of the $CaFe_2O_4$ and $Yb_3S_4$ compounds are rather similar as expected from the structural similarities. Between 10 and 14 Raman bands are observed, 14 being the number predicted in any one polarization direction. This is a common result from powder spectra of anisotropic crystals and suggests that vibrations along different crystallographic directions are in fact often nearly degenerate. Although some fine structure is resolved in the IR spectra, only a small fraction of the predicted number of modes is observed.

The greatest contrast between predicted and observed behavior occurs for the $MnY_2S_4$ structure type. Instead of 40 Raman bands, 33 of which are also IR active, figure 3 shows only three broad peaks. The infrared spectrum is a structureless broad absorption. The Raman lines are a factor of 10 broader than those of the $CaFe_2O_4$ structure and remain broad to liquid nitrogen temperature.

The Raman spectrum of $ZnGa_2S_4$ (Fig. 5) appears to be a superposition of some 7 sharp bands against a background of perhaps 5 rather broad bands. The IR spectrum appears as a single broad absorption with little resolved structure.

The rather striking contrast between the Raman spectra of the

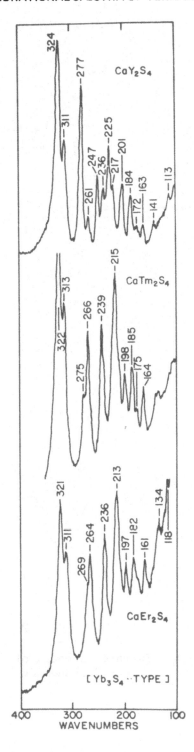

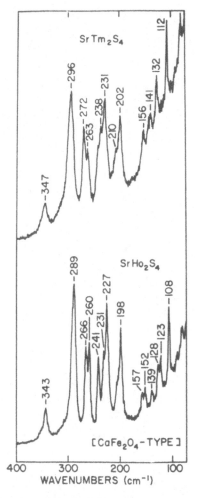

Figure 1. (Above) Raman spectra of compounds with CaFe₂O₄ type structure.

Figure 2. (Left) Raman spectra of compounds with Yb₃S₄ type structure.

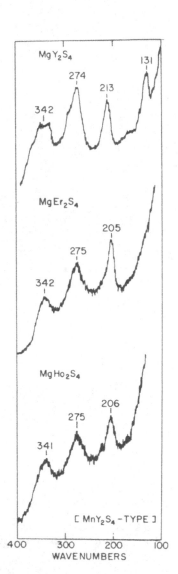

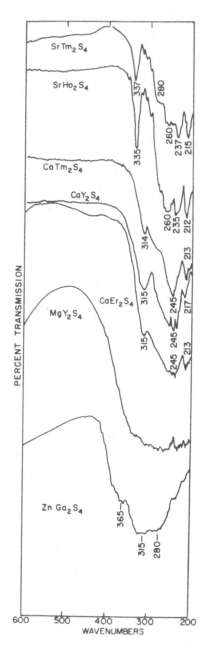

Figure 3. Raman spectra of
compounds with the
MnY$_2$S$_4$ type structure.

Figure 4.   Infrared spectra of
selected compounds

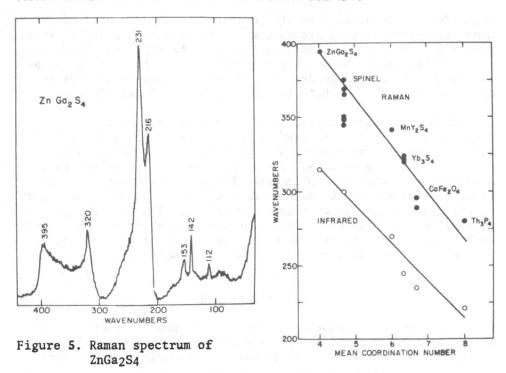

Figure 5. Raman spectrum of
ZnGa2S4

Figure 6. Plot of principal IR bands and highest frequency Raman
band against average coordination number

MnY2S4 compounds and compounds with the CaFe2O4 and Yb3S4 type
structures can be interpreted as evidence of pronounced cation disorder in the MnY2S4 structure as Flahaut (7) suggested, whereas
the CaFe2O4 and Yb3S4 compounds are highly ordered.  The absence
of line-narrowing with decreasing temperature is further evidence
that the disorder is structural and is locked in.  The remarkable
Raman spectrum of ZnGa2S4 is more difficult to interpret.  ZnGa2S4
is an ordered derivative of the sphalerite structure and requires
the ordering of one vacancy per three cations in addition to the
ordering between the cations themselves.  The contrast in line shape
between the sharp bands and the broader underlying bands suggests
that this sample is either a mixture of the highly ordered phase
with a partially disordered one or that the grains contain partially
disordered regions.  The fact that at least five broad bands can be
discerned is evidence that the disordered form is another intermediate structure and not sphalerite.  The sphalerite structure, a
diatomic crystal, would produce only a single broad band.

     The highest frequency bands in the Raman spectrum and the
center of the broad IR absorption envelope are a rough measure of
a characteristic stretching frequency for each structure.  Stretch-

ing modes are determined by atomic masses and by the bond force constants which in turn are determined by distances and coordination numbers.  The later are important because the effective force constant of any cation-anion pair will decrease as the cation charge is shared with more ligands.  The ternary sulfide structures are compared in figure 6.  Data for spinels and $Th_3P_4$ are from (5,2).

An average coordination number has been defined as [CN(M) + 2CN(Ln)]/3.  The Raman bands are better resolved and as a result there is more scatter when the highest frequency Raman peak is plotted against $\overline{CN}$.  The IR frequency was selected as the mean of the intense absorption band because some compounds have resolved fine structure and some do not.  The frequency shift with coordination number is much more systematic.  The mass effect can be seen in the range of Raman frequencies for a set of six spinel compounds.  Mass effects play a role in determining the frequencies but they are smaller than the coordination number effect.

## ACKNOWLEDGEMENTS

This work was supported by DARPA under sub-contract through the General Electric Company, Re-Entry and Environmental Systems Division.  We thank Patricia DeNoble for measuring the IR spectra.

## REFERENCES

1.  S.I. Boldish, P.L. Provenzano, L.E. Drafall and W.B. White. Proc. 11th Rare Earth Research Conf., Vol. 2, 938 (1974).
2.  P.L. Provenzano, S.I. Boldish and W.B. White.  Mat. Res. Bull. 12, 939 (1977).
3.  P.L. Provenzano and W.B. White.  Proc. 12th Rare Earth Research Conf.  Vol. 2, 522 (1976).
4.  S.I. Boldish and W.B. White. in The Rare Earths in Modern Science and Technology, G.J. McCarthy and J.J. Rhynne, eds.  Plenum Press,  New York, 607 (1978).
5.  S.I. Boldish and W.B. White.  J. Solid State Chem. 25, 121 (1978)
6.  J. Flahaut and P. Laruelle. in Progress in the Science and Technology of the Rare Earths, L. Eyring, ed., Pergamon Press, London, Vol. 3, 149 (1968).
7.  J. Flahaut. in Progress in the Science and Technology of the Rare Earths,  L. Eyring, ed., Pergamon Press, London, Vol. 3, 209 (1968).

# GROUP THEORETICAL EXAMINATION OF THE FAR INFRARED SPECTRA OF

$Tm_2(SO_4)_3 \cdot 8\,H_2O$ and $TmCl_3 \cdot 6\,H_2O$

Burton W. Berringer

Community College of Allegheny County

Boyce Campus, Monroeville, Pa. 15146

In 1972 the far infrared spectra of hydrated tripositive thulium sulfate and chloride were presented without analysis (1,2). The work was done to observe the spectra directly and support the validity of vibrational assignments made on the basis of vibronic spectra of these crystals observed in the visible region.

The aim of this work is to determine the number of lines expected in the infrared vibrational spectra. The correlation method of Fately et. al. (3,4) has been used to determine the number of infrared lines to be expected in the spectra of hydrated tripositive thulium sulfate and chloride crystals. The space group symmetry for these crystals has been reported (5) to be $C_{2h}^1$ for the chloride and $C_{2h}^6$ for the sulfate. The vibrations of the atoms or molecular units in the crystal modify the vibrations expected for a specific space group symmetry. This requires that the site symmetry be a subgroup of the space group of the crystal (3). To this end, one must consider the site symmetry of each atom or molecular unit in the Bravais cell and correlate the site symmetry with the space group symmetry.

The space group $C_{2h}^6$ of the hydrated sulfate, for example, can have the following site symmetries (4):

$$C_{6h}^2 : 4\ C_1(2);\ \underline{C_2(2)};\ \underline{C_1(4)}.$$

The coefficient indicates that there are four distinct $C_1$ sites in the $C_{2h}^6$ space group. Each of these sites has two equivalent atomic units associated with it. Not all four equivalent sites need be occupied but the number in parentheses must be the same as the number of equivalent units at the site. The underlined quantities correspond

to axes that can accomodate and infinite number of distinct sets with
the specified symmetry (4). Each set contains a number of units
corresponding to the number in parentheses. Thus, for this set of
site symmetries, there is an infinite number of distinct $C_2$ sites
each of which contains two equivalent units and an infinite number of
distinct $C_1$ sites each of which contains four equivalent units.

There are two $Tm_2(SO_4)_3 \cdot 8H_2O$ molecules per Bravais cell. The
correlation method suggests that the Tm atoms could lie on two distinct $C_2$ sites with two equivalent atoms per site. The $SO_4$ units
could, according to this scheme, occupy three distinct sites of $C_1$
symmetry each with two equivalent units. The water molecules could
occupy four distinct $C_1$ sites each with four equivalent units. The
irreducible representation of the vibration for this placement of
atoms in the crystal is

$$\Gamma^{vib} = 14\ A_g + 16\ B_g + 22\ A_u^{i.r.} + 24\ B_u^{i.r.}$$

This suggests that the crystal would have 22 vibrational species $A_u$
and 24 $B_u$ species which are infrared active. So 46 infrared lines
would be expected. If, on the other hand, the rare earth atoms lie
on a $C_s$ site as suggested by Stöhr, Seidel and Gruber(5) the irreducible representation is

$$\Gamma^{vib} = 19\ A_g + 20\ B_g + 16\ A_u^{i.r.} + 21\ B_u^{i.r.}$$

This representation will result if the Tm atoms lie on two distinct
$C_s$ sites each with two equivalent atoms; $SO_4$ on three distinct $C_2$
sites with two equivalent units per site; and water on four distinct
$C_i$ sites each with four equivalent units per site. Note in this
instance 16 $A_u$ vibrations and 21 $B_u$ vibrations would produce 37 lines
that are infrared active. The frequencies observed in the far infrared and optical spectra of the hydrated sulfate are presented in
Table I.

The space group symmetry of the hydrated thulium trichloride is
$C_{2h}^1$ (5). The possible site symmetries for this space group are (4)

$$C_{2h}^1 : 8C_{2h}(1);\ 4C_2\ (2);\ 2C_s(2);\ C_1(4)$$

For this crystal there are two molecules per Bravais cell. The con-
straints presented by the site symmetries suggest that the thulium
atoms occupy $C_2$ sites, four chlorine atoms bonded to the thulium (6)
occupy $C_1$ sites; two chlorine atoms bonded to hydrogen atoms occupy
$C_s$ sites and twelve water molecules occupy six distinct $C_2$ sites each
with two equivalent units. In this case the irreducible representation is

$$\Gamma^{vib} = 13\ A_g + 10\ A_u^{i.r.} + 16\ B_g + 19\ B_u^{i.r.}$$

Table I.   Vibrational Levels of Hydrated Thulium Chlorides and Sulfates (a)

| $Tm_2(SO_4)_3 \cdot 8H_2O$ | | $TmCl_3 \cdot 6H_2O$ | |
|---|---|---|---|
| Optically deduced Vibronic frequency $cm^{-1}$ | Observed IR Spectrum $cm^{-1}$ | Optically deduced Vibronic frequency $cm^{-1}$ | Observed IR Spectrum $cm^{-1}$ |
| 424 | | | |
| 390 | | 732 | |
| 383 | | 577 | |
| 363 | 355 | 517 | |
| 315 | | 419 | |
| 284 | 279 | 347 | |
| 275 | | 308 | 304 |
| 237 | 242 | 260 | 250 |
| 204 | 210 | 244 | 227 |
| 191 | 193 | 197 | 200 |
| 166 | 165 | 155 | 165 |
| 142 | 142 | 120 | 120 |
| 124 | 133 | 106 | 106 |
| 114 | 117 | 71 | 68 |
| 90 | 95 | 43 | 50 |
| 66 | 67 | | |
| 60 | 62 | | |
| 52 | 57 | | |
| 45 | 47 | | |
| 35 | | | |
| 26 | | | |
| 20 | | | |
| 17 | | | |
| 8 | | | |
| 2 | | | |

(a) From references 1 and 2.

From this it can be seen that the 10 $A_u$ vibrations and the 19 $B_u$ vibrations result in 29 infrared active lines. The vibrational frequencies observed in the far infrared and optical spectra of the hydrated trichloride are presented in Table I.

Examination of Table I shows that the number of infrared active vibrational lines predicted for crystals with $C_{2h}$ symmetry is much larger than the number of lines observed in either the vibronic spectra or the directly observed far infrared spectra. This is true for both and sulfate and trichloride. This observation could lead to the conclusion that the space group symmetry may not be $C_{2h}$ for these crystals.

In 1974, Stöhr, Seidel, and Gruber(5) reported spectroscopic evidence for higher symmetry effects in monoclinic thulium compounds. Based upon spectroscopic evidence, they suggested that the symmetry of the hydrated trichlorides is $D_6$ and that of the hydrated sulfates is $C_{6v}$. When the correlation method is applied to these higher symmetries interesting results are obtained.

The space group of the hydrated sulfate crystal is suggested to be $C_{6v}$. The possible site symmetries for this space group are

$$C_{6v}^2 : C_6(2); C_3(4); C_2(6); C_1(12)$$

The $Tm_2(SO_4)_3 \cdot 8H_2O$ molecules can be accomodated in the following way. Four thulium atoms occupy $C_3$ sites; six $SO_4$ units occupy $C_2$ sites; 12 water units occupy $C_1$ sites and 4 water molecules occupy two distinct $C_6$ sites with two equivalent units per site. The irreducible representation is

$$\Gamma^{vib} = 12\ A_1^{i.r.} + 13\ A_2 + 12\ B_1 + 12\ B_2 + 9E_1^{i.r.} + 4E_2$$

The 12 $A_1$ species and $9E_1$ species provide 21 infrared active vibrations which suggests one should observe 21 lines in the infrared spectrum.

The space group of the hydrated trichloride is suggested to be $D_6$. The possible site symmetries are

$$D_6^6 : 4D_3(2); 2C_3(4); 2C_2(6); C_1(12)$$

The $TmCl_3 \cdot 6H_2O$ molecules can be accomodated in the following way. Two thulium atoms occupy $D_3$ sites, six chlorine atoms occupy $C_2$ sites and twelve water atoms on two distinct $C_2$ sites with six equivalent units per site.

For this situation, the irreducible representation is

$$\Gamma^{vib} = 3\ A_1 + 6\ A_2^{i.r.} + 4\ B_1 + 6\ B_2 + 9\ E_1^{i.r.} + 10\ E_2$$

The 6 $A_2$ and 9$E_1$ vibrational species that are infrared active in this representation predict that there should be 15 infrared active lines in the spectrum of the hydrated trichloride.

Reference to Tables I and II shows much closer agreement between the number of optically observed vibrations and the number of vibrations predicted. The number of optically observed vibrations for the trichloride is 14 and the theory predicts 15 infrared active vibrations. Theory predicts 21 infrared vibrations for the sulfate and 25 vibrations are observed in the optical spectrum. Some of the vibrations of the sulfate observed vibronically may be vibrations that are not infrared active. The number of directly observed infrared vibrational frequencies in both types of crystal is less than the number of vibronically observed frequencies for several reasons. The first reason is that the infrared spectra were obtained at room temperature, so some of the vibrational frequencies could be masked by thermal motion of the molecules. The optical spectra were obtained at liquid nitrogen and/or liquid helium temperatures so the weaker vibrational transitions could be seen more easily. Another reason is that the infrared spectra were obtained in the range for 30 to 400 $cm^{-1}$.

Table II. Comparison of the Number of Observed Lines with the Number Predicted by this Analysis

|  | $Tm_2(SO_4)_3 \cdot 8H_2O$ | $TmCl_3 \cdot 6H_2O$ |
|---|---|---|
| Vibronic | 25 | 14 |
| Directly Observed IR | 14 | 9 |
| Predicted | 46[a] |  |
| For $C_{2h}$ Symmetry | 37[b] | 29 |
| Predicted |  |  |
| For Higher Symmetry | 21[c] | 15[d] |

(a) Tm on $C_2$ site for $C_{2h}^6$ space group

(b) Tm on $C_s$ site for $C_{2h}^3$ space group

(c) For $C_{6v}^2$ space group symmetry

(d) For $D_6^6$ space group symmetry

Examination of Table I will show that some of the vibronically observed frequencies lie outside the range of the spectrograph used.

In summary, this analysis suggests that the crystalline space group symmetry is somewhat higher than the $C_{2_h}$ symmetry suggested by earlier workers. According to this analysis the hydrated trichlorides appear to have $D_6^6$ or $P6_322$ space group symmetry with the thulium atoms occupying sites of $D_3$ symmetry, the chlorine atoms occupying sites of $C_2$ symmetry and the water molecules occupying two distinct sites of $C_2$ symmetry. The hydrated sulfates appear to have space group symmetry $C_{6v}^2$ or P6cc. In this case the thulium atoms occupy sites of $C_3$ symmetry, the sulfate groups occupy sites of $C_2$ symmetry, twelve water molecules occupy sites of $C_1$ symmetry and four water molecules occupy two distinct sites with $C_6$ symmetry each with two equivalent units per site.

In future work we will attempt to obtain evidence to support these suggestions. Among other possibilities we will compare the hydrated thulium sulfate spectrum with that of yttrium sulfate. We will also substitute deuterium oxide for waters of hydration.

## REFERENCES

1. B.W. Berringer, J.B. Gruber, and E.A. Karlow, Far Infrared Spectra of Hydrated and Anhydrous Tripositive Thulium Sulfate, J. Inorg. Nucl. Chem., 34 : 2084 (1972)
2. B. W. Berringer, J.B. Gruber, D.N. Olsen, and J. Stöhr, Far Infrared Spectra of Hydrated and Anhydrous Thulium Chloride, J. Inorg. Nucl. Chem., 34 : 373 (1972)
3. W.G. Fately, Neil T. McDevitt, and Freeman F. Bently, Infrared and Raman Selection Rules for Lattice Vibration: The Correlation Method, Appl. Spectrosc., 25 : 155 (1971)
4. William G. Fately, Infinite Number of Sites for Bravais Space Cell, Appl. Sectrosc., 27 : 395 (1973)
5. Joachim Stöhr, Ernst Seidel and John B. Gruber, Spectroscopic Evidence for Higher Symmetry Effects in Monoclinic $Tm_2(SO_4)_3 \cdot 8H_2O$, J. Chem. Phys., 61 : 4820 (1974) and references therein.
6. J. B. Gruber, USAEC Progress Report, RLO-2012-3, p. 109 (1969)

RADIATIVE TRANSITION PROBABILITIES AND STIMULATED CROSS-SECTIONS OF

$Nd^{3+}$ IN $3Ga_2S_3 \cdot La_2S_3$(GLS) AND $3Al_2S_3 \cdot La_2S_3$(ALS) GLASSES*

R. Reisfeld**, A. Bornstein**, J. Flahaut*** and A.M.
Loireau-Lazac'h***
**Department of Inorganic and Analytical Chemistry
The Hebrew University of Jerusalem, Israel
***Laboratoire de Chimie Minerale Structurale associé
au CNRS, Faculté des Sciences Pharmaceutiques et
Biologiques, 4 avenue de 1'Observatoire, 75006 Paris
Cedex 06, France

ABSTRACT

Radiative transition probabilities, oscillator strengths, radiative lifetimes and branching ratios of $Nd^{3+}$ in GLS and ALS glasses were calculated for levels ranging between $160cm^{-1}$ - 18,000 $cm^{-1}$. The matrix elements for the transitions were calculated in the intermediate coupling scheme using the measured energies in the glasses. The intensity parameters were obtained from these matrix elements and experimentally obtained oscillator strengths. The cross sections of the $^4F_{3/2} \rightarrow ^4I_{9/2}$, $^4I_{11/2}$ and $^4F_{3/2} \rightarrow ^4I_{13/2}$ were calculated and compared to the equivalent quantities in oxide glass. The similarity and difference in the radiative characteristics of $Nd^{3+}$ between oxide and chalcogenide glasses will be discussed.

The importance of chalcogenide glasses doped by $Nd^{3+}$ (1), $Ho^{3+}$ (2) and $Er^{3+}$ (3) has been emphasized. The qualitative behavior of these glasses has been described recently (4,5).

In the present work we have calculated the relevant optical properties of the ALS and GLS glasses and compared them with similar properties in oxide (6) and fluorophosphate glasses (7).

The spectroscopic properties were calculated by the use of the Judd-Ofelt theory (8,9) as described in ref. 10.

---

*Partically supported by U.S. Army Contract No. DAERO-76-G-006

The formulae used for the calculations are:

Line strength of electronic transitions

$$S_{ed}(aJ;bJ') = e^2 \Sigma\Omega_\lambda |<f^N[\gamma SL] \; J||U^{(\lambda)}||f^N[\gamma'S'L'] \; J'>| \qquad (1)$$
$$\lambda = 2,4,6$$

Line strength of magnetic transitions

$$S_{md}(aJ;bJ') = (e^2h^2/4m^2c^2)|<f^N[\gamma SL]J||L+2S||f^N[\gamma'S'L'] \; J'>|^2 \qquad (2)$$

For the calculation of the spontaneous transition probability

$$A(aJ;bJ') = [64\Pi^4\nu^3e^2/3hc^3(2J+1)] \; x \; [1/9n(n^2+2)^2S_{ed} + n^3S_{md}] \qquad (3)$$
$$n = \text{refractive index, for ALS} = 2.15, \text{ for GLS} = 2.5$$

The branching ratio

$$\beta ij = \frac{Aij}{\Sigma Aij} \qquad (4)$$

Experimentally the oscillator strength is obtained from the absorption spectrum by $f = 4.318 \times 10^{-9} \int \epsilon \; (\nu)d\nu$

Intensity parameters $\Omega_x$ are obtained from

$$f(aJ;bJ') = \frac{8\Pi mc\sigma}{3h(2J+1)} \frac{(n^2+2)^2}{9n} \; \Sigma\Omega_\lambda|<f^N \; [S,L] \; J||U^{(\lambda)}||f^N(S'L')J'>| \qquad (5)$$

In the least square fitting of the intensity paratmeters a statistical weight was given for various bands. This is especially important for the transition $^4I_{9/2} \to {}^4F_{3/2}$ as the lasing occurs from this level.

The $\Omega_x$ parameters and the predicted spectral intensity together with the matrix elements (11) are presented in Table I*. The measured and calculated oscillator strengths together with the absorption and intensity parameters are presented in Table II.

The integrated cross-sections were calculated from

$$\Sigma\sigma d\lambda = \frac{\lambda^4 A}{8\Pi cn^2} \qquad (6)$$

The radiative properties of $Nd^{3+}$ in chalcogenide glass compared to ED-2 glass and other oxide glasses are presented in Table III.

* for the transitions from $^4F_{3/2}$, $^4I_{15/21}$, $^4I_{13/2}$ and $^4I_{11/2}$, The complete set of levels will be published later.

Table I. Predicted Intensities and $U_x(2)$ matrix elements of $Nd^{3+}$ in GLS and ALS

| Transition | $U_2^2$ | $U_4^2$ | $U_6^2$ | GLS Energy of level cm⁻¹ | GLS $S_{ed}$ x10²⁰ cm² | GLS $S_{md}$ x10²² cm² | GLS A sec⁻¹ | GLS β | ALS Energy of level cm⁻¹ | ALS $S_{ed}$ x10²⁰ cm² | ALS $S_{md}$ x10²² cm² | ALS A sec⁻¹ | ALS β |
|---|---|---|---|---|---|---|---|---|---|---|---|---|---|
| $^4F_{3/2} \rightarrow$ | | | | | | | | | | | | | |
| $^4I_{15/2}$ | – | – | 0.0190 | 6139 | 0.08 | – | 39 | 0.0028 | 6075 | 0.13 | – | 33 | 0.0033 |
| $^4I_{13/2}$ | – | – | 0.2357 | 4016 | 1.03 | – | 1350 | 0.0993 | 4012 | 1.55 | – | 1110 | 0.1132 |
| $^4I_{11/2}$ | – | 0.1421 | 0.3772 | 2048 | 2.28 | – | 6120 | 0.4494 | 2048 | 3.17 | – | 4670 | 0.4759 |
| $^4I_{9/2}$ | – | 0.2298 | 0.0670 | 165 | 1.30 | – | 6110 | 0.4485 | 165 | 1.55 | – | 4000 | 0.4076 |
| $^4I_{15/2} \rightarrow$ | | | | | | | | | | | | | |
| $^4I_{13/2}$ | 0.0196 | 0.1186 | 1.4545 | 4026 | 7.04 | 71.00 | 62 | 0.2994 | 4012 | 10.30 | 71.00 | 46 | 0.2839 |
| $^4I_{11/2}$ | – | 0.0110 | 0.4172 | 2048 | 1.88 | – | 110 | 0.5283 | 2048 | 2.80 | – | 87 | 0.5372 |
| $^4I_{9/2}$ | – | 0.0001 | 0.0448 | 165 | 0.20 | – | 36 | 0.1723 | 165 | 0.30 | – | 29 | 0.1790 |
| $^4I_{13/2} \rightarrow$ | | | | | | | | | | | | | |
| $^4I_{11/2}$ | 0.0257 | 0.1352 | 1.2390 | 2048 | 6.20 | 93.30 | 52 | 0.3117 | 2048 | 9.06 | 93.30 | 41 | 0.3018 |
| $^4I_{9/2}$ | 0.0001 | 0.0136 | 0.4545 | 165 | 2.06 | – | 115 | 0.6883 | 165 | 3.06 | – | 94 | 0.6982 |
| $^4I_{11/2} \rightarrow$ | | | | | | | | | | | | | |
| $^4I_{9/2}$ | 0.0195 | 0.1072 | 1.1651 | 165 | 5.72 | 69.60 | 48 | 1.0000 | 165 | 8.38 | 69.60 | 38 | 1.0000 |

Table II. Oscillator strengths (f), radiative transition possibilities (A) and integrated cross-sections ($\Sigma\sigma d\lambda$) of the absorption bonds of $Nd^{3+}$ in GLS and ALS.

| Transition | GLS | | | | | | ALS | | | | | |
|---|---|---|---|---|---|---|---|---|---|---|---|---|
| | Energy of level $cm^{-1}$ | $\lambda$ nm | $f_{meas.}$ x10⁶ | $f_{cal.}$ x10⁶ | A $sec^{-1}$ | $\Sigma\sigma d\lambda$ x10²⁶cm³ | Energy of level $cm^{-1}$ | $\lambda$ nm | $f_{meas.}$ x10⁶ | $f_{cal.}$ x10⁶ | A $sec^{-1}$ | $\Sigma\sigma d\lambda$ x10²⁶cm³ |
| $^4I_{9/2} \rightarrow$ | | | | | | | | | | | | |
| $^4G_{9/2}$ | 18755 | 533 | - | 10.33 | 4220 | 2.54 | 18721 | 534 | 9.51 | 9.81 | 2980 | 2.43 |
| $^4G_{7/2}$ | | | 2.62 | | 10600 | | | | | | 7450 | |
| $^2G_{7/2}$, $^4G_{5/2}$ | 16843 | 593.5 | 52.15 | 51.59 | 9730 / 50200 | 15.78 | 16798 | 595 | 52.21 | 51.45 | 6700 / 37300 | 15.82 |
| $^2H_{11/2}$ | 15950 | 627 | 0.49 | 0.30 | 313 | 0.1 | 15895 | 629 | - | 0.32 | 243 | 0.11 |
| $^4F_{9/2}$ | 14505 | 689.5 | 0.75 | 1.08 | 924 | 0.44 | 14695 | 681 | - | 1.17 | 759 | 0.47 |
| $^4F_{7/2}$ | 13212 | 757 | 11.12 | 8.86 | 6250 | 4.68 | 13222 | 756 | 11.01 | 9.37 | 5090 | 4.8 |
| $^4S_{3/2}$ | | | | | 46.1 | | | | | | 32 | |
| $^2H_{9/2}$ | 12306 | 827 | 12.95 | 13.52 | 1490 | 8.25 | 12316 | 812 | 13.06 | 13.86 | 1210 | 7.90 |
| $^4F_{5/2}$ | | | | | 6820 | | | | | | 5120 | |
| $^4F_{3/2}$ | 11272 | 887 | 4.89 | 4.75 | 2440 | 4.6 | 11239 | 890 | 4.14 | 4.23 | 1600 | 2.88 |
| $^4I_{15/2}$ | 6139 | 1629 | 0.17 | 0.39 | 57.4 | 0.85 | 6075 | 1646 | - | 0.43 | 46.2 | 0.84 |
| $^4I_{13/2}$ | 4016 | 2490 | 1.97 | 2.60 | 161 | 13.1 | 4012 | 2493 | - | 2.89 | 132 | 14.55 |
| $^4I_{11/2}$ | 2048 | 4883 | - | 3.88 | 57.5 | 64.37 | 2048 | 4883 | - | 4.18 | 45.8 | 74.71 |

Table III. Radiative properties of neodymium in glasses as a frnction of network forming anion.
The last three parameters are for $^4F_{3/2} \rightarrow {}^4I_{11/2}$

| | Chalcogenide glass | | Phosphate | Borate | Germanate | Silicate | Tellurite | Aluminate | Titanate | Fluoro-phosphate |
|---|---|---|---|---|---|---|---|---|---|---|
| | GLS | ALS | LHG5 | H13 | H12 | ED2 | 162M | LG5 | 1.220 | 1.223 |
| $\tau_{rad}(^4F_{3/2})$, msee | 0.077 | 0.100 | 0.346 | 0.419 | 0.435 | 0.326 | 0.239 | 0.392 | 0.309 | 0.371 |
| $\lambda_p$, nm | 1.077 | 1.077 | 1.054 | 1.061 | 1.062 | 1.062 | 1.063 | 1.069 | 1.064 | 1.054 |
| $\Delta\lambda_{eff}$, nm | 22 | 22 | 25.5 | 36.8 | 34.7 | 34.0 | 28.9 | 43.1 | 38.6 | 27.2 |
| $\sigma_p(10^{-20}cm^2)$ | 7.95 | 8.2 | 3.9 | 2.2 | 1.9 | 2.9 | 2.9 | 1.8 | 2.5 | 3.5 |

While the branching ratio is slightly in favor of $^4F_{3/2} \to {}^4I_{11/2}$ (the lasing transition) the significant fluorescence of $^4F_{3/2} \to {}^4I_{13/2}$ should also be mentioned. This transition is responsible for the emission at 1.37 microns and is of importance as a light source for the IR region.

A striking property of chalcogenide glass is the cross-section of stimulated emission which is higher in chalcogenide glass as compared to ED-2 glass. It should be noted that this property of the glass enables its use as a laser material.

Thermal and mechanical stability of these glasses should also be mentioned. The glass transition temperature is $t_g = 530^oC$.

## REFERENCES

1. R. Reisfeld & A. Bornstein. Chem.Phys.Lett. 47:194 (1977).
2. R. Reisfeld, A. Bornstein, J. Flahaut, M. Guittard & A.M. Loireau-Lozac'h. Chem.Phys.Lett. 47:408 (1977).
3. R. Reisfeld & A. Bornstein. J.Noncryst.Solids 27:143 (1978).
4. A. Bronstein, J. Flahaut, M. Guittard, S. Jaulmes, A.M. Loireau-Lozac'h, G. Lucazeau & R. Reisfeld. The Rare Earths In Modern Science & Technology. Ed. G.J. McCarthy & J.J. Rhyne, Plenum 599(1978).
5. R. Reisfeld, A. Bornstein, J. Bodenheimer & J. Flahaut. J. Luminescence 18/19:253 (1979).
6. W.F. Krupke. IEEE J. Quant.Electron. QE-10:450 (1974).
7. R.R. Jacobs & J.M. Weber. IEEE J.Quant.Electron. OE-12:102 (1976).
8. B.R. Judd. Phys.Rev. 127:750 (1962).
9. G.S. Ofelt. J.Chem.Phys. 37:511 (1962).
10. R. Reisfeld & C.K. Jørgensen, Lasers and Excited States of Rare Earths. Springer-Verlag, Heidelberg 1977.
11. N. Spector, C. Guttel and R. Reisfeld. Optica Purs y Aplic. 10:197 (1979.

LANTHANIDE HYDROXIDE HOST LATTICE EFFECTS IN THE EPR SPECTRA OF
$Gd^{3+}$ IONS

H.A. Buckmaster, W.J. Chang and V.M. Malhotra

Department of Physics, The University of Calgary

Calgary, Alberta, Canada   T2N 1N4

## INTRODUCTION

The search to elucidate the mechanism(s) that produce the
observed zero-field splittings of the ground state for S-state ions
$[Mn^{2+}(3d^5)^6S; Gd^{3+}(4f^7)^8S]$ under the influence of crystalline electric fields of various symmetries has proven to be one of the most
intractable problems in EPR spectroscopy.  Some success has been
achieved for $Mn^{2+}$ but $Gd^{3+}$ has proven to be more difficult as has
been shown by Wybourne (1), Buckmaster *et al.* (2), Newman (3) and
Smith *et al.* (4).  It appears that alternative experimental
approaches should be attempted so as to yield new information on
the host lattice on the impurity S-state paramagnetic ion.  One
approach which has been attempted during recent years was to study
the $Gd^{3+}$ ion in various lanthanide host lattices with isomorphic
crystal structures.  The expectation was that the functional relationship between the largest spin-Hamiltonian crystalline electric
field parameters (usually $B_{20}$ or $b_2^0$) and some physical property of
the lattice (usually the ionic radius of the lanthanide host)
would provide this new insight.  These studies have failed to ful-
fill this expectation since it has been found that although the
relationship is linear as expected, the slope varies in sign with
some, as yet unknown, physical characteristic of each lattice.  In
some cases, $[LnF_3$ (5), $LnCl_3 \cdot 6H_2O$ (6), $Ln_2Mg_3(NO_3)_{12} \cdot 24H_2O$ (7),
$Ln(NO_3)_3 \cdot 6H_2O$ (8)$]$ the magnitude of $B_{20}$ decreases monotonically
with increasing ionic radius as expected from a point charge lattice
model while the opposite behaviour is observed in $[Ln(C_2H_5SO_4)_3 \cdot
9H_2O$ (4), $Ln_2(SO_4)_3 \cdot 8H_2O$ (9)$]$.  It is for this reason that it was
considered profitable to investigate the $Ln(OH)_3$ system.  Measurements of the EPR spectra of $Gd^{3+}$ ions in $La(OH)_3$ and $Y(OH)_3$ have
been report by Scott (10) and in $Eu(OH)_3$ by Cochrane *et al.* (11).

However, no systematic study of this system has been attempted.

## CRYSTAL STRUCTURE AND EXPERIMENTAL PROCEDURE

The lanthanide hydroxides (Ln ≡ La, Ce, . . . , Yb, Lu, Y) are believed to be isostructural with space group $P^63/m$ (12). They form hexagonal crystals which have two molecules per unit cell with each Ln ion surrounded by 9 OH⁻ radicals, of which six are at 0.242 nm and three at 0.254 nm. The $Ln(OH)_9^{6-}$ complex has $C_{3h}$ symmetry. The $Ln(OH)_3$(Ln ≡ La, Eu, Ho, Y) single crystals doped with ∿ 1 wt% $Gd(OH)_3$ were grown using a hydrothermal technique described by Maroczkowski et al. (13).

The measurements reported in this paper were performed at 293K using a 34.2 GHz EPR synchronous spectrometer with a cylindrical $TE_{011}$ mode cavity. The microwave frequency was stabilized and measured to $1:10^7$ and the magnetic flux density was stabilized and measured to $1:10^5$.

## RESULTS

The EPR spectra of $Gd^{3+}$ ions in $Ln(OH)_3$ (Ln ≡ La, Eu, Ho, Y) exhibited the characteristic 7-line fine structure, and the angular variation showed that the dominant crystalline electric field spin-Hamiltonian parameter was $B_{20}$. The phenomenological spin-Hamiltonian for an ion with fictitious spin $\overline{S}$ = 7/2 in a crystalline electric field of $C_{3h}$ symmetry has the form

$$H = \beta\overline{B}\cdot\underline{g}\cdot\overline{S} + B_{20}T_{20}(S) + B_{40}T_{40}(S) + B_{60}T_{60}(S) + B_{66}[T_{66}(S)+T_{6-6}(S)]$$

where β is the Bohr magneton (13.9960 GHz/T), $\overline{B}$ is the magnetic flux density, g is the spectroscopic splitting factor, $\overline{S}$ is the fictitious or free-ion spin angular momentum, $B_{\ell m}$ is the zero-field crystalline electric field parameter and $T_{\ell m}$ is a tensor operator (14).

Table 1 gives the values of the various spin-Hamiltonian parameters $g_{||}$ and $B_{\ell m}$. The values of the ionic radii were obtained from Templeton and Dauben (15) and corrected for systematic error using values of Greenwood (16). The values obtained by Scott (10) and Cochrane et al. (11) are appended for comparison.

Figure 1 is a graph of the spin-Hamiltonian parameter $B_{20}$(GHz) as a function of the ionic radius R(nm) of the lanthanide hosts Ln ≡ La, Eu, Y, Ho. This data satisfies a linear relationship with slope, $d|B_{20}|/dR$ = 3.63 GHz/nm if $B_{20}$(La) is not included. The slope is positive rather than negative as would be expected

TABLE 1

SPIN-HAMILTONIAN PARAMETERS IN Ln(OH)$_3$ AT 293K*

| Parameters | La(OH)$_3$ | | Eu(OH)$_3$ | | Y(OH)$_3$ | | | Ho(OH)$_3$ |
|---|---|---|---|---|---|---|---|---|
| | Scott 77K | Present Work 293K | Cochrane et al. 77K | Present Work 293K | Scott 300K | Scott 77K | Present Work 293K | Present Work 293K |
| $g_{\parallel}$ | 1.992 (1) | 1.99225 | 1.990 (1) | 1.98728 | 1.992 (1) | 1.992 (1) | 1.98880 | 1.99907 |
| $B_{20}$ | −0.67069(489) | −0.65855 | −0.35468(293) | −0.34856 | −0.33340(14) | −0.31919(147) | −0.33008 | −0.32938 |
| $B_{40}$ | −0.00192(42) | −0.00165 | −0.0025 (33) | −0.00211 | −0.00167(25) | −0.00182(17) | −0.00165 | −0.00138 |
| $B_{60}$ | 0.00007(7) | 0.00007 | 0.00009(5) | 0.00008 | 0.00007(4) | 0.00009(3) | 0.00008 | 0.00009 |
| RMS Error per point | | 0.00049 | | 0.00103 | | | 0.00016 | 0.00468 |

*$F_{\ell m}$ and RMS are in GHz.

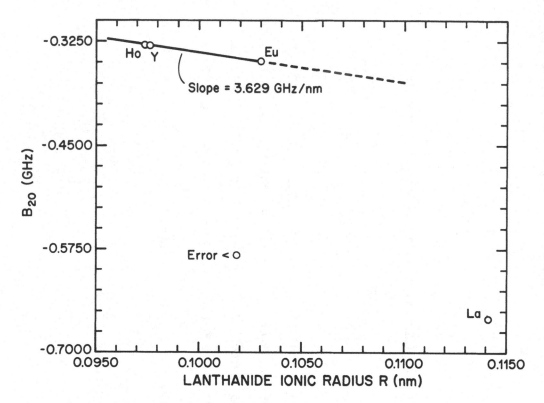

Figure 1.   Spin–Hamiltonian parameter $B_{20}$ describing the EPR spectra of $Gd^{+3}$ ion in $Ln(OH)_3$ (Ln = La, Eu, Y, Ho) as a function of the host's ionic radius.

from a point-charge model for the crystalline electric field. Consequently, it is confirmed that there exist two $C_{3h}$ symmetry systems, $Ln(C_2H_5SO_4)_3 \cdot 9 H_2O$ (4) and $Ln(OH)_3$ and one $C_{1h}$ symmetry system, $Ln_2(SO_4)_3 \cdot 8 H_2O$ (9), that behave contrary to the expectations of the point-charge model. It should be noted that a discontinuity exists in the host lattice data for $LnCl_3 \cdot 6 H_3O$ since there is a chemical and structural change for $Ln \equiv La$, Ce, Pr in $LnCl_3 \cdot 7 H_2O$, while for $Ln(NO_3)_3 \cdot 6 H_2O$ (8), a similar discontinuity exists for $Ln \equiv La$, Ce although there is no chemical change and no crystallographic information to indicate a change. Consequently it is suggested tentatively that $La(OH)$ may not have the same crystallographic structure as the other hydroxides. More EPR data on other $Ln(OH)_3$ is required as well as a detailed x-ray structure study. This paper demonstrates that EPR host lattice studies can be a useful tool in the determination of structural changes.

## ACKNOWLEDGEMENTS

The authors are indebted to NSERC and The University of Calgary for financial support. One of us (V.M.M.) is the recipient of a Shared Cost Postdoctoral Fellowship from The University of Calgary.

## REFERENCES

1. B. G. Wybourne, "Energy levels of trivalent gadolinium and ionic contributions to the ground state splitting," Phys. Rev. 148:317 (1966).

2. H. A. Buckmaster, R. Chatterjee and Y. H. Shing, "Ethylsulfate host lattice effects in the EPR spectra of Gd$^{3+}$ ions," Can. J. Phys. 50:991 (1972).

3. D. J. Newman, "Interpretation of S-state ion E.P.R. spectra," Advan. Phys. 24:793 (1975).

4. M. R. Smith, Y. H. Shing, R. Chatterjee and H. A. Buckmaster, "Ethyl sulfate host lattice effects in the EPR spectra of Gd$^{3+}$ ions. II," J. Magn. Reson. 26:351 (1977)

5. V. K. Sharma, "EPR of Gd$^{3+}$ in some rare earth trifluoride single crystals," J. Chem. Phys. 54:496 (1971).

6. V. M. Malhotra, H. D. Bist and G. C. Upreti, "Electron paramagnetic resonance of Gd$^{3+}$ in single crystals of some lanthanide tricholride hexahydrates," J. Magn. Reson. 27:439 (1977).

6a. S. K. Misra and G. R. Sharp, "EPR of Gd$^{3+}$ in trichloride

hexahydrates of Tb, Dy, Ho, Er, Tm and Yb, and systematics of
spin-Hamiltonian parameters in rare earth trichloride hexahydrates," J. Phys. C. 10:897 (1977).

7.  S. Misumi, T. Isobe and T. Higa, "Electron spin resonance of
    Gd(III) in single crystals of some lanthaoid(III) double-
    nitrates, $Ln_2Mg_3(NO_3)_{12} \cdot 24H_2O$," Nippon Kagaku Kaishi (Japan),
    11:2039 (1973).

8.  S. K. Misra and P. Mikolajczak, "Systematics of EPR spectra of
    $Gd^{3+}$ in rare earth trinitrate hexahydrate hosts," J. Chem. Phys.
    69:3093 (1978).

9.  V. M. Malhotra, H. D. Bist and G. C. Upreti, "Host lattice
    effects in the EPR spectra of $Gd^{3+}$ ion in some hydrated lan-
    thanide sulfates," J. Chem. Phys. 69:1919 (1978).

10. P. D. Scott, "Spin resonance of $Gd:Y(OH)_3$ and $Gd:La(OH)_3$,"
    J. Chem. Phys. 54:5384 (1971).

11. R. W. Cochrane, C. Y. Wu and W. P. Wolf, "Electron paramagnetic
    resonance of $Gd^{3+}$ pairs in rare-earth hydroxides," Phys. Rev.
    B 8:4348 (1973).

12. D. R. Fitzwater and R. E. Rundle, "Crystal structure of hydrated
    erbium, yttrium and praseodymium ethylsulfates," Zeitschrift
    für Kristallographie 112:362 (1959).

13. S. Mroczkowski, J. Eckert, H. Meissner and J. C. Doran, "Hydro-
    thermal growth of single crystals of rare earth hydroxides,"
    J. Cryst. Growth 7:333 (1970).

14. H. A. Buckmaster, R. Chatterjee and Y. H. Shing, "The applica-
    tion of tensor operators in the analysis of EPR and ENDOR
    spectra," Phys. Status Solidi (a) 13:9 (1972).

15. D. H. Templeton and C. H. Dauben, "Lattice parameters of some
    rare earth compounds and a set of crystal radii," J. Amer.
    Chem Soc. 76:5237 (1954).

16. N. N. Greenwood, "Ionic Crystals, Lattice Defects and
    Nonstoichiometry," Butterworths, London (1968) pp.40-41.

# FLUORESCENCE AND PHOSPHORESCENCE QUENCHING OF L-(-)-TRYPTOPHAN IN DMSO BY Eu$^{3+}$

J. Chrysochoos and V. Anantharaman

Department of Chemistry

University of Toledo, Toledo, Ohio 43606

## ABSTRACT

The fluorescence emission of tryptophan in DMSO at room temperature, with $\lambda_{max}$ = 355$\pm$1 nm, becomes weaker in the presence of Eu$^{3+}$ due to an enhancement of intersystem crossing via an external heavy atom perturbation. The corresponding phosphorescence spectrum of tryptophan, with maxima at about 407, 435 and 460 nm, also becomes weaker in the presence of Eu$^{3+}$ due to energy transfer from the $^3L_a$ or $^3L_b$-states of tryptophan to Eu$^{3+}$, leading to sensitized fluorescence monitored at 591$\pm$1 nm ($^5D_0 \rightarrow {}^7F_1$). The energy transfer rate constant is given by $k_{ET} \geq 3 \times 10^4$ M$^{-1}$ sec$^{-1}$.

## INTRODUCTION

The fluorescence lifetime and quantum yield of tryptophan depend very strongly upon the nature of the microenvironment in which tryptophan residues are located. Variations in the viscosity of the microenvironment (1), the polarity of the solvent (2) and the excitation wavelength (3,4,5) are accompanied by pronounced changes in the spectroscopic properties of the tryptophan moieties. In view of the role of tryptophan as a probe of the structural characteristics of many biomolecules, it is imperative that the spectroscopy of tryptophan in such microenvironments is thoroughly understood. The well defined spectroscopic properties of the lanthanide ions coupled with their magnetic properties make them ideal spectroscopic and magnetic probes of tryptophan in particular and of biomolecules in general. Some preliminary results on fluorescence and phosphorescence quenching of tryptophan by Eu$^{3+}$ will be discussed. Experimental details were given elsewhere (2).

RESULTS AND DISCUSSION

The fluorescence emission spectrum of tryptophan in DMSO has a maximum at 355 nm which is in agreement with the polarity of DMSO (2). The spectral distribution of the fluorescence spectrum is independent of the excitation wavelength. The fluorescence emission resulting from light excitation of tryptophan at 260 nm ($^1L_a \leftarrow {}^1A_O$) is relatively weak, characterized by a fluorescence lifetime of 17.6 nsec (6). On the other hand, a much stronger fluorescence emission is obtained under light excitation at 290 nm ($^1L_b \leftarrow {}^1A_O$) characterized by a shorter fluorescence lifetime (6) namely 8.3 nsec. The intensity of the fluorescence emission decreases in the presence of $Eu^{3+}$, although the spectral distribution does not change. Some typical results are plotted in Fig. 1(A) at $\lambda exc = 290\pm1$ nm.

The phosphorescence spectra of tryptophan in DMSO were recorded under similar excitation conditions at 77 K (liquid $N_2$). The phosphorescence spectrum consists of a structured band with components separated by 1468, 1436 and 1425 $cm^{-1}$, respectively (Fig. 1 B). A similar progression of 1350 $cm^{-1}$ was observed in the phosphorescence spectrum of indole (7) in EG-W at 77 K. The intensity of phosphorescence decreases in the presence of $Eu^{3+}$, whereas the corresponding spectral distribution remains unchanged. Very weak sensitized fluorescence arising from the $^5D_O$ state of $Eu^{3+}$ was observed at $[Eu^{3+}]$ higher than $8 \times 10^{-3}$ M at 77 K. Due to the weakness of both the phosphorescence of tryptophan and the sensitized fluorescence of $Eu^{3+}$ under these conditions, no correlation was established. However, the sensitized fluorescence of $Eu^{3+}$ is much stronger at room temperature in spite of the dissolved $O_2$. A typical spectrum is shown in Fig. 2. Although the sensitized fluorescence spectrum is weak, it is quite measurable. A correlation between the intensity of the fluorescence of tryptophan and the corresponding intensity of the sensitized fluorescence of $Eu^{3+}$ at room temperature is depicted in Fig. 3. It is quite apparent that the decrease in the intensity of fluorescence of tryptophan does not match the build-up of the sensitized fluorescence of $Eu^{3+}$. In fact the latter builds-up at a much slower rate. These preliminary observations are attributed to a heavy atom effect upon the intersystem crossing of excited tryptophan and energy transfer from the triplet state of tryptophan to $Eu^{3+}$. The results observed can be accounted for via the mechanism

$$^1A_O \xrightarrow{\text{290}\pm1 \text{ nm}} {}^1L_b$$

$$^1L_b \rightarrow {}^1A_O + h\nu_{fl}; \ k_{fl}$$

$$^1L_b \rightarrow ({}^3L_a, \ {}^3L_b); \ k_{isc}$$

$$^1L_b + Eu^{3+} \rightarrow ({}^3L_a, \ {}^3L_b); \ k_{isc}^{ind}$$

$$({}^3L_a, \ {}^3L_b) \rightarrow {}^1A_O + h\nu_{ph}; \ k_{ph}$$

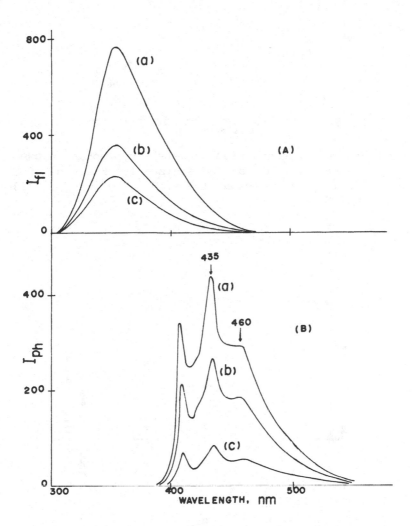

Figure 1: (A) Fluorescence spectra of L-(-)-Tryptophan in DMSO in the presence of $Eu^{3+}$ at room temperature. (B) Phosphorescence spectra of L-(-)-Tryptophan in DMSO in the presence of $Eu^{3+}$ at 77 K; $2 \times 10^{-4}$M Tryptophan; $\lambda_{exc}$ = 289+1 nm. (a) No $Eu^{3+}$, (b) $2 \times 10^{-3}$M $Eu^{3+}$, (c) $8 \times 10^{-3}$M $Eu^{3+}$. Fluorescence and phosphorescence spectra of L-(-)-Tryptophan have been corrected for the direct absorption of light by $Eu^{3+}$ in DMSO.

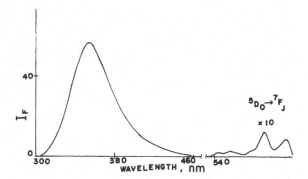

Figure 2:  Fluorescence spectrum of L-(-)-Tryptophan in DMSO in the presence of $Eu^{3+}$ and sensitized fluorescence of $Eu^{3+}$; $2 \times 10^{-4}M$ tryptophan; $1.6 \times 10^{-2}M$ $Eu^{3+}$; $\lambda_{exc} = 289\pm1$ nm; $T=22^{\circ}C$.

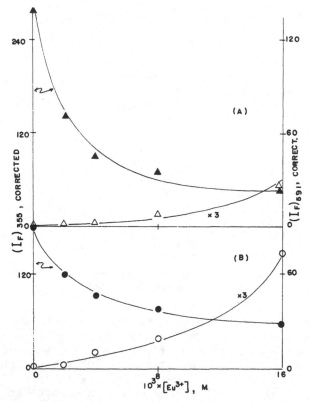

Figure 3:  Corrected fluorescence intensities of L-(-)-Tryptophan, $(I_F)_{355} \times (O.D)^{289}_{total}/(O.D)^{289}_{trypt}$, and of $Eu^{3+}$, $(I_F)_{591} \times (O.D)^{289}_{total}/(O.D)^{289}_{trypt}$, vs $[Eu^{3+}]$. (A) $3 \times 10^{-4}M$ Tryptophan; $\lambda_{exc} = 289\pm1$ nm, $T=22^{\circ}C$ (B) $2 \times 10^{-4}M$ Tryptophan; $\lambda_{exc} = 289\pm1$nm; $T=22^{\circ}C$.

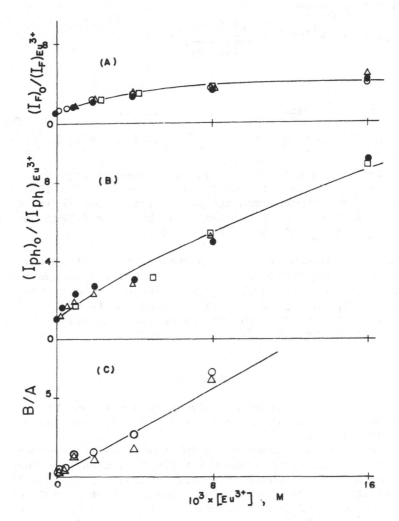

Figure 4:  Fluorescence and phosphorescence intensity ratios of L-(-)-Tryptophan in DMSO vs $[Eu^{3+}]$. (A) $(I_F)_o/(I_F)Eu^{3+}$ vs $[Eu^{3+}]$; $\lambda_{exc} = 289\pm1$ nm; $\lambda_{fl}=355\pm1$ nm; T=22°C (B) $(I_{ph})_o/(I_{ph})Eu^{3+}$ vs $[Eu^{3+}]$; $\lambda_{exc}=289\pm10$ nm; $\lambda_{ph}=435\pm1$ nm; 77K  (C) $(I_{ph})_o(I_F)Eu^{3+}/(I_{ph})Eu^{3+}(I_F)_o$ vs $[Eu^{3+}]$

■    $1 \times 10^{-4}$M Tryptophan          ●    $2 \times 10^{-4}$M Tryptophan

▲    $3 \times 10^{-4}$M Tryptophan          ○    $4 \times 10^{-4}$M Tryptophan

$$(^3L_a, \ ^3L_b) + Eu^{3+} \ (^7F_0) \rightarrow {}^1A_0 + Eu^{3+} \ (^5L_J, \ ^5G_J), \ k_{ET}.$$

This mechanism leads to

$$(I_F)_0/(I_F)Eu^{3+} = 1 + (\frac{k_{isc}^{ind}}{k_{isc}+k_{fl}}) \ [Eu^{3+}] \tag{1}$$

and

$$\frac{(I_{ph})_0/(I_{ph})Eu^{3+}}{(I_F)_0/(I_F)Eu^{3+}} = \frac{k_{isc}}{k_{isc} + k_{isc}^{ind} \ [Eu^{3+}]} + \frac{k_{isc} \ k_{ET}}{k_{ph}(k_{isc} + k_{isc}^{ind} \ [Eu^{3+}])} [Eu^{3+}] \tag{2}$$

Plots according to Eqs. (1) and (2) are shown in Fig. 4. Based upon $k_{fl} + k_{isc} = 1.2 \times 10^8$ sec$^{-1}$, a value $k_{isc}^{ind} \leq 7 \times 10^{10}$ sec$^{-1}$ is obtained. Furthermore, from the phosphorescence lifetime of tryptophan (6) in EG-W, namely 6.8 sec, and the value $\phi fl/\phi ph \approx 4.0$ the value $k_{ET} \geq 3 \times 10^4$ M$^{-1}$ sec$^{-1}$ is obtained.

The exact origin of the phosphorescence is not known. Work is underway regarding the dependence of the phosphorescence lifetime of tryptophan upon the excitation wavelength. Direct population of the $^1L_a$ and $^1L_b$ of tryptophan via the excited states of Eu$^{3+}$, by eliminating the participation of the $^1L_a$ and $^1L_b$ state of tryptophan is also being studied.

## REFERENCES

1.  I. Weinryb, "The Effect of Solvent Viscosity on the Fluorescence of Tryptophan Derivatives," Biochem. Biophys. Res. Comm. 34, 865 (1969).
2.  J. Chrysochoos, "Fluorescence Spectra of L-(-)-Tryptophan in Solution. Dependence Upon the Excitation Light and the Concentration of the Substrate", Mol. Photochem. 5, 1 (1973).
3.  N. Mataga, Y. Torihashi and K. Ezumi, "Electronic Structures of Carbazole and Indole and the Solvent, Effects on the Electronic Spectra", Theoret. Chim. Acta 2, 158 (1964).
4.  H. Zimmermann and N. Joop, "Polarization of Electronic Bands of Aromatics", Z. Elektrochem. 65, 61 (1970).
5.  P. S. Song and W. E. Kurtin, "A Spectroscopic Study of the Polarized Luminescence of Indoles", J. Am. Chem. Soc., 91 4892 (1969).
6.  W. X. Balcavage and T. Alväger, "On the Fluorescence Lifetime of Tryptophan in Proteins", Mol. Photochem. 7, 309 (1976).
7.  D. Muller, E. Ewald and G. Durocher, "Isothermal and Photosensitized Emissions of Indole and Tryptophan in Two Polar Matrices at 77 K", Can. J. Chem. 52, 407 (1974).

NEAR RESONANCE INTERACTIONS BETWEEN STARK COMPONENTS OF THE $^6H_{5/2}$

AND $^4G_{5/2}$ STATES OF Sm$^{3+}$ IN POCl$_3$:SnCl$_4$

J. Chrysochoos and P. Tokousbalides

Department of Chemistry

University of Toledo, Toledo, Ohio 43606

## INTRODUCTION

Light excitation of Sm$^{3+}$ ions in POCl$_3$:SnCl$_4$ leads to fluorescence emission arising exclusively from the lowest excited state of Sm$^{3+}$, namely the $^4G_{5/2}$-state. The fluorescence lifetime of the $^4G_{5/2}$-state of Sm$^{3+}$ in POCl$_3$:SnCl$_4$ is fairly long (1), approaching 3.3 msec at very low [Sm$^{3+}$], in contrast to the relatively short lifetimes of Sm$^{3+}$ ($^4G_{5/2}$) observed in such solvents as CH$_3$OH, C$_2$H$_5$OH, C$_3$H$_7$OH and H$_2$O. Values of $\tau_{fl}$ equal to 10.5, 10.0, 9.0 and < 3.0 µsec have been reported for Sm$^{3+}$ in these latter solvents (2). The fluorescence lifetime and quantum yield of the $^4G_{5/2}$-state of Sm$^{3+}$ in POCl$_3$:ZrCl$_4$ and POCl$_3$:TiCl$_4$ decrease with increasing [Sm$^{3+}$] due to the self-quenching (1,3) of the $^4G_{5/2}$ by the $^6H_{5/2}$-state of Sm$^{3+}$. In addition, the fluorescence is quenched by the solvent, $k_{sol}$, whose maximum value is 320 sec$^{-1}$ (4). The temperature dependence of the self-quenching process of the $^4G_{5/2}$-state of Sm$^{3+}$ will be discussed in this article. Experimental details regarding this study were given elsewhere (2,3,5).

## RESULTS AND DISCUSSION

The intensity of the fluorescence emission arising from the $^4G_{5/2}$-state of Sm$^{3+}$ in POCl$_3$:SnCl$_4$ was monitored at 597$\pm$1 nm ($^4G_{5/2} \rightarrow {}^6H_{7/2}$) using an excitation flash (100 µsec duration) at 402$\pm$1 nm. The decay of the fluorescence intensity was found to be exponential at all [Sm$^{3+}$] and temperatures. The value of the fluorescence lifetime, $\tau_{fl}$, decreases with [Sm$^{3+}$]. Plots of $\tau_{fl}^{-1}$ vs [Sm$^{3+}$] were used to determine the value of $k_{Sm3+}$ at several temperatures. Such values are summarized in the first column of Table 1.

Table 1.  Values of the Fluorescence Quenching Rate Constant, $k_{Sm^{3+}}$, of the $^4G_{5/2}$-state of $Sm^{3+}$ in $POCl_3:SnCl_4$

| T, $^\circ$C | $k_{Sm^{3+}}$ <br> $(M^{-1}sec^{-1})$ | $(k_{Sm^{3+}})_{corr} = k_{Sm^{3+}})\{N(^6H_{5/2})_{lowest}/N(^6H_{5/2})_{upper}\}$ <br> $(M^{-1}sec^{-1})$ |
|---|---|---|
| -15 | $5.0 \times 10^2$ | $2.8 \times 10^3$ |
| -10 | $4.6 \times 10^2$ | $2.5 \times 10^3$ |
| - 2 | $4.7 \times 10^2$ | $2.45 \times 10^3$ |
| +11 | $5.3 \times 10^2$ | $2.70 \times 10^3$ |
| 15 | $5.9 \times 10^2$ | $2.8 \times 10^3$ |
| 21 | $5.95 \times 10^2$ | $2.7 \times 10^3$ |
| 30 | $6.1 \times 10^2$ | $2.7 \times 10^3$ |
| 40 | $7.3 \times 10^2$ | $3.0 \times 10^3$ |
| 49 | $7.45 \times 10^2$ | $3.0 \times 10^3$ |
| 55 | $8.0 \times 10^2$ | $3.1 \times 10^3$ |

The fluorescence emission of $Sm^{3+}$ can be accounted for via the following mechanism

$$Sm^{3+}(^6H_{5/2}) \xrightarrow{\lambda = 402 \pm 1 \text{ nm}} (Sm^{3+})^* \rightsquigarrow Sm^{3+}(^4G_{5/2})$$

$$Sm^{3+}(^4G_{5/2}) \rightarrow Sm^{3+}(^6H, ^6F) + h\nu_{fl}; \ k_{fl} \qquad (1)$$

$$Sm^{3+}(^4G_{5/2}) + Sm^{3+}(^6H_{5/2}) \rightarrow 2 \ Sm^{3+}(^6H, ^6F); \ k_{Sm^{3+}} \qquad (2)$$

$$Sm^{3+}(^4G_{5/2}) + solvent \rightarrow Sm^{3+}(^6H, ^6F); \ k_{solv} \qquad (3)$$

This mechanism leads to

$$\tau_{fl}^{-1} = k_{fl} + k_{solv} + k_{Sm^{3+}}[Sm^{3+}] \qquad (4)$$

Plots of $\log_{10}k_{Sm^{3+}}$ vs $1/T(K^{-1})$ can be employed to determine $E_A$. Such plots are shown in Fig. 1(A) leading to an apparent value of $E_A$ equal to 1.3 kcal/mole.  The dependence of $k_{fl} + k_{solv}$ upon the temperature is depicted in Fig. 1(B).

Reaction (2) can be rewritten as follows taking into consideration the energy levels of $Sm^{3+}$ in crystals (6)

$$Sm^{3+}(^4G_{5/2}) + Sm^{3+}(^6H_{5/2}) \rightarrow 2 \ Sm^{3+}(^6F_{9/2}) \qquad (5)$$

$$Sm^{3+}(^4G_{5/2}) + Sm^{3+}(^6H_{5/2}) \rightarrow Sm^{3+}(^6F_{5/2}) + Sm^{3+}(^6F_{11/2}) \qquad (6)$$

$$Sm^{3+}(^4G_{5/2}) + Sm^{3+}(^6H_{5/2}) \rightarrow Sm^{3+}(^6F_{11/2}) + Sm^{3+}(^6F_{5/2}) \qquad (7)$$

The energy mismatch between the remaining states in the ground state manifold as $Sm^{3+}$ and the $^4G_{5/2}$-state is too large to be

bridged by the participation of the host.   Processes (5) through
(7) involve coupling of the following transitions:   ${}^4G_{5/2} \rightarrow {}^6F_{9/2}$
with ${}^6F_{9/2} \leftarrow {}^6H_{5/2}$, ${}^4G_{5/2} \rightarrow {}^6F_{5/2}$ with ${}^6F_{11/2} \leftarrow {}^6H_{5/2}$ and ${}^4G_{5/2} \rightarrow$
${}^6F_{11/2}$ with ${}^6F_{5/2} \leftarrow {}^6H_{5/2}$, respectively.   Some of the transitions
involved cannot be studied by direct observation of the corre-
sponding spectra.   However, the corresponding energies can be obtained indirectly via emission and near infrafred absorption
spectra of $Sm^{3+}$ in $POCl_3:SnCl_4$.

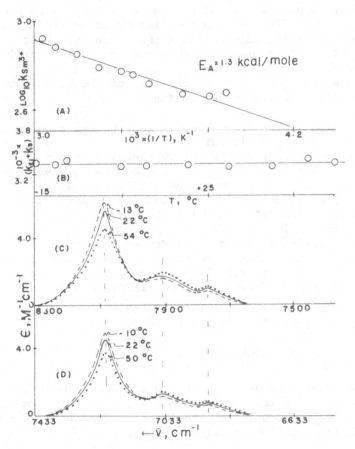

Figure 1:   (a)  Variation of $Log_{10}k_{Sm}{}^{3+}$ with $1/T$ $(K^{-1})$.   (B)  Variation of $k_{fl} + k_s$ with $T$ (°C); $4 \times 10^{-2}M$ $Sm^{3+}$ in $POCl_3$:
$SnCl_4$ (10:1 v/v); $\lambda_{exc}=402\pm1$ nm; $\lambda_{fl}=597\pm1$ nm   (C)  Near
infrared absorption band of $Sm^{3+}$ in $POCl_3$:$SnCl_4$ at several
temperatures (${}^6F_{7/2} \leftarrow {}^6H_{5/2}$)   (D)  Near infrared absorption band of $Sm^{3+}$ in $POCl_3:SnCl_4$ at several temperatures
(${}^6F_{5/2} \leftarrow {}^6H_{5/2}$).

The near infrared spectra of $Sm^{3+}$ in $POCl_3:SnCl_4$ consist of a series of bands attributed to the $(^6H, {}^6F) \leftarrow {}^6H_{5/2}$ transitions of $Sm^{3+}$ (5). Each band consists of three components separated by about 160 and 150 $cm^{-1}$, respectively. At lower remperatures the main component gains intensity at the expense of the other two components. Two such typical absorption bands attributed to the ${}^6F_{7/2} \leftarrow {}^6H_{5/2}$ and ${}^6F_{5/2} \leftarrow {}^6H_{5/2}$ transitions are shown in Figs. 1(C) and (D). The spectral characteristics of four of the most distinctive near infrared absorption bands are summarized in Table 2. Since all the bands considered are associated with the ${}^6H_{5/2}$-state (initial) of $Sm^{3+}$ but with different final states, the differences observed are attributed to the splitting of the ${}^6H_{5/2}$-state of $Sm^{3+}$ by the solvation sphere, into three components separated by 160 and 150 $cm^{-1}$, respectively in $POCl_3:SnCl_4$. The effect of the temperature is attributed to differences in the Boltzmann population of the three components. This is also verified by the temperature independence of the total absorption envelope, given in terms of the integral $\int \varepsilon(\lambda)d\lambda$, in spite of the pronounced changes of the intensities of the components. Some of these results are summarized in Table 3.

Combination of the near infrared and the emission spectra lead to energy determinations illustrated in Fig. 2. It is apparent that process (5) corresponds to the energy ranges 8,192 to 9,123 $cm^{-1}$ and 8,772 to 9,524 $cm^{-1}$, respectively. At very low temperatures the limit of the ${}^4G_{5/2} \rightarrow {}^6F_{9/2}$ approaches 8,192 $cm^{-1}$ whereas that of ${}^6F_{9/2} \leftarrow {}^6H_{5/2}$ approaches 9,524 $cm^{-1}$ leading to maximum energy mismatch of about 1,300 $cm^{-1}$. Under such conditions process (5) becomes very improbable. At higher temperatures the energy of the ${}^4G_{9/2} \rightarrow {}^6F_{9/2}$ shifts to higher energies whereas that of the ${}^6F_{9/2} \leftarrow {}^6H_{5/2}$ shifts to lower ones. The energy mismatch diminishes, near resonance conditions are established and process (5) becomes more effective. Similar effects take place in processes (6) and (7),

$${}^4G_{5/2} \rightarrow {}^6F_{5/2}; \quad \Delta E = 10,108 \text{ to } 11,048 \text{ cm}^{-1}$$
$${}^6F_{11/2} \leftarrow {}^6H_{5/2}; \quad \Delta E = 10,204 \text{ to } 10,869 \text{ cm}^{-1}$$

and

$${}^4G_{5/2} \rightarrow {}^6F_{11/2}; \quad \Delta E = 6,705 \text{ to } 7,645 \text{ cm}^{-1}$$
$${}^6F_{5/2} \leftarrow {}^6H_{5/2}; \quad \Delta E = 6,802 \text{ to } 7,463 \text{ cm}^{-1}$$

Maximum energy mismatches of about 760 $cm^{-1}$ make these two processes ineffective at low temperatures. However, at higher temperatures near resonance conditions are established. Therefore, fluorescence self-quenching of $Sm^{3+}$ gains significance.

The population of the highest energy component of the ${}^6H_{5/2}$-state of $Sm^{3+}$ rather than the total $[Sm^{3+}]$, is more closely associated with the self-quenching process. Let $N({}^6H_{5/2})_{upper}$ and $N({}^6H_{5/2})$ lowest represent the appropriate Boltzmann distributions leading to the corrected value of $k_{Sm^{3+}}$

$$(k_{Sm^{3+}})_{corr} + (k_{Sm^{3+}})_{app} \text{ X } (N({}^6H_{5/2})_{lowest}/N({}^6H_{5/2})_{upper}) \quad (8)$$

Table 2: Spectral characteristics of the near infrared absorption bands of $Sm^{3+}$ in $POCl_3:SnCl_4$, $POCl_3:ZrCl_4$ and $POCl_3:TiCl_4$; $Sm^{3+}$ = 1 x $10^{-2}$M, T=22°C. The frequencies tabulated represent peak maxima.

| Transition | $POCl_3:SnCl_4$ $[POCl_3]/SnCl_4] = 10.34$ | | | $POCl_3:ZrCl_4$ $[POCl_3]/[ZrCl_4] = 45$ | | | $POCl_3:TiCl_4$ $[POCl_3]/[TiCl_4] = 10.7$ | | |
|---|---|---|---|---|---|---|---|---|---|
| | $\bar{\nu}(cm^{-1})$ | $\Delta\bar{\nu}(cm^{-1})$ | $\varepsilon(M^{-1}cm^{-1})$ | $\bar{\nu}(cm^{-1})$ | $\Delta\bar{\nu}(cm^{-1})$ | $\varepsilon(M^{-1}cm^{-1})$ | $\bar{\nu}(cm^{-1})$ | $\Delta\bar{\nu}(cm^{-1})$ | $\varepsilon(M^{-1}cm^{-1})$ |
| $^6F_{1/2} \leftarrow ^6H_{5/2}$ | 6281 | | 0.20 | 6273 | | 0.46 | 6275 | | 0.55 |
| | | 162 | | | 152 | | | 143 | |
| $^6F_{3/2} \leftarrow ^6H_{5/2}$ | 6443 | | 0.90 | 6425 | | 1.0 | 6418 | | 1.45 |
| | 6544 | | 0.60 | 6551 | | 0.90 | 6564 | | 1.30 |
| | | 165 | | | 147 | | | 135 | |
| $^6F_{5/2} \leftarrow ^6H_{5/2}$ | 6709 | | 1.90 | 6698 | | 2.00 | 6699 | | 2.50 |
| | 6920 | | 0.30 | 6920 | | 0.50 | 6944 | | 0.75 |
| | | 147 | | | 140 | | | 123 | |
| | 7067 | | 1.20 | 7060 | | 1.60 | 7067 | | 2.00 |
| | | 166 | | | 155 | | | 145 | |
| $^6F_{7/2} \leftarrow ^6H_{5/2}$ | 7233 | | 4.60 | 7215 | | 3.70 | 7212 | | 4.70 |
| | 7794 | | 6.60 | 7797 | | 0.80 | 7812 | | 1.55 |
| | | 149 | | | 139 | | | 124 | |
| | 7943 | | 1.40 | 7936 | | 1.80 | 7936 | | 2.20 |
| | | 161 | | | 148 | | | 144 | |
| | 8104 | | 5.30 | 8084 | | 4.90 | 8080 | | 5.70 |

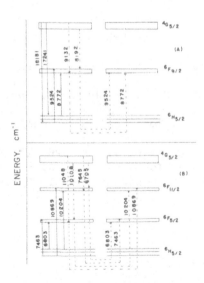

Figure 2: Energy levels participating in the fluorescence self-quenching of $Sm^{3+}$ in $POCl_3:SnCl_4$. The energy values assigned are based upon near infrared and emission spectra of $Sm^{3+}$.

Table 3.  Integrated Absorption Bands of $4\times10^{-2}M$ $Sm^{3+}$ in $POCl_3:SnCl_4$
(10:1 v/v) in the Near Infrared, Corrected for the Temp-
erature.  $10^6$ x $\int \varepsilon(\lambda)d\lambda M^{-1}$.

| T(°C)/ Transition | $^6F_{3/2}\leftarrow^6H_{5/2}$ (14,700–15,200 A) | $^6F_{5/2}\leftarrow^6H_{5/2}$ (13,400–14,700 A) | $^6F_{7/2}\leftarrow^6H_{5/2}$ (11,900–13,100 A) | $^6F_{9/2}\leftarrow^6H_{5/2}$ (10,500–11,400 A) | $^6F_{11/2}\leftarrow^6H_{5/2}$ (9,200–9,800 A) |
|---|---|---|---|---|---|
| 50   | 4.2 | 11.9 | 13.8 | 6.1  | 0.96 |
| 45   | 4.3 | 11.7 | 13.4 | 7.05 | 0.94 |
| 40   | 4.2 | 11.9 | 13.6 | 7.3  | 0.97 |
| 35   | 4.3 | 11.6 | 13.6 | 7.25 | 0.99 |
| 30   | 4.6 | 11.9 | 13.5 | 7.15 | 0.98 |
| 24.5 | 4.1 | 11.4 | --   | 7.3  | 0.89 |
| 15   | 4.3 | 11.8 | --   | 7.1  | 0.81 |
| 5    | 4.3 | 11.0 | --   | 7.4  | 0.81 |
| -6   | 4.2 | 10.9 | --   | 7.3  | 0.84 |
| -10  | 4.4 | 11.2 | --   | 7.3  | 0.82 |

Values of $(k_{Sm^{3+}})_{corr}$ are summarized in the second column of Table 1.
They are virtually temperature independent.  Consequently, the temperature dependence of the fluorescence self-quenching of $Sm^{3+}$ is
partly due to varying contributions of the Stark components of the
$^6H_{5/2}$-state of $Sm^{3+}$.


                              REFERENCES

1.  P. Tokousbalides and J. Chrysochoos, J. Phys. Chem. 76, 3397
    (1972).
2.  N. A. Kazanskaya and E. B. Sveshnikova, Opt. Spectrosc. 28, 376
    (1970).
3.  J. Chrysochoos and P. Tokousbalides, Spectrosc. Lett. 6, 435
    (1973).
4.  J. Chrysochoos, J. Chem. Phys. 70, 3264 (1979).
5.  J. Chrysochoos and P. Tokousbalides, J. Lumines.  (In Press).
6.  G. H. Dieke and H. M. Crosswhite, Appl. Optics 2, 675 (1963).

# OXIDATION OF CERIUM STUDIED BY PHOTOELECTRON SPECTROSCOPY

A. Plateau and S.-K. Karlsson

Department of Physics and Measurement Technology
Linköping University
S-581 83 Linköping, Sweden

Cerium is, except for europium, the most reactive rare earth metal. The present paper concerns photoemission measurements (XPS Mg K$\alpha$ and UPS, He I and He II) on evaporated cerium films after various oxygen exposures.

The aim of the work was to study the growth and the composition of the oxide film thus formed. After higher oxygen exposures (> 100 L, 1 L = $10^{-6}$ torr sec) the emission due to the metallic Ce 5d6s band vanishes, in both XPS and UPS, while the emission due to the Ce 3d and 4p levels (XPS) shows satellite structures, similar to those reported on some Ce(III)-compounds. Contrary to previous reports (1,2) we thus draw the conclusion that a protective layer- if existent - is thicker than the probing depth, determined by the electron escape length.

As an example of the results obtained, we present in Fig. 1 some XPS spectra of the 3d doublet of cerium. The emission due to the 3d doublet, fairly simple in the case of the clean metal (A), becomes a rather involved satellite structure when cerium is oxidized. Cerium becomes trivalent when oxidized at room temperature, and tetravalent when oxidized at elevated temperatures, as can be concluded from UPS and XPS measurements of the valence band (3) performed on the same films. Thus, the satellite spectra (C) and (D) are characteristic for $Ce_2O_3$ and $CeO_2$, respectively. The similarity of the spectra (D) and (F) is obvious, the apparent binding energy shift of (F) may be due to charging effects. The slight discrepancy between the spectra (C) and (E) is interpreted as being due to a surface contamination of the $Ce_2O_3$ powder grains by $CeO_2$, as a result of keeping the powder exposed to air prior to experiments. This interpretation is supported by spectrum (B), which shows a decaying $CeO_2$ phase at the surface immediately after exposture to oxygen.

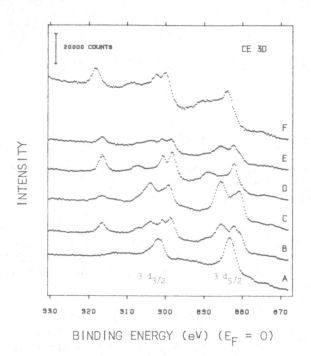

Fig. 1: XPS (hν = 1253.6 eV) spectra of the 3d doublet of cerium recorded under different conditions: A, immediately after evaporation, B, immediately after an exposure to 1000-L oxygen at room temperature, C, the same film as in B a few hours later, D, after further exposure to 1000-L oxygen with the Ce film heated to approximately 600° C, E, $Ce_2O_3$ powder, F, $CeO_2$ powder. All measurements were performed at room temperature.

REFERENCES

1) R.C. Helms and W.E. Spicer, Appl. Phys. Lett. 21: 237 (1972).

2) A. Plateau, L.I. Johansson, A.L. Hagström, S.-E. Karlsson and S.B.M. Hagström, Surf. Sci., 63: 153 (1977).

3) A. Plateau and S.-E. Karlsson, Phys. Rev. B, 18: 3820 (1978).

# SOFT X-RAY APPEARANCE POTENTIAL SPECTROSCOPY OF $^{70}$Yb[*]

D. Chopra and G. Martin

Physics Department, East Texas State University

Commerce, Texas 75428, USA

## ABSTRACT

The soft x-ray appearance-potential spectra (SXAPS) of
ytterbium (Yb) in $Yb_2O_3$ have been obtained in the 160-600 eV range
with a nondispersive spectrometer, which measures the derivative of
the total x-ray intensity of the anode surface as a function of
energy of incident electrons. The spectra represent the differential excitation probability of 4d core levels and give detailed
information regarding the self-convolution of the density of unoccupied conduction band states in ytterbium ions in the surface
region. The $N_{4,5}$ spectrum is a complex structure superimposed on
a broad band of 12.7 eV base width. The broad band exhibits two
peaks above the threshold, 3.1 eV apart. The spectral features
have been related to the exchange coupling between the 4f electrons
and 4d vacancies in Yb.

## INTRODUCTION

Soft x-ray appearance-potential spectroscopy (SXAPS) has been
regarded as an important core level spectroscopic technique for investigating the surface characteristics of solids (1-3). The
surface and near-surface properties of the solid material are of
considerable importance in modern technology since they govern
their interaction with the environment. The surface properties of
a solid are not amenable to theoretical predictions as are the bulk
properties. Considerable effort has, therefore, been made in the
development of a great variety of surface characterization techniques. Much of the interest in SXAPS has been centered in investigating core electron binding energies and conduction band

densities of states of the atoms in the surface region. In SXAPS
a fast electron (100-2000 eV) is used to excite an electron from
an inner core level, leaving the system in a final state which has
a core hole and up to two additional electrons in the conduction
band. The total x-ray yield produced when the core hole de-excites
is then measured as a function of incident electron energy and represents a measure of the cross section for exciting the core hole.
The information obtained is similar to that usually derived from
soft x-ray absorption (SXA) and isochromate measurements but is
much more sensitive to the surface region. The purpose of the
present paper is to investigate the SXAPS measurements of ytterbium
(Yb). The 4f subshell in metallic Yb is reported to be full with a
$4f^{14}6s^2$ configuration. Ytterbium metal has a divalent character,
but in compounds, Yb shows trivalent properties with a $4f^{13}$ configuration. Ytterbium is, therefore, particularly interesting for the
investigation of the spectral features associated with the so-called
filling of the 4f subshell in rare earths; i.e., the filled $4f^{14}$
compared to the $4f^{13}$ configurations.

## EXPERIMENT

The appearance-potential spectrometer which had been described
in detail elsewhere (4) consists simply of a vacuum diode and photoelectron collector which measures the soft x-ray fluorescence of an
electron-bombarded sample. The electron source was a tungsten
filament heated by 20V, 400Hz pulses having a duty cycle of eight
percent to improve the resolution by minimizing the energy spread
of the emitted electrons. The emission current was usually a few
mA. As the target was operated at relatively low potentials, it
was deemed necessary to regulate the emission current against fluc-
uations due to space charge limitation. This was accomplished by
operating the electron source in the close proximity of the sample
and also by using an emission control circuit which maintained
constant emission current irrespective of the sample potential.
The sample was screened from the photoelectron collector with a
grid which is maintained at -50 V relative to the filament to prevent the thermal electrons from reaching the detector assembly.
The appearance-potential spectrum was obtained by superimposing a
small oscillation of 1.0 $V_{p-p}$ on the target potential and detecting
the resulting synchronous variations in photocurrent with a
phase-lock amplifier. The derivative of the x-ray yield as a function of the incident electron energy was thus directly obtained
and plotted on an x-y recorder. The vacuum was kept on the order
of $10^{-9}$ Torr. A polycrystalline sample (10 x 10 x 0.75 mm) of
high purity Yb was mechanically polished, degreased in isopropyl
alcohol, sputtered and annealed to ensure a clean surface for
examination.

## RESULTS AND DISCUSSION

The soft x-ray appearance-potential spectra of N levels of Yb were obtained in the 160-600 eV range. Only the $N_5$ level exhibits a significant structure. Figure 1. shows the $N_{4,5}$ SXAPS of Yb in metal oxide. The spectrum consists of two positive peaks, B and C superimposed on a continuum and an undershoot. The continuum extends 12.7 eV above the threshold and results from the autoionization of $4d^9 4f^{N+1}$ to $4d^9 4f^N \epsilon f$ configurations where $\epsilon f$ is the continuum state. The principal peaks, 3.1 eV apart are well resolved and accompanied by a step-like threshold, A and a low shoulder peak, D on the high energy side. The binding energy of $N_5$ core level has been determined to be 179.5 ± 1.0 eV as compared to 184.9 ± 1.3 eV reported by Bearden and Burr (5) and 186 eV from photoelectron spectroscopy measurements (6). The main peaks seem to precede the excitation potential of the $N_5$ threshold and no contribution to the spectrum lies in the region corresponding to the $N_4$ threshold.

Trebbia and Colliex (7) have investigated the 4d excitation spectra of Yb by means of energy loss experiments and report a single maximum at 181 ± 1 eV which agrees satisfactorily with our principal peak B at 183.5 eV. Because of their moderate vacuum and poor energy resolution they were unable to detect additional structures. Fomichev et al. (8) report a main peak at 179.6 eV preceded by a weak secondary maximum at 171.3 eV in their $N_{4,5}$ level SXA spectra. The measurements of Fomichev et al. were performed at a pressure of $10^{-5}$ Torr which is relatively high for the rare earths which oxidize easily. Because of the vacuum conditions for many of these measurements, it is quite probable that their spectra are representative of the metal oxide.

We have not been able to keep Yb surface in pure metallic state long enough to obtain SXAPS from it. The sample oxidizes too rapidly even under a working vacuum of $10^{-9}$ Torr and exhibits a K level oxygen peak. The 4f levels in Yb are not localized and are, therefore, easily affected by oxidation. The $N_{4,5}$ region showed two peaks, 3.1 eV apart which are indicative of the tripositive ytterbium ion. It appears that a protective thin layer of $Yb_2O_3$ is formed on a clean Yb surface. The protective layer is at most a few monolayers thick. Since the surface sensitive SXAPS techniques probes a limited number of surface layers, the spectrum is likely to be representative of the metal oxide, $Yb_2O_3$. To verify the influence of oxygen contamination, the sample was heavily sputtered with argon ions and the spectrum recorded immediately afterwards. The $N_{4,5}$ SXAPS of Yb showed a decrease in intensity accompanied by a corresponding reduction in the oxygen K peak. After a few hours at $10^{-9}$ Torr, the intensities built up to previous levels. Similar changes in intensities were observed as a result of heating the sample to elevated temperature for several hours. This would be

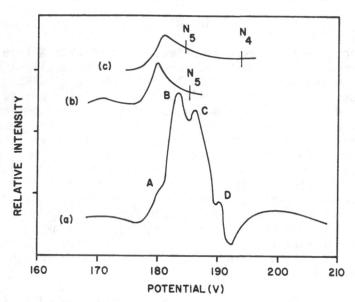

Fig. 1.    (a)  Ytterbium $N_{4,5}$ soft x-ray appearance potential spectrum from 160 to 210 eV (electron current 6 mA, modulation
potential 1.0 $V_{p-p}$).   The derivative of the photocurrent
(I) relative to electron accelerating potential (V) is
plotted vs. V.
(b)  Ytterbium $N_{4,5}$ soft x-ray absorption spectrum taken
from Ref. 8.
(c)  Ytterbium $N_{4,5}$ electron energy loss spectrum taken
from Ref. 7.

consistent with the assumption that a protective metal oxide layer
is formed even under ultra high vacuum conditions.  To provide further evidence that the spectrum is characteristic of the metal
oxide, the sample was exposed to 300-L oxygen.  There was no noticeable change in the $N_{4,5}$ SXAPS features of Yb and the relative intensities of the two peaks appeared to be independent of additional oxygen exposure, indicating that the oxygen on the metal is chemisorbed
instead of physisorbed and the thickness of the protective film remains essentially the same.  Physisorbed oxygen atoms are weakly
bound and would have been desorbed by prolonged exposure to incident electron current of a few mA.  For this reason it is believed
that only chemisorbed oxygen is detected as an oxygen K peak.

Sugan (9) has interpreted the $N_{4,5}$ multipeaked SXA spectra
spanning over 10-20 eV above the threshold, in the case of rare
earths, in terms of the atomic-like $4d^{10} 4f^N \rightarrow 4d^9 4f^{N+1}$
(N = o for La and N = 13 for Lu) transitions. The multiplet splitting in the excited $4d^9 4f^{N+1}$ configuration is caused by the extensive

overlap between the partially filled 4f subshells and 4d vacancies. The 5s and 5p electrons screen the 4f and lower subshells from the environment of the solid. The 4f electrons in rare-earth metals, therefore, behave like well localized atomic states. No edge structure is predicted since 5f and higher orbitals do not overlap 4d wave functions. It is, therefore, a reasonable assumption to treat the present solid state measurements in terms of the pure atomic model. The spectra of tripositive rare-earth ions are considered dependent on the degree of filling of the 4f subshell with increasing Z. When the 4f subshell becomes half full, there is a discontinuous change in the spectrum. Following the filling of the 4f subshell, the regularity in the spectra will again be interrupted. Metal Yb with a full 4f subshell cannot be expected to show any structures corresponding to $4d^{10}\ 4f^{N} \rightarrow 4d^{9}\ 4f^{N+1}$ transition (10). However, for $Yb_2O_3$ the excitation of a 4d electron allows the complete filling of the 4f subshell. As a result the exchange interaction between the paired electron distribution in the filled 4f subshell and the 4d hole vanishes leading to minimal or no splitting in the excited $4d^{9}\ 4f^{N+1}$ configuration. The decrease in exchange interaction may also be assumed due to the increase in the binding energy of the 4d electrons. It can be understood easily why, in our experimental data the spectrum is not rich in structure in contrast to spectra of other heavy rare earths with unfilled 4f subshell configurations. (8).

In SXAPS, however, the relative x-ray intensity is determined by the product of transition probabilities of incident and core electrons. The incident electron probes the spatially extended states in the conduction band of the solid. The core electron performs a transition to the excited state of the crystal, such as in SXA. The excited state for rare earths is localized consistent with Wendin's two-density of states model (11). Smith et al. (12) have calculated the SXAPS of La and Ce using the reported experimental isochromate and SXA data. The two density of states model calculation can explain only a part of the observed features. Because of the lack of such experimental data for Yb we have not been able to compare the observed spectra with theoretical predictions. However, for the electronic configuration of Yb oxide, it would be appropriate to consider the two-densities of states to be non-interacting. SXAPS would then represent the situation where the projectile electron will go into free electron-like states above the Fermi energy independent of the ion with electron-hole pair. In principle, the core electron can be excited to unoccupied 4f and 6p states. In the metal oxide where hybridization occurs, the transition to 5d and 6d states also becomes possible. However, due to the intensity of the sharp peaks observed in the present experiment, it is logical to assign them the 4d → 4f transition. The difference between SXAPS and SXA must account for the extra electron which is not present in the final state configuration of SXA. Since the relative transitions

are shielded by the centrifugal potential barrier and the 5s and 5p
subshells, the SXAPS and SXA structure should have one to one correspondence since they reflect the similar excited states of tripositive Yb ions.  The present SXAPS measurements correlate the experimental SXA data of Fomichev et al. and Trebbia et al. very well
as shown in Fig. 1.

It is possible that structures of the spectra correspond to the
$4d^{10} 4f^N \rightarrow 4d^9 4f^{N+1}$ transitions for the tripositive rare-earth ions
in the solid state lattice.  The simple one-electron model fails to
explain the 4d spectra of rare earths where localized, atomic-like
excitations are involved.  The nonavailavility of the values of
energies of the final $4d^9 4f^{N+1}$ configurations precludes a quantitative understanding of the spectra.  To arrive at a single model in
the case of rare earths a great effort, both theoretical and experimental, is still necessary.

## REFERENCES

\*     Work supported by the Robert A. Welch Foundation and an East
      Texas State University Grant.

1.    R.L. Park, Phys. Today 28, 52 (1975); Surf. Sc. 48. 80 (1975).

2.    R.L. Park and J.E. Houston, J. Vac. Sc. Technol. 11, 1974.

3.    A.M. Bradshaw, Surf, Defect. Prop. Solids, 3, 153 (1973).

4.    D. Chopra, H. Babb and R. Bhalla, Phys. Rev. B 14, 5231 (1976).

5.    J.A. Bearden and A.F. Burr, Rev. Mod. Phys. 39, 125 (1967).

6.    T.A. Carlson, Photoelectron and Auger Spectroscopy (Plenum Press,
      New York-London, 1975).

7.    P. Trebbia and C. Colliex, Phys. Stat. Sol. (b) 58, 523 (1973).

8.    V.A. Fomichev, T.M. Zimkina, S.A. Gribovskii and I.I. Zhukova
      Soviet Phys. Solid State 9, 1163 (1967).

9.    J. Sugar, Phys. Rev. B5, 1985 (1972).

10.   C.S. Fadley and D.A. Shirley, Phys. Rev. A2, 1109 (1970).

11.   G. Wendin, Proc. Intern. Conf. Vacuum Ultraviolet Radiation
      Physics, eds. E. Koch, R. Haensel and C. Kunz, Hamburg 1974
      (Vieweg Pergamon, 1974).

12.   R.J. Smith, M. Piacentini, J.L. Wolf and D.W. Lynch Phys. Rev.
      B 14, 3419 (1976).

# LUMINESCENCE PROCESSES IN Nb DOPED YVO$_4$

Dilip K. Nath

General Electric Company, Lighting Business Group

1099 Ivanhoe Road, Cleveland, Ohio 44110

## INTRODUCTION

Impurities like Nb$^{5+}$ or Ta$^{5+}$ (1) and Tb$^{3+}$ (2) in minor or trace quantities in the lattice of YVO$_4$(Eu) are known to increase the efficiency of Eu$^{3+}$ emission under 3650Å but not under 2537Å. Efforts have been made to explain the luminescence processes associated with the said impurities without systematically investigating the effect of these minor impurities on the luminescence characteristics of self activated YVO$_4$. In this paper, the absorption, excitation and emission spectra as well as life time of Nb doped YVO$_4$ will be presented to discuss the role played by these impurities in the luminescence processes of self-activated YVO$_4$.

## Experimental

Phase pure specimens were synthesized by standard solid state techniques. The phosphors were polycrystalline white powders with a size median 8-10µ.

Absorption and emission spectra were obtained from a modified Cary 14 spectrophotometer. The fluorescence spectra were measured by a spex 1402 monochromator with a low pressure Hg lamp and filter to isolate 2537 or 3650Å line. Fully corrected fluorescence spectra were obtained with a dedicated mini-computer. Life time were obtained with a box car integrator using TRW pulsed deuterium lamp with a pulse width approximately 10 nano-seconds.

## Results and Discussion

### 1.  Diffuse   Reflectance Spectra:

The absorption spectra of Nb doped $YVO_4$ showed a non-additive shift of the $VO_4^{3-}$ charge transfer edge towards longer wavelength indicating the formation of a complex between $NbO_4^{3-}$ and $VO_4^{3-}$ ions in their ground state (Figure 1).  Verification of the absence of any $V^{4+}$ or $Nb^{4+}$ was done by examining EPR spectra at room temperature and $77^oK$.  DeLosh et al (3) reported a similar shift of the absorption edge in $YV_{.7}P_{.3}O_4$ and $Y_{.98}Tb_{.02}V_{.7}P_{.3}O_4$ and referred to an interaction of Tb and V and not to Tb alone because the band was absent in $YPO_4$:Tb which also has the same zircon structure.  A similar absence of this absorption band in $YPO_4$:Nb indicates the non-additive change is not due to Nb alone but due to an inter-action between $NbO_4^{3-}$ and $VO_4^{3-}$ ions.  In Figure 2 the concentration dependence of the absorption strength (log k/s) showed a monotonic increase until the solid solubility limit of 2 atom% Nb is reached.  The lowering of absorption strength above 2 atom% Nb is due to the presence of a secondary phase $3Y_2O_3 \cdot Nb_2O_5$ isostructural with $3Y_2O_3 \cdot Sb_2O_5$ (4).

### 2.  Excitation and Emission Spectra and Lifetime:

The excitation and emission spectra of $YVO_4$, niobium doped $YVO_4$ and for comparison $YNbO_4$ are shown in Figure 3. The characteristic $VO_4^{3-}$ excitation band at 3200Å did not show any new structure when doped with Nb.  The emission spectra, on the other hand, showed a band at 4720Å in addition to the characteristic $VO_4^{3-}$ band at 4330Å.  The ratio of the intensities of these two bands did not change with Nb concentration.  The origin of these bands cannot be assigned to any transition due to $Nb^{5+}$ alone because the absorption and emission spectral overlap did not change by doping niobium in $YVO_4$.  Also a similar $VO_4^{3-}$ emission spectrum was obtained by doping $YVO_4$ with 0.01 atom% $Eu^{3+}$ as shown in Figure 4.

According to the configuration coordinate diagram (Figure 5) of Hsu and Powell (5), the lowest excited level $^1A_1(^1A_1)$ should emit through the transitions to the two crystal field components of the ground state namely, $^1E(^1T_1)$ and $^1A_2(^1T_1)$.  The reported literature shows the existence of the band at about 4330Å for the transition $^1A_1(^1A_1) \rightarrow ^1A_2(^1T_1)$ but no band at 4720Å due to the transition $^1A_1(^1A_1) \rightarrow ^1E_1(^1T_1)$ as evidenced in the present investigation.

The maximum efficiencies of $VO_4^{3-}$ emission in Nb doped YVO₄ were found to be 20% and 30% higher than pure YVO₄ under 2537 and 3650Å respectively, corresponding to an optimum Nb concentration of 0.01 atom% (Figure 6). The absorption strength of Nb doped YVO₄ was higher than pure YVO₄ under 3650Å but remained the same under 2537Å. Since the spectral overlap does not change, the process which increases the quantum efficiency of $VO_4^{3-}$ emission in Nb doped YVO₄ is not a radiative type.

The effect of Nb concentration on the relative yield under 2537Å and non-radiative life time of $VO_4^{3-}$ emission is shown in Table 1. The term $T_n'/T_n$ was derived from the afterglow equation as:

$$T_n'/T_n = \frac{\eta' T}{\eta T'} \quad \ldots\ldots\ldots\ldots (1)$$

Table 1. Effect of Nb Concentration on Relative Yield and Nonradiative Life Time

| Nb Concentration (Atom%) | Life Time $T_{obs}(10^{-6}$ sec) | Relative Yield Under 2537Å $\eta*$ | $\dfrac{T_n'}{T_n}$ |
|---|---|---|---|
| 0 | 17 | 0.19 | 1.00 |
| 0.002 | 21 | 0.21 | 1.37 |
| 0.01 | 23 | 0.23 | 1.62 |
| 0.02 | 21 | 0.21 | 1.37 |
| 0.10 | 13 | 0.17 | 0.67 |

\* Based on $YP_{.99}V_{.01}O_4$, $\eta = 1.00$

The results show a maximum of 1.62 times increase in the nonradiative life time of $VO_4^{3-}$ emission at a concentration of 0.1 atom% Nb.

The quenching rate of this nonradiative process can be expressed as:

$$I(t) = I_0 \exp - t \left(\frac{1}{T_0} + W\right) \quad \ldots\ldots (2)$$

If $\dfrac{1}{T_{obs}}$ = Observed decay time, then

$$\frac{1}{T_{obs}} = \frac{1}{T_o} + W \quad \ldots \ldots \ldots \quad (3)$$

where $\frac{1}{T_o}$ =    On-site de-excitation rate (radiative and nonradiative)

W    =    Quenching rate

The relative yield for $VO_4^{3-}$ emission is:

$$\eta = \frac{T_{obs}}{T_o} \quad \ldots \ldots \ldots \ldots \ldots \quad (4)$$

The quenching rate W can be deduced from equation (3) and (4) as:

$$W = \frac{1}{T_{obs}} (1 - \eta) \quad \ldots \ldots \ldots \ldots \quad (5)$$

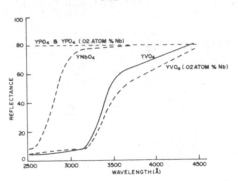

Figure 1. Diffuse Reflectance Spectra of $YVO_4$, $YNbO_4$, $YVO_4(Nb^{5+})$, $YPO_4$, and $YPO_4(Nb^{5+})$

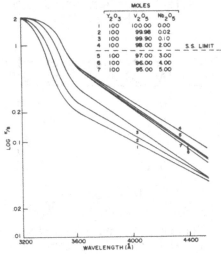

Figure 2. Effect of $Nb^{5+}$ Concentration on the Absorption Strength of $YVO_4$

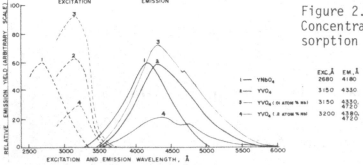

Figure 3. Excitation and Emission Spectra of $YNbO_4$, $YVO_4$, and Nb Dope and Nb Doped $YVO_4$

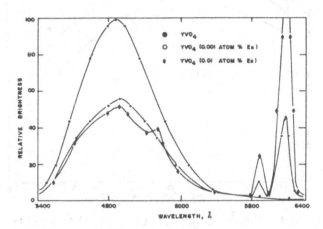

Figure 4. Effect of Eu$^{3+}$ Doping on VO$_4{}^{3-}$ Emission in YVO$_4$

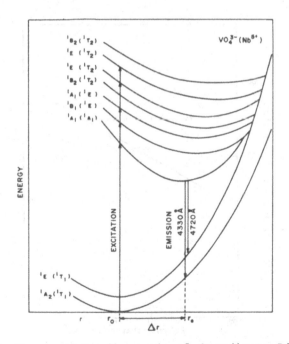

Figure 5. Proposed Condigurational Coordinate Diagram of
VO$_4{}^{3-}$ Ions in YVO$_4$, (HSU and Powell)

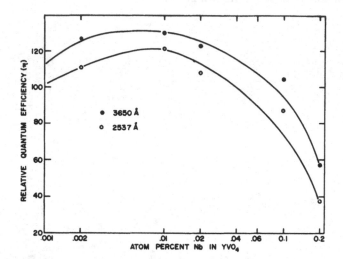

Figure 6. Effect of $Nb^{5+}$ Concentration on the Relative Quantum
Efficiency of $VO_4^{3-}$ Emmission in $YVO_4$ ($YVO_4$
Relative Yield 100)

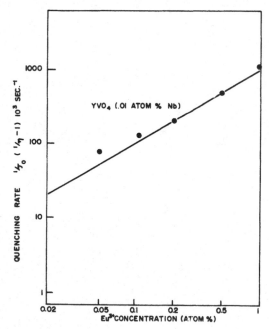

Figure 7 Plot of Quenching Rate Vs. $Eu^{3+}$ Concentration

If C is the molar concentration of "killers", the molar transfer rate (W/C) is obtained from the slope of Figure 7. The results as given in Table 2 show a change in quenching and molar transfer rate on doping Nb in YVO$_4$, the molar concentration of "killers" on the other hand remaining unchanged.

Table 2.  Effect of Nb Doping on Quenching and Molar Transfer Rate and Molar Concentration of "Killers" in YVO$_4$.

| Nb in YVO$_4$ (Atom%) | On-Site-Rate $1/T_0(10^3\text{sec}^{-1})$ | Quenching Rate $W(10^3\text{sec}^{-1})$ | Molar Transfer Rate $W/C(10^6\text{sec}^{-1})$ | Concentration of "Killers" |
|---|---|---|---|---|
| 0 | 11 | 48 | 200 | $2.4\times10^{-4}$ |
| 0.002 | 10 | 38 | | |
| 0.01 | | 33 | 140 | $2.3\times10^{-4}$ |
| 0.02 | | 38 | | |
| 0.10 | | 64 | | |
| 100.0 | 46 | 0.05 | | |

The increase in the relative quantum efficiency of the VO$_4^{3-}$ emission under 2537Å in Nb doped YVO$_4$ is, therefore, neither due to a reduction in the concentration of "killers" nor to an increase in the diffusion length as observed in the isomorphous substitution of PO$_4^{3-}$ ions in VO$_4^{3-}$ sites. Most probably, the NbO$_4^{3-}$ ions act as potential wells to hinder the migration of energy to a quenching sink.

## ACKNOWLEDGEMENTS

The author would like to thank Dr. R. L. Bateman of LRTSO, General Electric Company, for spectroscopic measurements and helpful discussions and Dr. J. A. Parodi of LRTSO, General Electric Company for EPR measurements. Also appreciated are helpful discussions by Prof. F. E. Williams of the University of Delaware and sample preparation by Mrs. C. Roland of Q&CPD General Electric Company.

## REFERENCES

1. .T. Kano and Y. Otomo, U.S. Pat. No. 3,586,636, June (1971).

2.  S. Faria and E. J. Mehalchick, J.E.C.S., 121, (1974), 305.

3.  R. G. DeLosh, T. Y. Tien, E. F. Gibbons, P. J. Zachmanidis and H. L. Stadler, J. Chem. Phy., 53 (1970), 681.

4.  D. K. Nath, Inorg. Chem 9 (1970), 2714.

5.  C. Hsu and R. C. Powell, J. Lum., 10 (1975), 273.

# ELECTROLUMINESCENCE OF RARE-EARTH DOPED CADMIUM FLUORIDE

J. F. Pouradier, G. Grimouille and F. Auzel

Centre National d'Etudes des Télécommunications, 196 rue

de Paris, 92220 Bagneux, France

Rare-earth doped cadmium fluoride ($CdF_2:Ln^{3+}$) has been recognized as a promising electroluminescent (E.L.) material for several years (1) because of its large transparency (0.2 $\mu$m $< \lambda <$ 12 $\mu$m) and its ability to be converted to a semi-conducting state ($\rho \sim 1\Omega$ cm is readily available). This note is devoted to results obtained on $CdF_2:Ln^{3+}$ crystals and their E.L.

We have grown $CdF_2:Ln$ crystals (Ln = Ho, Er, Tm, Yb) by a Bridgman Stockbarger method, using a graphite crucible and heated teflon atmosphere (P $\sim 10^5$ Pa). Undoped $CdF_2$ had first been refined by melting in dry nitrogen and in a teflon atmosphere, reducing near UV absorption to less than 1 $cm^{-1}$ for $\lambda >$ 0.205 $\mu$m. As-grown doped crystals were found to be n-type conducting without any subsequent annealing process. Red and infrared absorptions (giving the crystals a blue color) connected with this conducting state have been measured at room temperature (0.5 $\mu$m $< \lambda <$ 3 $\mu$m) and found to follow a power law

$$\alpha\,(\lambda) = \alpha(\lambda)\left[\frac{\lambda}{\lambda_o}\right]^n$$

with n increasing with Ln concentration (n = 3.7 for x = Ho/(Ho+Cd) $\leqslant$ 1% and n = 4.2 for x = 5%). Such a wavelength dependence is hardly in accordance with previous statements attributing (infra) red absorption to hydrogenic level ionization (2) in which case one should measure n $\leqslant$ 3.5 (3). Using an ion gun technique we have achieved E.L. structures with Au contacts on $CdF_2:Ln^{3+}$. Good ideality factors could be measured in many cases (n $\sim$ 1.05 to 1.1). When applying continuous or pulsed reversed bias of 15 to 20 volts, we get E.L. at wavelengths characteristic of $Ln^{3+}$. E.L. intensities

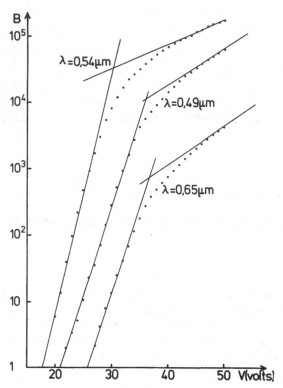

Figure 1.   Electroluminescence intensity (relative units) versus
            applied voltage for 0.49 μm, 0.54 μm and 0.65 μm emis-
            sion groups in $CdF_2:Ho^{3+}$.


(B) have been recorded versus bias (V), for various emission lines.
Our results (Fig. 1) show that B(V) curves are similar for 0.49 μm
and 0.65 μm of $Ho^{3+}$ (both originating from $^5F_3$) and different for
0.54 μm ($^5F_4 - ^5S_2$).  It then clearly follows that direct excitation
of each level takes place in $CdF_2:Ln^{3+}$, any process leading to a
highly excited $Ln^{3+}$ state followed by nonradiative decay to the
emitting level being ruled out as it should result in similar B(V)
relations for any emission.


REFERENCES

1.  J. M. Langer, Semiconductor Sources of Electromagnetic Radia-
    tion, Proc. Intern. Autumn School on Semicond. Optoelectronics,
    Cetniewo, 1975, M. Herman, Ed.  Polish Scient. Pub., Warszawa,
    p. 411 (1976).
2.  F. Moser, D. Matz and S. Lyu, Phys. Rev. 182:808 (1969)
3.  Y. A. Rozneritsa and V. D. Prodan, Sov. Phys. Semicond. 11:1306
    (1977).

RARE-EARTH PERMANENT MAGNETS:   DEVELOPMENT TRENDS AND THEIR

IMPLICATIONS FOR THE INDUSTRY

Karl J. Strnat*

University of Dayton, School of Engineering

Dayton, Ohio 45469

## ABSTRACT

The various types of rare earth-cobalt magnets and their properties are briefly reviewed.  A survey of applications is given,
with emphasis on commercially successful devices.  The use of rare-
earth magnets is often economical, despite their high cost.  Such
magnets are now used in electric watches and clocks, industrial
motors, computer printers, etc., and they are seriously considered
for large electrical machines and even mass-produced consumer items,
including automotive devices.

The competitive situation of rare-earth magnets relative to
other permanent magnets is discussed.  The consequences of limited
samarium supply and the recent cobalt crisis on projected use patterns of the different RE magnet types are considered.  The apparent
possibilities for further RE magnet development are reviewed and
some specific suggestions made.

## DIGEST

Basic research on rare earth-transition metal intermetallics
between the late 1950's and mid-60's led to the prediction of a new
type of permanent magnets based on light rare earth-cobalt alloys.
Intense development efforts that began in 1967, and still continue,
brougnt about a whole family of new magnet materials with properties
far superior to conventional magnets.  Several alternative alloy and

---

* Visiting Scientist at University of California, San Diego, EE and
  CS Department in 1978/79.

magnet production methods evolved.  About 30 companies worldwide
now manufacture rare earth-cobalt permanent magnets (REPM).

The prototype of the REPM materials, sintered $SmCo_5$ was initially very expensive and consequently used only in devices for military and space applications.  However, Sm and alloy prices dropped
rapidly and the choice of materials widened to include alloys with
Pr, Ce, La, mischmetal (MM) and other rare-earth substitutions for
the Sm, as well as partial replacement of the cobalt by Fe.  This
and the increasing production volume helped reduce magnet cost.  It
was also realized that, in certain device categories, as little as
1/5 to 1/20 of the magnet volume is needed, compared to Alnico or
ferrite designs.  This in turn means device miniaturization, secondary savings of other materials, often a design simplification, and
cheaper device fabrication.  A whole-system-redesign approach can
bring additional economies.  As a consequence, REPM have become the
magnets of choice in some mass-production applications for which
they were first considered too expensive.  Examples are computer
line printers, small electric motors for business machines, horse-
power-size industrial motors, magnetic couplings, and drives for
wristwatches and clocks.

There are classes of devices in which the use of permanent magnets became truly feasible only with REPM:  Magnetic bearings, very
flat loudspeakers, automotive starter motors and alternators are in
this category.  Now, the use of REPM in very large electric motors
and generators is being advocated as potentially economical with regard to initial systems cost as well as energy use.  It is probable
that we shall see revolutionary changes in the design of electrical
machinery as a consequence of the introduction of REPM.

The REPM have created a market for the previously nearly use-
less metal samarium.  But now, the rapid growth of applications
threatens to cause a demand in excess of the limited production capacity and raw-material supply rate for this moderately rare element
within the next few years.  The recent crisis in the world cobalt
supply, with consequent steep price rises and allocations has tentatively prevented an explosive expansion of the demand for the new
magnets.  It has caused postponement of some potentially very-large-
scale commercial REPM uses in the automotive industry and elsewhere.
This supply problem is mostly of political origin and there is hope
that it can be resolved in a few years.

Because of the materials supply situation and the projected
growth of REPM use, certain changes in the production patterns of the
magnet industry seem inevitable, and certain material development
projects have acquired new urgency.  The almost exclusive dependence
on samarium has to be reduced, and the cobalt use must be minimized
as far as possible.

Choosing from the already commercially available REPM types, this can, e.g., be done by shifting from $SmCo_5$ to the new $Sm_2(Co, Fe)_{17}$-based magnets with still higher energy density. But this is not nearly enough. Magnets of $(MM,Sm)Co_5$, $(Sm,Pr)Co_5$, $Ce(Co,Cu, Fe)_{5-7}$ and similar alloys are now in limited production, and they should find increasing utilization.

Polymer- or soft-metal-bonded rare-earth powder magnets ("matrix magnets") are likely to be more economical in materials use, magnet and device production than their sintered counterparts in the many cases where their lower energy density and poorer thermal stability are acceptable. Present commercial bonded magnets based on $SmCo_5$ with epoxy resins, EVA copolymers or other organic binders, have from 5 to 12 MGOe energy product. They have been used in quartz watches, fractional horsepower motors, phonograph pickups and tonearm supports, headphones, flat speakers, camera shutters, magnetic locks, pump drives, switches and magnetic jewelry. For the latter, a cheaper but less stable $(Sm,MM)Co_5$ is also in use. In the laboratory, $Sm_2(Co,Fe)_{17}$-based matrix magnets have been made with 16.5 MGOe, equivalent to the energy of the best sintered $MMCo_5$ or lower-grade $SmCo_5$ magnets.

In order to best serve the magnet-using industry and to promote the continued rapid growth of REPM applications, it is important that much materials R&D now be directed toward lowering the magnet cost per energy unit, and also toward broadening the raw material supply base, without severely compromising the magnetic property advantages characteristic of the REPM. For use in those cases where highest remanence and energy product are essential, researchers must strive to further increase the saturation and coercivity of the 2-17 alloys, e.g., by increasing their content of Fe and Mn. Further efforts must be made to replace the Sm in these as much as possible by more abundant rare earths. In the lower-energy magnet regime, the design engineers will have to redesign machines and devices destined for mass production such that they can use $MMCo_5$ and Ce-based precipitation-hardened magnets instead of $SmCo_5$. Since the properties of these at elevated temperatures are simply too poor for some applications, modified "$MMCo_5$" alloys optimized for specific devices must also be developed. Promising approaches involve the increased use of La, Nd and Pr, optimizing the Ce content, and making various compromises regarding the ratio of Sm to the other rare earths. Such development efforts can be justified where the potential materials consumption is large, e.g., for automotive applications.

Bonded magnets having higher maximum operating temperatures than the present range ($60^o$ to $100^oC$ quoted by different producers) need to be developed. Good long-term stability of remanence and coercivity at 125 to $150^oC$ is a realistic objective. In the area of production methods for polymer-matrix magnets, lower-cost alternatives to

the presently practiced diepressing in a magnetic field, such as
extrusion pressing and injection molding, should be introduced.
Bonded magnets of better magnetic stability and physical integrity
are possible through the use of a soft-metal binder.  Higher energy
density in matrix magnets can be achieved by utilizing 2-17 alloy
powders and by more efficient particle packing.  Economic production
methods that employ these principles have yet to be developed.

The most important boosts to the commercial utilization of the
REPM at this time will have to come from the magnet manufacturers
and from the chemical and metallurgical industry supplying their raw
materials.  Magnet and alloy production are now moving from small
pilot plants, that were attached to the research laboratories, into
larger, dedicated production facilities at several companies.  This
will help to lower the manufacturing costs and improve product consistency.  Increased production yield and efficient scrap recycling
will be necessary, particularly in view of the high cobalt price and
the potential samarium shortage.  The rare-earth industry must prepare to manufacture mischmetal and certain tailor-made "Magmisch"
blends of rare earths (or their magnet alloys with cobalt) with much
closer composition control and consistency than earlier, non-magnetic technological applications of mixed rare earths have required.
The reason is that, as the Sm content of the alloys is reduced, the
coercive-force safety margin is lost, while the "windows" for such
process parameters as sintering and heat treating temperatures, oxygen content of the powder, etc., become narrower.  Poor control of
the mischmetal chemistry and of the rare earth-to-transition metal
ratio can make an economic magnet production impossible.

Putting the REPM in perspective to the development of permanent
magnets in general, we can perceive the following trends:  The use
of all magnets will grow steadily, as replacement for electromagnets
and in new applications made possible by ferrites and REPM.  Due to
the cobalt shortage, alnico will quickly lose ground.  In less de-
manding applications--the large majority--it will be replaced by
ferrites.  In high-performance, smaller devices (but perhaps also in
large electric machines), the shift will be to the REPM, which make
much more efficient use of the available cobalt.  Two newly developed magnet materials, the ductile Fe-Cr-Co alloys and Mn-Al-C, may
also gain importance at the expense of alnico.  The Fe-Cr-Co offers
properties like the lesser alnico grades at somewhat lower Co content, while the Mn-Al-C contains no Co at all.  Its drawback is a
very low Curie point (275-300$^{\circ}$C vs. 450$^{\circ}$ for ferrites, >700$^{\circ}$ for
alnico and Sm-Co) and the consequent strong temperature dependence
of $B_r$ and $H_c$.  Neither of the new materials is likely to compete
with the REPM for the same applications.

## CONCLUSIONS

The rare earth-cobalt magnets have become a viable commercial product whose use is likely to increase rapidly throughout the next decade. It has been estimated that the REPM may in 5 years account for 15 to 20% (by price) of the total world market for all permanent magnets. This total market is also growing steadily due to many novel applications for magnets which the introduction of, first, the ferrites, and now the rare-earth magnets has opened up. Thus there will be an increasing demand from the magnet industry for all the samarium that can be economically produced, but also for La, Ce, Pr, some Nd and Y, and for high-quality mischmetal of reproducible and often modified composition. There will also be a small demand for the heavy rare earths, Gd through Er, which are used in some special magnet grades for temperature compensation.

## GENERAL BIBLIOGRAPHY

1. Proceedings of the International Workshop on Rare Earth-Cobalt Permanent Magnets and Their Applications. (Exclusively papers on all aspects of REPM technology.)
   a) K. J. Strnat, Ed., 2nd Workshop, Dayton, Ohio, 1976.
   b) K. J. Strnat, Ed., 3rd Workshop, San Diego, California, 1978.
   c) H. Kaneko and T. Kurino, Eds., 4th Workshop, Hakone, Japan, 1979; Society of Non-Traditional Technology, No. 2-8, 1-Chome, Toranomon, Minato-Ku, Tokyo 105, Japan.
2. IEEE Transactions on Magnetics. Proceedings issue of the INTERMAG Conferences from MAG-7, No. 3 (1971) through MAG-14, No. 5 (1978). (Contain many papers on REPM technology.) The Institute of Electrical and Electronic Engineers, 345 E. 47th Street, New York, NY 10017, USA.
3. F. Kornfeld, Ed., Goldschmidt informiert. (Company journal. Two special issues on REPM metallurgy and technology. In English.) a) 4/75, No. 35 (1975); b) 2/79, No. 48 (1979). Th. Goldschmidt AG, 43 Essen 1, Postfach 17, West Germany (BRD).
4. Proc. 2nd Conference on Advances in Magnetic Materials and Their Application (1976). (Several individual papers concerned with REPM and their properties.) Inst. of Electrical Engineers, Savoy Place, London, WC2, England, UK.
5. C. D. Graham, Jr., G. H. Lander and J. J. Rhyne, Eds., AIP Conference Proceedings. A series of books containing the proceedings of the US Conference on Magnetism and Magnetic Materials. (Contains many papers on REPM physics, metallurgy and technology.) a) No. 5 (1972); b) No. 10 (1973); c) No. 18 (1974); d) No. 24 (1975); e) No. 29 (1976). American Institute of Physics, Inc., 335 East 45th Street, New York, NY 10017, USA.

6.  J. J. Becker and J. C. Bonner, Eds., <u>Journal of Applied Physics</u>,
    <u>Vol. 80</u>, <u>No. 3</u>, <u>Part II</u> (1979).  Proceedings of the 24th Annual
    Conference on Magnetism and Magnetic Materials.  (Contains pa-
    pers on REPM physics, metallurgy and technology.)  Amer. Inst.
    of Physics, NY, NY 10017, USA.

7.  H. Zijlstra  Ed., <u>Proceedings of the 3rd European Conference on</u>
    <u>Hard Magnetic Materials</u> (1974).  (Contains papers on REPM phys-
    ics, metallurgy and technology.)  Bond voor Materialenkennis,
    P. O. Box 9321, Den Haag, The Netherlands.

8.  "<u>Rare Earth-Cobalt Permanent Magnets--A Comprehensive Overview</u>."
    Multi-client report issued on subscription basis by Wheeler
    Associates, Inc., Elizabethtown, Kentucky 45701, USA. 1979.

9.  J. P. Fort, "<u>Availability of Rare Earths for the Rare Earth-</u>
    <u>Cobalt Permanent Magnet Market</u>."  Rhodia Inc., POB 125, Mon-
    mouth Junction, New Jersey 08852, USA.  (Company rept.)

10. J. J. Cannon, "Review of the Rare Earth Industry 1978," Ann.
    Rev. of the Metal and Mineral Industry.  <u>Engineering and Mining</u>
    <u>Journal</u>, March 1979, McGraw-Hill Book Co., New York, NY.

11. J. H. Jolly, "Rare Earth Elements and Yttrium," in <u>U.S. Bureau</u>
    <u>of Mines Bulletin 667</u>, <u>Mineral Facts and Problems</u>, 1975.  Super-
    intendent of Documents, US Government Printing Office, Washing-
    ton, D.C. 20402.

12  S. F. Sibley, "Cobalt", <u>Mineral Commodity Profiles</u>, <u>MCP-5</u>, July
    1977.  US Bureau of Mines, Washington, D.C. 20241.

13. P. Scheidweiler, "<u>Cobalt Availability</u>," Societe Generale des
    Minerais S.A., rue du Marais, 31, B-1000, Brussels, Belgium.
    November 1977.  (Company report.)

# THE DIVERSITY OF RARE-EARTH MINERAL DEPOSITS AND THEIR GEOLOGICAL DOMAINS

E. Wm. Heinrich and Ralph G. Wells

Dept. Geology & Mineralogy, University of Michigan,
Ann Arbor, MI   48109 and Colt Industries, Pittsburgh,
PA   15230

## ABSTRACT

Concentrations of rare-earth (RE) minerals are represented by an unusually great diversity of geological types of deposits, particularly when compared to the variety of deposits of such elements as, for example, Pb, Zn and Cu.  In part this diversity stems from the fact that RE elements not only form their own minerals in various groups (oxides, multiple oxides, fluorides, carbonates, phosphates, silicates) but also appear widespread vicariously in numerous minerals of many other elements (oxides, multiple oxides, fluorides, silicates).

RE mineral deposits occur in all three geological domains: 1) igneous-hydrothermal, 2) sedimentary and 3) metamorphic, in each of which the deposits were formed under specific and peculiar genetic controls.  Igneous-hydrothermal deposits, which are the most diverse, are genetical dichotomous:  1) those related to calc-alkalic rocks, and 2) those related to alkalic rocks.  Analogously, sedimentary deposits occur in both clastic and non-clastic types.  Groups of genetically diverse deposits are concentrated in geographically restricted areas or provinces.  Examples of United States RE provinces include 1) the Blue Ridge, Piedmont and Coastal Plain of the southeastern states, 2) eastern Idaho-southwestern Montana, 3) north-central Colorado, 4) south-central Colorado, 5) San Bernardino County, California and 6) southeastern Alaska.  The ratio Y/Y + lanthanides (Ln) and the composition of the lanthanides in RE minerals, which are strongly controlled by crystallo-chemical factors, also are influenced markedly by differences in the geological environment under which the deposit formed.

## INTRODUCTION

The purpose of this paper is to summarize and systematize published and new information on the geology of RE mineral deposits. RE mineral concentrations occur in a great number of different geological types of deposits, among which are representatives of all three geological domains:  igneous-hydrothermal, sedimentary and metamorphic.

## MINERALOGY

In part the diversity of RE deposits is a reflection of the complex mineralogy of the RE elements.  They not only form their own minerals in a variety of groups (oxides, multiple oxides, fluorides, carbonates, phosphates and silicates), but they also occur vicariously in numerous minerals of many other elements (oxides, multiple oxides, fluorides, silicates), in isomorphous substitution, particularly for Ca but also for Th, U and Zr.  Rare earths are essential elements in over 100 minerals, and occur in lesser amounts in substitution in many more.  Inasmuch as the total range of stabilities of all of the RE minerals is very broad, deposits can be formed under a wide variety of geological environments.  Despite this unusually diverse mineralogical representation, only a very few RE species are economically important because of abundance, high RE content and processability.  The two most important are monazite (Ce, La, etc.,Y, Th) $PO_4$, and bastnaesite, (Ce, etc.) $FCO_3$.  Others of lesser economic significance include xenotime, $YPO_4$, apatite, $(Ca, RE)_5 (PO_4)_3F$, and brannerite, $(V, Ca, Fe, Th, Y)_3 Ti_5 O_{16}$.

## RE PROVINCES

Geographically restricted areas in which RE deposits are concentrated are defined as <u>provinces</u>.  Usually a province is time-independent, containing a genetic variety of deposits of different geological ages.  Among the well defined RE provinces in North America are:

1.  Southeastern United States (Virginia, North Carolina, South Carolina, Georgia, Florida).  Precambrian, Cretaceous and Tertiary deposits.  Types:  disseminations in metamorphic rocks, pegmatites, all types of placers, phosphorite.

2.  North-central Colorado.  Precambrian and Tertiary.  Types: granites, pegmatites, metasomatic allanite dissemination, xenotime and monazite migmatites, radioactive RE fluorite veins.

3.  South-central Colorado.  Precambrian, Cambrian, Tertiary. Types:  granites, pegmatites, magnetite-perovskite lenses,

carbonatites, fenites, monazite veins, RE apatite-molybdenite breccia, radioactive fluorite veins.

4.   Idaho-Montana border. Silurian, Permian, Cretaceous, Tertiary, Pleistocene. Types: disseminations in granite, pegmatites, carbonatites, Th veins, fluorite replacements, sulfide-parasite veins, marine phosphorite, monazite placers, euxenite placers.

Igneous-hydrothermal deposits represent the most complex group, both in mineralogy and geology as shown in Table 1.  To a large extent their distribution depends on the prior emplacement of plutons, either calc-alkalic or alkalic, anomolously enriched in RE elements. Intrusion sites of alkalic plutons are closely guided by large-scale lineaments.

Table 1.   Classification of RE Deposits - Igneous-Hydrothermal

| Type | RE Mineralogy | Examples |
|---|---|---|
| **Calc-Alkalic Association** | | |
| 1.  Disseminations in granitoid plutons | Monazite, xenotime, allanite, RE-sphene | Several phases of Idaho batholith; Silver Plume granite of Colorado |
| 2.  Granite pegmatites | Monazite, xenotime, RE-fluorite, microlite, uraninite, uranothorite | South Platte, CO; Llano Co., Texas; Haliburton-Bancroft, Ont. |
| 3.  Metasomatic disseminations | | |
| a)  Metamorphic rocks and granite | Allanite | North Central, Colorado |
| b)  In para-pyroeuxenites | Uranothorianite | Madagascar |
| 4.  Veins and replacement deposits | | |
| a)  Tin-tungsten veins | Monazite | Bolivia; South Africa |
| b)  Copper-thorium veins | Monazite (20-75%!) | Steenkampskraal, South Africa |
| c)  Barite-sulfide-quartz veins | Bastnaesite | Karonge, Urundi |
| d)  Davidite veins | Davidite | South Australia; Mavuzi, Mozambique |
| e)  Fluorite veins and replacement breccias | Uranothorite | Jamestown, Colorado |
| **Alkalic Association** | | |
| 5.  Disseminations in alkalic plutons | Pyrochlore, thorite | Teria District, Nigeria |
| 6.  Pegmatites | | |
| a)  Alkalic granite | Thorite, Niobian rutile, zircon | Mt. Rosa, Colorado |
| b)  Syenitic | Uraninite, RE-sphene, allanite | Haliburton and Renfrew Counties, Ontario |
| c)  Nepheline syenitic | Allanite, sphene, apatite, tritomite, melanocerite, thorite, pyrochlore, xenotime, rinkite, steenstrupine, lovchorrite, loparite | Langesundfjord, Norway; Southern Greenland; Kola Peninsula, USSR |
| 7.  Alkalic-carbonatitic complexes | | |
| a)  Magnetite-perovskite lenses | Perovskite | Iron Hill, Colorado |
| b)  Carbonatites | Bastnaesite | Mtn. Pass, California |
| c)  Fenites | Perovskite, pyrochlore, thorogummite, brockite, RE-miserite | McClure Mtn., Colorado |
| 8.  Veins and replacement deposits | | |
| a)  Quartzose-thorium veins (several types) | Thorite, monazite, brockite | Lemhi Pass, Idaho |
| b)  Fluorite replacement bodies | Bastnaesite | Gallinas Mtns., New Mexico |

The mineral assemblages of sedimentary RE deposits, summarized in Table 2, are much simpler than those of the igneous-hydrothermal group and depend, in the clastic group, on the survivability of such resistates as monazite and xenotime to the destructive actions of weathering and transportation. The accumulation of these re-sistates is a long, complex, and repetitive process involving usu-ally several recyclings into progressively younger sediments to an ultimate marine placer deposit. The precambrian metastable branner-ite placers of Blind River, Ontario, are postulated to have formed during a time when the earth's atmosphere was significantly lower in oxygen.

Table 2.  Classification of RE Deposits – Sedimentary

| Type | RE Mineralogy | Examples |
|---|---|---|
| Clastic Sediments | | |
| 1. Eluvial and colluvial deposits (in saprolite) | Monazite | Patrick and Henry Counties, VA |
| 2. Alluvial (fluviatile placers) | a) Monazite, xenotime<br>b) Euxenite | a) Costal plain of Georgia, S.C.; Cascade, Idaho<br>b) Bear Valley, Idaho |
| 3. Littoral (beach) placers | Monazite, zircon | Jacksonville, FL; Bahia, Brazil; Travancore, India; Ceylon |
| 4. Consolidated (fossil) placers | a) Monazite<br>b) Brannerite | a) Goodrich quartzite, Palmer, MI<br>b) Blind River, Ontario |
| 5. Adsorptions on clay minerals | Clay minerals | Pennsylvania sediments, Mid-continental USA |
| Non-Clastic Sediments | | |
| | | Idaho and Montana; Florida |
| | | Dale Co., Ala., Garland Co., Arkansas |
| | | Paradox Basin, Utah |

RE mineral assemblages in regional metamorphic deposits (see Table 3), are the most simple and are based on the geochemical cy-cle for monazite.(1) Detrital monazite in clastic sediments be-comes unstable during low-grade metamorphism and the RE's enter the structure of other minerals. With increasing grade of metamorphism, monazite again becomes stable and reforms, attaining its maximum abundance in rocks most intensely metamorphosed. Also, the percen-tage of Th in monazite increases with increasing grade of metamor-phism. With migmatization, monazite is further concentrated.

Table 3.    Classification of RE Deposits - Metamorphic

| Type | RE Mineralogy | Examples |
|---|---|---|
| **In Regional Metamorphic Rocks** | | |
| 1. Disseminations in medium to high-grade ortho and para metamorphic rocks | Monazite | Monazite belt in Piedmont of VA, NC and SC |
| 2. Veins, pods, dissemination concentrations in migmatites | a) Monazite<br>b) Xenotime<br>c) Brannerite | a) Kulyk Lake, Saskatchewan; Puumala, Finland; North-central Colorado<br>b) Music Valley, CA<br>c) San Bernadino Mtns., CA |
| **In Contact Metamorphic Rocks** | | |
| 3. Hematite skarns | Allanite, cerite, fluocerite, lanthanite, tornebohmite, bastnaesite | Bastnas, Sweden |
| 4. Magnetite skarns | RE-apatite, doverite, xenotime, bastnaesite, monazite | Mineville, N.Y; Dover, NJ |
| 5. Uraninite Skarns | Allanite, stillwellite | Mary Kathleen, Queensland, Australia |

Other provincial concentrations occur in San Bernardino County, California, migmatitic deposits and carbonatites, and in southeastern Alaska, carbonatites and various types of RE veins. Malawi, with disseminations in alkalic rocks, numerous carbonatites very rich in RE's (Chilwa Island, Tundulu, Kangankunde), and phosphorite, is an example of a foreign province.

## RE FRACTIONATION AND GEOLOGY

Both the ratio Y/Y + Ln and the distribution percentages of the lanthanides in RE minerals are strongly controlled by crystallochemical factors. The causes for partitioning include 1) ionic radius differences, 2) the detailed geometry of the crystal structures, 3) basicity differences, 4) oxidation states and 5) stability of complexes.[2] However, these variations also are markedly influenced by differences in the geological environment under which the deposit formed. In general, silica-rich igneous rocks favor the concentrations of heavy lanthanides and yttrium, whereas igneous rocks low in silica and high in carbonate favor concentrations of light lanthanides.[3] Data on the distributional variation of RE elements with paragenesis have been determined for pegmatitic monazites,[4] for monazite and apatite,[3] for sphene,[5] and for bastnaestite. [6] Two general studies on the geology, mineralogy, geo-chemistry, etc. of RE deposits are reported by Olson and Staatz (7) and by Heinrich. (8)

REFERENCES

1.    Overstreet, The Geologic Occurrence of Monazite, U.S.
      Geol. Survey Prof. Paper #530 (1967).

2.    Wells, R. G.  Light Lanthanoid Partitioning in Rocks and
      Minerals Containing Rare Earths, The Rare Earths in Modern
      Science and Technology, G. J. McCarthy and J. J. Rhyne (Eds.),
      Plenum Press, New York, pp 253-258 (1978).

3.    Fleischer, M. and Altschuler, Z. S.  The Relationship of the
      Rare-Earth Composition of Minerals to Geological Environment.
      Geochim. Cosmochim Acta, 33, 725-732 (1969).

4.    Heinrich, E. Wm., Borsup, R. A. and Levinson, A. A.  Rela-
      tionships Between Geology and Composition of Some Pegmatitic
      Monazites.  Geochim. Cosmochim. Acta 19, 222-231 (1960).

5.    Fleischer, M.  Relation of the Relative Concentrations of
      Lanthanides in Titanite to Type of Host Rocks.  Am. Mineral,
      63, 869-873  (1978-A).

6.    Fleischer, M.  Relative Proportions of the Lanthanides in
      Minerals of the Bastnaesite Group.  Canad. Mineral, 16,
      361-363  (1978-B).

7.    Staatz, M. H. and Olson, J. C.  Thorium.  In, "United States
      Mineral Resources."  U.S. Geol. Survey Prof. Paper 820,
      468-476  (1973).

8.    Heinrich, E. Wm.  Economic Geology of the Rare-Earth
      Elements.  Mining Mag. 98(5), 265-273  (1958).

# QUALITY CONTROL ON AN INDUSTRIAL SCALE AT THE LA ROCHELLE RARE EARTHS PLANT

Pierre Melard

Rhone Poulenc Industries, Division Chimie Fine

La Rochelle Plant, B.P. 2049, 17010  LA ROCHELLE

Rare earth compounds are used in very diversified applications : luminescence, nuclear reactors, optical glasses, glass polishing, ceramics, catalysis, permanent magnets, metallurgy, energy storage and bubble memories.

They not only must meet very exacting chemical specifications, but must also possess well defined physical characteristics. In addition some specific products require particular properties because of the type of applications in which they will be used ; these properties are not easily measured by routine analyses. Rather, they must be evaluated by specific tests relating to the comptemplated end-use.

Precise and constant adjustment in production is necessary for the manufacture of high-quality, reproducible rare earth compounds . Rare earth separation by liquid/liquid extraction is in itself a guarantee of quality since this is a continuous and well-mastered technology. Automatic analyzers are installed to control the production line in the plant, to check key parameters in an on-line fashion and to assure optimization of the adjustments. Moreover, numerous analyses are carried out in the laboratory to constantly follow in a step by step fashion the evolution of the product.

These imperatives imply the existence of a fairly large laboratory fitted with modern industrial equipment for inorganic chemistry and highly skilled personnel. 35 chemists are charged with this responsibility at our La Rochelle plant.

Customers require a high degree of purity – up to 5 and 6 N – for some applications. For elements such as chromium, copper, iron one should bear in mind that the determination of the degree of purity at ppm and ppb levels is a difficult, lengthy and costly undertaking.

It has been proved by using radio-active tracers, that separation by liquid/liquid extraction allows one to attain a degree of purity as high as is desired. The real limitations to the development of ultra-pure products that is those purer than 6 N derive not from process shortcomings, but rather from inadequate analytical techniques and there, improvements are being made.

The evolution of each rare earth is followed throughout its manufacture. This is achieved by testing intermediate products, on-line analysis and laboratory analysis. We will speak of each again briefly.

## TESTING OF INTERMEDIATE PRODUCTS

Continuous manufacturing processes are best attended to (and at least cost) by means of on-line analyzers.

The operating principles of these analyzers were defined by taking into account the particular characteristics of rare earths.

Once the incoming ores are crushed, a free acidity analyzer of original design permits us to feed the liquid/liquid extraction batteries with a constant acidity solution.

Molecular absorption spectra of aqueous rare earth solutions (fig 1) which possess very narrow absorption bands are particularly adapted to on-line spectrophotometry.

The stability of the front of the rare earths separation in each liquid/liquid extraction battery is thus permanently monitored.

Rare earth cations which exist in two oxidation levels require the use of electrochemical methods. We have on-line coulometers and polarographs.

We chose an X-Ray fluorescent spectrometer to analyze inter-mediate compounds ; this is a quick and satisfactory method for measuring levels between 100 and 1 000 ppm. Our Philips PW 1410 is equiped with two detectors, one with a NaI scintillator, the other with a gaseous flux.

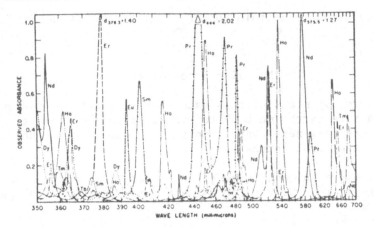

Fig. 1

A sodium tetraborate pellet is prepared. In the minerals, we thus regularly measure for silica, titanium, zirconium and phosphorus.

Of course analysis sensitivity could be increased by making more concentrated pellets. Matrix effects would increase however.

We also use rather frequently potentiometric methods for analyzing intermediary products : chloride traces, acidity in the presence of ionic precipitants, acide mixtures of various strengths. The use of ionic electrodes is particulary interesting for the study of effluent problems.

Neutronic activation is not used in our plant but is practised in our Research Center to determine rare earth elements in the different compositions.

CONTROL OF FINISHED PRODUCTS
CHEMICAL & PHYSICAL CHARACTERISTICS

Let us now examine the determinations to be made on separated rare earths. We will use gadolinium oxide as an example of what has to be done to guarantee a 3 or 4 N oxide having clearly defined physical characteristics.

Chemical Analysis

As already stated, the complete chemical analysis of a pure rare earth oxide uses all the instrumental inorganic methods available very often to the extreme limit of their sensitivity.

We carry out a certain number of measurements on manufactured powders.

## Emission Spectrometry

We have, by means of this excellent test method, the ability to record rare earths on a photographic plate. The resulting spectras are in general very complex and possess thousands of lines of almost identical intensity.

We measure in gadolinium all rare earths except cerium, but including yttrium. We work from 1 ppm for ytterbium to 50 ppm for Tb, Nd, Pr.

Some non rare earths are measured : Cu 0,5 ppm ; Co 3 ppm ; Ni 5 ppm ; Pb 10 ppm ; Cr 3 ppm ; $SiO_2$ 50 ppm.

A 200-fold enrichment can also be effected to increase method sensitivity. Achieved on exchange resins, its allows us to measure rare earth concentration similar to those seen by mass spectrometry. Radioactivity tracers can be used to follow the progress of products on the resin.

We thus measure : $La_2O_3$  0,1 ppm ; $CeO_2$  0,4 ppm ; $Pr_6O_{11}$ 0,4 ppm ; $Nd_2O_3$  0,1 ppm ; $Sm_2O_3$  0,05 ppm ; $Eu_2O_3$  0,02 ppm.

However, this method is long, unpractical and costly and is not used in routine analysis.

## Optical fluorescence (X-Ray Excitation)

This method which is very specific and rather limited in the rare earth field yields excellent results.

Gadolinium oxide must be removed before other rare earths are measured because the crystalline structure of the sample must be clearly defined and constant so as to insure good reproducibility.

Europium can be measured down to 1 ppb by this method. Similarly Pr, Sm, Eu, Tb, Dy can be measured to levels below  1 ppm.

## Mass Spectrometry

This method allows one to attain ppm precision for most rare earths and thus applies to very pure products (4-6 N). This is not an universal method however, since certain elements register poorly. Each run in long, costly and therefore not well suited to routine controls. The apparatus is available in our Research Center and is used essentially for statistical controls.

## Molecular Absorption Spectrometry

The BECKMAN 5230 apparatus with a band width of 0,2 nm permits us to analyze at very low levels. Hence, in gadolinium one can determine : Nd 1 ppm ; Pr 2 ppm ; Sm 3 ppm ; Ho 2 ppm ; Er 2 ppm ; compared to the usual emission spectrometry performances of 10 to 50 ppm.

Non-rare earth impurities (iron and silica) are determined by colorometry. Iron through an ortho-phenanthroline complex is seen and measured down to 0,5 ppm and silica through a silico-molybdic complex is measured down to 1 ppm.

## Spectrofluorimetry

Cerium and terbium are measured with this method in gadolinium solutions. For cerium the excitation wavelength 254 nm and the measurement wavelength is at 350 nm. We can reach 0,3 ppm in a perchloric acid medium in a 40 g/1 $Gd_2O_3$ oxide equivalent solution. In the case of terbium we use 100 g/1 of $Gd_2O_3$ in perchloric acid medium. The sensitivity is 10 ppm of terbium.

## Atomic Absorption Spectrometry

We use this method particularly for determination of europium. It is mainly used to follow non-rare earth impurities.

Alkali and alkaline earths can also be detected at the ppm level ; chromium can be determined down to 100 ppb in 5 g/1 solution $Gd_2O_3$.

## Plasma Spectrometry

We have just acquired this apparatus from INSTRUMENTS S.A. (Jobin-Yvon) and are running tests using this new method. We think that to get good results we have to at least use a 2,5 kw generator, ultrasonic nebulisation, a plasma generator and a dispersion system with a theoretical resolution of 100 000 at least.

The hight temperature (6 000° K) should limit possible chemical interferences and should occasion more precise measurements than emission spectrometry. Our first trial on an yttrium matrix allowed us to attain the following walues :

$La_2O_3$   :   0,2 ppm
$Eu_2O_3$   :   0,1 ppm
$Er_2O_3$   :   1   ppm

## Physical Characteristics

The chemical purity is a necessary guarantee of quality but is not always sufficient to insure good end-use results.

### Specific Surface Area

This characteristic has a close correlation with oxide reactivity. Two methods are in use in our La Rochelle plant :

- Static Volumetric analysis with a Micromeritics 2200 an 2205
- Dynamic volumetric analysis. This is a chromatographic determination in which an adsorbable gas diluted by a non adsorabable gas is pushed through a column of the product to be assayed.

We work on Perkin-Elmer and Monosorb Quantachrome absorption meters. Specific surface of our rare earth oxides are generally smaller than 20 m2/g.

### Particle Size

Different methods are used, depending on what we want to find out, from sieving to Sub-Sieve-Sizer Fisher, Coulter-Counter Z 4 and Sedigraph 5000 D Micromeritics. The Sub-Sieve-Sizer Fisher is used between 0,2 and 50 µm. It is based on the principle of air permeability. Reproducible results by this method are very dependent on prior compaction of the powder. Results can be very different from one lab to another and can sometimes create problems.

The Coulter-Counter is well known and suitable for powders of particle size larger than 1 μm. The powder must be dispersed before-hand by ultrasonic vibration ; frequency and duration of the vibration must be specified.

We prefer, to the methods already mentioned, the Sedigraph 5000 D which is very useful for particles below 1 μm and which is based on Stoke's sedimentation law. The vertical displacement of the cell containing the suspension in front of a low-energy X-Ray beam allows us to carry out the measurement in less than half an hour.

## Powder Density

The absolute volume of rare earth powders is measured with a helium pycnometer. This method can also give the porosity of an object when the form and the total volume are known. For bulk density we use the usual DIN or ASTM standards.

## Radio Activity

All rare earth ores contain radioactive elements from the thorium family. Our solvent extraction process gives a very good separation of non-rare earth elements and especially of radio-active elements. Trivalent actinium is the most difficult to separate.

The radioactivity of each isotope can be followed either with a NaI scintillator doped with thallium, or with a Ge-Li detector. Very long measurement times allow one to identify all isotopes and to determine radioactivity levels inferior to one picocurie. With this method we have shown that Rhone-Poulenc's Cerox polishing powders have less radioactivity than any other product on the market. This level is around 10 picocuries and we know which isotopes are responsible for it.

However, for gadolinium oxide which is used to make phosphors for X-Ray intensifying screens, the residual radioactivity is so weak that it cannot be determined by the usual method and we have to develop a special test. A gadolinium oxide sample is mixed with a phosphor and the mixture is put into contact for two weeks with X-Ray film. The phosphor is excited by the radioactivity and the resulting spots of light on the film can be measured. We also use this test to follow the lanthanum quality for intensifying screens in medical X-Rays.

END-USE TESTS

The characteristics, analytical and physical, investigated cannot always ensure that the product will have good performances in a given end-use and it is sometimes difficult to find a "middle-of-the-road" language with the users. They generally consider a physical effect such as brightness, light absorption and so on...; on the other hand the rare earth producers speak of ppm impurities. A synthesis between the two points of view is sometimes difficult to achieve. To overcome this problem we try to translate the physicist's vocabulary into chemical terms by means of end-use tests. So as to give to our customers a product adapted to their needs, we have to conduct application research on phosphorescence, enamelling, glass polishing. An application laboratory has been created for this mission. Its objective is to work with methods similar to those employed by our customers. This approach has helped to bring better products on the market and in some cases we have been able to assist our customers to solve some of their problems.

1.  Glass polishing powders have been one of the first areas to benefit from these tests. Our Cerox and Opaline products are systematically tested on machines and under conditions identical to those used by the industrial user. Our machines were chosen to cover all the different ways of polishing in use in the industry. Thus, for example, high precision felt pads are studied on Coburn machines. High-speed polishing using polyurethane pads are studied on a CMV ICM 7 machine. Polishing for prescription lenses Pellon pads are studied on an Autoflow machine.

We study, for each type of glass, for each type of product and for each polishing product, the best working conditions : suspension concentration, rotation speed, pressure etc.

One can thus compare the relative efficiencies of different polishing powders on every usual pad.

The mechanism for glass polishing can be determined by Electron Spectroscopy Chemical Analysis. By changing the angle of impingement of an X-Ray beam on polished glass one can study the variations in the composition of a glass on a 10 to 50 Å layer. We can correlate this data with that obtained with an X EDAX microprobe. By this method we have been able to locate K and Ca at the edge of a defect and on the surface of the glass we have determined the concentration gradient of these two elements along a line intersecting these defects. We have thus observed a lower content of potassium and calcium at the periphery of the defect, than in the center.

Finally with the ESCA method we have studied the concentration of these elements on the surface of a CEROX particle.

2.    Cerium and praseodymium oxides with specific characteristics are used in the enamel industry. They are studied in the application laboratory in sintered compounds which are specially designed to obtain maximum sensitivity. Enamel is applied either to metal sheets or to bricks called "biscuits". We measure the chromatic components L, a, b, which afford an insight into the performance of the oxides and permit us to evaluate the influence of impurities such as iron, neodymium, praseodymium.

3.    Glass coloration and discoloration problems and the influence of impurities such as the chromium in lanthanum are studied with the help of a glass furnace.

4.    Our firm has developed different techniques to study a large variety of rare earth applications in electronics. We are not involved in the production of phosphors, however, we have studied phosphors to know what is the quality of the oxides we have to produce to afford good performance to our industrial customers. We use ultraviolet or X-Ray excitation and determine luminescence under conditions defined depending on the application : cathode tube, lamps, radiology.

The purpose of these studies is to correlate the chemical and physical characteristics of the rare earth oxides to their performance in the end-use.

We have for example, studied the following problems :

- correlation between the characteristics of yttrium oxide and the cathodoluminescent properties of the yttrium/ trivalent europium oxysulphide phosphor use as a red component in colour T.V.

- determination of residual radioactivity in lanthanum and gadolinium oxides which are used in X-Ray phosphors for intensifying screens.

CONCLUSION
IMPACT OF THE COST OF THE ANALYSIS
ON THE PRICE OF THE PRODUCT

We have seen the various possibilities for determining the quality of an oxide by analytical techniques and end-use tests. All these techniques are often costly and time-consuming. We have already said that this requires the activity of 35 technicians in our La Rochelle plant. The quality control laboratory in our plant runs several hundred determinations per day. The cost of these analyses correspond, on average, to about 4 % of the cost of the product. However, analytical costs of high purity rare earths can reach about 10 %.

The demand for purer compounds for even more sophisticated applications is increasing. The effect of analytical costs on rare earth manufacturing cost will at best remain stable : in some cases it will increase in direct proportion to the purity level required.

We would like to emphasize that the development of rare earth compounds of higher purity is not dependent on either production capacity or technological problems of separation. We believe that liquid/liquid extraction has no theoretical limits in terms of purity but rather is related to present analytical possibilities. Current analytical technology does not allow us to offer a guarantee of purity higher than 6 N in a routine industrial manner. This is one of our main concerns and corresponds to one of our major research activities in the La Rochelle plant as well as in the Central Research Laboratory in Paris.

# TRENDS IN THE INDUSTRIAL USES FOR MISCHMETAL

I. S. Hirschhorn

Ronson Metals Corporation

Newark, NJ   07105

Ronson Metals Corporation, which was founded in 1915, has witnessed all of the changes which have occurred in the uses of mischmetal (mixed rare earth metals) over the last sixty-four years. It is illuminating to note currently the drastic changes which have occurred in the last decade as compared with the present and as projected over the next ten years.  These patterns are seen in Table 1.

In 1968, ductile iron and magnesium alloys dominated the picture.  By 1978, steel was by far the major use for mischmetal, with ductile iron a distant second.  Our forecast for 1988 promises new industrial uses, mainly for magnets and special alloys, the latter probably falling mostly into the category of heat resistant steels.

Not included in Table 1 is the oldest use of mischmetal, namely the production of lighter flints which contain about 75-80% mischmetal.  An interesting derivative of this use occurred during the years 1971/1972.

Mischmetal and alloys of mischmetal have unique burning characteristics which make them very useful in ordnance applications. When ignition of a mischmetal alloy particle is achieved by high energy explosion or impact, the metal will burn until completely consumed.  As a result, uses for this material were developed in the form of shell linings, aircraft penetrator devices, tracer bullets, bomblet markers and other similar applications.

Most of these uses were developed on a laboratory or pilot production scale.  However, during the Vietnam war, large tonnages of a mischmetal magnesium alloy were consumed in the form of a liner for a 40mm shell.

Table 1

Changes in the Industrial Use Pattern of Mischmetal

|                    | 1968 Actual | 1978 Actual | 1988 Forecast |
|--------------------|-------------|-------------|---------------|
| Ductile Iron Etc.  | 56%         | 13%         | 15%           |
| Magnesium Alloys   | 20          | 1           | 1             |
| Steel              | 17          | 85          | 65            |
| Magnets            | -           | -           | 9             |
| Special Alloys     | -           | -           | 6             |
| Miscellaneous      | 7           | 1           | 4             |

Currently, there is little known R & D taking place in the military establishment or in ordnance oriented companies. A metallurgical breakthrough, substantially improving ductility (for adaptability to certain applications) and/or improving resistance to corrosion (for long-term shelf life), could add a tremendous impetus to the use of mischmetal and its alloys for incendiary and pyrophoric applications.

The major current use for mischmetal is in the production of HSLA steels. This is a direct result of the "energy crisis." Low sulfur steels of exceptional strength require the addition of small quantities of mischmetal for control of the shape of residual sulfide inclusions. The benefits are due mainly to the replacement of manganese sulfide inclusions, which elongate during rolling, with rare earth oxysulfide inclusions, which retain their spherical form at steel processing temperatures. This change results in greater resistance to cracking under stress and to more uniform strength properties in both longitudinal and transverse directions. Mischmetal also acts to reduce hydrogen-induced cracking in steels exposed to hydrogen sulfide and other sources of hydrogen. This is important in steels used in deep-well drilling for oil and gas, and for steels in general which are used below the surface of the earth.

HSLA steels to which mischmetal has been added are now widely used in the automotive industry, in bridges, off-the-road vehicles, in high-tension towers, in roll-over bars, in off-shore drilling platforms (wherever light but strong structures are required); in the production of large diameter pipe for gas pipelines, in the production of "oil country goods" such as casings and other tubular products, and in other applications which require both high strength and great resistance to stresses under arctic and other difficult environments.

Major new uses for mischmetal in the next decade are expected to be based on the following developments:

## RARE EARTH-COBALT MAGNETS

General Motors recently announced development of a new process for fabrication of thin curved sections of rare earth-cobalt magnets (2). These have been considered for use in automotive equipment to reduce size and weight of electric motors (3). The new manufacturing process offers the potential for maximizing the use of expensive cobalt magnetic materials for DC motors, by producing thin arc segments which do not require diamond grinding and do not involve waste of material or breakage during fabrication. To be practical for use in cars, cobalt magnets must be low-cost as well as low-weight. This may be accomplished in part by substituting mischmetal for most of the samarium in rare earth-cobalt magnet alloys without significant sacrifice of motor performance (1). Ferrite permanent magnet motors are already widely used in cars for air conditioning blowers, windshield wiper motors, rear window defoggers and heater blowers. Since the new rare earth-cobalt magnets have an energy product five times greater than that of ferrites and much better resistance to demagnetization, they provide an opportunity to reduce size and weight in cars thereby greatly increasing their fuel efficiency.

The greatest deterrent to this application is the current high cost of cobalt. As a result, efforts are under way to use metals such as manganese, copper, iron, titanium and chromium to replace as much of the cobalt as possible (4). In any case, use of the rare earth metals maximizes the magnetic energy obtainable from cobalt. This should eventually lead to practical large-scale automotive applications.

## CONTINUOUSLY CAST STEELS

Several European steel producers have already successfully used mischmetal in the form of steel-clad wire in continuous casting. Residual sulfide has been effectively converted to globular form and reports indicate that the rate of solidification has been changed with rather sharp reduction in internal cracking of the slab. Metallurgical studies are now under way to determine the mechanism responsible for the results obtained.

By making mischmetal wire additions to the mold, which is downstream of the tundish nozzle, clogging of the nozzle cannot occur as it often does when ladle additions are made. Secondary reactions with refractory linings are avoided altogether and by use of protective shrouding, reoxidization due to exposed turbulence of the tapping stream is effectively prevented. Wire feeding rates are calculated on the basis of residual sulfur and shape-change requirements to yield the desired properties in the finished steel (5).

Mischmetal additions in almost all HSLA steels are presently made either in the ladle or in the ingot mold (5). When the steel industry develops a need, as expected, to manufacture HSLA steels by continuous casting, then use of mischmetal in wire form should prove to be advantageous for the reasons stated above.

## FERRITIC STAINLESS STEELS

Additions of mischmetal have been made to certain austenitic stainless steels for almost 30 years to make them forgeable. In recent years, research has been done to improve ferritic stainless steels, which are much less costly, by argon-oxygen-decarburization. This results in better corrosion resistance, particularly at elevated temperatures and in improved formability (7). Additions of 0.03 to 0.1% mischmetal typically improve oxidation resistance up to about 1100°C (6).

## VERMICULAR AND NODULAR GRAPHITE CAST IRONS

Vermicular graphite cast irons have properties between gray and ductile irons (9). They may be produced by adding mischmetal to aid in desulfurization and to promote shape change of the graphite in these irons. High thermal conductivity and good machinability, with about 80% of the strength of fully nodular cast iron, are achieved with additions such as 0.10% of mischmetal to iron containing .012% sulfur (8). Typical applications are transmission housings, automotive engine exhaust manifolds, eccentric gears, ingot molds and brake drums. These applications show a good growth trend.

The rising cost of magnesium in recent years should lead to increasing use of mischmetal for treatment of both vermicular and fully nodular cast irons. Since use of mischmetal provides the additional benefit of elimination of environmental problems arising from magnesium pyrotechnics and oxide fumes, good growth in demand is likely in this area.

## FREE-MACHINING STEEL

According to recent Japanese research, use of less than 0.2% mischmetal to promote formation of spheroidal graphite in steels containing 0.20-0.90% carbon makes it possible to produce free-machining steels, thus avoiding poisonous atmospheres in screw machine shops resulting from use of existing free-machining steels (10).

Screw machine steels containing sulfur have poor directional properties and those containing lead, sulfur, tellurium or selenium are not hot-rollable or hot-workable. Free-machining steels containing carbon, which has been graphitized by mischmetal additions, are reported to be both hot-workable and hot-rollable and to have good directional properties as well.

## ELECTROSLAG REMELTED STEELS

The main benefit of electroslag remelting is the isotropic properties developed in the steels (11). These steels have cleanliness comparable to vacuum melted steels, very good homogeneity, excellent forgeability, and improved toughness. Since research in this relatively new field has already shown improved Charpy V-notch toughness and ductility resulting from maintenance of spherical residual sulfides, it is probable that mischmetal will find new uses in ESR steels.

## ENERGY STORAGE DEVICES

It is well-known that $LaNi_5$ will, at room temperature and low pressures, store as much hydrogen as an equal volume of liquid hydrogen and will release hydrogen safely on demand. Furthermore, storage of gaseous hydrogen in conventional forged steel cylinders at 2300 psi takes three times as much space as storage of the same quantity of hydrogen in $LaNi_5$. Since hydrogen holds tremendous promise as the prime gaseous fuel of the future, there is a good probability of growth in the use of intermetallic compounds related to $LaNi_5$ in which mischmetal replaces lanthanum as a means of achieving greater cost effectiveness.

Although the foregoing remarks have been intentionally abbreviated, it is hoped that members of the research community will nevertheless gain from them an insight into the possibilities of the wider use of mischmetal and be stimulated to do the kind of research which will result in making these forecasts of future uses a reality.

## REFERENCES

1. B. R. Patel, Proceedings of the Third International Workshop on Rare Earth-Cobalt Permanent Magnets (San Diego, CA, June 1978). Mischmetal Rare Earth Magnet Tradeoffs in Automotive Accessory Motors.

2.  W. F. Jandeska  _Ibid_  Fabricating Rare Earth-Cobalt Magnets in Thin Arc Segments for Light Weight DC Motors.

3.  K. Grayzel  _American Metal Market (July 24, 1978)_  Curved Cobalt Magnets made by General Motors.

4.  Albert Mari  _American Metal Market (November 27, 1978)_  Panel Says Cobalt Crunch Spurring Substitution Moves.

5.  I. S. Hirschhorn  _ASM/ATAC Meeting (Cleveland, Ohio, September 26, 1978)_  Trends in the Use of Mischmetal.

6.  H. Brandis and R. Oppenheim  _Offenlegungsschrift 21 61 954 (1973)_  Ferritischer Hitzebestandiger Stahl.

7.  _Iron Age, (September 20, 1973)_  What's Behind the Push for New Ferritic Stainless?

8.  Michael J. Lalich  _Foundry M&T  (March 1978)_  Effects of Rare Earths on Structure and Properties of Cast Iron.

9.  Michael J. Lalich and S. J. LaPresta  _Foundry M&T, (September 1978)_  Uses of Compacted Graphite Cast Irons.

10. T. Yokokawa, et al  _Deutsche Offlengungsschrift 24 61 520 (1975)_  Automatenstahl.

11. Robert R. Irving, _Iron Age, (October 9, 1978)_  ESR Raises the Quality of Plate Steels, Hollow Rounds.

# CATALYSIS USING RARE EARTH INTERMETALLICS

A. A. Elattar and W. E. Wallace

Department of Chemistry, University of Pittsburgh

Pittsburgh, PA 15260

## ABSTRACT

Catalytic studies of the reaction of CO and $H_2$ to form hydro-carbons involving $LaNi_5$ and other members of the $RNi_5$ series have been extended to include the $RCo_{5+\delta}$ systems, the $LaNi_x$ systems in which $x \neq 5$, the $ThNi_x$ systems with $x \neq 5$ and the $LaCo_x$, $HoCo_x$ and $ThCo_x$ systems. The $RCo_{5+\delta}$ systems show Arrhenius-type temperature dependence of the rate with a pronounced compensation effect, i.e., log of the pre-exponential factor varies linearly with the activation energy. In the $LaNi_x$ and $ThNi_x$ systems the most activity is observed when $x = 2$. For $LaCo_x$ systems, activity increased monoton-ically with increasing $x$ whereas with $HoCo_x$ systems activity decreas-ed monotonically with increasing $x$. $ThCo_x$ systems exhibited a U-shaped activity curve. The highest activity was observed with $Th_7Co_3$ for which 91.4% of CO was converted in a single pass through the catalyst bed at 350°C.

## INTRODUCTION

Extensive studies have been made in this laboratory of inter-metallic compounds containing rare earths and actinides combined with d-transition metals in regard to the catalytic formation of $NH_3$ from $N_2$ and $H_2$ (1) and the formation of hydrocarbons and alcohols from CO and $H_2$ (2-4). It has been shown that the catalytic activity is not a feature of the intermetallic compound but is instead characteristic of products which form when the intermetallic compound is exposed to $N_2$ and $H_2$ or CO and $H_2$. In the first instance the compound is par-tially transformed (1) into rare earth nitride and d-transition metal, e.g., Fe, Co, Ni or Cu, whereas in the second instance rare earth or actinide oxide is formed along with elemental Fe, Co, Ni or

Cu (2-4). The procedure can be viewed as a novel way of making supported catalysts. However, these new catalysts have been shown (5) to exhibit substantially higher activity than catalysts formed by conventional wet chemical techniques. Previous work has been concerned largely with catalysts formed from $RNi_5$ systems (R = a rare earth or Th). In the present work these studies have been extended to the $RCo_5$ systems, to $RNi_x$ where $x \neq 5$ and to various $ThNi_x$ intermetallics. The latter systems are included to ascertain the influence of varying concentration of Ni. Catalysis experiments using $LaCo_x$, $HoCo_x$ and $ThCo_x$ were also carried out. The thorium compounds were included because the decomposition products of $ThNi_5$ have been found to be exceedingly active as methanation catalysts.

## PROCEDURE AND RESULTS

Most of the experimental techniques employed have been described in previous publications from this laboratory (4,5). In the present work the $CO/H_2$ ratio employed was 1:1 for the cobalt-containing systems and 1:3 for the systems containing nickel. The methanation reaction was examined in a single pass microreactor operated under steady state conditions. The synthesis mixture gas purified by passage through a molecular sieve and liquid nitrogen was admitted in the system at 1 atm total pressure. The effluent gas after the reaction was monitored by gas chromatography (Chromasorb 102). The catalytic activity of each catalyst for methane formation was experimentally determined from the rate of CO consumption. X-ray diffraction measurements on the used catalytic materials showed transformations along the lines described in the Introduction. Accompanying this transformation and carbon deposition formation is a large increase in surface area. The very large surface area of $HoCo_{5.5}$ may be due to extensive carbon formation on the surface (see Table 1). A similar effect was observed for $Fe_3Si$, in which there was a very large unexplained increase in surface area (6).

The rate of formation at low yields followed an Arrhenius type expression. The logarithm of the pre-exponential factor A varies linearly with the measured activation energy (Fig. 1) for the series of $RCo_{5+\delta}$* systems studied. Thus these systems exhibit a well-defined compensation effect.

A number of Ni-containing intermetallics with varying Ni/R ratio were studied to ascertain the effect of this ratio on activity. Results are shown in Fig. 2. The activity is found to be maximal in both $LaNi_x$ and $ThNi_x$ compounds when x = 2. The effect of the Co/R

---

*For the light rare earths $\delta$ = 0; with increasing atomic number $\delta$ increases to a value $\delta$ = 1 at R = Er. The stable 1:5 phase actually contains a stoichiometric excess of Co.

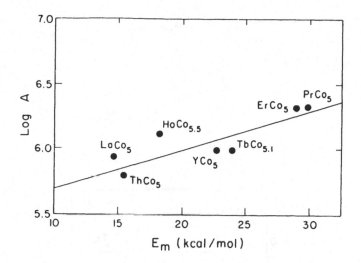

Fig. 1  Compensation plots for $RCo_5$ systems when used in the methanation reaction.  A = pre-exponential factor.  $E_m$ = measured activation energy.

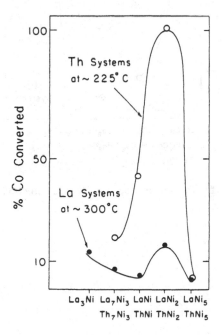

Fig. 2  Conversion of CO and $H_2$ to $CH_4$ over La-Ni and Th-Ni systems.

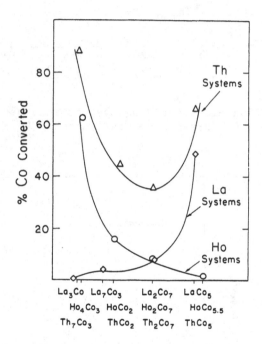

Fig. 3   Conversion of CO and $H_2$ to $CH_4$ over La-Co, Ho-Co and Th-Co systems.

Table 1.  Surface Areas of $RCo_{5+\delta}$ Catalysts

| Catalyst | Surface Area[a] $(m^2/g)$ | | Metallic Surface Area[b] $(m^2/g)$ |
|---|---|---|---|
| | Fresh Sample | Used Sample | |
| $YCo_5$ | 0.23 | 64.9 | 1.88 |
| $LaCo_5$ | 0.96 | 50.9 | 2.24 |
| $PrCo_5$ | 0.25 | 10.0 | 1.7 |
| $SmCo_5$ | 1.5 | 37.7 | 1.45 |
| $GdCo_5$ | 0.11 | 9.1 | 0.55 |
| $TbCo_{5.1}$ | 1.5 | 18.5 | 0.3 |
| $HoCo_{5.5}$ | 0.14 | 246.0 | 0.58 |
| $ErCo_6$ | 0.29 | 8.30 | 0.19 |
| $ThCo_5$ | 0.03 | 42.7 | 2.0 |
| Co powder | 0.22 | 0.26 | - |

a.  Measured by Ar adsorption.
b.  Measured by CO chemisorption.

ratio on activity has been investigated using three systems - $LaCo_x$, $HoCo_x$ and $ThCo_x$.  The behavior (Fig. 3) is different in that reactivity is not maximal when x = 2.  Instead, a U-shaped curve is found with $Th_7Co_3$ exhibiting the highest activity - 91.4% of CO converted in a single pass at 350°C.

Surprisingly, contrasting behavior was exhibited by the $LaCo_x$ and $HoCo_x$ systems in that activity monotonically increased and decreased, respectively, for these two systems.  The reason for this differing behavior is as yet not clear.

## REFERENCES

1.  T. Takeshita, W. E. Wallace and R. S. Craig, J. Catal. 44:236 (1976).
2.  V. T. Coon, T. Takeshita, W. E. Wallace and R. S. Craig, J. Phys. Chem. 80:1878 (1976).
3.  A. Elattar, T. Takeshita, W. E. Wallace and R. S. Craig, Science 196:1093 (1977).
4.  A. Elattar, W. E. Wallace and R. S. Craig, in "The Rare Earths in Science and Technology," eds. G. J. McCarthy and J. J. Rhyne, Plenum Press, N.Y. (1978), p. 87.
5.  A. Elattar, W. E. Wallace and R. S. Craig, Advances in Chemistry Series, in press.
6.  H. Imamura and W. E. Wallace, J. Phys. Chem., to appear July, 1979.

# THE HYDROGENATION OF ETHYLENE OVER LANTHANUM TRANSITION METAL PEROVSKITES

Dennis Anderson and W. J. James

Department of Chemistry and Graduate Center for Materials

Research, University of Missouri-Rolla, Rolla, MO  65401

## INTRODUCTION

In recent years, considerable interest has been expressed in the possible use of rare-earth transition metal perovskites as catalysts and electrodes. (1-2)  A theory has been proposed by Wolfram and Morin (3) to explain the catalytic activity of these compounds with respect to symmetry-forbidden reactions.  With these points in mind, the study of the hydrogenation of ethylene over lanthanum transition metal perovskites was undertaken to correlate the catalytic activity with the Fermi level of the catalysts.  The results of this study are presented herein.

## EXPERIMENTAL

Reagent grade lanthanum sesquioxide and transition metal oxides were used in preparing these compounds.  The reagents were mixed in stoichiometric proportions by grinding with mortar and pestle. Samples were fired in Pt-lined crucibles at conditions reported in Ref. 4, quenched in their firing atmospheres and ground to approximately one micron for testing.  The crystal structure and the presence of impurity phases were determined by x-ray powder techniques.  No detectable impurities were present.  The magnetic susceptibility at room temperature was determined via a Faraday balance.  The surface areas were approximated by employing scanning electron microscopy.

The surface properties of these compounds were examined by using ESCA and Auger electron spectroscopy.  The work functions were obtained by measuring the energy of the Fermi level relative to the

impurity carbon ls signal. The carbon ls signal was taken to be
at 284 ev.

A plug-flow microreactor was employed to determine the kinetic
data. C.P. grade ethylene and hydrogen were mixed via flow control-
lers. In most instances, a prepared reaction mixture (99.2% $H_2$,
0.184% $C_2H_4$) was used. The reaction mixture was allowed to flow
through the reactor such that the yield was less than 10% by volume.
The reaction temperature was confined to a range of 175–300°C,
measured to ±0.2C. A gas chromatograph was used to analyze the
products.

## RESULTS AND DISCUSSION

The catalytic activity measurements reveal some very interesting
results. Figure 1 is representative of the results obtained for the
entire series. The empirical rate expression for the anomalous
behavior shown in Figure 1 (□) is:

$$\text{Rate} = k \; p_{H_2} \; p_{C_2H_4}^{x} \; \text{moles/sec/meters}^2 \tag{1}$$

where x is a linear function of temperature; x is unity at high
temperature and zero at low temperature. By changing the temperature range over which x varies, the various curves in Figure 1 can
be duplicated.

This type of rate expression implies that a Freundlich-type
isotherm for ethylene adsorption is operating. In this study, the
isotherm introduced by Sips (5) has been used to represent the ethylene adsorption.

$$\theta_{C_2H_4} = K_1 \; p_{C_2H_4}^{x} / (1 + K_1 \; p_{C_2H_4}^{x}) \tag{2}$$

A reasonable reaction mechanism is the Twigg-Rideal treatment
(6) which involves the following steps:

$$C_2H_4(g) + S \underset{k_{-1}}{\overset{k_1}{\rightleftharpoons}} C_2H_4 \cdot S \tag{3}$$

$$H_2(g) + C_2H_4 \cdot S + S \underset{k_{-2}}{\overset{k_2}{\rightleftharpoons}} C_2H_5 \cdot S + H \cdot S \tag{4}$$

$$C_2H_5 \cdot S + H \cdot S \overset{k_3}{\rightarrow} C_2H_6(g) + 2S \tag{5}$$

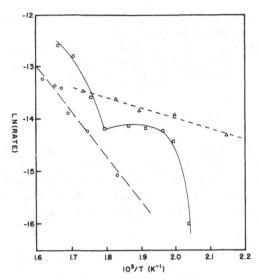

Figure 1.    Reaction rate (moles/sec/m$^2$) versus 1/T:    (o) LaScO$_3$ outgassed in Helium at 300°C; (□) LaScO$_3$ previously subjected
to the hydrogenation reaction and subsequently outgassed;
(Δ) LaScO$_3$ allowed to reach a steady-state at 300°C prior
to testing.

where S is a surface site and the k's are rate constants.  If the
rate determining step is the adsorption of hydrogen, Eq. 4, then
the resultant rate expression is:

$$\text{Rate} = k_2 K_1 \, P_{H_2} \, P^x_{C_2H_4} / (1 + K_1 \, P^x_{C_2H_4})^2 \tag{6}$$

The rate expression is similar to a Langmuir-Hinshelwood expression
and is consistent with all literature previously reported. (7-11)

The activation energy for this reaction can be approximated
using a classical approach.  The rate determining reaction (Eq. 4)
can be expressed in two parts:

$$\tfrac{1}{2}H_{2\,(g)} + C_2H_4 \cdot S \rightarrow C_2H_5 \cdot S \tag{7}$$

$$\tfrac{1}{2}H_{2\,(g)} + S \rightarrow H \cdot S \tag{8}$$

It is assumed that the activation energy is given by the sum of the
activation energies for each part of this concerted process.  Equation 7 has an activation energy which should be essentially constant
for this series of catalysts.  The activation energy of the last
step was approximated using a Lennard-Jones 6-12 potential.

The adsorption of hydrogen can be described in two ways; it adsorbs as a proton on a p-type surface, or as an anion on an n-type surface. This procedure is treated in more detail in the literature. (12,13) Figure 2 illustrates the potential energy diagram for the positive ion. The energy, $E_1$, is given by:

$$E_1 = \frac{1}{2} E_D - \phi + E_I \tag{9}$$

where $E_D$ is the dissociation energy of the hydrogen molecule, $\phi$ is the work function of the catalysts, and $E_I$ is the ionization energy of the hydrogen atom. The work function is present, since the electron from the hydrogen is transferred to the hole in the catalyst. The heat of adsorption, $E_2$, is given by:

$$E_2 = e^2/4r_{H^+} \tag{10}$$

where $e^2/4r_{H^+}$ is the image energy resulting from the electrostatic attraction of the hydrogen ion to its image in the surface. (12) $E_3$ is the heat of adsorption of $H_2$ on the surface.

Using these relationships for the energies, the potential energy curves can be determined for each absorbing species. The activation energy, $E_a$, is approximated by the intersection of these curves. Reasonable values for the quantities defined above were chosen, and the activation energy for the reaction was calculated as a function of the work function. These are listed in Table 1.

For p-type adsorption the activation energy decreases as the work function increases. On the other hand, for n-type adsorption the activation energy decreases as the work function decreases. Also, note that the activation energy is much less for n-type adsorption than for p-type.

It should be noted here that hydrogen adsorption does not proceed as ions, as can be seen from the endothermic heat of adsorption of the hydrogen ion relative to the hydrogen molecule. The process therefore involves a mixture of covalent and ionic bonding. The ionic cases were chosen for simplicity of calculation. Upon going to a more covalent model, the activation energies decrease dramatically but the same relationship between the work function and the activation energies remains.

Figure 3 shows the variation of the work function and the catalytic activity versus the number of d-electrons. The $d^0$ to $d^6$ catalysts are p-type, hence there should be a direct correlation between the work function and the catalytic activity, which indeed there is. The $d^7$ and $d^{10}$ catalysts are n-type and therefore exhibit an inverse relationship of catalytic activity to the work function.

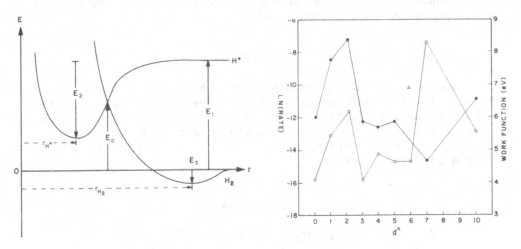

Figure 2.   Potential energy diagram for hydrogen adsorption on a
            single atom surface.

Figure 3.   Comparison of reaction rates at 477 K to work function
            of each compound:   (o) Reaction rate (moles/sec/m$^2$);
            (Δ) Reaction rate after additional reduction; (•) Work
            function (ev).

Table 1.   Activation Energy for Hydrogen Absorption as a Function
           of Work Function.

| Work Function (ev) | Activation Energy (ev) | |
|---|---|---|
|  | p-type | n-type |
| 0 | 13.64 | 0.62 |
| 1 | 12.69 | 1.46 |
| 2 | 11.73 | 2.34 |
| 3 | 10.79 | 3.23 |
| 4 | 9.84 | 4.13 |
| 5 | 8.91 | 5.05 |
| 6 | 7.97 | 5.97 |
| 7 | 7.03 | 6.90 |
| 8 | 6.11 | 7.83 |
| 9 | 5.18 | 8.76 |
| 10 | 4.45 | 9.71 |

The triangle (Δ) in Figure 3, representing additional reduction of
$LaCoO_3$, provides additional insight into the workings of this model.

This reduction causes the $LaCoO_3$ to become n-type, which predicts an increase in the reaction rate.

## CONCLUSIONS

The results and theories presented here form a consistent model. The reaction mechanism and rate expression are consistent with most data available for the hydrogenation of ethylene over various catalysts. We have shown that the reaction rate can be maximized over perovskite catalysts by choosing an n-type compound with a low work function.

## REFERENCES

1. D. B. Meadowcroft, Nature 226:847 (1970).

2. W. F. Libby, Science 171:499 (1971).

3. T. Wolfram and F. J. Morin, Appl. Phys. 8:125 (1975).

4. D. Anderson, "The Hydrogenation of Ethylene over Lanthanum Transition Metal Perovskites," University Microfilms, 106 pp. (1979).

5. R. Sips, J. Chem. Phys. 16:490 (1948).

6. E. K. Rideal and G. H. Twigg, Proc. Roy. Soc. London A171:55 (1939).

7. H. Zur Strassen, Z. Physik. Chem. A169:81 (1938).

8. R. N. Pease, J. Am. Chem. Soc. 45:1196 (1923).

9. A. Farkas and L. Farkas, J. Am. Chem. Soc. 60:22 (1938).

10. O. Tayoma, Rev. Phys. Chem. Jap. 11:153 (1937).

11. H. Koh and R. Hughes, J. Cat. 33:7 (1974).

12. J. M. Thomas and W. J. Thomas, "Introduction to the Principles of Heterogeneous Catalysis," Academic Press, pp. 17-32 (1967).

13. P. H. Emmett, "Catalysis," Vol. III, Reinhold, pp. 315-317 (1954).

# A NEW SUPERIONIC CONDUCTOR: (LaO)AgS

Marcel Palazzi, Claude Carcaly and Jean Flahaut

Laboratoire de Chimie Minérale Structurale, associé au

CNRS - Faculté de Pharmacie, 4 av. de l'Observatoire,

75270 Paris Cedex 06, France

## ABSTRACT

The new compound (LaO)AgS is prepared by the reaction of $Ag_2S$ and $La_2O_2S$ in sealed evacuated ampule, between 580 and 640°C, in the presence of a small quantity of iodine. It decomposes at 750°C into the two constituents.

The powder diffractogram shows a tetragonal cell, space group $P4/_n$ or $P4/_n$ mm, with a = 4.066Å, c = 9.095Å; Z = 2. Due to the absence of convenient single crystals, the structure is still unknown. However, from the previously known structural relations which are observed in the oxysulfides formed by rare earths and a second cation, it appears that this structure has a sheet arrangement, with alternating (LaO) and (AgS) sheets. Moreover, electron microscopy studies show a behavior that was previously observed with Ag β-alumina. With the impact of the electron beam, Ag threads rapidly grow from the crystal perpendicularly to the edges. Such a behavior shows the high mobility of the Ag, and suggests the existence of a sheet structure.

The total electrical conductivity is measured by the complex impedence method. The ionic part of the conductivity is obtained by the same method, with a $RbAg_4I_5$ blocking electrode. The results are compared with e.m.f. measurements using a Ag/(LaO)Ag/S.C. cell. The conductivity is found to be purely ionic, with $\sigma = 10^{-3}$ to $10^{-1}\Omega^{-1}cm^{-1}$ between 25 and 250°C. The activation energy is 0.20 eV.

This compound can be used as a specific electrode relative to the Ag and S ions, in the solution concentration range $10^{-2}M$ to $10^{-6}M$.

   This work will be published in the Journal of Solid State
Chemistry.

STEM CHARACTERIZATION OF THE SYNTHETIC RARE EARTH MINERALS IN

NUCLEAR WASTE "SUPERCALCINE-CERAMICS"

John G. Pepin and Gregory J. McCarthy[a], Materials Re-
search Laboratory, The Pennsylvania State University,
University Park, PA 16802; and D. R. Clarke, Rockwell
Science Center, Thousand Oaks, CA 91360

INTRODUCTION

"Supercalcine-ceramics" are a type of nuclear waste form made
by modifying the compositions of high-level liquid wastes with
selected additives so that after drying and firing, an assemblage
of mutually compatible crystalline phases is produced (1,2).
Wherever possible, these phases are designed to be synthetic versions of nature's most stable minerals, of which the sedimentary
resistates are the most desirable (3).  A class of prototype
supercalcine-ceramics are under development for the high-level
liquid wastes that would result from reprocessing of nuclear power
plant spent fuel.  In such wastes, the rare earths, taken as a
group, are the major constituents. McCarthy (4) has previously
addressed the crystal chemistry of the rare earths in nuclear waste
forms.  In this paper we describe the behavior of the rare earths
in two prototype supercalcine-ceramics.

The rare earths (RE) have been designed to crystallize in the
following mineral-like phases:

Apatite structure solid solution ($A_{ss}$):  nominal stoichiometry:
$AE_2RE_8(SiO_4)_6O_2$, AE = Ca, Sr; silicate analogs of the hexagonal phosphate apatite minerals.

Monazite structure solid solution ($M_{ss}$):  nominal stoichiometry:  $REPO_4$; common rare earth mineral and important

[a]Address after September 1, 1979:  Departments of Chemistry and
Geology, North Dakota State University, Fargo, ND 58102.

sedimentary resistate.

Fluorite structure solid solution ($F_{ss}$): nominal stoichiometry: $(U,RE,Ce,Zr)O_{2\pm x}$; familiar structure type for lanthanide and actinide dioxides; found in nature as uraninite
and cerianite. This phase has cubic symmetry in U- and Ce-
rich versions, but is also found with tetragonal symmetry in
Zr-rich versions.

X-ray diffraction studies of the fired ceramics have confirmed
crystallization of these RE phases and crystal chemical modeling
of nominal stoichiometry phases has added insight into the roles
of RE in the various phases (4-6). Microchemical characterization
of individual grains with a scanning transmission electron microscope (STEM) has been undertaken to determine the partitioning of
the RE's and to identify other ions that crystallize in the
synthetic RE minerals.


EXPERIMENTAL

Formulation and preparation of the two prototype supercalcine-
ceramics used in this study are described in detail elsewhere (4-
6). They are assemblages of 8-9 crystalline phases (and at least one
minor glassy intergranular phase) that were prepared by first
calcining largely liquid-phase-mixed batches of 26 oxide components
followed by pressing of the batches into pellets and firing in air
at 1200°C for two hours.

The 3 or 4 synthetic RE minerals constitute about 60 wt % of
the fired ceramic. The discussions that follow will focus on the
elements RE, U and Zr. The RE's are both controlling elements in
formation of two of the synthetic minerals ($A_{ss}$ and $M_{ss}$) and crystal
chemical analogs of the radionuclides Am and Cm. Uranium is the
controlling element in $F_{ss}$ and a model for much of the crystal
chemistry of Np and Pu. Zirconium has been found in all three
phases. The molar ratios of these elements in the two specimens are
61.3 RE, 20.5 U and 18.2 Zr where RE in

|              | La  | Ce  | Pr  | Nd   | Sm  | Gd   | Y   |
|--------------|-----|-----|-----|------|-----|------|-----|
| PSU-SPC-2+U = | 4.1 | 8.8 | 3.9 | 12.6 | 2.9 | 26.6 | 2.4 |
| PNL-SPC-4+U = | 2.5 | 8.8 | 0.5 | 22.8 | 0.3 | 26.4 | 0.0.|

The first specimen had the distribution of RE's typical of LWR
fission products. The modified distribution of the second resulted
from the need to use less expensive RE mixes in the Kg quantity
engineering-scale preparations at Battelle, Pacific Northwest Laboratories (PNL). Note that the feed compositions were modified to
yield the appropriate Ce and Gd concentrations.

The STEM specimens were prepared from the sintered ceramics by grinding and then Ar ion thinning. They were mounted on Cu TEM grids. The instrument was a Philips EM-400 with a STEM attachment operated at 120 kv. When individual crystals were located in the TEM mode, the instrument was switched to STEM and static beam x-ray microanalysis made. For these RE phases, spectra were taken in two parts: a 0-20 keV spectrum for total chemistry, and a 20-40 keV scan for resolution of individual RE's. The latter analyses were performed only for specimen PSU-SPC-2+U. The electron probe diameter used for these analyses was 200Å, so that sampling of multiple crystals and excitation of x-rays from neighboring crystals was minimized.

## RESULTS AND DISCUSSION

The microstructure of a typical supercalcine-ceramic is illustrated in Fig. 1. Most of the individual crystals are less than 1 µm in cross section. It was this typically submicrometer crystal size that precluded use of more routine microchemical tools (SEM and electron microprobe) and required STEM with its much smaller electron probe diameter. In Fig. 1 some of the $A_{ss}$ crystals can be easily identified by their hexagonal cross section. There are also laths of $A_{ss}$ in this photomicrograph. The dark oval crystals are $F_{ss}$. No $M_{ss}$ crystals are included. A typical 0-20 keV x-ray energy spectrum is shown in Fig. 2.

Results of the microanalyses are listed in Table 1. X-ray unit cell data, abstracted from ref. (6), are also included. Note that the stoichiometries are only nominal and have yet to be established by quantitative analyses.

The $A_{ss}$ phase was confirmed to be an alkaline earth-rare earth silicate with only Zr and a small amount of Fe as substituting ions. The larger unit cell size in PNL-SPC-4+U results from a smaller Ca/Sr ratio (1:1 $vs$. 3:1 in PSU-SPC-2+U); the higher concentration of larger $Sr^{2+}$ ions gives a larger unit cell. No P was detected even though phosphate-silicate apatite solid solutions are well known and readily synthesized (4). Except for Ce, the relative concentrations of the RE's parallel their availability. In earlier research we synthesized all $AE_2RE_8(SiO_4)_6O_2$ apatites for AE = Ca, Sr and RE = La, Pr, Nd, Sm and Gd by the same methods used to prepare the supercalcine-ceramics (4). Cerium as $Ce^{4+}$ would be expected to concentrate in $F_{ss}$. Charge-balanced (Ca+Zr) substitution for $RE^{3+}$ is the likely model for Zr incorporation in $A_{ss}$. To test this hypothesis, syntheses of $Ca_{3+x}Nd_{8-2x}Zr_x(SiO_4)_6O_2$ were attempted. Phase pure apatite could be prepared up to x = 1. The small amount of Fe is probably replacing Si and can also contribute to charge balance for Zr substitution. In some of the x-ray spectra, a very small U signal was noted. It was not certain whether this

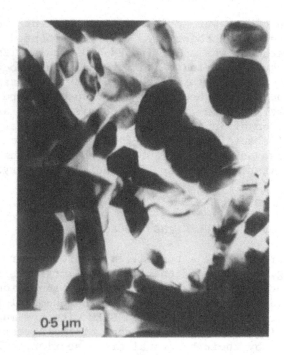

Figure 1.   TEM photomicrograph of PNL–SPC–4+U.

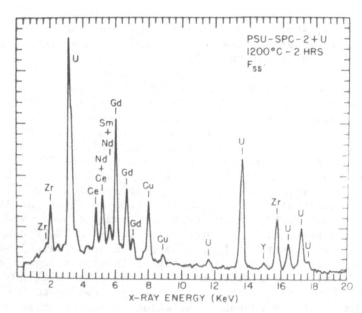

Figure 2.   X-ray energy spectrum of an $F_{ss}$ crystal in PSU–SPC–2+U (Cu peaks are from the Cu grid).

Table 1. Microchemistry and unit cell parameters[a] for the apatite, fluorite and monazite structure phases.

| Specimen | Apatite (hexagonal, P6$_3$/m) AE$_2$RE$_8$(SiO$_4$)$_6$O$_2$ | Fluorite (cubic, Fm3m) (U,RE,Ce,Zr)O$_{2\pm x}$ | Monazite (monoclinic, P2$_1$/n) REPO$_4$ |
|---|---|---|---|
| PSU-SPC-2+U | $a_o$ = 9.480(4) <br> $c_o$ = 6.957(8) <br> $v^o$ = 541.5 <br><br> Si,RE,Ca,Sr,Zr <br> (Gd>Nd>Pr>La>Ce>Y) | $a$ = 5.339(3) <br> $v^o$ = 152.2 <br><br><br> U,RE,Zr <br> (Gd>Ce>Nd>Sm,Y) | $a_o$ = 6.750(8) <br> $b_o$ = 6.998(8) <br> $c_o$ = 6.433(8) <br> $\beta^o$ = 103.98 (10) <br> V = 294.9 <br> RE,P <br> (Gd>Nd>La,Pr) |
| PNL-SPC-4+U | $a_o$ = 9.505(4) <br> $c_o$ = 6.975(4) <br> $v^o$ = 545.7 <br><br> Si,RE,Ca,Sr,Zr,Fe <br> (Gd>Nd)[b] | $a$ = 5.347(3) <br> $v^o$ = 152.9 <br><br><br> U,RE,Zr <br> (Gd>Ce>Nd)[b] | $a_o$ = 6.761(6) <br> $b_o$ = 6.950(8) <br> $c_o$ = 6.439(8) <br> $\beta^o$ = 103.72 <br> V = 293.9 |

[a] $a_o$, $b_o$, $c_o$ in Å; $\beta$ in degrees; V in Å$^3$.      [b] RE analyses incomplete.

was coming from the $A_{ss}$ crystals or excitation of U in adjoining $F_{ss}$ crystals. In any case, the U concentration in $A_{ss}$ would be very small.

The $F_{ss}$ phase was made up of the oxides of U, RE and Zr (see Fig. 2). Cerium was concentrated in $F_{ss}$, presumably because Ce$^{4+}$ is favored in this dioxide structure type. No La or Pr was detected and the Nd concentrations relative to Gd appeared to be smaller than their relative abundances in the bulk ceramic. This suggests that the $F_{ss}$ phase preferentially incorporates the smaller RE$^{3+}$ ions. Incorporation of the smaller Zr$^{4+}$ ion brings the cell parameter substantially below the 5.40–5.50Å range expected for UO$_{2+x}$-RE$_2$O$_3$-CeO$_2$ solid solutions. The oxidation state of U is unknown, but it is assumed to be largely U$^{+4}$ due to crystal chemical stabilization in $F_{ss}$ phase. Detailed phase equilibria studies of the systems UO$_{2+x}$-RE$_2$O$_3$-CeO$_2$-ZrO$_2$ now underway will model the solid solution behavior in $F_{ss}$ and allow the determination of the oxidation state of U in $F_{ss}$ as a function of $F_{ss}$ composition.

Crystals of the $M_{ss}$ phase were identified in one of the specimens. The larger RE's (La, Pr, Nd) preferentially crystallized in this phase. This observation is consistent with the x-ray data where the cell parameters fall between those of PrPO$_4$ and NdPO$_4$.

A Zr-rich phase was noted in PSU-SPC-2+U.  It was enriched in Ce relative to the bulk RE composition and also contained U, Gd and Nd.  It is almost certainly the tetragonal $ZrO_2$-rich phase found in other supercalcine-ceramics (2,5,6), but this symmetry cannot be confirmed because it is present below the level of detection by x-ray diffraction methods.

In summary, synthetic RE minerals isostructural with apatite, fluorite and monazite have been identified in these supercalcine-ceramics and an additional minor $ZrO_2$-rich phase containing RE's has also been noted.  Each of the phases shows some partitioning of individual RE's.  Further details of the RE distributions and the exact stoichiometries of the phases will require quantitative STEM microanalysis.

## ACKNOWLEDGEMENTS

This research was supported by a U.S. Office of Education Fellowship to JGP and by the U.S. Department of Energy through the Battelle, Pacific Northwest Laboratories.

## REFERENCES

1.  G. J. McCarthy and M. T. Davidson, Am. Ceram. Soc. Bull. 54, 782-786 (1975).
2.  G. J. McCarthy, Nucl. Technol. 32, 92-104 (1977).
3.  G. J. McCarthy, W. B. White and D. E. Pfoertsch, Mat. Res. Bull. 13, 1239-1245 (1978).
4.  G. J. McCarthy, in Proc. 12th Rare Earth Res. Conf., Vail, CO, C. E. Lundin, Ed., pp. 665-676 (July 1976).
5.  J. M. Rusin, M. F. Browning and G. J. McCarthy, Scientific Basis for Nuclear Waste Management, C. J. McCarthy, Ed., pp. 169-180, Plenum, NY (1979).
6.  G. J. McCarthy, J. G. Pepin and D. R. Clarke, Proc. Intl. Symp. Ceramics Nucl. Waste Management, T. D. Chikalla and J. E. Mendel, Eds., U.S.D.O.E. (in print).

# RARE-EARTH HAFNIUM OXIDE MATERIALS FOR MAGNETOHYDRODYNAMIC (MHD) GENERATOR APPLICATION

D. D. Marchant and J. L. Bates

Pacific Northwest Laboratory

P.O. Box 999, Richland, WA  99352

Several ceramic materials based on rare-earth hafnium oxides have been identified as potential high-temperature electrodes and low-temperature current leadouts for open cycle coal-fired MHD generator channels.  The electrode-current leadouts combination must operate at temperatures between 400 and 2000K with an electrical conductivity greater than $10^{-2}ohm^{-1}cm^{-1}$.  The electrodes will be exposed to flowing (linear flow rates up to 100 m/s) potassium seeded coal combustion gases (plasma core temperatures between 2400–3200K) and coal slag.  During operation the electrodes must conduct direct electric current at densities near 1.5 amp/cm$^2$.  Consequently, the electrodes must be resistant to electrochemical decompositions and interactions with both the coal slag and potassium salts (e.g., $K_2SO_4$, $K_2CO_3$).  The current leadout materials are placed between the hot electrodes and the water-cooled copper structural members and must have electrical conductivities greater than $10^{-2}ohm^{-1}cm^{-1}$ between 1400 and 400K.  The current leadouts must be thermally and electrochemically compatible with the electrode, copper, and potassium salts.  Ideally, the electrodes and current leadouts should exhibit minimal ionic conductivity.

Several materials have been suggested as possible candidates for electrodes.  These include doped lanthanum chromites (1,2), iron doped magnesium alumina spinel (1,2), stabilized zirconia (1,2), doped yttrium chromites (3), and rare-earth hafnium oxides (4,5). This paper focuses on rare-earth hafnium oxides.

Several rare-earth hafnium oxides have adequate electrical conductivity at temperatures above 1100K; however, at lower temperature the conductivity is not adequate for direct contact with the copper.  Consequently, current leadout materials are needed.  Several rare-earth hafnium oxide $M_xO_y$ ceramics ($M_xO_y$ represents a highly conductive oxide) have been developed which have adequate electrical conductivity and are thermally and electrochemically compatible with the rare-earth hafnium oxide electrodes (3).

## Fabrication

Sinterable hafnium oxide rare-earth oxide powders (BET surface areas around 14 $m^2$/g) were coprecipitated from neutralized dilute acidic solutions of the cations using a process similar to that of Dole, et al. (6). The acidic solutions were made from dissolving the rare-earth oxides in concentrated boiling HCl or $HNO_3$ and the hafnium oxychloride in water. Neutralized to a pH of 7.5 with dilute $NH_4OH$, the precipitate was washed with acetone and toluene, dried and air calcined for 2-4 hrs between 1100-1300K. The powders were ground through -200 mesh or ball milled with steel balls for 6-8 hrs, then cold die pressed at 82.7 MPa and isostatically pressed at 154 MPa and sintered in vacuum between 2173 and 2273K for as-ground powders or 1900-2100K for the ball milled powders. The sintered samples were partially reduced (dark blackish color) and were reoxidized (buff color) by air at 1773K.

The limited equilibrium phase data for $Re_2O_3$-$HfO_2$ appears similar to the better known analogous $Re_2O_3 \cdot ZrO_2$ systems. When added to $HfO_2$, the larger rare earth cations tend to form pyrochlore structures (7,8) and the smaller cations tend to form fluorite solid solutions (7,9). The predicted crystallographic structures of the rare earth cations were obtained after heat treatment. For most of the compositions, the predicted structures developed after the low temperature calcine.

## Electrical Conductivity

The DC electrical conductivity using a 4 probe technique was measured in air to 1620K for several of the compositions. The range of values are shown in Figure 1 along with data for other potential electrode materials. Included in the Figure is data for the current lead out and undoped $HfO_2$ (12). The electrical conductivity of the current lead out materials are between $10^{-1}$ and $10^{-2}$ohm$^{-1}$cm$^{-1}$ at 300K. Of the electrode compositions, the highest electrical conductivity are those with the fluorite phase containing less than 30 mol% $Re_2O_3$. The results are consistant with data on $Er_2O_3$-$HfO_2$ (10) and $Y_2O_3$-$HfO_2$ (11).

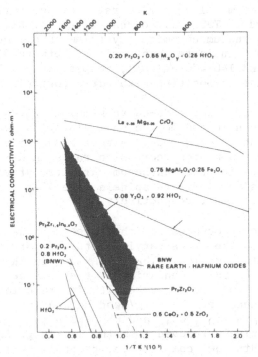

Figure 1.  Electrical conductivities of rare-earth hafnium oxides
           compared with other hafnates, zirconates and some
           potential MHD electrodes.


## Electrochemical Corrosion

The electrochemical corrosion of several rare-earth hafnium
oxides was measured in molten coal slag/alkali seed using laboratory test methods (4).  An anode and cathode were suspended in the
molten slag/seed electrolyte and a direct current passed between
these electrodes.  Aluminum oxide sleeves were placed around the
electrodes to direct the current through the ends of the electrodes.
A platinum voltage probe equidistant from each electrode measured
the electric potential of the anode and cathode.  Tests were conducted in molten $K_2SO_4$ (at 1373K) and synthetic Montana "Rosebud"
(MR-1) coal slag (at 1720-1730K) containing $K_2O$.  The composition
of the MR-1 was 42.4% (weight percent) $SiO_2$, 18.8% $Al_2O_3$, 12.9% CaO,
6.9% $Fe_2O_3$, 4.1% MgO, 0.7% $TiO_2$, 13.4% $K_2O$, 0.4% $Na_2O$, and 0.2% $P_2O_5$

The corrosion rates were determined by geometry changes and
material loss measured after metallographic preparation.  These
enhanced rates provided a relative corrosion resistance for comparing different materials but cannot be related directly to corrosion
rates anticipated in an MHD channel.

The electrochemical corrosion rates of the rare-earth hafnium oxides are listed in Table 1. Comparative corrosion rates in molten MR-1 coal for lanthanum chromites, yttrium chromites, and iron containing magnesium aluminate spinel varied between 30-280 µg/coul, 13-160 µg/coul and 140-470 µg/coul, respectively. In contrast, the better hafnia-base compositions exhibited corrosion rates <10 µg/coul.

The corrosion of the rare-earth hafnium oxides in molten $K_2SO_4$ were much less than in the molten coal slags, Table 1. Little potassium penetration occurred in either anode or cathode. For comparison, the electrochemical corrosion rates in $K_2SO_4$ tested under similar conditions for the lanthanum chromite, yttrium chromite and iron doped magnesium aluminate spinels were 45-500 µg/coul, 5-30 µg/coul and 60-140 µg/coul, respectively.

The corrosion processes for the rare-earth hafnium oxides were very similar, with the rates varying with the rare earth additions. In a typical test (Test 163: $Yb_{0.64}Hf_{0.36}O_2$ in MR-1 slag), slag interacts with the grain boundaries of the cathode causing grain separation. Continued reaction results in grain loss into the slag. The reaction products of the slag/seed-hafnia interaction consists of $K_{0.15}Si_{0.32}Al_{0.03}Yb_{0.13}O_{2+x}$ and $Ca_{0.16}Si_{0.17}Al_{0.16}Yb_{0.38}O_{2+x}$.* The Yb was selectively removed from the matrix grains, ultimately leaving a $Yb_{0.41}Hf_{0.59}O_2$ composition, probably a fluorite phase. No other products were observed, except for some metallic iron in the slag near the cathode surface, probably resulting from the charge transfer reaction between the cathode and the slag.

Table 1. Electrochemical Corrosion of Several
Rare-Earth Hafnia Compositions

| Composition | Electrolyte | Current Density, $A/cm^2$ | Corrosion Rate, µg/coul Anode | Cathode |
|---|---|---|---|---|
| $Tb_{0.31}Hf_{0.69}O_2$ | $K_2SO_4$ | 1.6 | 1-5 | 1-5 |
| $Tb_{0.31}Hf_{0.69}O_2$ | MR-1 | 1.6 | 3 | 5 |
| $Pr_{0.04}Yb_{0.08}Tb_{0.17}Hf_{0.71}O_2$ | $K_2SO_4$ | 0.5-1 | 1-5 | 1-5 |
| $Pr_{0.04}Yb_{0.08}Tb_{0.17}Hf_{0.71}O_2$ | MR-1 | 0.8 | 5-10 | 3-6 |
| $Yb_{0.09}Pr_{0.27}Hf_{0.64}O_2$ | MR-1 | 1 | 11 | 43 |
| $Pr_{0.04}Yb_{0.08}Tb_{0.17}Hf_{0.71}O_2$ | MR-1 | 0.7-1.2 | 12 | 16 |
| $Tb_{0.31}Hf_{0.69}O_2$ | MR-1 | 1 0.5 | 50-82 | 25-53 |
| $Yb_{0.64}Hf_{0.36}O_2$ | MR-1 | 1.2 | 55 | 99 |
| $Tb_{0.18}Hf_{0.82}O_2$ | MR-1 | 1.2 | 3-6 | -- |
| $Pr_{0.2}Hf_{0.8}O_2$ | $K_2SO_4$ | 0.3 | 12-24 | 12-24 |
| $Tb_{0.18}Hf_{0.82}O_2$ | MR-1 | 1.1 | 16-32 | 126 |
| $Eu_{0.06}Sm_{0.11}Hf_{0.83}O_2$ | MR-1 | 1 | 5-14 | 200-400 |

*The compositions determined by quantitative scanning electron microscopy and microprobe (SEM-EDX). The oxygen analysis is an estimate since oxygen cannot be detected directly.

The electrochemical processes appear strongly dependent upon the electrochemical and electrical character of the slag/seed electrolyte, e.g., the ionic and electronical transport. Passage of a dc current through the molten electrolytes results in ionic migration toward the anode and cathode. In molten potassium salts, $K^+$ migrates toward the cathode and $CO_3^=$ or $SO_4^=$ migrates to the anode. In complex coal slag/seed, the specific migrating ions, the degree of slag ionicity and the electronic-ionic transfer mechanism across the slag-oxide interfaces are not known. It is assumed that cations ($K^+$, $Ca^{++}$, $Na^+$) in coal slag carry a majority portion of the current in the "Rosebud" slag. The anion transfer between the anode and slag is more complex and could involve the decomposition of silicate ion ($SiO_4^=$) and the formation of $O_2$ gas.

The oxygen or other gases formed at the anode surface can accelerate material loss by 1) channeling of electric current resulting in increased current density, 2) increasing the oxygen chemical potential, 3) removal of reaction products and movement of unreacted slag to surface, and 4) mechanical erosion.

Cation migration from anode results in a cation depletion zone in the slag at the anode surface and decreases the electrical conductivity as indicated by the 3-fold increase in electric potential in the slag at the anode. This may have increased the potential for slag or anode decomposition.

Cation buildup in the slag near the cathode surface increases the charge buildup. Electrochemical reactions probably involve the reduction in oxygen activity resulting in the reduction of other slag species, i.e., iron. Potassium and calcium rich silicates were found in the cathode slag but not at the anode. These potassium rich slags also interact with the cathode grain boundaries (which are nominally high in silica) and ultimate loss in grain integrity.

Certain rare earths in hafnia appear to be more reactive in the electrochemical molten coal slag environmental than others. In MR-1 coal slag, preliminary results suggest that the reactivity of Eu>Sm>Yb>Pr>Y>Tb. In addition, the fluorite structures appear to be more resistant to electrochemical corrosion than the pyrochlore structures. The rare earths tend to be more electrochemically reactive than the hafnium in both molten slag and alkali salts. The terbia-hafnia composition appears to be the most stable.

## ACKNOWLEDGMENTS

The authors express appreciation to D. I. Boget, J. E. Coleman and W. M. Gerry for conducting the electrochemical experiments, physical property measurements, and SEM-EDX examination. This work supported by the Department of Energy under contract EY-76-C-06-1830, Division of MHD.

REFERENCES

1. J. B. Heywood, W. T. Morris, and A. C. Warren, in: Open-Cycle MHD Power Generation, J. B. Heywood and G. J. Womack, ed., Pergamon Press, New York (1969).

2. S. J. Schneider, D. P. R. Frederikse, G. P. Telegin and A. I. Romanov, in: Open-Cycle Magnetohydrodynamic Electrical Power Generation, M. Petrick and B. Ya. Shumyatsky, ed., Argonne National Laboratory, Argonne (1978).

3. D. D. Marchant and J. L. Bates, 18th Symposium Engineering Aspects of Magnetohydrodynamics, Butte, Montana, D.1.5.1-D.1.5.8 (1979).

4. D. D. Marchant, C. W. Griffin and J. L. Bates, 17th Symposium Engineering Aspects of Magnetohydrodynamics, Stanford University, Stanford, California D.5.1-D.5.5 (1978).

5. L. H. Cadoff, B. R. Rossing, D. D. Marchant and J. L. Bates, Fourth US-USSR Colloquium on Magnetohydrodynamic Electric Power Generation, Washington D.C. CONF-781009, 689-711 (1978).

6. S. L. Dole, R. W. Scheidecker, L. R. Shiers, M. F. Berard and O. Hunter, Jr., Materials Science and Engineering, 32:227-281 (1978).

7. D. R. Wilder, J. D. Buckley, D. W. Stacy and J. K. Johnstone, Editions du Centre National de La Recherche Scientifigue No. 205, 335-44 (1972).

8. M. V. Kravchinskaya, A. K. Kuznetsov, P. A. Tikhonov and E. K. Koehler, Ceramurgia International 4(1):14-18 (1978).

9. R. W. Schneidecker, D. R. Wilder and H. Moeller, J. American Ceramic Society 60(11-12):501-504 (1977).

10. J. K. Johnstone, Ph.D. Thesis, Iowa State University, Ames, Iowa (1970).

11. J. D. Scheiltz, Ph.D. Thesis, Iowa State University, Ames, Iowa (1970).

12. N. M. Tallen, W. C. Tripp and R. W. Vest, J. American Ceramic Society 50(6):279-283 (1967).

# MAGNETISM OF HYDROGENATED RARE EARTH INTERMETALLIC COMPOUNDS[*]

W. E. Wallace

Department of Chemistry, University of Pittsburgh.

Pittsburgh, PA  15260

## ABSTRACT

Rare earth intermetallic compounds containing the 3d-transition metals are very useful in the study of the effects of hydrogen on th magnetism of Mn, Fe, Co and Ni.  This cannot be done directly except by making use of rather high pressures since these 3d metals do not absorb significant amounts of hydrogen at atmospheric pressure; however, hydrogen introduction can be accomplished at moderate pressures using intermetallic compounds (vide infra).  As an illustration of the effect of hydrogenation on 3d transition metals, Baranowski (1) has recently reported on the magnetism of Mn when placed under a pressure of some thousands of atmospheres of hydrogen.  He finds that Mn, an antiferromagnetic metal, becomes ferromagnetic with a Curie temperature near room temperature.  This alteration in behavior is consistent with the current concept of the electronic makeup of the 3d metals, advocated most strongly by Stearns (2), to the effect (1) that the d electrons divide into two categories, localized and itinerant and (2) that the concentration of the latter determines whether a metal orders ferromagnetically or antiferromagnetically. According to the point of view of Stearns Fe, Co and Ni have few itinerant 3d-electrons, $\sim 0.05$ per atom, and they order ferromagnetically whereas Mn has a larger number which leads to antiferromagnetic ordering.  According to the point of view espoused by the author

---

*Work assisted by a grant from the Army Research Office and acknowledgement is made to the donors of the Petroleum Research Fund administered by the American Chemical Society for partial support of the preparation of this review.

hydrogenation leads to a decrease in the number of itinerant d-electrons, since hydrogen effectively absorbs electrons, and it is this absorption which transforms Mn into a ferromagnetic material.

In rare earth intermetallics containing 3d-transition metals, exchange originating with the transition metals is usually dominant (3,4) and in the Y, La, Lu and Th-containing systems it is the sole interaction. Accordingly, the influence of hydrogenation on the electronic makeup of and exchange involving Mn, Fe, Co and Ni can be studied using intermetallics containing these elements combined with rare earths or actinides. Large pressures of hydrogen are not needed to achieve a significant hydrogen concentration. Instead the chemical affinity of rare earths for hydrogen is exploited to introduce large amounts of hydrogen. This considerably simplifies the experimental procedure.

$Th_7Fe_3$ is a Pauli paramagnet (5) and a superconductor (6). Electron transfer from Th to Fe fills the Fe 3d band. When this material is hydrogenated, hydrogen absorbs electrons from the d-band, Fe regains its moment and the material becomes ferromagnetic with a Curie temperature near room temperature. In the process, superconductivity is quenched. Thus hydrogenation effectively transforms a superconductor into a ferromagnet.

Studies (7-15) of the $RFe_3$, $RFe_2$, $RCo_2$ and $RCo_3$ systems show that exchange is invariably weakened by hydrogenation. The iron moment, measured by neutron scattering studies, is either left unchanged (in $ErFe_2$-H) (11) or rises (in $HoFe_2$-H) (13). In contrast, the cobalt moment invariably falls. This is probably due to the fact that the cobalt moment is largely induced (16); hence as exchange is weakened, its moment drops.

$GdNi_2$ is a system in which rare earth exchange is dominant. The Ni d-band is filled by electron transfer from Gd and Ni becomes non-magnetic (17). Hydrogenation in this case also shows (18) a weakening of exchange, the Curie temperature falling from 81 K for the host metal to 8 K for the hydride $GdNi_2H_4$.

Hydrogenation of $Y_6Mn_{23}$ and $Th_6Mn_{23}$ produces diametrically opposite effects (19). These two compounds exhibit ferromagnetism and Pauli paramagnetism, respectively. Hydrogenation of $Th_6Mn_{23}$ transforms it into a ferromagnet whereas the yttrium compound becomes a Pauli paramagnet when hydrided. Thus, hydrogenation "switches on" ferromagnetism in one case and "switches it off" in the other.

Replacement of Mn in $Y_6Mn_{23}$ by Fe sharply reduces $T_c$, the Curie temperature (20). Exchange is thus weakened by a decrease in the number of itinerant d electrons in this system brought on when Fe replaces Mn. The decrease in $T_c$ or the destruction of ferromagnetism, accompanying hydrogenation is therefore expected since hydrogen absorbs d electrons and hence also decreases the concentration of itinerant d electrons.

The effects noted with the various compounds studied involving Mn, Fe, Co and Ni are all consistent with the notion that hydrogen withdraws electrons from the 3d band.

## REFERENCES

1. B. Baranowski, paper presented at the International Meeting on Hydrogen in Metals held in Münster, Germany, March 6-9, 1979.
2. M. B. Stearns, Physics Today 31:34 (1978).
3. See W. E. Wallace, Rare Earth Intermetallics, Academic Press, Inc., New York (1973), Chapters 10, 11 and 12.  See also W. E. Wallace, Ber. Bunsenges, Physik. Chem., to appear.
4. See, for example, T. Tsuchida, S. Sugaki and Y. Nakamura, J. Phys. Sol. Japan 39:340 (1975) for an analysis of $GdCo_2$ showing the relative importance of the Gd-Gd, Gd-Co and Co-Co interactions.
5. S. K. Malik, W. E. Wallace and T. Takeshita, Solid State Commun. 28:359 (1978).
6. B. T. Matthias, V. B. Compton and E. Corenzwit, J. Phys. Chem. Solids 17:130 (1961).
7. S. K. Malik, T. Takeshita and W. E. Wallace, Mag. Lett. 1:33 (1976).
8. K. H. J. Buschow, Solid State Commun. 19:421 (1976).
9. K. H. J. Buschow and A. M. Van Diepen, ibid., 19:79 (1976).
10. A. M. Van Diepen and K. H. J. Buschow, ibid., 22:113 (1977).
11. J. J. Rhyne, S. G. Sankar and W. E. Wallace, in The Rare Earths in Science and Technology, eds. G. J. McCarthy and J. J. Rhyne, Plenum Press, N.Y. (1978), p. 63.
12. J. J. Rhyne, G. E. Fish, S. G. Sankar and W. E. Wallace, J. de Physique, in press.
13. G. E. Fish, J. J. Rhyne, S. G. Sankar and W. E. Wallace, J. Appl. Phys. 50:2003 (1979).
14. F. A. Kuijpers, Ph.D. Thesis, Technische Hogeschool, Delft (1973).
15. S. K. Malik, W. E. Wallace and T. Takeshita, Solid State Commun. 28:977 (1978).
16. B. Bleaney, in Proceedings of the 3rd Rare Earth Research Conference, K. S. Vorres, ed., Vol. 2, p. 499, Gordon and Breach, New York (1964).  See also ref. 3, p. 155.
17. See ref. 3, chapter 10.
18. S. K. Malik and W. E. Wallace, Solid State Commun. 24:283 (1977).
19. S. K. Malik, T. Takeshita and W. E. Wallace, ibid., 23:599 (1977).
20. See ref. 3, chapter 12.

# LOW TEMPERATURE HEAT CAPACITY STUDIES ON HYDROGEN ABSORBING INTERMETALLIC COMPOUNDS

T. Takeshita, G. Dublon,[*] O. D. McMasters and
K. A. Gschneidner, Jr.
Ames Laboratory-DOE and Department of Materials
Science and Engineering
Iowa State University, Ames, IA 50011

## ABSTRACT

Low temperature heat capacities of some ternary Haucke compounds [$ThNi_{5-x}Al_x$ (x = 0 to 3), $YNi_{5-x}Al_x$ (x = 0 to 1.5) and $LaNi_{5-x}Cu_x$ (x = 0 to 5)] were studied. The electronic specific heat constants of these compounds decrease as Al or Cu is substituted for Ni. The Debye temperatures show some variations but not in a systematic manner for the aluminum substitution. In the aluminum substituted ternaries, there is no correlation between the hydrogen absorption capacity with the electronic specific constant and the Debye temperature except for $ThNi_2Al_3$. In the pseudo-binary system, $LaNi_{5-x}Cu_x$, the stability of the metal hydride may be related to the rigidity of the metal lattice, i.e. softer the metal lattice the more stable the metal hydride. This is in agreement with our previous observations.

## INTRODUCTION

In the previous communication, low temperature heat capacit results of some representative Haucke compounds were reported (1). It was observed that the density of states at the Fermi energ appears to have little correlation with the stability of their hy-drides. But the rigidity of the metal lattice may be an important parameter, along with the unit cell volume and electron concentration, relative to their ability to retain hydrogen in their lattice. Preliminary results of the low temperature heat capacity studies of the ternary compounds, $YNi_{5-x}Al_x$ (x = 0 to 1.5),

---

[*]Present address: Harvard University, Division of Applied Sciences, Pierce Hall, Cambridge, Massachusetts 02138

$ThNi_{5-x}Al_x$ (x = 0 to 3), and $LaNi_{5-x}Cu_x$ (x = 0 to 5), are reported in this study.

These alloys were chosen since the Al substitution for Ni in $YNi_5$ and $ThNi_5$ increases the stability of the metal hydrides (2), and the Cu substitution for Ni in $LaNi_5$ lowers the plateau pressures (i. e. increases the stability) of the metal hydrides (3). A preliminary result on the $ThNi_{5-x}Al_x$ system has been presented elsewhere (4). The stability of metal hydrides can be discussed in terms of the plateau pressure of the hydrogen gas in equilibrium with them. $ThNi_5$ and $YNi_5$ do not form stable hydrides under mild conditions such as room temperature and 100 atmospheres pressure of hydrogen gas. But a small amount of Al substitution for Ni drastically lowers the plateau pressure (2). On the other hand, the Cu substitution for Ni in $LaNi_5$ lowers the plateau pressure (3) but not as strongly as Al substitutions.

## EXPERIMENTAL

The arc-melted $YNi_{5-x}Al_x$ and $ThNi_{5-x}Al_x$ samples were annealed at $\sim 800°$ C for three weeks, and the $LaNi_{5-x}Cu_x$ samples were annealed at 700° C for six weeks or longer. The metallographic examination showed all samples to be single phase except $YNi_{3.5}Al_{1.5}$ and $ThNi_2Al_3$, in which small amounts of second phases were seen, less than 2% by an areal analysis. Debye Scherrer X-ray powder diffraction patterns ($CuK\alpha$ radiation) showed no extraneous lines for any of the samples. The extrapolated lattice constants were obtained by using the Nelson-Riley method.

The low temperature (1 - 20 K) isothermal type calorimeter used was checked by measuring the heat capacity of the 1965 Calorimetry Conference copper standard. The measured electronic specific heat constant and the Debye temperature of this sample was in agreement with literature values within 0.8%.

The low temperature heat capacity, $C_p$, of a normal metal can be expressed as $C_p = \gamma T + \beta T^3$, where $\gamma$ is the electronic specific heat constant and $\beta$ is related to the Debye temperature ($\theta_D$) by the expression, $\beta = (12 \pi^4 R/5) (1/\theta_D)^3$, where R is the gas constant (R = 8.317 J/g -at. K). The results of heat capacity measurements were analyzed in terms of the above relationships, since the plot of $C_p/T$ vs $T^2$ is linear at temperatures below $\theta_D/50$. A least square fit program was used to obtain the coefficients, $\gamma$ and $\beta$ ($\theta_D$).

## RESULTS AND DISCUSSION

The heat capacity data, hydrogen absorption characteristics and unit cell volumes of the systems studied are given in Table 1. The lattice constants of the Al substituted ternaries increase as Al replaces Ni in $YNi_5$ and $ThNi_5$, and consequently the unit cell

TABLE 1. Hydrogenation properties, unit cell volumes and low temperature heat capacity results of ternary Haucke compounds *

| Compound | Plateau pressure 25°C (atm) | H formula unit | $\gamma \left(\dfrac{mJ}{g\text{-at. } K^2}\right)$ | $\theta_D$ (K) | Unit cell volume ($A^3$) |
|---|---|---|---|---|---|
| ThNi$_5$ | >200 | - | 6.38 | 365.5 | 83.31 |
| ThNi$_{4.5}$Al$_{.5}$ | >100 | - | 3.69 | 359.2 | 85.80 |
| ThNi$_{4.0}$Al$_{1.0}$ | ~ 7 | 2.5 | 2.27 | 404.5 | 87.88 |
| ThNi$_{3.5}$Al$_{1.5}$ | ~ 1 | 3.0 | 2.58 | 376.4 | 90.44 |
| ThNi$_{3.0}$Al$_{2.0}$ | ~ 0.1 | 2.7 | 2.50 | 393.5 | 91.91 |
| ThNi$_{2.5}$Al$_{2.5}$ | - | - | 2.24 | 395.8 | 94.62 |
| ThNi$_{2.0}$Al$_{3.0}$ | - | - | 1.41 | 566.6 | 98.18 |
| YNi$_5$ | >100 | - | 6.07 | 449.8 | 80.81 |
| YNi$_{4.5}$Al$_{0.5}$ | 1 | 4.5 | 4.48 | 451.0 | 84.34 |
| YNi$_{4.0}$Al$_{1.0}$ | 0.2 | 3.75 | 1.97 | 449.4 | 86.70 |
| YNi$_{3.5}$Al$_{1.5}$ | < 0.01 | 3.25 | 2.29 | 473.1 | 90.02 |
| LaNi$_5$ | 2.5 | 6.2 | 6.08 | 351.0 | 86.80 |
| LaNi$_4$Cu | 1.6 | 5.6 | 5.30 | 337.4 | 88.80 |
| LaNi$_3$Cu$_2$ | 1.0 | 3.8 | 2.69 | 349.3 | 90.08 |
| LaNi$_2$Cu$_3$ | .65 | 3.6 | 1.43 | 322.9 | 90.62 |
| LaNiCu$_4$ | .43 | 3.4 | 1.68 | 295.0 | 92.77 |
| LaCu$_5$ | .30 | 3.0 | 2.40 | 258.9 | 95.41 |

*The standard deviation in $\gamma$ and $\theta_D$ is less than 0.2%.

volumes increase also with the increase of Al content. A similar trend is also noted in the LaNi$_{5-x}$Cu$_x$ system from x = 0 to x = 5 (Table 1).

R(Ni$_{5-x}$Al$_x$ Alloys

It is noted that the Al substitution for Ni in YNi$_5$ and ThNi$_5$ lowers the electronic specific heat constants substantially in a comparable manner, but changes in Debye temperatures do not appear to vary in a systematic way. A sharp decline of the electronic specific heat constant occurs with the substitution of a small amount of Al (up to about x = 1), and the change after this value of x is gradual. This result indicates that the nearly filled d-bands of Ni are filled with electrons from Al up to about x = 1 and with x > 1 the electron levels near the Fermi surface are mainly s- and p-like in character. This behavior is quite similar to that found for Al and Cu additions to pure Ni metal (5). The hydrogen contents in the hydrides of those ternaries show some decrease with increasing x for x > 1, but the stability of these hydrides increases drastically as the Al content increases except ThNi$_2$Al$_3$ where no hydrogen absorption occurs. From these results, it appears that there is little correlation between the density of states at the Fermi surface and the hydrogen absorption capacity of those ternary Haucke compounds.

The change in Debye temperature due to the Al substitution in $ThNi_5$ and $YNi_5$ is relatively small except $ThNi_2Al_3$ where a sharp increase is seen. There is also an anomalous increase of the Debye temperature at the composition $ThNi_4Al$. The substitution of Ni by Al occurs preferentially in 3(g) sites in $ThNi_5$ (2), and the composition of $ThNi_4Al$ produces a superlattice in which the Al atom occupies the center position of the 3(g) sites in a unit cell ($\frac{1}{2}$ $\frac{1}{2}$ $\frac{1}{2}$) and this ordering is consistent with an increase in the lattice rigidity. The unit cell of $ThNi_2Al_3$ consists of alternate layers of planes containing only Th and Ni atoms and planes containing only Al atoms, i.e., all Ni atoms on 3(g) sites of $ThNi_5$ have been completely replaced with Al atoms in $ThNi_2Al_3$. This complete replacement of Ni atoms by Al atoms leads to a significant stiffening of the lattice and the large observed increase in $\theta_D$. In general there is no correlation between the hydrogen absorption characteristics and the Debye temperature, except for the $ThNi_2Al_3$ alloy where the hydrogen absorption in the $Th(Ni_{5-x}Al_x)$ abruptly ceases when x = 3 due to the sudden increase in lattice rigidity as a result of the complete filling of the 3(g) sites. This lattice stiffening is the dominant factor in the hydrogen absorption characteristics of $ThNi_2Al_3$ and overcomes the favorable size factor (see below). As noted before, the unit cell volume of these alloys increases as the Al content increases, which correlates well with the stability of metal hydrides except $ThNi_2Al_3$ which is discussed above.

## La($Ni_{5-x}Cu_x$) Alloys

The $LaNi_{5-x}Cu_x$ system shows a substantial decrease in the electronic specific heat constant as x increases from 0 to 3 and a small increase from x = 3 to x = 5, while the Debye temperature decreases gradually from x = 0 to x = 5 except for a small peak at x = 2. The cause of this increase in the Debye temperature at this composition is not known. It may indicate some ordered substitution. Furthermore, if it occurs, the Cu substitution for Ni does not seem to be preferential as in the case of Al substitution. Both the decrease in the lattice rigidity ($\theta_D$) and the increase in unit cell volume as the copper content increases is consistent with the increase of the stability of the hydride in the $LaNi_{5-x}Cu_x$. This behavior is in good agreement with the previous observation (1). Finally, it is seen that the electronic specific heat constant decreases as Ni is replaced with Cu, and it does not correlate with the observed hydride stability. Again we conclude that the density of states at the Fermi surface is not important with regard to the stability of metal hydrides.

ACKNOWLEDGMENT

This work was supported by the U.S. Department of Energy, Office of Basic Energy Sciences, Division of Materials Sciences. The authors thank J. Holl for his assistance in carrying out these experiments.

REFERENCES

1. T. Takeshita, K. A. Gschneidner, Jr., D. Thome and O. D. McMasters, to be submitted for publication (1979).
2. T. Takeshita and W. E. Wallace: J. Less-Common Metals, 55:61 (1977).
3. J. Shinar, L. Jacob, D. Davidov and D. Shaltiel, Int. J. Hydrogen Energy, to be published (1979).
4. K. A. Gschneidner, Jr., T. Takeshita, B. J. Beaudry, O. D. McMasters, S. M. Taher, J. C. Ho, G. B. King and J. B. Gruber, J. Phys. (Paris) Colloq. C-5: 114 (1979).
5. K. P. Gupta, C. H. Cheng and P. A. Beck, Phys. Rev. 133:A203 (1964).

# NEUTRON SCATTERING STUDIES OF HYDRIDES OF THE LAVES PHASE RARE EARTH COMPOUNDS $RFe_2$

G. E. Fish* and J. J. Rhyne
National Bureau of Standards, Washington, DC  20234

T. Brun, P. J. Viccaro, D. Niarchos, B. D. Dunlap, and
G. K. Shenoy
Argonne National Laboratory[+], Argonne, IL  60439

S. G. Sankar* and W. E. Wallace
University of Pittsburgh**, Pittsburgh, PA  15260

We have used neutron scattering to study the structure and
magnetic ordering of a series of hydrides and deuterides of the
rare earth-iron compounds $RFe_2$.  The parent $RFe_2$'s have the cubic
close packed Laves phase (C15) structure for R heavier than Nd and
order ferrimagnetically at $\sim$600 K with the full free ion moment on
each R and $\sim 1.6\mu_B$/Fe at saturation.  The stable hydride phases with
$\sim$2 and $\sim$3.5 H(D) per formula unit are known from x-ray and neutron
diffraction to retain the C15 cubic structure, with lattice parameters increased by $\sim$5% and $\sim$7%.  The $RFe_2H_4$ phase is rhombohedrally
distorted and is not ordered at 4.2 K (1).

Figure (1) shows the magnetization of both Er and Fe sublattices
obtained by neutron scattering for $ErFe_2$, $ErFe_2D_2$, and $ErFe_2D_{3.5}$.
The pattern illustrated is common to all the $RFe_2H_2$ and $RFe_2H_{3.5}$'s
studied (2).  The Fe moment either is unchanged or is raised by a
slight amount (comparable to experimental error), while the coherent
R moment drops substantially.  The lowering of $T_c$ in the hydrides
signals a reduction in the Fe-Fe exchange which is known to dominate
in the parent $RFe_2$'s.  The R-Fe exchange, weak in $RFe_2$, is further
lowered, particularly in the case of $ErFe_2D_{3.5}$, where the Er sublattice disorders well below the overall $T_c$.  Sublattice magnetizations in these hydrides have also been inferred from measurement of
hyperfine fields using $^{57}$Fe, $^{161}$Dy, and $^{166}$Er Mossbauer spectroscopy
(1,3).  These measurements, unlike the neutron data, probe the local
moment and not a coherent average.  They reveal an Fe moment in
agreement with the present values, but the full free ion moment on
R.  This indicates that the weakened exchange coupling to R and the
spatially varying anisotropy field produced by random H site occu-

pancy allow the colinear R spin alignment seen in RFe$_2$ to be broken.
The resulting "fanning" of R spins gives a progressively lower observed R moment (defined as the net projection of the R spin along
the Fe spin direction) as the H content increases.

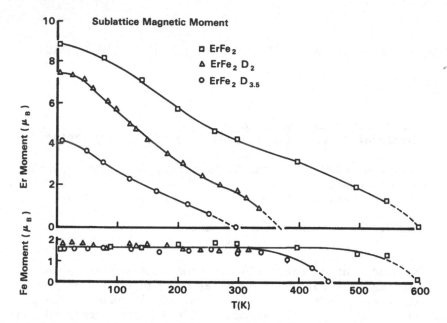

Figure 1.  Magnetization of Er and Fe sublattices in ErFe$_2$, ErFe$_2$D$_2$,
and ErFe$_2$D$_{3.5}$.  Curie temperatures (T$_c$) for ErFe$_2$ and ErFe$_2$D$_{3.5}$ were
determined by bulk magnetization.  For ErFe$_2$D$_2$, desorption begins
below T$_c$, so moments were determined using a smooth extrapolation
of diffraction peak intensity to the nuclear limit.

## REFERENCES

*Present address, Allied Chemical Corp., Morristown, NJ.
+Supported by the U. S. Department of Energy.
**Supported by Pet. Res. Fund of ACS and NSF CHE77/27252.

1.  P. J. Viccaro, G. K. Shenoy, B. D. Dunlap, D. G. Westlake and
    J. F. Miller, Journal de Phys., 40, C2-198 (1979).

2.  J. J. Rhyne, G. E. Fish, S. G. Sankar, and W. E. Wallace, Journal
    de Phys., 40, C5-209 (1978) and G. E. Fish, J. J. Rhyne,
    S. G. Sankar, and W. E. Wallace, J. Appl. Phys., 50, 2003
    (1979).

3.  P. J. Viccaro, J. M. Friedt, D. Niarchos, B. D. Dunlap,
    G. K. Shenoy, A. T. Aldred, and D. G. Westlake, J. Appl. Phys.,
    50, 2051 (1979).

# KINETICS OF DESORPTION OF HYDROGEN FROM $RFe_2H_n$ Systems[*]

H. Imamura and W. E. Wallace

Department of Chemistry, University of Pittsburgh

Pittsburgh, PA   15260

## ABSTRACT

The kinetics of desorption of hydrogen from $GdFe_2H_{3.2}$, $TbFe_2H_{3.3}$, $HoFe_2H_{3.3}$, $ErFe_2H_{3.3}$ and $TmFe_2H_{2.7}$ were measured over a range of temperature extending from 200 to 230 K. The desorption rate was measured using two methods:  (1) by following the increase in gaseous pressure in a closed system and (2) by determining the rate of formation of $C_2H_6$ from $C_2H_4$. The initial desorption rate in each case conformed to the expression $-\dfrac{dC}{dt} = kC$, where $k$ = the rate constant and C is the concentration of hydrogen in the compound. The observed behavior is accounted for by assuming that the rate is controlled by interfacial diffusion of monatomic hydrogen. $k$ varies with temperature according to the Arrhenius expression with activation energies ranging from 6.2 to 7.8 kcal/mole.

## INTRODUCTION

It has been shown that many rare earth intermetallic compounds reversibly absorb large amounts of hydrogen. The dissolved hydrogen exists in the atomic state and occupies interstitial positions in the parent compound (1-4). Recently, because of their large absorption capacity along with high absorption or desorption rates much attention has been directed to these intermetallics. The physical properties of intermetallic-hydrogen systems have been extensively

*Acknowledgement is made to the Donors of the Petroleum Research Fund, administered by the American Chemical Society, for the support of this work.

examined from the viewpoint of thermodynamics (1-4), crystallographics (1-3,5,6) and magnetism (7). The kinetic features of the process have also been studied (8-10), but less extensively.

The absorption process is accomplished by bulk penetration of the hydrogen adsorbed dissociatively on the absorber surface and the reverse desorption process must include the recombination reaction of atomic hydrogen. Thus the hydrogen-absorbing intermetallics can essentially activate hydrogen molecules on the surface and, accordingly, they effectively catalyze hydrogenation reactions of olefins or dienes (11,12).

In this work, we studied the desorption kinetics of hydrogen from $RFe_2H_n$ (R = Gd, Tb, Ho, Er and Tm) by using two methods: (I) by determining the rate of pressure increase in a closed system as a conventional method and (II) by determining the rate of formation of $C_2H_6$ from $C_2H_4$.

## EXPERIMENTAL

The intermetallic compounds were prepared by techniques which are standard in this laboratory - melting of the component metals in a water-cooled copper boat under an atmosphere of purified argon.

Before kinetic measurement for the intermetallics, it was necessary to activate the samples. This was accomplished by heating the samples at 300°C under vacuum ($\sim 10^{-5}$ torr), exposing to hydrogen, cooling gradually to room temperature and then slowly heating to 300°C under vacuum. This treatment was repeated until hydrogen absorption took place without difficulty.

The measurement of the desorption rate was made with about 0.15 g of each sample thus prepared. Prior to every measurement the compounds were outgassed at 280°C to $\sim 10^{-5}$ torr for 3 hrs, exposed to hydrogen at pressure up to 20mm Hg cooled slowly to -78°C. The composition of each compound was determined from this pressure drop of hydrogen in the closed system. The hydrides formed had the compositions $GdFe_2H_{3.2}$, $TbFe_2H_{3.3}$, $HoFe_2H_{3.3}$, $ErFe_2H_{3.3}$ and $TmFe_2H_{2.7}$. After the hydrogenated samples were briefly evacuated to remove hydrogen in the gas phase, the increase in pressure was continuously measured by a Pirani gauge or a manometer.

The hydrogenation reaction of ethylene was studied in a flow reactor with about 0.5 g of $RFe_2H_n$ catalysts. After the preparation of $RFe_2H_n$ at -78°C, the reactor was briefly evacuated to remove hydrogen in the gas phase and then a mixture gas of ethylene and helium as a carrier gas (He/$C_2H_4$ = 7.9) was introduced into the system at 1 atm total pressure. The reacting gas was collected from the system into a gas sampler and transferred to gas chromatography.

H$_2$, D$_2$ and HD in an exchange reaction were analyzed by mass spectrometry.

## RESULTS AND DISCUSSION

Desorption rates were measured using GdFe$_2$H$_{3.2}$, TbFe$_2$H$_{3.3}$, HoFe$_2$H$_{3.3}$, ErFe$_2$H$_{3.3}$ and TmFe$_2$H$_{2.7}$. These contained 1-2 matom of H. Typical time courses of the desorption from the hydrogenated samples are shown in Fig. 1. According to dependence of the hydride concentration on the desorption rate plots of log C against time (Fig. 2) are linear, indicating that the desorption process in this experiment is describable by first order kinetics. The temperature coefficient of the first order rate constant (Fig. 3) gives for the apparent activation energies for GdFe$_2$H$_{3.2}$, TbFe$_2$H$_{3.3}$, HoFe$_2$H$_{3.3}$, ErFe$_2$H$_{3.3}$ and TmFe$_2$H$_{2.7}$, 7.8, 7.0, 7.4, 7.1 and 6.2 kcal/mol, respectively.

The desorption process essentially consists of bulk diffusion, injection into the interfacial region (13), interfacial diffusion, release from the interfacial region perhaps by recombination and finally the release into the gas phase. An alternative is the release from the interfacial region as atoms and recombination on the surface. To ascertain whether the latter could be rate-determining, an exchange reaction of H$_2$ and D$_2$ (H$_2$/D$_2$ = 1) was undertaken over ErFe$_2$H$_{3.3}$. The reaction easily occurred at -196°C and was found to

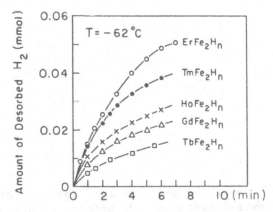

Fig. 1.  Time dependence of the desorption of hydrogen from several hydrides.

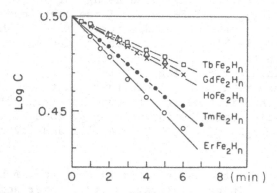

Fig. 2.  Plot of log C versus time for several hydrides.  n in the
formula $RFe_2H_n$ is taken as a measure of C.

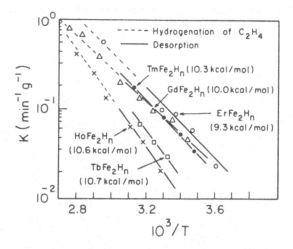

Fig. 3.  Dependence of the rate constant for extraction of hydrogen
from several hydrides.  Quantities in parentheses are
effective activation energies.

be first order.  The exchange rate at -196°C was observed to be much larger than the desorption rate extrapolated to -196°C.  This establishes that surface recombination on Fe (vide infra) and the final process, i.e., hydrogen release into the gas phase, are not rate-determining.  Moreover, taking into consideration that the desorption process followed the first order kinetics, it seems unlikely that the diffusion in the bulk is a rate-determining step.  Accordingly, for the present study it can be deduced that the rate of the desorption process is likely controlled by the interfacial diffusion of the absorbed hydrogen to the surface.  Moldovan, Smith and Wallace (14) have shown that the surface of $ErFe_2$ consists of a film of $Er(OH)_3$, iron[*] and an oxide of iron.  Diffusion through this film seems rate-determining, in contrast with the situation for the $LaNi_5$-H system in which release $f$ om the interfacial region is rate-controlling (13).

In the work of Soga et al. (11,12) it was found that the hydrogenation rate of $C_2H_4$ at -78°C was nearly equal to the desorption rate.  The results obtained can be explained by assuming the mechanism that the migration process of hydrogen through the surface film is rate-determining.

To investigate the mechanism of the present desorption process in further detail, the hydrogenation reaction of ethylene over $RFe_2H_n$ was carried out by introducing ethylene and helium as a carrier gas.  Slight reaction occurred around room temperature.  Surprisingly, the hydrogenation reaction readily occurred at -78°C over $RFe_2H_n$ by admitting a mixture of ethylene and hydrogen.  This phenomenon was significantly different from that observed by Soga et al. in which it was observed that $C_2H_4$ reacted more readily with the hydride than with gaseous $H_2$.  The present behavior is probably due to irreversible changes of the adsorbed ethylene, e.g., carbon deposit as a result of dehydrogenation on the surface, which retards hydrogen escape from the hydride.  This was confirmed by the desorption measurement from $RFe_2H_n$ treated with ethylene.  For the hydrogenated samples exposed to ethylene the desorption and hydrogenation rates were approximately equal (Fig. 3), indicating that the hydrogenation reaction proceeded in the same mechanism as that proposed by Soga et al.

The desorption rate from the hydrogenated samples was larger by over two orders of magnitude than that from the same samples treated with ethylene.  However, the desorption process from both the untreated and treated samples obeyed first order kinetics.  This implies that the desorption rate is probably controlled by the diffusion of the absorbed hydrogen in the interfacial region.

---

[*]Probably the fast $H_2$-$D_2$ exchange takes place on this surface.

REFERENCES

1.  J. H. N. Van Vucht, F. A. Kuijpers and H. C. A. N. Bruning, Philips Res. Repts. 25, 133 (1970).
2.  T. Takeshita, W. E. Wallace and R. S. Craig, J. Inorg. Chem. 13, 2282 (1974).
3.  F. A. Kuijpers, Philips Res. Repts., Suppl. 2 (1973).
4.  H. H. van Mal, ibid., Suppl. 1 (1976).
5.  P. Fischer, A. Furrer, G. Busch and L. Schlapbach, Helv. Phys. Acta 50, 421 (1977).
6.  A. F. Andresen in Hydrides for Energy Storage, A. F. Andresen and A. J. Maeland, Eds., Pergamon Press, 1978, p. 61.
7.  W. E. Wallace, "Magnetic Properties of Metal Hydrides and Hydrogenated Intermetallic Compounds," Springer-Verlag (1977).
8.  O. Boser, J. Less-Common Metals 46, 91 (1976).
9.  S. Tanaka, J. D. Clewley and T. B. Flanagan, J. Phys. Chem. 81, 1684 (1977).
10. A. Goudy, W. E. Wallace, R. S. Craig and T. Takeshita, Advances in Chemistry Series 167, 312 (1978).
11. K. Soga, H. Imamura and S. Ikeda, J. Phys. Chem. 81, 1762 (1977).
12. K. Soga, H. Imamura and S. Ikeda, J. Catal. 56, 119 (1979).
13. W. E. Wallace, R. F. Karlicek, Jr. and H. Imamura, J. Phys. Chem. 83, 1708 (1979). See also paper in this conference.
14. A. G. Moldovan, H. K. Smith and W. E. Wallace, unpublished work.

# DECOMPOSITION KINETICS OF PLUTONIUM HYDRIDE *

John M. Haschke, Jerry L. Stakebake

Rockwell International, Rocky Flats Plant

P. O. Box 464, Golden, CO 80401

## INTRODUCTION

Actinide hydrides are frequently employed as intermediates for preparing the nitrides or the powdered metals and, therefore, are particularly important compounds of elements employed as nuclear fuels. Of the three actinides (Th, U, Pu) of interest to the nuclear-reactor industry, the least is known about the properties of the plutonium-hydrogen system.

Earlier studies have shown that the complex behavior of plutonium closely resembles that of the lanthanides in that it reacts with hydrogen to form a $CaF_2$-type dihydride, a cubic hydride of variable composition between $PuH_2$ and $PuH_3$, and a $LaF_3$-type trihydride.[1-3] Thermodynamic data based on PTX equilibrium measurements have been included in these reports, but kinetic data for plutonium hydride, $PuH_x$, are quite limited. Rate data have been reported for the reaction of massive and powdered Pu.[4-6] The absence of kinetic data for the decomposition process is somewhat surprising, since the only practical method for preparing powdered plutonium is based on thermal decomposition of the hydride.[7,8] The present study was undertaken to investigate the dissociation kinetics of $PuH_x$ both in the two-phase region (0<X<1.95) and in the nonstoichiometric region (1.95<X<3.01).

## EXPERIMENTAL

Electrorefined $\alpha$-phase plutonium (461 wppm impurity) and $\delta$-stabilized plutonium with 1.0 wt % Ga were used in these tests. The Ga alloy, which contained 0.13 wt % impurities, was used

exclusively after initial tests demonstrated that the hydrides
from alloyed and unalloyed samples behaved identically. Ultra-pure
hydrogen used in the experiments was purified with $UH_3$. $PuH_x$ was
prepared and dehydrided in Al or quartz buckets on two vacuum
microbalance systems, a Cahn Model RG and a Cahn Model 100. $H_2$
was admitted to the systems at controlled temperatures and pres-
ures to form $PuH_x$ samples in the range $2.7 < X < 3.0$. The mass of $H_2$
evolved during thermal decomposition was measured as a function of
time, temperature and $H_2$ pressure.


## RESULTS AND DISCUSSION

### The $Pu-PuH_{1.95}$ Region

Equilibrium Measurements.  Static equilibrium pressure data
were measured in a temperature range (349-498 °C) which was below
the 600-800 °C range of the previous study.(2)  Results show that
the equilibrium hydride composition is approximately $PuH_{1.95}$ and
that the vaporization reaction is best described by Eqn. 1.

$$1.026 \ PuH_{1.95} (s) \rightleftarrows 1.026 \ Pu(s) + H_2(g). \tag{1}$$

The least squares refinement of the $\ell nP$ ($H_2$, torr) vs $1/T$ data
yields a slope of $(-19150 \pm 850)$ and an intercept of $(22.9 \pm 1.21)$.
For Eqn. 1, $\Delta H°_{697} = 38.1 \pm 1.7$ kcal/mol and $\Delta S°_{697} = 32.3 \pm 1.7$
cal/deg mol.  The enthalpy of formation calculated for $PuH_{1.95}$ at
697K, $-37.1 \pm 1.7$ kcal/mol, is in excellent agreement with the results reported by Mulford and Sturdy(2) for the hydride prepared
from  α-plutonium.

Kinetic Measurements.  Decomposition rates for $PuH_{1.95}$ were
obtained from linear mass-loss isotherms.  Since the decomposition
rates defined by their slopes are independent of the $PuH_{1.95}$ to Pu
ratio, the kinetics of the reaction are best described as zero
order.  An evaluation of the data obtained with both balance systems show that the absolute decomposition rate, K, is influenced
by vacuum system conductance, sample container material, and sample configuration.  For a given set of conditions, the rates
measured for hydride prepared from α-Pu were in excellent agreement with those of samples prepared from the δ-stabilized alloy.
K is apparently independent of the hydride surface area, which
varied from 0.10 to 0.30 $m^2/g$.  The effects of temperature and
vacuum system conductance are shown by the Arrhenius results in
Fig. 1.  The data points indicated by open circles were obtained
with the vacuum system at maximum conductance; those marked by
solid symbols were obtained with the system at a reduced conductance.  Although the absolute rate is reduced by decreasing the
pumping speed, the energies of activation are identical.  The
average $E_a$ obtained from the data sets in Fig. 1 and from a third

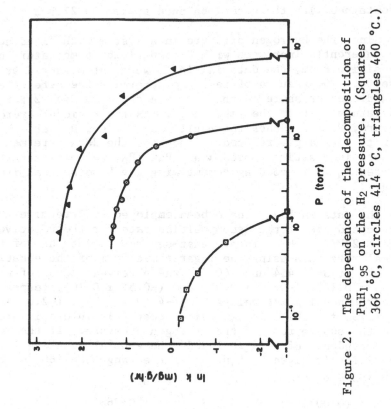

Figure 2.   The dependence of the decomposition of PuH₁.₉₅ on the H₂ pressure. (Squares 366 °C, circles 414 °C, triangles 460 °C.)

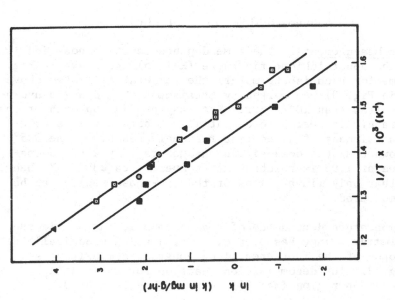

Figure 1.   Arrhenius results for the decomposition of PuH₁.₉₅. (Open symbols are for maximum vacuum conductance; solid symbols are for reduced conductance.)

set obtained with the second balance system is $27.3 \pm 1.4$ kcal/mol.

Since the hydrogen pressure in a system with fixed conductance inherently increases with decomposition temperature or decomposition rate, the data in Fig. 1 were not measured at the same pressure. The results of tests to determine the effects of residual hydrogen pressure on the rate are shown in Fig. 2 for three temperatures within the test range. As the residual hydrogen pressure, P, approaches the equilibrium value, $P_0$, at $\ell nk = -\infty$, the rate is sharply reduced. Since all the data were measured in pressure insensitive ranges where $P \leq 0.1 P_0$, we believe that the reported $E_a$ is in good agreement with the hypothetical isobaric value.

The data in Fig. 2 have been employed to determine the pressure dependence of the decomposition rate for the relative pressure range $0.2 < P/P_0 < 0.6$. Linear least-squares refinements of the isothermal rate data using the logarithmic form of the equation $K = AP^n$ follow. At 366, 414 and 460 °C, the observed values of n are $(-0.48 \pm 0.13)$, $(-0.49 \pm 0.07)$ and $(-0.55 \pm 0.10)$, respectively; the corresponding $\ell nA$ values are $(-4.49 \pm 1.16)$, $(-2.02 \pm 0.43)$ and $(-1.03 \pm 0.50)$. It is obvious that K is inversely proportional to the square root of the hydrogen pressure. If the temperature dependence of A is included, the general equation for decomposition is given by Eqn. 2 for the range 366-460 °C, P in torr and K in mg/g hr.

$$K = (0.00369 \ °C - 1.349)P^{-1/2}. \quad (°C > 366) \tag{2}$$

### The Nonstoichiometric Hydride

Kinetic Measurements. The time dependence of composition isotherms for the nonstoichiometric range (X>1.95) are shown in Fig. 3. Zero time for each isotherm (cf. the temperature listed above each curve in Fig. 3) was marked by attainment of a dynamic hydrogen pressure less than $10^{-3}$ torr. The asymptotic approach of the lower temperature isotherms to essentially constant values coincides with attainment of a constant residual pressure. The 305° isotherm shows that the composition changes rapidly until a constant X-value of 1.95 is reached. At temperatures ≥389°, X changes so rapidly that only linear rates of the two-phase region can be accurately measured.

The temperature dependence of the decomposition rate in the nonstoichiometric range has been quantified using graphically determined slopes of the isotherms along constant-composition sections of Fig. 3. The decomposition reaction for the hydride in the solid solution region (X>1.95) is described in Eqn. 3.

$$PuHx(s. \; soln.) \xrightleftharpoons{} Pu(s. \; soln.) + X/2 \; H_2(g).$$ (3)

Use of the formulation "Pu(s. soln.)" implies that the condensed product is a hydride that is more metal-rich than the reactant. Values of $E_a$ (in kcal/mol) obtained by Arrhenius analysis of the rate data at constant X-values are: 26.0 at 2.10; 20.5 at 2.20; 17.5 at 2.30; 11.2 at 2.40; 8.3 at 2.50; and 7.8 at 2.60. These results show that $E_a$ is inversely proportional to X over the accessible composition range.

Preparative Methods. Another result derived from the decomposition isotherms in Fig. 3 is important to the preparative chemistry of the nonstoichiometric hydride. For the 25-250 °C range, a graph of the terminal composition vs. isotherm temperature is linear with a slope of $(-3.300 \pm 0.007 \; X \; 10^{-3}$ X-units/°C) and an intercept of $(2.78 \pm 0.01)$. Compositions are accurately predicted by this relationship only if the $PuH_x$ is freshly prepared.

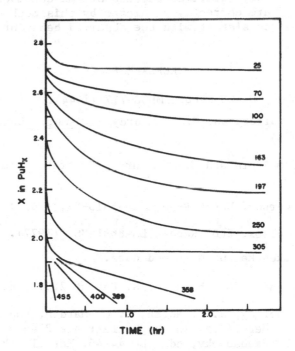

Figure 3. Decomposition Isotherms of Nonstoichiometric

CONCLUSIONS

A closer examination of the kinetic data for decomposition of $PuH_{1.95}$ provides insight into a possible mechanism for the hydriding and dehydriding reactions of plutonium. The fact that the rate of the hydriding reaction, $K_H$, is proportional to $P^{1/2}$ (5,6) and the rate of the dehydriding process, $K_D$, is inversely proportional to $P^{1/2}$ suggests that the forward and reverse reactions proceed by opposite paths of the same mechanism. The $P^{1/2}$ dependence of hydrogen solubility in metals is characterestic of the dissociative absorption of hydrogen;(9) i.e., the reactive species is atomic hydrogen. It is reasonable to assume that the rates of the forward and reverse reactions are controlled by the surface concentration of atomic hydrogen, $[H_s]$, that $K_H = c'[H_s]$, and that $K_D = c/[H_s]$, where $c'$ and $c$ are proportionality constants. For this surface model, the pressure dependence of $K_D$ is related to $[H_s]$ by the reaction $[H_s] \rightleftarrows 1/2 \ H_2(g)$ and by its equilibrium constant $K_e = [H_2]^{1/2}/[H_s]$. In the pressure range of ideal gas behavior, $[H_s] = K_e^{-1}(RT)^{-1/2}P^{1/2}$ and the decomposition rate is given by $K_D = cK_e(RT)^{1/2}P^{-1/2}$. For an analogous treatment of the hydriding process with this model, it can be readily shown that $K_H = c'K_e^{-1}(RT)^{-1/2}P^{1/2}$. The inverse pressure dependence and direct temperature dependence of the decomposition rate given by Eqn. 2 are correctly predicted by this mechanism which we believe is most consistent with the observed behavior of the Pu-H system.(10)

## REFERENCES

1.   I. B. Johns, USAEC Report MDDC-717 (1944).

2.   R. N. R. Mulford and G. E. Sturdy, J. Amer. Chem. Soc., 77: 3449 (1955).

3.   R. N. R. Mulford and G. E. Sturdy, J. Amer. Chem. Soc., 78; 3897 (1956).

4.   D. F. Bowersox, USAEC Report LA-5515-MS, (1974).

5.   D. F. Bowersox, ERDA Report LA-6681-MS, (1977).

6.   J. L. Stakebake, unpublished data.

7.   S. Fried and H. Baumbach, U. S. Patent 2,915,362, Dec. 1, 1959.

8.   M. J. F. Notley, J. M. North, P. G. Mardon, and M. B. Waldron, in "Powder Metallurgy in the Nuclear Age-Plansee Proceedings 1961," F. Benesovsky, ed., pp 44-45, Metallwerk Plansee AG, Reutte, Austria, 1962.

9.   S. Dushman, "Scientific Foundations of Vacuum Technique," J. M. Lafferty, ed., ch. 8 revised by F. J. Norton, John Wiley and Sons, New York, 1962.

10.  J. M. Haschke and J. L. Stakebake, DOE Report RFP-2878.
*    Funded by U.S. Department of Energy.

# THE KINETICS OF HYDROGEN ABSORPTION BY RARE EARTH INTERMETALLICS

W. E. Wallace, R. F. Karlicek, Jr. and H. Imamura

Department of Chemistry, University of Pittsburgh

Pittsburgh, PA 15260

Intermetallic compounds containing the rare earths in chemical union with d-transition metals absorb large amounts of hydrogen rapidly and reversibly (1). Hydrogenated $LaNi_5$, for example, at room temperature releases $\sim$90% of its hydrogen in a five-minute period (2). $ErFe_2$ absorbs hydrogen at 25 C to 95% of saturation in approximately a minute (3). The factors responsible for the rapid absorption or release of hydrogen have heretofore not been adequately revealed. Since hydrogen is absorbed dissociatively, it is clear that a sequence of events of considerable complexity is involved in the sorption process. Obviously dissociation of $H_2$ at the surface is involved, but until now it has not been clear whether this is rate determining for absorption.

Since dissociation and recombination take place at the surface, it is obviously of interest to have detailed information about the surface features of the solid – the elemental composition, valence states of the surface species, etc. This information is beginning to be provided. Several investigators (4-6) have shown that the surface of $LaNi_5$ (approximately the top 50 Å) consists of Ni plus $La_2O_3$ or $La(OH)_3$. Since the oxide and Ni do not absorb hydrogen in quantity, access to the underlying $LaNi_5$ must take place along the $La_2O_3$-Ni interface. The absorption process involves the following: (1) chemisorption of $H_2$ on the exposed Ni, (2) cleavage of the hydrogen bond in the chemisorbed molecular hydrogen, (3) diffusion along the $Ni-La_2O_3$ interface and (4) solution in $LaNi_5$.

Evidence indicates that each of these steps for $LaNi_5$ and its overlayer is rapid and none is rate determining. Soga, Imamura and Ikeda (7) have studied the reaction $H_2 + D_2 \rightarrow 2HD$ over $LaNi_5$ and found it to be $\sim 10^3$ faster at room temperature than the sorption rate of $H_2$ into $LaNi_5$.  These same investigators found (8) that when $C_2H_4$ is flowed over hydrogenated $LaNi_5$, hydrogen is removed much more rapidly ($\sim 10^2$ times faster) than by depressurization. Detailed analysis of the results of Soga, Imamura and Ikeda made elsewhere (9) shows that intergranular diffusion is so rapid that it cannot be rate determining for the hydrogenation of $LaNi_5$.  In that analysis it was concluded that transfer of atomic hydrogen from Ni into the intergranular space (or vice versa) is the rate-determining step for $H_2$ sorption by $LaNi_5$.

Desorption of hydrogen from $LaNi_5H_x$ is second order (9,10). The desorption of hydrogen from $R_2Co_7H_x$ systems is also second order (11) and the behavior of these systems is generally similar to $LaNi_5H_x$.  Although the surface features of the $R_2Co_7$ systems have not been established, it seems likely that they possess an outer layer of $R_2O_3$ [or $R(OH)_3$] and Co and hence the sorption process is similar to that for $LaNi_5$.  Auger spectroscopy (AES) results for $RCo_5$ systems show (12) a surface consisting of $R_2O_3$ [or $R(OH)_3$] and Co.  AES results for $ErFe_2$ show a surface consisting of $Er_2O_3$ or $Er(OH)_3$, Fe and an oxidized form of Fe.  The differing surface features of $RFe_2$ lead to differing kinetic features, as is indicated in an investigation of the sorption of hydrogen by $RFe_2$ systems (R = Tb, Ho, Er, Tm) reported elsewhere in this conference (13).

## REFERENCES

1.  W. E. Wallace, R. S. Craig and V. U. S. Rao, in "Advances in Chemistry Series," in press.
2.  J. H. N. Van Vucht, F. A. Kuijpers and H. C. A. M. Bruning, Philips Res. Repts. 25:133 (1970).
3.  D. M. Gualtieri, K. S. V. L. Narasimhan and T. Takeshita, J. Appl. Phys. 47:3432 (1976).
4.  H. C. Siegmann, L. Schlapbach and C. R. Brundle, Phys. Rev. Lett. 40:972 (1978).
5.  A. G. Moldovan, Ph.D. Thesis, University of Pittsburgh, 1978
6.  Th. von Waldkirch and P. Zürcher, Appl. Phys. Lett. 33:689 (1978).
7.  K. Soga, H. Imamura and S. Ikeda, Nippon Kagaku Kaishi 9:1304 (1977).
8.  K. Soga, H. Imamura and S. Ikeda, J. Phys. Chem. 81:1762 (1977).
9.  W. E. Wallace, R. F. Karlicek, Jr. and H. Imamura, J. Phys. Chem. 83:1708 (1979).
10. O. Boser, J. Less-Common Metals 46:91 (1976).
11. A. Goudy, W. E. Wallace, R. S. Craig and T. Takeshita, in "Advances in Chemistry Series" 167:312 (1978).
12. A. G. Moldovan and W. E. Wallace, unpublished measurements.
13. H. Imamura and W. E. Wallace, this conference.

# THERMODYNAMIC AND STRUCTURAL PROPERTIES OF LaNi$_{5-x}$Mn$_x$ COMPOUNDS AND THEIR RELATED HYDRIDES

C. Lartige, A. Percheron-Guégan and J. C. Achard, ER209,
C.N.R.S., 1, A. Briand, 92190 – Meudon, France
and
F. Tassett, Institut Laüe Langevin, 156X, 38042 Grenoble
Cédex, France

## INTRODUCTION

It has been shown from the linear correlation between the hydrides equilibrium pressure and the lattice dimensions of the original alloys, that substituting nickel atoms in LaNi$_5$ compounds by larger atoms leads to more stable hydrides. This is a good way to extend the working temperature range of electrodes (1,2).

Unfortunately, this type of substitution was often accompanied by a reduction of the hydrogen capacity (3,4). For manganese substitution, on the contrary, this reduction is negligible (2,5,6).

The present work reports the existence of the solid solution LaNi$_{5-x}$Mn$_x$ up to x = 2.2 at 800°C, and describes the absorption-desorption isotherms of the related hydrides between 25 and 200°C.

The substitutional sites occupied by the manganese atoms in the intermetallic compounds as well as the interstitial sites of deuterium in their deuterides have been determined by neutron diffraction.

The comparison of these results with those for aluminum substitution (7-9), involving also an increase of the hydrides, is made based on the structural factors acting on the stability and composition of the hydrides.

## INTERMETALLIC COMPOUNDS

The compounds were prepared by induction melting the pure components (La 99.9%, Ni 99.9%, Mn 99.9%) under vacuum in a water

585

cooled copper crucible.  The stoichiometry and homogeneity of compounds were checked by metallographic examination and microprobe analysis.

The composition of the alloys corresponds to the formula $LaNi_{5-x}Mn_x$, with x increasing from 0 to 2.5 by steps of 0.25.  The melting point of the compounds (Mp) decreases rapidly with the manganese concentration:  Mp > 1150° for $X_{Mn}$ ≤ 1.25; 1150°C ≤ Mp ≤ 900°C for 1.5 ≤ $X_M$ < 2; 8000 < Mp < 900°C for 2 ≤ $X_{Mn}$ < 2.5.  As a consequence the limit of substitution was determined on the samples annealed at 800°C and then quenched to room temperature.

X-ray diffraction on powders shows that the compounds are monophase and retain the hexagonal $LaNi_5$ structure ($P6/_mmm$) up to $X_{Mn}$ = 2.2.  Beyond this concentration a second phase appears and replaces progressively the first one.  The composition of this phase is sensibly $La_2(Ni+Mn)_{17}$, and its structure is hexagonal with the $P6_3/mmc$ space group.

The introduction of manganese in the lattice leads to an increase of the lattice parameters, as shown on Fig. 1.  The cell volume increases linearly with increasing manganese concentration following the equation:  $V = 86.82 + 4.61 X_{Mn} Å^3$.

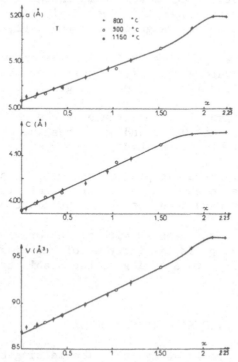

Figure 1.  Variation of the lattice dimensions versus $X_{Mn}$ in the intermetallic compounds.

Neutron diffraction measurements were performed on powdered samples of LaNi$_{4.5}$Mn$_{0.5}$, LaNi$_4$Mn and LaNi$_3$Mn$_2$ compounds using the D1B multidetector instrument at I.L.L., Grenoble ($\lambda$ = 1.282Å, 0.5° < $\theta$ < 67° by steps of 0.05°, up to sin$\theta$/$\lambda$ = 0.72Å$^{-1}$). The structures of the intermetallic compounds (Table 1) have been determined using the standard Rietveld technique (10). The structural model (11) is basically the CaCu$_5$ structure (12) with the possible substitution of a few lanthanum atoms (S1a) by pairs of nickel atoms in the position 2e of the P6/$_m$mm space group. The Mn atoms occupy mainly the 3g site in the plane z = 1/2 (S 2). A small proportion (nS1) of those occupying the 2c site in the plane z = 0 decreases regularly with the manganese concentration (19% for X$_{Mn}$ = 0.5, 14% for X$_{Mn}$ = 1 and 9% for X$_{Mn}$ = 2).


HYDRIDES


The hydrides were obtained by exposing samples to a hydrogen pressure of about 40 bar at room temperature. Samples were subjected to several adsorption-desorption hydrogen cycles before measurement, in order to obtain activated powders with a homogeneous grain size.

The pressure composition isotherms were determined between 25 and 200°C from incremental adsorption-desorption measurements (3). The adsorption-desorption isotherms at 60°C shown in Fig. 2 show that the increase of the manganese concentration X$_{Mn}$ in the intermetallic compounds leads to a large decrease in the plateau pressure of the re-related hydride from 6.4 atm for LaNi$_5$ to 0.1 atm for LaNi$_4$Mn. The estimated value for LaNi$_3$Mn$_2$ is 10$^{-5}$ atm. The logarithm of the plateau pressures of the hydrides varies linearly with the manganese concentration. This variation is represented by the following equations: LnP$_{25°}$ = -6.05 x$_{Mn}$ + 0.48; LnP$_{60°}$ = -5.81 x$_{Mn}$ + 1.91; LnP$_{200°}$ = -3.61 x$_{Mn}$ + 4.90. On the other hand, the hydrogen content changes very little with manganese substitution. As a matter of fact the

Table 1. Intermetallic Compounds LaNi$_{5-x}$M$_x$

| Compounds | La$_{1-s}$Ni$_{5+2s}$ | LaNi$_{4.5}$Mn$_{0.5}$ | | | LaNi$_4$Mn | | | LaNi$_3$Mn$_2$ | | |
|---|---|---|---|---|---|---|---|---|---|---|
| a Å | 5.016 | 5.045 | | | 5.089 | | | 5.174 | | |
| c Å | 3.983 | 4.022 | | | 4.082 | | | 4.145 | | |
| (n) S La | 0.017 (3) | 0.010 (3) | | | 0.005 (2) | | | 0.017 (2) | | |
| (n) S 1 | | 0.059 (16) | | | 0.128 (12) | | | 0.155 (20) | | |
| (n) S 2 | | 0.256 (22) | | | 0.779 (13) | | | 0.643 (9) | | |
| B La | 0.46 (4) | 0.63 (6) | | | 0.72 (5) | | | 0.53 (8) | | |
| B S 1 | 0.61 (4) | 0.76 (4) | | | 0.98 (4) | | | 0.93 (5) | | |
| B S 2 | 0.44 (2) | 0.56 (3) | | | 0.71 (3) | | | 0.59 (8) | | |
| Composi-tion | La 0.98  Ni 5.03 | La 0.99 | Ni 4.71 | Mn 0.315 | La 0.995 | Ni 4.1 | Mn 0.91 | La 0.983 | Ni 3.236 | Mn 1.798 |
| R Factor | 4.4% | 5% | | | 6.4% | | | 5.98% | | |

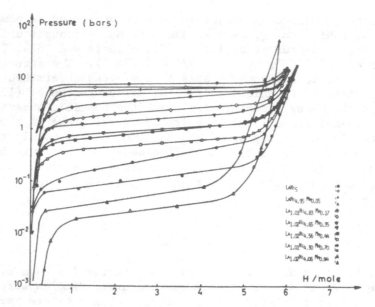

Figure 2.    Adsorption-desorption isotherms at 60°C for $LaNi_{5-x}Mn_x$ compounds.

capacity at room temperature increases up to 6.6 H/mole for $X_{Mn}$ = 0.3 and then decreases to 5.8 H/mole for $X_{Mn}$ = 2.

The enthalpy of formation of the hydrides increases linearly with $X_{Mn}$ from -7.48 Kcal/mole $H_2$ for $LaNi_5$ to -12.1 Kcal/mole $H_2$ for $La_{1.01}Ni_{4.16}Mn_{0.84}$ following the equation:

$$\Delta H° = -7.5 - 5.5 \, X_{Mn} \pm 0.2 \text{ Kcal/mole } H_2.$$

The entropy can be considered as being almost constant in this series; -28 ± 3 cal/K/mole $H_2$ to be compared with -30 ± 6 cal/K/mole $H_2$ for $LaNi_5$.

The x-ray diffraction patterns of the hydrides maintained at room temperature under 1 atm pressure of hydrogen can be indexed in the same hexagonal structure of the original alloys.  The relative increase of the unit cell volume of the hydrides with respect to those of the metallic compounds is 20% for $LaNi_4MnH_6/LaNi_4Mn$, 17% for $LaNi_3Mn_2H_{5.9}/LaNi_3Mn_2$, and both are slightly smaller than for $LaNi_5H_6/LaNi_5$ (25%).

Neutron measurements of the hydrides were made in the same experimental conditions as for intermetallic compounds.  Based on the

Table 2. Hydrides of Intermetallic Compounds

| Compounds | LaNi$_{4.5}$Mn$_{0.5}$D$_{6.6}$ | | | | LaNi$_4$MnD$_{5.9}$ | | | | LaNi$_3$Mn$_2$D$_{5.95}$ | | | |
|---|---|---|---|---|---|---|---|---|---|---|---|---|
| a (Å)  c (Å) | 5.387 | 4.289 | | | 5.406 | 4.329 | | | 5.451 | 4.344 | | |
| B La | 1.47 | | | | 1.53 | | | | 0.50 | | | |
| B S 1  B S 2 | 1.44 | 1.41 | | | 1.0808 | 1.53 | | | 1.29 | 2.28 | | |
| B$_D$ f | 3.66 | | | | 1.5 | | | | 1 | | | |
| B$_D$ o | 2.96 | | | | 2.23 | | | | | | | |
| B$_D$ n | 4.60 | | | | 3.17 | | | | 1.35 | | | |
| B$_D$ m | 0.16 | | | | 4.66 | | | | 3.20 | | | |
| B$_D$ h | | | | | 1.50 | | | | | | | |

| sites | x | y | z | n (D) | x | y | z | n (D) | x | y | z | n (D) |
|---|---|---|---|---|---|---|---|---|---|---|---|---|
| f | 0.225 | 0.451 | 0.316 | 1.28 | 0.5 | 0 | 0 | 0.35 | | | | |
| o | 0.466 | 0 | 0.102 | 2.78 | 0.232 | 0.464 | 0.319 | 0.81 | | | | |
| n | 0.138 | 0.276 | 0.5 | 2.205 | 0.461 | 0 | 0.105 | 2.48 | | | | |
| m | 0.333 | 0.666 | 0.408 | 0.34 | 0.132 | 0.264 | 0.5 | 2.06 | 0.445 | 0 | 0.118 | 1.70 |
| h | | | | | 0.333 | 0.666 | 0.40 | 0.2 | 0.127 | 0.255 | 0.5 | 2.05 |

| R Factor | 5.92% | | | | 6.74% | | | | 7.91% | | | |
| Composition | La 0.986 | Ni 4.68 | Mn 0.34 | D 6.6 | La 0.988 | Ni 4.12 | Mn 0.91 | D 5.9 | La 0.98 | Mn 3.25 | Ni 1.80 | D 4.79 |

previous neutron diffraction study of $LaNi_5H_6$ (14) several space groups such as P31m and P3 were tested. The best agreement with the experimental data was obtained with the more symmetrical space group $P6/_mmm$ for the three compounds: $LaNi_{4.5}Mn_{0.5}D_{6.6}$, $LaNi_4MnD_{5.9}$, $LaNi_3Mn_2D_{5.95}$. For the last compound, poor agreement results were obtained due to the occurrence of short range order. Among the interstitial sites available for deuterium a preferential occupaton exists for the 6m and 12n sites as can be seen in Table 2.

## DISCUSSION

By comparison of the results on the two systems $LaNi_{5-x}Mn_x$ and $LaNi_{5-x}Al_x$ (13) some remarks can be made:

- Although aluminum atoms (atomic radius Rat = 1.43Å) are larger than manganese (Rat = 1.37Å), both induce a comparable increase of the cell volume ($V_{LaNi_4Al}$ = 90Å$^3$, $V_{LaNi_4Mn}$ = 92Å3) when substituted for nickle atoms.

- Aluminum atoms can only enter in the 3g sites (z = 1/2) whereas the manganese replaces both nickel in the 3g sites and in the 2c sites (z = 0) leading to a much more disordered structure.

- For the deuterides, in spite of a very similar value of the cell volume of corresponding intermetallics ($LaNi_4Mn$, $LaNi_4Al$), the hydrogen concentrations are quite different ($LaNi_4MnH_6$, $LaNi_4AlH_{4.8}$). In the case of aluminum substitution only the 6m and 12n sites are occupied. In the case of manganese, these two sites are occupied to about the same extent. In addition the o, f, h sites become slightly occupied. It can be noted that the larger m and n sites in both compounds are the most filled, being 33% and 25% respectively.

## REFERENCES

1. G. Bronoël, J. Sarradin, A. Percheron-Guégan, and J. C. Achard, Int. J. Hydrogen Energy 1:251 (1976).
2. A. Percheron-Guégan, J. C. Achard, J. Sarradin and G. Bronoël, French Patents N° 7516160 Mai 1975, N° 7706138, Mars 1977, N° 7723812, Aout 1977.
3. J. C. Achard, A. Percheron-Guégan, H. Diaz and F. Briaucourt, 2nd Int. Cong. Hydrogen in Metals, Paris (1977), 1 E 12.
4. G. Busch, L. Schlapbach, and W. Toeni, 2nd Int. Cong. Hydrogen in Metals, Paris (1977), 1 E 12.
5. A. Percheron-Guégan, J. C. Achard, J. Sarradin and G. Bronoël, Hydrides for Energy Storage Geilo, August (1977), p. 485, Eds. A. F. Andresen and A. J. Maeland.
6. C. E. Lundin and F. E. Lynch, Int. Report Denver Research Institute, July (1977).

7.  H. Diaz, A. Percheron-Guégan, and J. C. Achard, 2nd World Hy-
    drogen Energy Conference Zurich, August (1978).
8.  J. C. Achard, F. Givord, A. Percheron-Guégan, J. L. Soubeyroux
    and F. Tasset, Colloque C.N.R.S., Physique des terres rares à
    l'état métallique St Pierre du Chartreuse, Sept. (1978).
9.  M. H. Mendelsohn, D. M. Gruen, and A. E. Dwight, J. Less Common
    Met. 63:193 (1979).
10. G. M. Rietveld, Acta Cryst. 22:151 (1967).
11. J. Schweizer and F. Tasset, J. Less Common Met. 18:245 (1965).
12. J. H. Wernick, Acta Cryst. 12:662 (1959).
13. J. C. Achard, P. Germi, C. Lartigue, A. Percheron-Guegan and
    F. Tasset, VI Int. Conf. on Solid Compounds of Transition Ele-
    ments, Stuttgart, June (1979).
14. P. Fischer, A. Furrer, G. Busch and L. Schlapbach, Helv. Phys.
    Acta 50:421 (1977).

# THE EFFECT OF GROUP III A AND IV A ELEMENT SUBSTITUTIONS (M) ON THE HYDROGEN DISSOCIATION PRESSURES OF LaNi$_{5-x}$M$_x$ HYDRIDES*

Marshall H. Mendelsohn and Dieter M. Gruen

Chemistry Division, Argonne National Laboratory

Argonne, IL   60439

Ternary modifications of the hydrogen absorbing alloy LaNi5 have been found to have a large effect on the hydrogen dissociation pressure of the corresponding hydride (1-4). Recently, we have studied a series of alloys LaNi$_{4.6}$M$_{0.4}$ (M = Al, Ga, In, Sn) to establish the effect of various group III A and IV A element substitutions on the dissociation pressures of their corresponding hydrides (5). In this paper, we extend those results to include the alloys LaNi$_{4.6}$Si$_{0.4}$, LaNi$_{4.6}$Ge$_{0.4}$ and LaNi$_{4.4}$B$_{0.6}$. Further dissociation pressure measurements have been made on the alloy LaNi$_{4.6}$Ga$_{0.4}$ as well as two additional indium alloys. The data obtained for the three indium alloys appear to follow a linear correlation of lnP or $\Delta$G with the amount of substituted element as has been previously observed for Al (1,6).

## EXPERIMENTAL

The experimental procedures and apparatus were essentially the same as described earlier (7). The initial hydrogen absorption (i.e. "activation procedure") for the alloys in this study was performed at hydrogen pressures of from 150-400 PSIA.

## RESULTS

Desorption isotherms for the hydrides LaNi$_{4.6}$Si$_{0.4}$H$_y$ and LaNi$_{4.6}$Ge$_{0.4}$H$_y$ are shown in Fig. 1 for temperatures of 30° and 60°C. Isotherms were also obtained at 40°C but are not shown.

---

*Work performed under the auspices of the U.S. Department of Energy.

An alloy of approximate composition $LaNi_{4.4}B_{0.6}$ was prepared and
its x-ray powder diffraction pattern was indexed as hexagonal
with a = 5.09 ± 0.01 Å and c = 3.69 ± 0.01 Å. A more detailed
report on this newly discovered ternary alloy will be published
elsewhere (8). Desorption pressures were measured at five temperatures and at approximately the same composition for the hydride
$LaNi_{4.4}B_{0.6}H_y$. From the data shown in Table I, the enthalpy and
entropy of transition were calculated to be -8.0 kcal/mole $H_2$ and
-23.4 cal/deg-mole $H_2$, respectively.

Two indium alloys of nominal composition $LaNi_{4.8}In_{0.2}$ and
$LaNi_{4.7}In_{0.3}$ were also prepared. However, as will be explained in
the discussion section, the actual compositions of the hexagonal
$AB_5$-type phases were $LaNi_{4.93}In_{0.07}$ and $LaNi_{4.85}In_{0.15}$ as determined
from their x-ray powder diffraction patterns. For $LaNi_{4.93}In_{0.07}$,
a = 5.031 ± 0.002 Å and c = 3.991 ± 0.001 Å, while for
$LaNi_{4.85}In_{0.15}$, a = 5.039 ± 0.003 Å and c = 4.018 ± 0.003 Å. Desorption isotherms for the indium samples were measured at three
temperatures and their plateau pressures are given in Table II.
Also given in Table II are plateau pressures measured for the
hydride of $LaNi_{4.6}Ga_{0.4}$. The thermodynamic and crystallographic
data for all of the group III A and IV A substituted alloys
studied to date are summarized in Table III. The enthalpies and
entropies given are for the transition of α-phase alloy to
β-phase hydride.

Figure 2 shows the approximately linear correlation observed
for both the Al and In substituted systems of lnP versus amount of
substituted element.

## DISCUSSION

The stoichimetries $LaNi_{4.93}In_{0.07}$ and $LaNi_{4.85}In_{0.15}$ were
obtained by measuring the cell volumes accurately and then reading
the In concentration from a plot previously obtained of cell
volume vs In concentration (9). The presence of an unknown second

Table I.

| Composition (H atom/mole $LaNi_{4.4}B_{0.6}$) | Temperature (°C) | $H_2$ Pressure (atm) |
|---|---|---|
| 0.76 | 30.0 | 0.210 |
| 0.76 | 35.0 | 0.261 |
| 0.76 | 40.0 | 0.326 |
| 0.74 | 45.0 | 0.395 |
| 0.74 | 50.0 | 0.479 |

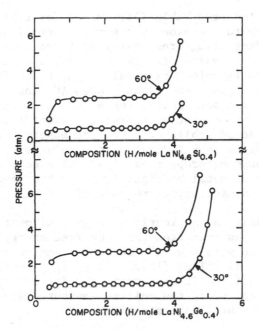

Figure 1.  Pressure vs composition isotherms for the hydrides
           of LaNi$_{4.6}$Si$_{0.4}$ and LaNi$_{4.6}$Ge$_{0.4}$.

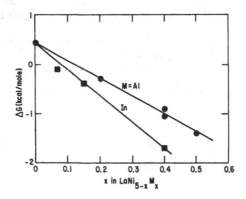

Figure 2.  Free energy vs x (x = 0-0.5) for the hydrides of
           LaNi$_{5-x}$Al$_x$ and LaNi$_{5-x}$In$_x$.

phase was noted by the appearance of 2 or 3 weak lines in the x-ray powder diffraction pattern of the annealed alloys.

As can be noted from Table III, except for $LaNi_{4.4}B_{0.6}$, all the alloys follow the previously observed approximately linear correlation of decreasing free energy of formation of $AB_5$ hydrides with increasing cell volume (1,6). However, a detailed examination of the data reveals a reversal of the trend for the two hydride pairs $LaNi_{4.6}Al_{0.4}H_y$ – $LaNi_{4.6}Ga_{0.4}H_y$ and $LaNi_{4.6}Si_{0.4}H_y$ – $LaNi_{4.6}Ge_{0.4}H_y$. For both pairs, there occurs a change in the electronic structure of the substituted element from empty 3d orbitals to filled 3d orbitals. Thus the observed reversal in hydride free energies may indicate the involvement of specific metal-hydrogen bonding properties in addition to the previously noted importance of metal-hydrogen (10) and hydrogen-hydrogen (11) distances on the dissociation pressure of the alloy hydrides.

The alloy $LaNi_{4.4}B_{0.6}$ clearly does not follow the above noted trend. Sandrock has suggested that the free energy may be more closely correlated with the "a" lattice parameter of the alloy rather than the cell volume (12). $LaNi_{4.4}B_{0.6}$ would be consistent with this correlation. In any event, the further explanation of this notable exception to the cell volume correlation awaits a more precise structural determination.

Although no extensive measurements were made of absorption isotherms, some data points taken in order to compare the magnitude of hysteresis of the Group III A and IV A substituted hydrides with the parent hydride $LaNi_5H_6$ are shown in Table IV. The reason for the reduction in hysteresis in some of the alloys compared to $LaNi_5$ is not clear, but is an important benefit in some proposed practical applications of metal hydrides (13).

Table II.

| Composition | Temperature (°C) | Pressure (atm) |
|---|---|---|
| $LaNi_{4.93}In_{0.07}H_{2.5}$ | 30 | 1.37 |
| $LaNi_{4.93}In_{0.07}H_{2.5}$ | 40 | 2.20 |
| $LaNi_{4.93}In_{0.07}H_{2.5}$ | 60 | 4.83 |
| $LaNi_{4.85}In_{0.15}H_{2.25}$ | 30 | 0.78 |
| $LaNi_{4.85}In_{0.15}H_{2.25}$ | 40 | 1.17 |
| $LaNi_{4.85}In_{0.15}H_{2.25}$ | 60 | 2.42 |
| $LaNi_{4.6}Ga_{0.4}H_{2.5}$ | 30 | 0.49 |
| $LaNi_{4.6}Ga_{0.4}H_{2.5}$ | 40 | 0.745 |
| $LaNi_{4.6}Ga_{0.4}H_{2.5}$ | 60 | 1.71 |

Table III.

| Alloy | $\Delta H_{\alpha \to \beta}$ $\dfrac{kcal}{mole\ H_2}$ | $\Delta S_{\alpha \to \beta}$ $\dfrac{cal}{mole\ H_2\text{-}deg}$ | $\Delta G$ for 20° | Alloy Cell Vol.(Å) |
|---|---|---|---|---|
| LaNi$_{4.4}$B$_{0.6}$ | −8.0 | −23.4 | −1.1 | 82.4 |
| LaNi$_{4.6}$Al$_{0.4}$ | −8.7 | −26.1 | −1.05 | 87.9 |
| LaNi$_{4.6}$Ga$_{0.4}$ | −8.4 | −26.2 | −0.7 | 88.1 |
| LaNi$_{4.93}$In$_{0.07}$ | −8.4 | −28.3 | −0.1 | 87.5 |
| LaNi$_{4.85}$In$_{0.15}$ | −7.6 | −24.5 | −0.4 | 88.4 |
| LaNi$_{4.6}$In$_{0.4}$ | −9.5 | −26.5 | −1.7 | 91.0 |
| LaNi$_{4.6}$Si$_{0.4}$ | −8.5 | −27.3 | −0.5 | 86.9 |
| LaNi$_{4.6}$Ge$_{0.4}$ | −8.2 | −26.5 | −0.4 | 87.8 |
| LaNi$_{4.6}$Sn$_{0.4}$ | −9.2 | −26.2 | −1.5 | 91.0 |

Table IV.

| Alloy | Temperature (°C) | H/mole alloy | P$_{abs}$ (atm) | P$_{des}$ (atm) | P$_a$/P$_d$ |
|---|---|---|---|---|---|
| LaNi$_{4.6}$Al$_{0.4}$[a] | 30 | 3.09 | 0.31 | 0.28 | 1.11 |
| LaNi$_{4.6}$In$_{0.4}$ | 30 | 1.96 | 0.097 | 0.095 | 1.02 |
| LaNi$_{4.4}$B$_{0.6}$ | 25 | 0.76 | 0.159 | 0.157 | 1.01 |
| LaNi$_{4.6}$Si$_{0.4}$ | 30 | 1.58 | 0.73 | 0.67 | 1.09 |
| LaNi$_{4.6}$Si$_{0.4}$ | 40 | 1.23 | 1.11 | 1.04 | 1.07 |
| LaNi$_{4.6}$Ge$_{0.4}$ | 30 | 1.32 | 0.85 | 0.78 | 1.09 |
| LaNi$_{4.6}$Ge$_{0.4}$ | 40 | 1.97 | 1.31 | 1.21 | 1.08 |
| LaNi$_{4.6}$Sn$_{0.4}$[b] | 20 | 2.41 | 0.078 | 0.079 | 0.99 |
| La$_{1.05}$Ni$_{4.6}$Sn$_{0.4}$[b] | 32 | 2.38 | 0.162 | 0.156 | 1.04 |
| LaNi$_5$[c] | 20 | 3.0 | 2.0 | 1.6 | 1.25 |

[a] from ref. 6
[b] two different samples
[c] from ref. 14

Acknowledgment.  We wish to thank Dr. A. E. Dwight for preparing some of the alloy samples.

References

1. M. H. Mendelsohn, D. M. Gruen and A. E. Dwight, Nature, 269, 45 (1977).

2. J. C. Achard, A. Percheron-Guegan, H. Diaz, F. Braincourt, and F. Denany, 2nd Int'l. Cong. on Hydrogen in Metals, Paris, France, June, 1977.

3. T. Takeshita, S. K. Malik and W. E. Wallace, J. Solid State Chem. 23, 271 (1978).

4. G. D. Sandrock, 2nd World Hydrogen Energy Conf., August 1978.

5. M. H. Mendelsohn, D. M. Gruen, and A. E. Dwight, Mat. Res. Bull., 13, 1221 (1978).

6. M. H. Mendelsohn, D. M. Gruen, and A. E. Dwight, J. Less-Common Metals, 63, 193 (1979).

7. M. H. Mendelsohn, D. M. Gruen, and A. E. Dwight in The Rare Earths in Modern Science and Technology; Plenum Press: New York, N.Y., 1978.

8. M. H. Mendelsohn, D. M. Gruen and G. D. Sandrock, to be submitted for publication.

9. A. E. Dwight, private communication.

10. E. S. Machlin, private communication.

11. A. C. Switendick, Sandia Laboratories, Report No. SAND 78-0250, April, 1978.

12. G. D. Sandrock, Proc. 12th IECEC, 1977.

13. D. M. Gruen, M. H. Mendelsohn and I. Sheft, Solar Energy, 21, 153 (1978).

14. F. A. Kuijpers and H. H. van Mal, J. Less-Common Metals, 23, 395 (1971).

# HYDROGEN COORDINATION NUMBER, VOLUME, AND DISSOCIATION PRESSURE

# IN LaNi$_5$ SUBSTITUTED HYDRIDES

F. L. Carter
Naval Research Laboratory
Washington, D.C. 20375, U.S.A.

J. C. Achard and A. Percheron-Guegan
Chimie Metallurgique des Terres Rares
CNRS, 1 pl. A. Briand
92190 Meudon, France

## INTRODUCTION

The concepts of Polyhedral Atomic Volume PAV (1-3), generalized coordination numbers CN, partial coordination numbers PCN, and PCN coefficients (4,5) are employed to elucidate further the insight proposed by Lundin, Lynch, and Magee (6) relating hydrogen decomposition plateau pressures to interstitial hole volumes in intermetallic compounds. Recent powder neutron diffraction studies (7,8) indicated the location of nickel substitutes and deuterium in several sites in such compounds as LaNi$_4$AlD$_4$ and LaNi$_4$MnD$_6$ related to the space group P6/mmm. By applying Pauling's Metallic Radii PMR (9) in combination with the above concepts to a further refinement of the diffraction data (7) for these compounds a more detailed analysis of the chemical characteristics of the deuterium sites is obtained (see also Mendelsohn, Gruen, and Dwight (10)). We further note that lanthanum recovers its PAV volume during the hydriding process. Finally we indicate several needed studies.

## METHOD

By ordering the nickel substitutes Al and Mn the PAV cells are calculated as the space filling Voronoi (11) (or Wigner-Seitz) cells but with the difference that the planes perpendicular to the internuclear axes bisect the distance between the outer radii of the atoms. These radii are Pauling's metallic radii R1 (9) obtained in a self-consistent manner employing the "neutral cell" approach (see 1-3). The advantage of PMR over the hard sphere model of Lundin

et al. (6) is that one obtains valence and charge estimates as well
as radii.  However, intermetallic compound formation leads to
dramatic volume contraction in the rare earth volume (1-3) and a
correspondingly larger d character and smaller Rl in the PMR self-
consistent calculations.  Uncertainty in the proper Rl for H or
deuterium is also a problem.  The use of PAVs also accounts for all
but 0.2% of the unit cell volume compared to hard sphere models that
typically leave 15-35% of the cell volume unassigned.

The generalization of the CN and PCN coefficient concepts can
make use of any measure of bonding interaction $A_{ij}$ between atom i
and its j neighbors provide that $A_{iT} = \Sigma \, A_{ij}$ (over j) is finite.  If
atom i has K different kinds of neighbors then it has been shown
(4,5) that

$$1/CN_i = \sum_j (A_{ij})^2/A_{iT}^2 = \sum_k^K f_{ik}^2/PCN_{ik} \qquad (1)$$

where all the PCN coefficients $f_{ik}$ sum to one, $1 = \Sigma \, f_{ik}$ (over k
kinds).  By further application of Eq. (1) it can be readily demonstrated that PCN and their coefficients can be combined as in Eq.
(2) to form new $PCN_{in}$

$$f_{in}^2/PCN_{in} = \Sigma \, f_{ik}^2/PCN_{ik} \qquad (2)$$

where $f_{in} = \Sigma \, f_{ik}$ and the sum is over the desired atoms.  The interactions $A_{ij}$ can refer to various measures, for example, bond energies,
bond orders, overlap integrals, etc.  We will make use of $CN_B$ referred
to bond orders and $CN_V$ referred to the pyramidal volume of atom i
having atom i as the apex and the common PAV polygon face between
atom i and atom j as the pyramid  base.

## RESULTS

The primary results are indicated in Tables 1 and 2.  In Table
1 we show the dramatic PAV decrease for La upon formation of $LaNi_5$
(see Refs. 1 and 3 for other examples) and how the La PAV increases
back toward the elemental volume (12) upon Al and Mn substitution and
then hydriding.  Also shown in Table 1 are the trigonal and icosa-
hedral nickel, Ni(T) and Ni(I), and Al and Mn volumes and $CN_V$.
The valences for the nickels and managanese are normal in the alloys,
as per $LaNi_4Mn$ - 4.7 (Ni-T), 4.5 (Ni-I), and 4.2 (Mn); for the
hydrides the valences vary appreciably, but within expected ranges.
Aluminum however has an unrealistically high calculated valence, 5.5
to 6.0, in both the alloys and the hydrides.  Thus, the assumption of
unrelaxed atomic positions is poor for these Al compounds.  This is
also observed as a low PAV for Al, Table 1.

TABLE 1.  METAL PAV's

| | La | | Ni(T) | | Ni(I) | | M | | |
|---|---|---|---|---|---|---|---|---|---|
| | Vol.* | $CN_V$ | Vol.* | $CN_V$ | Vol.* | $CN_V$ | | Vol.* | $CN_V$ |
| $LaNi_5$ | 25.20 | 19.53 | 11.96 | 11.34 | 12.53 | 11.97 | | – | – |
| $LaNi_4Al$ | 26.26 | 19.47 | 12.18 | 11.33 | 12.78 | 11.96 | Al | 14.47 | 11.98 |
| LaNi4Mn | 27.29 | 19.49 | 12.40 | 11.33 | 13.01 | 11.98 | Mn | 13.42 | 11.98 |
| $LaNi_4AlD_{4.85}$ | 33.23 | 23.42 | 12.18 | 12.52 | 12.72 | 14.22 | Al | 14.31 | 14.22 |
| $LaNi_4MnD_{5.9}$ | 35.29 | 22.58 | 13.13 | 12.58 | 13.76 | 14.22 | Mn | 14.16 | 14.22 |
| Elements | 37.17 | | 11.13 | | 11.13 | | Al | 16.60 | |
| | | | | | | | Mn | 12.21 | |

*Volumes in $Å^3$

Table 2 gives the PAV's, CNs, and PCNs for deuterium in various Pb/mmm type sites in the substituted compounds where total site occupation is indicated by n(D), based on neutron diffraction data.

We note that PCN coefficients based on bond order $f_B$ show surprising differences between sites of similar symmetry and volume (unprimed sites have high f(D-M) character).  Also note that the largest occupation n(D) at 12n is not at the site of largest D PAV (6m and 6m').  Table 2 gives both the number of the faces and vertexes of the D PAV cells and their CN based on pyramidal volume $CN_V$ and bond order $CN_B$.  The greatest occupation sites show high $CN_B$ = $\sim 4.5$ in $LaNi_4MnD_{5.9}$.  Figure 1 indicated the various PAV cells for Mn hydride of 12n site occupation and Figure 2 compares the deuterium PAV cells for the various type of sites of Table 2.  Note that while the geometric center $\bowtie$ of the metal PAV cells coincide closely with the atomic positions $+$ that is not the case generally with the deuterium cells of Figure 2.

## DISCUSSION AND CONCLUSIONS

The large reexpansion of the La PAV is undoubtedly a very important factor in hydride formation.  The substitution of larger metals like Al and Mn for Ni assist in this process and lead to the significant lowering of the hydrogen equilibrium pressure observed (13).  However, the composition range associated with the first plateau is decreased with increasing Al content (13) but not Mn content.  For aluminum this is consistent with the preferred occupation of the primed sites, i.e., D avoiding either the smaller (8%) sites or avoiding bonding with Al.  In contrast to the large rigid Al atom, Mn is more accommodating since it can change valence and size.

By associating the various sites with imaginary pure phases like $LaN_4MnD_3$ (3f sites) and assuming the free energy of the various sites to be equal we note that the size considerations of Lundin et al. (6)

TABLE 2.  DEUTERIUM POSITIONS IN SUBSTITUTED $LaNi_4MD_x$ *

| Cpd. | No./Position | n(D) | Face Vert. | Vol. | $CN_V$ | $CN_B$ | D-La $f_B$ | D-La $PCN_B$ | D-M $f_B$ | D-M $PCN_B$ | D-D $f_B$ | D-D $PCN_B$ |
|---|---|---|---|---|---|---|---|---|---|---|---|---|
| $LaNi_4AlD_{4.85}$ | | | | | | | | | | | | |
| | 4  6 m | 2.06 | 8/12 | 1.120 | 5.79 | 3.83 | 0.215 | 2.00 | 0.394 | 1.00 | .106 | 2.00 |
| | 2  6 m' | | 8/12 | 1.191 | 5.89 | 4.29 | .244 | 2.00 | 0 | 0 | .121 | 2.00 |
| | 2 12 n | 2.74 | 6/8 | 0.814 | 5.63 | 4.32 | .074 | 1.57 | .267 | 1.00 | .132 | 1.00 |
| | 4 12 n' | | 6/8 | 0.881 | 5.56 | 4.34 | .080 | 1.57 | 0 | 0 | .109 | 1.00 |
| $LaNi_4MnD_{5.9}$ | | | | | | | | | | | | |
| | 1  3 f | 0.31 | 6/8 | 1.192 | 6.00 | 2.84 | .081 | 2.00 | 0.089 | 2.00 | 0 | 0 |
| | 2  3 f' | | 6/8 | 1.213 | 5.99 | 2.81 | .081 | 2.00 | 0 | 0 | 0 | 0 |
| | 4  4 h | 0.18 | 8/12 | 0.738 | 4.52 | 4.79 | .003 | 3.00 | 0.247 | 1.00 | .126 | 1.00 |
| | 4  6 m | 2.01 | 8/12 | 1.284 | 5.94 | 4.59 | .251 | 2.00 | 0.309 | 1.00 | .150 | 2.00 |
| | 2  6 m' | | 8/12 | 1.302 | 5.96 | 4.68 | .257 | 2.00 | 0 | 0 | .154 | 2.00 |
| | 2 12 n | 2.54 | 6/8 | 0.953 | 5.59 | 4.50 | .085 | 1.35 | .217 | 1.00 | .162 | 1.00 |
| | 4 12 n' | | 6/8 | 0.971 | 5.57 | 4.49 | .087 | 1.35 | 0 | 0 | .165 | 1.00 |
| | 2 12 o | 0.85 | 6/8 | 1.056 | 4.37 | 3.28 | .062 | 1.03 | .370 | 1.00 | 0 | 0 |
| | 2 12 o' | | 7/10 | 1.056 | 4.38 | 3.33 | .062 | 1.03 | .366 | 1.00 | .009 | 1.00 |
| | 2 12 o" | | 7/10 | 1.078 | 4.41 | 3.38 | .064 | 1.03 | .006 | 1.00 | .009 | 1.00 |

*M is Al or Mn

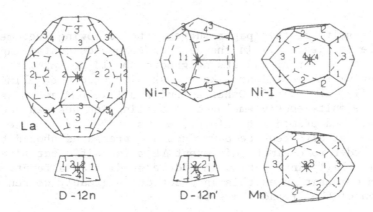

Figure 1:  The PAV cells are illustrated for an ordered arrangement of $LaNi_4MnD_6$ with the deuterium in 6 of the 12n positions.  The numbers in the face centers indicate the neighbors; 1-La, 2-Ni(T), 3-Ni(I), 4-Mn(I), and blank-D.  Note that in this D arrangement all direct Ni(T) - Ni(T) is eliminated, the same occurs for 3D in the 3f positions.

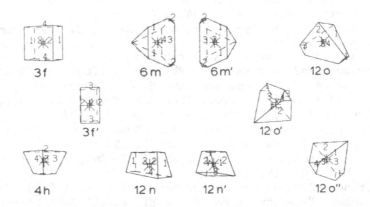

Figure 2:  The wide variety of deuterium PAV cells possible in $LaNi_4MnD_6$ are illustrated.  The primed notation indicates decreasing D-Mn contact.  The primed and unprimed cells are usually very similar but are shown in different orientations.

suggest a much larger composition range than the n(D) indicates, the configurational entropies of Gruen and Mendelsohn (14) notwithstanding.

The results of this paper suggest the following recommendations for future studies. (1) In the case of compounds containing Al, relaxed positions for Al and its neighbors should be permitted. (2) Low deuterium PAV positions like the 4h site should be eliminated in favor of (1), above. (3) Other techniques like NMR and Mossbauer should be simultaneously employed to distinguish between occupation of unprimed and primed sites (e.g., 6m vs. 6m'). (4) Deuterium site occupations at different temperatures and pressures should be determined in order to estimate different $\Delta H$'s for different sites. (5) The Ni(T) - Ni(T) interaction should be studied as a function of deuterium composition as this interaction is greatly weakened with occupation of the 3f and 12n sites.

## REFERENCES

1.  F. L. Carter, Proc. 9th Rare Earth Res. Conf., ed., D. E. Field, Blackburg, Va., Vol. 2 (Oct. 1971) 617.
2.  F. L. Carter, T. L. Francavilla and R. A. Hein, Proc. 11th Rare Earth Research Conf., eds., J. M. Haschke and H. A. Eick, Traverse City, Michigan, (Oct. 1974), 36.
3.  F. L. Carter, "Atomic Volume Contraction in Rare Earth Nickel Intermetallics as a Function of Partial Coordination Number Coefficient," this conference.
4.  F. L. Carter, J. Less-Common Metals 47, 157 (1976).
5.  F. L. Carter, Acta Cryst., B34, 2962 (1978).
6.  C. E. Lundin, F. E. Lynch, and C. B. Magee, J. Less-Common Met., 56 19 (1977).
7.  J. C. Achard, F. Givord, A. Percheron-Guegan, J. L. Soubeyroux, and F. Tasset, J. Physique Colloq., 40 C5-218 (1979).
8.  A. L. Bowman, J. L. Anderson, and H. G. Nereson, eds., C. J. Kevane and T. Moeller, Proc. 10th Rare Earth Res. Conf., Carefree, Arizona, (April 30 - May 4, 1973) 458.
9.  L. Pauling, "Nature of the Chemical Bond," 3rd ed. (Cornell Univ. Press, Ithaca, New York, 1960), 398 ff.
10. M. H. Mendelsohn, D. M. Gruen, and A. E. Dwight, J. Less-Common Met., 63, 193 (1979).
11. G. Voronoi, J. Reine und Angew. Math., 134 198 (1908).
12. J. Donohue, "The Structure of the Elements," John Wiley and Sons, New York USA, 1974.
13. J. C. Achard, A. Percheron-Guegan, H. Diaz, F. Briaucourt, and J. Demany, Second International Congress on Hydrogen in Metals, Paris, June 1977.
14. D. M. Gruen and M. H. Mendelsohn, J. Less-Common Met., 55, 149 (1977).

# NEW HYDRIDIC RARE EARTH COMPOUNDS

G. L. Silver
Monsanto Research Corporation
Mound Facility*
Miamisburg, Ohio

Modern textbooks dealing with the chemistry of the lanthanide elements state that the elemental lanthanides, with a few exceptions, dissolve in dilute aqueous acids with the formation of the trivalent lanthanide cation and hydrogen gas. One exception that has been reported is the red coloration imparted to dilute hydrochloric acid solutions containing a chip of dissolving samarium metal. This red coloration has been attributed to the divalent samarium cation. But careful observation of specimens of elemental lanthanides actually being dissolved in dilute acids may show unexpected results. For example, a faint brown coloration can sometimes be observed above a similarly dissolving chip of neodymium metal. There is nothing in the reported chemistry of neodymium that indicates a brown coloration under such circumstances, and the possibility of the formation of divalent neodymium cations is suggested by analogy. Also by analogy, the rapid disappearance of the brown coloration further suggests oxidation of the supposed divalent neodymium cations by hydrogen ions. To diminish the speed of the supposed oxidation reaction, the dissolution experiment may be repeated in a solution with a diminished concentration of hydrogen ions, as in a solution of a weak acid. Repeating the dissolution experiment in a solution of warm, dilute ($1\underline{M}$) acetic acid has been found to produce an unexpected brown precipitate (1). Other acids which produced unexpected precipitates were monochloro-, dichloro-, trichloroacetic, propionic, and propiolic. Formic, methyllactic, benzoic, butyric, and isobutyric acids did not yield precipitates. The colors of the precipitates produced from acetic

---

*Mound Facility is operated by Monsanto Research Corporation for the U. S. Department of Energy under Contract No. DE-AC04-76-DP00053.

acid solution were black for La and Ce, brown for Pr and Nd, and yellow for Sm and heavier lanthanides. (Eu produced no precipitate.)

Analyses of the precipitates from several preparations were not consistent, and did not correspond to the compositions of any simple lanthanide compounds. The precipitates were reducing agents toward many aqueous oxidants and, when decomposed in a solution of $D_2SO_4$ and $DCl$, gave a gas with a predominant proportion of HD (2, 3). The proton NMR spectrum of one of these compounds suggested the presence of hydridic hydrogen. The precipitates were decomposed (with gas evolution) by solutions of strong acids, and this decomposition reaction was noticeably accelerated by sulfate ions. A comparison of the x-ray powder diffraction patterns of the new compounds with the patterns of fluorite-structured $Pr_6O_{11}$ and $PrH_2$ suggests that the new compounds are formed in the fluorite crystal system, since the patterns of all of these compounds are similar (2,3). It is generally recognized that commercial rare earth metals are often contaminated with rare earth oxides, $M_2O_3$, which may also occur in the fluorite crystal system. The x-ray diffraction data may thus be explained as undissolved oxide $M_2O_3$, an explanation which fails to account for the differential behavior of the acids mentioned above, or for the reducing properties of the compounds. An alternative explanation, consistent with both x-ray and chemical data, is that a mixed oxide-hydride phase is formed during dissolution. Schematically, this phase may be represented as $M_2O_2(pH+qX)$ where X is an appropriately sized anion from the parent acid, H is hydride, and p+q=2. In this representation, substitution of H and X occurs randomly into two of the crystal vacancies which remain after removal of one oxygen anion from the fluorite-structured oxide, resulting in both charge balance and structural stabilization (4). The irreproducibility of the carbon, hydrogen, and metal analyses may reflect this process of random substitution.

## REFERENCES

1.  G. L. Silver, J. Inorg. Nucl. Chem., 30, 1735 (1968).

2.  G. L. Silver, P. W. Seabaugh, W. H. Smith, and R. R. Eckstein, Hydridic Nature of Rare Earth Preparations with Reducing Properties, USAEC Report MLM-1648 (August 8, 1969), 13 pp.

3.  G. L. Silver, Further Studies on New Hydridic Rare Earth Compounds, USAEC Report MLM-1914 (May 16, 1972), 11 pp.

4.  T. Moeller, The Chemistry of the Lanthanides, Reinhold Publishing Corp., New York, 1963, p. 64.

# SUGGESTIONS FOR NEW FOUR-LEVEL LASERS

Christian K. Jørgensen

Départment de Chimie minérale, analytique et appliquée

Université de Genève, CH 1211, Geneva 4

## THE PROBLEM

The theory of stimulated emission of coherent light proposed by Einstein 1917 is based on thermodynamic arguments accepting Planck's constant, but is eight years earlier than Schrödinger's equation, and hence does not involve detailed insight in electronic structure. On the other hand, any material object opaque at a given photon energy emits the standard continuous spectrum (only dependent on T). The development of lasers (1) has clearly shown that it is much more suitable to use a partly transparent four-level system, though the first laser in the visible range was the ruby, a three-level system with $E_1$ (defined below) coinciding with $E_0$. In a four-level laser, the energy is pumped into the higher level $E_3$ corresponding to a strong transition from the ground state $E_0$. By radiationless (or much less frequently, radiative) decay of $E_3$, a small concentration of the metastable (relatively long-lived) $E_2$ is built up. The absolutely necessary condition for laser action is population inversion (excluded in thermal equilibrium) where the instantaneous concentration of $E_2$ is higher than of $E_1$ to which the light-producing transition takes place. Whereas a dramatic high power is needed to obtain a denser population of $E_2$ than $E_0$, the Boltzmann population proportional to $(2J_n+1)\exp(-(E_n-E_0)/kT)$ can be negligible if $E_1$ is situated more than 14 kT above $E_0$ ($k=0.7$ cm$^{-1}$) since $\exp(-14) \sim 10^{-6}$. An additional condition is rapid decay of $E_1$ to $E_0$.

## SPECIFIC ROLE OF THE LANTHANIDES

Gaseous atoms (or $M^+$) obtained in electric discharges are
transparent in broad regions of the spectrum, and the cascading
down from $E_2$ to $E_1$ can easily maintain a population inversion.
Nevertheless, condensed matter (crystals, glasses and liquids) and
perhaps also gaseous molecules present many advantages, some related to more practical <u>energy transfer</u> (1,2) for supplying $E_3$ and
others derived from the less pronounced deactivation by collision
in viscous materials. Actually, the terawatt lasers used since
January 1978 in the SHIVA apparatus (for inducing thermonuclear
reactions in imploding spheres consisting mainly of deuterium and
tritium) in the Livermore Laboratories in California consist of
glass containing neodymium(III). The emitted line is the near
infra-red transition from the $^4F_{3/2}$ of this $4f^3$ system (at 11500
$cm^{-1}$) to the first excited J-level $^4I_{11/2}$ situated 2000 $cm^{-1}$ (about
10 kT at room temperature) above $E_0(^4I_{9/2})$. It is not trivial to
obtain emission from $E_2$ to an intermediate $E_1$ rather than to $E_0$
though this desired behavior is also shown by $4f^6$ europium(III)
where the transitions from the long-lived $^5D_0$ at 17200 $cm^{-1}$ above
$E_0(^7F_0)$ are stronger to $^7F_2$ (as known from the red cathodolumines-
cence in color television) and to $^7F_0$. The counteracting factor is
the emission probability being intrinsically proportional to
$(E_2-E_1)^3$ as pointed out by Einstein. However, it is not a strin-
gent requirement for laser action that the radiative decay of $E_2$
goes exclusively to one $E_1$ though it is beneficial if the branching
ratio is above a-half. In practice, most $E_2$ are unsuitable because
of non-radiative decay, and an important criterium is the largest
feasible ratio between $(E_2 - E_1)$ and the highest phonon energy
among the fundamental vibrational frequencies of the material (1,2).
A few four-level lanthanide lasers use a higher-lying sub-level of
the lowest J-level as $E_1$. Because of the local (very weak) deviation from spherical symmetry of the $4f^q$ system, the J-levels are
each split in at most (2J+1) sub-levels (for even q) or $(J+\frac{1}{2})$(odd q)
but rarely more than 500 $cm^{-1}$, and one has to be quite lucky with
the branching ratio, and may have to cool the solid to achieve
Boltzmann depopulation of $E_1$. The classical cases of luminescent
lanthanides are concentrated around the half-filled shell (q=5,6,7,
8,9) because these systems present a large gap between the two
closest adjacent J-levels in the visible, but in glasses with suffi-
ciently low phonon energies even $4f^{11}$ erbium(III) and $4f^{12}$ thulium
(III) show strong narrow-band luminescence from several excited
J-levels (3).

By the same token as the energies of the J-levels are described
(4,5) to a good approximation by the three parameters (Slater-
Condon-Shortley or Racah) of interelectronic repulsion and by the
Landé parameter of spin-orbit coupling, it is possible to calculate
the radiative transition probabilities from $(J_1, S_1, L_1)$ to another

($J_2$, $S_2$, $L_2$) of the $4f^q$ levels (with a typical precision of 30 percent) using only three host-dependent Judd-Ofelt parameters (1,6).
This theory was originally applied to the intensities of the narrow
absorption bands due to transitions to excited J-levels, but it works
nearly as well for luminescence, with resulting good agreement with
observed radiative life-times and branching ratios. The physical
mechanism behind the Judd-Ofelt treatment is rather enigmatic. The
parameter $\Omega_2$ varies strongly with the chemical bonding, and produces
pseudo-quadrupolar hypersensitive transitions, possibly connected
with the local inhomogeneity of the dielectric. On the other hand,
as already pointed out by Broer, Gorter and Hoogschagen in 1945, $\Omega_4$
and $\Omega_6$ (formally corresponding to 16- and 64-polar transitions) are
quite large, even in the rather electrovalent (1,5) aquo ions and
fluorides. In Russell-Saunders coupling ($S_1 = S_2$) the only general
selection rule is $|J_1 - J_2| \leq 6$ but the $\Omega_t$ parameters multiply matrix
elements of $U^{(t)}$ which can be evaluated once for all (4) in a given
M(III).

## ADVANTAGES OF ENERGY TRANSFER

The energy level $E_3$ originally receiving the energy may belong
to the lanthanide emitting the spectral line. Thus, the intense
$4f^8 \rightarrow 4f^7 5d$ transitions (in the ultra-violet) of terbium(III) may
populate $^5D_3$ or $^5D_4$ then decaying with high yield to the seven $^7F_J$
levels (groundstate $^7F_6$). This mechanism is quite general for M(II)
substituted in fluorite-type or other crystals, and laser action
has been reported for M=Sm, Dy, Ho, Er and Tm. However, with exception of $4f^6$ samarium(II), these transitions occur in the rather far
infra-red. One would expect electron transfer bands (1) to be
equally effective, but though they feed luminescence of Eu(III) and
other species, these absorption bands are much broader than $4f \rightarrow 5d$
transitions, and their propensity toward radiationless decay may be
connected with highly different minima of the potential surfaces.
The qualitative discussion of the concomitant effects of Franck and
Condon's principle usually is restricted to one "breathing mode"
scaling all internuclear distances by the same factor, but when N
nuclei vibrate, (3N-6) independent distances occur, and we may fear
surprises in the (3N-5) dimensional space describing the variation
of the energy.

Energy transfer (1,2) can be observed from Tb(III), or from the
$4f \rightarrow 5d$ transition in cerium(III), to other trivalent lanthanides.
The most readily obtained evidence is strong, new bands in the excitation spectrum. Like the isoelectronic mercury atom, thallium(I),
lead(II) and bismuth(III) in condensed matter show two strong transitions in the ultra-violet. Though it can be discussed (7) how
exactly the orbitals involved are 6s and 6p-like, these species are
very efficient (1,2) in energy transfer to M(III). However, also

much weaker transitions can be useful in undiluted crystals, such as the first excited (quartet) state of $3d^5$ in manganese(II) fluoride known for life-time in the msec range and great tendency to energy migration, and Weber (8) found energy transfer from $3d^3$ chromium (III) to M=Nd, Ho, Er, Tm and Yb in the perovskites $Y_{1-x}M_xAl_{1-y}Cr_yO_3$. The strong electron transfer bonds of vanadates, molybdates and tungstates are known to transfer energy to incorporated lanthanides, as well as the (very weak) electron transfer band of the uranyl ion to Eu(III) in glasses(1).

If we want to use the luminescence produced by energy transfer to lanthanides in four-level lasers, we prefer co-operative states, where the terminal level $E_1$ of the light emission cannot be formed directly. The <u>excimer</u> <u>lasers</u> (such as noble-gas monohalides) constitute a quite interesting analogy. Their groundstate is a dissociative potential curve without minimum (like the first triplet state of $H_2$) and hence, any excited level $E_2$ is born with inverted population. The electron transfer band (9) of $Yb_xLu_{1-x}PO_4$ shows a branching, some emission centered around 24300 $cm^{-1}$ corresponding to formation of the excited level $^2F_{5/2}$ of ytterbium(III), to be compared with 34500 $cm^{-1}$ producing the groundstate $^2F_{7/2}$. As reviewed (1), green luminescence from $Yb_{2-x}Gd_xO_3$ from $Yb_{1-x}Gd_xPO_4$ corresponds to the lowest excited level $^6P_{7/2}$ of Gd(III) at 32000 $cm^{-1}$ forming $^2F_{5/2}$ of Yb(III) at 10000 $cm^{-1}$. The <u>antiferromagnetic</u> coupling corresponding to super-exchange between two oxide-bridged Yb(III) in undiluted $Yb_2O_3$ and YbOF is so strong that bands (with very low oscillator strengths $\sim 10^{-9}$) can be (10) detected between 20500 and 21850 $cm^{-1}$ corresponding to simultaneous excitation of two Yb(III) by the same photon, much like the pair-spectra of Cr(III) in ruby (in the near ultra-violet) characteristized by the intensity being proportional to the square of x in $Cr_xAl_{2-x}O_3$. Such co-operative effects depend on the square of overlap integrals, and decrease very rapidly (exponentially) with increasing distance between the two ions containing a partly filled shell. In practice, the "super-exchange" is conducted via the bridging closed-shell ligand.

It would be very attractive to construct a four-level laser around the energy subtraction between the first excited level of Gd(III) and of Yb(III). The primordial requirement is a lot of luck, but we may help by providing as short GdOYb or GdSYb distances as possible. If we work with transparent samples of high optical quality (no bubbles, inhomogeneities nor parasitic absorption) oscillator strengths of the desired $(E_2-E_1)$ in the range $10^{-9}$ to $10^{-8}$ should be sufficient for overwhelming population inversion of $E_1$. A much wider selection would be available of energy transfer between any species having strong absorption bands and a closely adjacent, trivalent lanthanide. In the latter case, we could select a photon energy out of a very broad band, much like in dye lasers (of the rhodamine 6G type). However, the major difficulty is to persuade the lanthanide to choose one of its excited J-levels. One approach may be to exploit the formation (11) of resultant S values by coupling between $S_1$ and $S_2$ of the antiferromagnetic sub-systems, and to trap the lanthanide in a very unusual S value. Unfortunately, the lowest levels with S two units smaller than of the groundstate (such as quartets of $4f^7$ and triplets of $4f^6$ and $4f^8$) are always within spin-orbit distance from other S values. An example would be $GdCo(CN)_6$ forming the (otherwise rather inaccessible) quintet state of the cobalti cyanide anion. Praseodymium(III) is fairly close to Russell-Saunders coupling, and $^3P$ is separated by 14000 $cm^{-1}$ from the closest other triplet levels. If the strongly absorbing species had the "ambition" to form Pr(III) in a triplet state, collective emission might occur at 21000 $cm^{-1}$ lower wave-numbers than the energy difference to the $^3H_4$ groundstate. Much like $5f^2$ uranium(IV) aquo ions, Pr(III) has no light absorption in a broad window starting after $^3P_2$ (say above 24000 $cm^{-1}$) and would be transparent for "co-operative luminescence" falling in this interval. A much more moderate shift is the excited level $^2F_{7/2}$ of cerium(III) at 2200 $cm^{-1}$ shown by Kröger to provide a secondary maximum at lower energy in the 5d broad-band luminescence. However, this shift is still 10 kT and might help a four-level laser in a transparent part of the spectrum. The theory for energy transfer (1) is, unfortunately not in a very mature state at the moment, and there has been a general tendency for the observed quantities to be one or more orders of magnitude larger than predicted.

It is not an entirely trivial proposition that any excited state, which does not decay in some other way, luminesces. Our "normally behaving" colored materials shown very efficient non-radiative decay, in some cases connected with photochemical reactions. The experimental investigators have not been very keen on looking for luminescence in the near infra-red, and it is for instance not known whether the pronounced "quenching" of luminescence by iron(III) is entirely non-radiative, or is accomplished by emission close to 10000 $cm^{-1}$ from the lowest quartet state. It is not devastating for a four-level laser to have a low transition probability of the $(E_2-E_1)$ type, if there are no seriously competing alternatives. If it can be obtained, it is an undoubted advantage to have $E_1$ high above the groundstate $E_0$ as long as the decay of $E_1$ to $E_0$ does not become slow. As soon as the three Judd-Ofelt parameters can be determined from absorption spectra for M(III) in a given environment, it is feasible (1,3) to calculate reasonably accurate transition probabilities from one J-level to another of M(III), without any need for direct determination. This situation is much more favorable than in atomic spectra, where the oscillator strengths of the electric dipole-allowed transitions have their "serious" origin, and hence behave individually.

The specific application of trivalent lanthanides in four-level lasers is a striking case of the imitation of monatomic entities by systems containing a partly filled 4f shell.  This behavior is much better understood now that $4f^q \rightarrow 4f^{q-1}$ ionization energies can be measured (12) in photo-electron spectra of $MF_3$, $M(IO_3)_3$, $M_2O_3$, MSb and in the metallic elements and alloys (13).  The physical mechanism for the exceptional "atomic character" of the 4f shell has been discussed recently (14,15) but for our purposes, the major aspect is the conservation of a definite pattern of J-levels (which can be classified in S,L-terms, but showing perceptible effects of deviation from Russell-Saunders coupling) in lanthanides, allowing the oxidation state z to be defined (7) as (Z-K) where K is the integer giving the number of electrons in a Kossel isoelectronic series (14).  This statement does not imply exclusive electrovalent bonding since the total symmetry (16) of the closed shells, and of the filled molecular orbitals in general, is the neutral element of Hund vector-coupling not modifying the number of states (and their symmetry types) of the partly filled shell.

## REFERENCES

1.  R. Reisfeld and C. K. Jørgensen, Lasers and Excited States of Rare Earths, Springer, Berlin (1977).
2.  R. Reisfeld, Structure and Bonding, 22:129 (1975) and 30:65 (1976).
3.  R. Reisfeld and Y. Eckstein, J. Chem. Phys., 63:4001 (1975).
4.  W. T. Carnall, in Handbook on the Physics and Chemistry of Rare Earths (ed.: K. A. Gschneider and LeRoy Eyring), Vol. 3, Chapter 24, North-Holland, Amsterdam (1979).
5.  C. K. Jørgensen, ibid, Vol. 3, Chapter 23.
6.  R. D. Peacock, Structure and Bonding, 22:83 (1975).
7.  C. K. Jørgensen, Oxidation Numbers and Oxidation States, Springer, Berlin (1969).
8.  M. J. Weber, J. Appl. Phys., 44:4058 (1973).
9.  E. Nakazawa, Chem. Phys. Letters, 56:161 (1978).
10. H. J. Schugar, E. I. Solomon, W. L. Cleveland and L. Goodman, J. Amer. Chem. Soc., 97:6442 (1975).
11. C. K. Jørgensen, Modern Aspects of Ligand Field Theory, North-Holland, Amsterdam (1971).
12. C. K. Jørgensen, Structure and Bonding, 13:199 (1973) and 22:49 (1975).
13. M. Campagna, G. K. Wertheim and E. Bucher, Structure and Bonding, 30:99 (1976).
14. C. K. Jørgensen, Adv. Quantum Chem., 11:51 (1978).
15. C. K. Jørgensen, Israeli J. Chem., in press.
16. C. K. Jørgensen, Archives des Schieces (Genève), in press.

# MAGNETIC PROPERTIES OF THE MINI-LASER MATERIAL : THE CHLORO-APATITE OF NEODYNIUM

M. Guillot[+], M. Fadly[++], H. Le Gall[++], H. Makram[++]

[+]Laboratoire Louis Néel, Laboratoire Propre du CNRS,
associé à l'USMG, 166 X, 38042 Grenoble-Cedex, France
[++]CNRS, Laboratoire de Magnétisme et d'Optique des
Solides, 92190 Meudon Bellevue, France

## INTRODUCTION

In rare earth laser materials, the rare earth ion enters
either as a dopant or as a constituant ; in the second type, a
high number of ions per unit length can be obtained therefore
a high cavity gain is achieved. Such is the case for the chloroapatite of neodynium $Nd_2Na_2Pb_6(PO_4)_6Cl_2$ (ClAP : Nd) where

direct laser effect experiments have been recently performed (1).
In this work, the magnetic properties of ClAP : Nd are reported
in the temperature range 2-300 K ; all the experiments have been
performed on powders.

## CRYSTALLOGRAPHIC STRUCTURE AND PREPARATION OF ClAP : Nd

Apatites have the general formula $M_{10}(AO_4)_6B_2$ in which
M stands for a divalent metal ion as $Ca^{2+}$, $Sr^{2+}$, $Ba^{2+}$, $Pb^{2+}$ and
$Cd^{2+}$, A for $P^{5+}$, $As^{5+}$, $V^{5+}$, $Mn^{5+}$, $Si^{5+}$ and $Ga^{5+}$ and B for $F^-$,
$(OH)^-$ and $Cl^-$.

They crystallize in a hexagonal lattice (two formulae units
per cell) (2). The ten metal ions are distributed among two nonequivalent crystallographic sites. The prototypic apatite is the
natural calcium fluoro apatite $Ca_{10}(PO_4)_6F_2$ (its structure is

given on the figure 1). Four Ca ions ($Ca_I$) are surrounded by
three oxygen triangles with the larger triangle at the same
height along z axes as $Ca_I$. In the $Ca_{II}$ site, each ion is surrounded by six $O^{-2}$ ion and one $F^-$ ion ; so each $F^-$ is coplanar
with a triangle of $Ca_{II}$ ions. In the rare earth substituted
fluoroapatite and in the chloroapatite of calcium, a distorsion
of the hexagonal structure is found and the rather large $Cl^-$ ions

which are located approximately at z = 0 and z = 1/2 are no more coplanar with the triangle of Ca$_{II}$ ions (3) ; in the fluoroapatite the F$^-$ positions are z = 1/4 and z = 3/4).

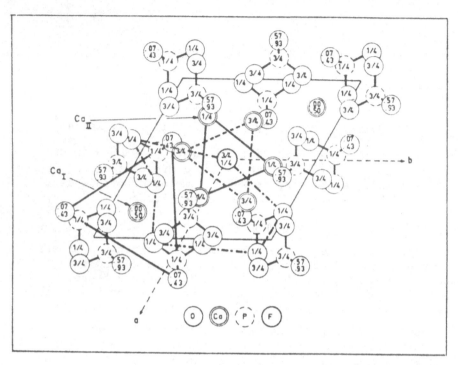

Fig. 1 - View of Ca$_{10}$(PO$_4$)F$_2$ structure projected on ab plane.

The numbers shown are z coordinates of the atoms.
Ca$_I$ has z coordinates 0.00 and 0.50 ; Ca$_{II}$ has z coordinates 1/4 and 3/4 (ref. 2).

The ClAP : Nd has been prepared in the powder form by reaction as follows :

$$2 \text{ NaCl} + 6 \text{ PbO} + \text{Nd}_2\text{O}_3 + 6 \text{ NH}_4\text{H}_2\text{PO}_4$$

$$\rightarrow \text{Na}_2\text{Nd}_2\text{Pb}_6(\text{PO}_4)_6 \text{ Cl}_2 + 6 \text{ NH}_3 + 9 \text{ H}_2\text{O}$$

The products are mixed in stoichiometric proportions and heated in a platinum crucible at 623 K during 3 hours to eliminate the ammonia and the water (water elimination must be complete to avoid the formation of hydroxyapatite) ; the reaction is achieved by many heatings and crushings during 3 hours in the 920 - 1120 K temperature range (4).

Until now, we are unable to determine the exact atomic positions in ClAP : Nd ; nevertheless X ray powder diagrams show us that the structure remains hexagonal without distorsion

(space group $P6_3/m$ ; the lattice parameters of the compound
are a = 9.912 Å and c = 7.252 Å (5). These values have been
confirmed by a Weissenberg camera photographs on a small single
crystal. The positions of the rare earth ion cannot be determined
unambiguously but it is reasonable to believe that the substitution of $Nd^{3+}$ ion in the Ca sites proceeds in a disordered way and
that these ions are statistically distributed among all the
cation positions of the $P6_3/m$.

## RESULTS AND DISCUSSION

At  first the initial magnetic susceptibility, χ, has been
measured using a translation balance in an applied field H up to
5 kOe in the 4.2 - 300 K temperature range. Secondly magnetization
measurements in field up to 26 kOe were performed in the helium
temperature range.

The temperature evolution of the reciprocal susceptibility
is plotted on figure 2. Firstly, a linear variation of $\chi^{-1}$ is
found only for T > 140 K ; it is given by a Curie Weiss law :
$1 / \chi = (T - \theta_p) / C$ where C = 3.30 e.m.u. °K/mole and $\theta_p$=-35.5 K
This experimental value of the Curie constant is in very good
agreement with the value (3.28) which can be calculated from two
free $Nd^{3+}$ ions. Such a result leads us to conclude that the

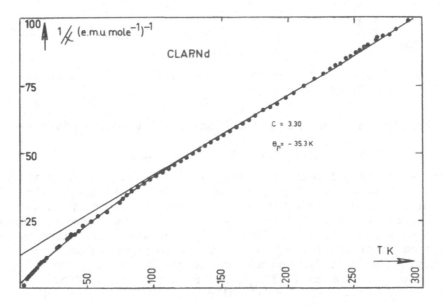

Fig. 2 - Reciprocal susceptibility for ClAP : Nd in the 4.2-300 K
        temperature range.

concentration of $Nd^{3+}$ ions is practically equal to the theoretical value (2 $Nd^{3+}$ per unit formula). For T<140 K, the Curie constant is decreasing as the temperature decreases. At 4.2 K, C is equal to 1.4 e.m.u. mole$^{-1}$. Such a behaviour implies that the crystalline field effects must be taken into account.

The magnetization curves, M(H) (figure 3) which have been obtained at low temperatures confirm that the crystalline field influences the magnetic properties. At different temperatures, M

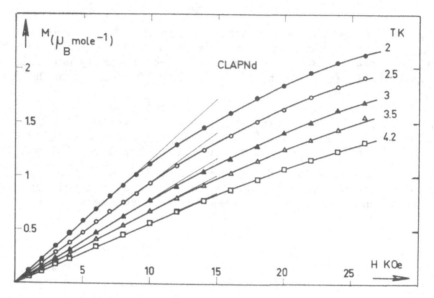

Fig. 3 - Magnetization versus applied field for ClAP : Nd at low temperature.

can be written as an unique function of H/T which is given by :

$$M \ (\mu_B \ mole^{-1}) = 2 \ N \ m_0 \ tanh \ m_0 H/kT$$

where k is the Boltzmann constant, N the Avogadro number and $m_0$ a constant equal to 1.3 Bohr magneton mole$^{-1}$. As the experimental variations are fitted to the curve calculated with the preceeding relation (figure 4), we deduce that the lowest energy level of the $Nd^{3+}$ ion is a Kramers doublet which is well separated from other levels. This doublet may be represented by a fictitious spin S'=1/2 and a magnetic g' factor equal to 2.6 ; its saturation moment (1.3 $\mu_B$), is smaller than the free ion value (3.27 $\mu_B$). Such a result allows us to deduce that the most part of $\theta_p$ (-35.3 K) is not determined by superexchange terms between magnetic ions but only by crystalline field effects. Such a conclusion is confirmed by the following results which have been obtained recently on ClAP : Ho and ClAP : Er ; the reciprocal susceptibilities of these chloroapatites follow a Curie Weiss law in the whole 2-300 K temperature range with $\theta_p$ equal to 2 K (details on these results will be given in a forthcoming paper).

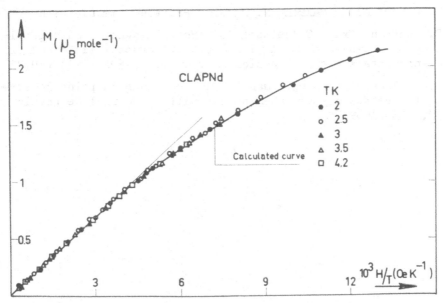

Fig. 4 - M versus H/T (in Oe/degree).

Therefore it seems reasonable to attribute the high value of $\theta_p$ (in ClAP : Nd) to crystalline field effects. P. Boutron (6) (7) has established that the crystal field effect has no influence on the susceptibility of a powder when there is only one kind of magnetic ion in one crystalline site ; so we can conclude that the $Nd^{3+}$ ions are distributed among the two cation sites of the hexagonal lattice.

## REFERENCES

1. M. Fadly, J. Ostorero, H. Makram, J.C. Michel and F. Auzel, Single crystal growth from the fluxed melt of new neodynium doped material for mini-laser.
   The Rare Earths in Modern Science and Technology, edited by Gregory J. McCarthy and J.J. Rhyne - Plenum Press. New York.

2. P.E. Mackie and R.A. Young, Location of Nd dopant in Fluorapatite, $Ca_5(PO_4)_3$ F : Nd, J. Appl. Cryst. 6, 26, (1973).

3. I. Mayer, R.S. Roth and W.E. Brown, Rare Earth Substituted Fluoride-Phosphate Apatites, Journ. of Solid State Chemistry, 11, 33-37 (1974).

4. J.C. Michel, D. Morin et F. Auzel, Intensité de Fluorescence et durée de vie du niveau $^4F_{3/2}$ de $Nd^{3+}$ dans une chloroapatite fortement dopée.
   C.R. Acad. Sci., <u>281</u>, 445-448 (1975).

5. J. Ostorero, H. Makram, M. Herpin, Private Communication.

6. P. Boutron, Exact Calculation of the Paramagnetic Susceptibility of a single crystal with arbitrary crystal field and exchange interactions, Physical Review B, <u>7</u>, 3226-3238, (1973).

7. P. Boutron, Anisotropie magnétique au-dessus du point d'ordre et paramètres d'environnement cristallin, Journal de Physique, <u>30</u>, 413 (1969).

# NECESSARY NUMERICAL CRITERIA FOR SOME PHYSICAL PROPERTIES OF LOW-QUENCHING NEODYMIUM MATERIALS

F. Auzel

Centre National d'Etudes des Télécommunications

196 rue de Paris, 92220 Bagneux, France

## ABSTRACT

The energy matching criterion for $Nd^{3+}$ $(^4F_{3/2}) \to Nd^{3+}(^4I_{15/2})$ energy transfer is shown to be in close relationship with the crystal field strength. Small-quenching conditions are derived as a function of lattice cohesive energy (or hardness) and melting point.

A generalization to some other ions such as $Eu^{3+}$ and $Tb^{3+}$ is also considered, giving an additional proof of the important role played by energy "ill-matching" conditions in stoichiometric laser materials.

## INTRODUCTION

With the advent of $NdPP(NdP_5O_{14})$ as a minilaser material a great deal of research work has been recently devoted to the search for other materials with similar low-quenching properties (1,2). As often in material research, we have to use "cut and try" methods based upon intuitive feelings and a posteriori measurements of the required behaviour. I would like to present here a first step toward an a priori approach to low-quenching materials.

This step is based upon recent recognition of two facts (3): (i) the importance of energy matching conditions in the energy transfer involved in quenching

$$Nd^{3+}(^4F_{3/2}) + Nd^{3+}(^4I_{9/2}) \to 2Nd^{3+}(^4I_{15/2});$$

(ii) the existence of an invariant with respect to host, namely the

619

energy difference $(e_2-e_1) \simeq 430$ cm$^{-1}$ between the energy differences $e_2$ and $e_1$ for the involved "free ions" levels, respectively ($^4F_{3/2}-^4I_{15/2}$) and ($^4I_{15/2}-^4I_{9/2}$); see Fig. 1. Because $(e_2-e_1)$ is negative, quenching is essentially forbidden in the "free ion" state. The crystal field strength, removing degeneracies, gives existence to a zero or positive energy difference $(E_{2M}-E_{1m})$ (Fig. 1) which by resonance provides a path for concentration quenching.

Recalling briefly the role of crystal field for $Nd^{3+}$, we shall link physical properties with crystal field, ending with a generalization to the quenching of $Eu^{3+}$ ($^5D_1$) and $Tb^{3+}$ ($^5D_3$).

### FIELD STRENGTH AND ENERGY MATCHING CONDITIONS

In order to describe by a single parameter the largest amount of splitting obtained regardless of rare-earth ion site symmetry, a crystal field strength scale parameter has been introduced as (3):

$$Nv = \left[ \sum_{k \neq 0, q} \left( \frac{4\pi}{2k+1} B_q^k \right)^2 \right]^{1/2} \simeq \left[ \sum_{k \neq 0, q} \left( B_q^k \right)^2 \right]^{1/2}$$

where $(4\pi/2k+1)^{1/2}$ is the normalization factor which can be neglected in a first approximation except when $k \neq 4,6$ only.

The $B_q^k$ are the usual crystal field parameters in Wybourne notation. $Nv$ is found to describe well the maximum splitting for different site symmetries as long as the considered $B_q^k$ in $Nv$ are the same as those necessary to the degeneracy removal of a given J level. That occurs when: $2J \gtrless k$ for $k = 2,4,6$,

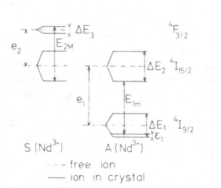

Figure 1. "Free-ion" and crystal levels $^4F_{3/2}$, $^4I_{15/2}$, $^4I_{9/2}$ of $Nd^{3+}$

In $Nd^{3+}$, the field strength sufficient to give a $(E_{2M}-E_{1m}) \geq 0$ is found to be (3): $Nv \geq 1800$ cm$^{-1}$ giving a necessary criterion for small quenching as $Nv \leq 1800$ cm$^{-1}$. This limit translated into, for instance, the $^4I_{9/2}$ splitting is $\Delta E(^4I_{9/2}) < 470$ cm$^{-1}$, $\Delta E(^4I_{9/2})$ being easily obtained from the absorption of level $^2P_{1/2}$ at temperature $\geq 300°K$.

## RELATION BETWEEN FIELD STRENGTH AND GENERAL PHYSICAL PROPERTIES

Because the crystal field strength at the rare-earth ion site contributes to the cohesive energy of the lattice, we shall look at two physical properties of solids, namely hardness and melting point, in order to obtain predictions prior to the measurement of optical spectra.

Scratch hardness in Mohs scale is converted to Plendl's absolute scale (4) from which the lattice cohesive energy U is calculated. Results are presented in Table 1. U is well known to be due essentially to coulombic forces. For binary ionic solids it is given as

Table 1. Hardness, Cohesive Energy, Crystal Field, and Melting points for different Matrix.

| Matrix | z | H*<br>Mohs | U/V**<br>kcal/mole<br>cm$^{-3}$ | U<br>kcal/mole | U/z<br>kcal/mole | $N_{V}$<br>cm$^{-1}$ | Tm,d<br>d°C |
|---|---|---|---|---|---|---|---|
| NdCl$_3$ | 3 | 2.7 | 11 | 679 | 226 | 928 | 810 |
| NdF$_3$ | 3 | 4.5 | 57,6 | 1780 | 593 | 1841 | 1374 |
| Nd$_2$O$_3$ | 6 | 5 | 82 | 4500 | 750 | 2200 | 1900 |
| NdP$_5$O$_{14}$ | 28 | 6 ⊥domains | 79 | 11507 | 411 | 1406 | ≈1000 |
| LiNdP$_4$O$_{12}$ | 24 | 5 | 55 | 7627 | 318 | 1474 | 975 |
| NdAl$_3$(BO$_3$)$_4$ | 24 | 8 | 127 | 14094 | 587 | 1630 | 1170 |
| Na$_2$Nd$_2$Pb$_6$(PO$_4$)$_6$Cl$_2$ | 50 | 3 | 16 | 5910 | 118 | 1236 | 1030 |
| NaNd(WO$_4$)$_2$ | 16 | 5 | 82 | 8014 | 501 | 1820 | 1250 |
| KNd(WO$_4$)$_2$ | 16 | 4.5 | 58 | 5620 | 351 | 1390 | 1075 |
| YAlO$_3$:Nd | 6 | 8.7 | 259 | 6260 | 1043 | 2760 | 1875 |
| Y$_3$Al$_5$O$_{12}$:Nd | 24 | 8.25 | 133 | 17350 | 723 | 3575 | 1970 |

*H is measured hardness on Moh's scale.
**U/V is absolute hardness on Plendl's scale.

$$\bar{U} = \frac{Z_1 \ Z_2 q^2 \ A}{r} \qquad\qquad [1]$$

where A is the Madelung constant, $Z_1$ and $Z_2$ are the ionic charges, r the distance between ions, and q the electronic charge.  If $Z_1$ is taken as the $Nd^{3+}$ charge and $Z_2$ a neighbouring charge, Nv is of the form:

$$\sum_i \frac{Z_{2i} \ q}{r_i}.$$

This induces us to look for a relation:

$$U/z \ \alpha \ Nv$$

where z is the number of bonds in a binary compound such as $NdCl_3$, $NdF_3$, or $Nd_2O_3$.  Figure 2 gives a plot of U/z versus Nv showing reasonable proportionality.  Trying to generalize this result to multicomponent solids such as stoichiometric neodymium materials, we find that Fig. 2 shows also a good fit for them if z is taken as the number of bonds per molecule (z = maximum valence x number of ions of such valence).  Solids for which neodymium cannot be a constituent give more erratic points because hardness varies rapidly with $Nd^{3+}$ concentration.

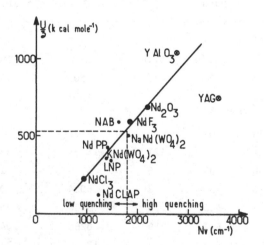

Figure 2.   Cohesive energy per bond versus crystal field for:
⊝ binary $Nd^{3+}$ compounds; • multicomponent $Nd^{3+}$ compounds; (x) $Nd^{3+}$ doped compounds.

Though a general theory of melting points of solids ($T_m$) does not exist yet, we have been looking for a relation between $T_m$ and Nv. We found (3) that $T_m \simeq 0.48 \, N_v + 300$ (correlation coefficient of 0.88).

Translation of the $N_v$ criterion shows that small quenching is to be found for materials with:

$$U/z < 500 \text{ kcal mole}^{-1} \text{ per bond (see Fig. 2)}$$

and

$$T_m < 1200°C.$$

Equation [1] shows that r should be kept large and that means large ionic radius for material components. Comparison between $NaNd(WO_4)_2$ and $KNd(WO_4)_2$ (see Table 1) is meaningful in this instance; the latter is predicted to have a lower self-quenching. For a given cohesive energy, the number of bonds should be large. This can explain why all minilaser materials known to date have a large number of component ions (4 on the average) except $NdCl_3$ which is of very low cohesive energy ($\simeq 680 \text{ kcal mole}^{-1}$).

## GENERALIZATION TO $Eu^{3+}$ AND $Tb^{3+}$ DOPED PHOSPHORS

Concentration quenching of $Eu^{3+}$ ($^5D_0$) and $Tb^{3+}$($^5D_4$) levels are known to be small in any matrix, due to the large energy gap to the next lower level (5,6) and for this reason cannot be considered as typical for concentration quenching studies. In contrast, $Eu^{3+}$ ($^5D_1$) and $Tb^{3+}$ ($^5D_3$) are respectively quenched through the following energy transfers:

$$Eu^{3+}(^5D_1) + Eu^{3+}(^7F_0) \rightarrow Eu^{3+}(^5D_0) + Eu^{3+}(^7F_3 \text{ or } ^7F_2)$$
$$Tb^{3+}(^5D_3) + Tb^{3+}(^7F_0) \rightarrow Tb^{3+}(^5D_4) + Tb^{3+}(^7F_0 \text{ or } ^7F_1).$$

"Free ions" energy differences ($e_2-e_1$) involved in such transfers are found to be positive except for $^5D_1 \rightarrow {}^7F_3$ for which ($e_2-e_1$) $\simeq$ $-120 \text{ cm}^{-1}$. The splitting of $^7F_3$ is found to be described by $\Delta E(^7F_3) \simeq 0.112 \, N_v$ and $\Delta E(^5D_1)$ is taken as $\simeq 30 \text{ cm}^{-1}$ (2J = 2 case); this gives a condition for the field: $N_v < 890 \text{ cm}^{-1}$. From Table 1 this condition is never realized for usual hosts, even $LaCl_3$. It means that small quenching inorganic materials are not likely to be discovered for $Eu^{3+}(^5D_1)$ and $Tb^{3+}(^5D_3)$ within the limits of our description. This can explain why $EuP_5O_{14}$ and $TbP_5O_{14}$ show very little or no emission from $^5D_1$ and $^5D_3$ though they were claimed to be small quenching materials (7,8) (in fact for $^5D_0$ and $^5D_4$, which, as said above, is not pertinent).

## CONCLUSION

From a resonant condition, several numerical criteria have been derived for splitting, crystal field, cohesive energy per bond and melting point. The method can be generalized to any ion, but quenching properties have to be analyzed level by level and cannot be generalized to other levels, ions or matrices. The strong quenching of $Tb^{3+}$ $(^5D_3)$ (7), for instance in $TbP_5O_{14}$ in which the Tb-Tb distance could not be much shorter than the Nd-Nd distance in $NdP_5O_{14}$, is an additional proof of the primordial role of resonance over the shortest distance one, although the latter is the hypothesis generally put forward (1).

Of course a resonance criterion cannot give by itself any information about absolute quantum yield; low site summetry and small phonon coupling have also to be considered in order to obtain fully "necessary and sufficient" criteria for small quenching materials.

## REFERENCES

L.   S. R. Chinn, H. Y-P. Hong and J. W. Pierce, "Laser Focus," 64 (May 1976).
2.   F. Auzel, Proceedings of the 2nd International School on Semiconductors Optoelectronics, CETNIEWO (1978), Polish Scientific Publishers, Warsaw (to be published).
3.   F. Auzel, Mat. Res. Bull. 14, 223 (1979).
4.   J. N. Plendl and P. J. Gielisse, Phys. Rev. 125, 828 (1962).
5.   L. G. Van Uitert and R. R. Soden, J. Chem. Phys. 32, 1687 (1960).
6.   G. E. Peterson and P. M. Bridenbaugh, J. Opt. Soc. Am. 53, 1129 (1963).
7.   B. Blanzat, J. P. Denis and J. Loriers, Proceedings of the 10th Rare Earth Conference 2, 1170 (1973).
8.   C. Brecher, J. Chem. Phys. 61, 2297 (1974).

STARK LEVEL IDENTIFICATION OF THE 2.8 µm LASER TRANSITION IN

Er:YLF

L. Esterowitz, R.E. Allen, M.R. Kruer and R.C. Eckardt

Naval Research Laboratory

Washington, D.C. 20375

In this paper we describe the results of an investigation of the laser and spectroscopic properties of the $^4I_{11/2} \rightarrow {}^4I_{13/2}$ transition of $Er^{3+}$:LiYF$_4$(YLF). This laser transition has previously been observed in $CaF_2$[1,2,3], $GdAlO_3$[4], $Y_3Al_5O_{12}$[5], $LuAl_5O_{12}$[6], $Er_3Al_5O_{12}$[6], $YAlO_3$[7], and $Y_{0.8}Gd_{0.2}ScO_3$[7]. The best laser performance reported in previous studies was achieved in 30% Er:YAG[8].

A simplified energy level diagram relevant to the four level laser operation of $Er^{3+}$:YLF is shown in Fig. 1. Stimulated emission at 2.8094 µm is observed from the $^4I_{11/2} \rightarrow {}^4I_{13/2}$ transition following flashlamp excitation of an Er:YLF rod at room temperature. The $Er^{3+}$ ion absorbs strongly at many wavelengths shorter than one micron where the emission spectra of high current Xe flashlamps are very intense. The energy absorbed from the flashlamp at these wavelengths must be rapidly transferred to the $^4I_{11/2}$ level in order to achieve efficient laser performance. In pulsed fluorescence studies at room temperature, all states between 19,000 and 27,000 cm$^{-1}$ relax in less than 10 µsec to $^4S_{3/2}$. This energy is then transferred at a slower rate to the lower lying energy levels. The energy transfer processes in Er:YLF may occur through several modes including radiative decay, nonradiative relaxation with generation of phonons, phonon assisted radiative transfer or cross relaxation resonant transfer. The relative rates of the various relaxation modes affect both laser efficiency and crystal heat loading.

The measured fluorescent lifetimes for the $^4I_{13/2}$, $^4I_{11/2}$, $^4S_{3/2}$ and $^2H_{11/2}$ levels of $Er^{3+}$ in YLF are listed in Table I for erbium concentrations of 2, 5 and 10%. Table I also lists the build-up times measured for the initial and terminal manifolds for the 2.81 µm transition. The decay of the $^4S_{3/2}$ manifold is

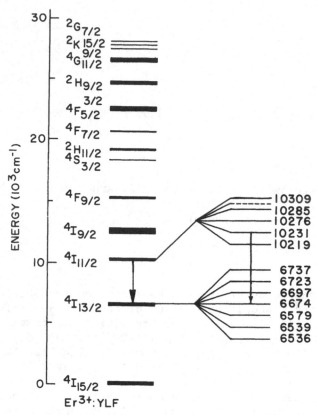

Figure 1.  Simplified energy level diagram of Er$^{3+}$:YLF showing the 2.81 μm laser transition.

Table 1.  Fluorescent lifetimes and build-up times as a function of Er$^{3+}$ concentration for the energy levels participating in cross relaxation.

| Er Concentration \ Level | Fluorescent Lifetimes (msec) | | | | Build-up Times (msec) | |
|---|---|---|---|---|---|---|
| | $^3H_{11/2}$ | $^4S_{3/2}$ | $^4I_{11/2}$ | $^4I_{13/2}$ | $^4I_{11/2}$ | $^4I_{13/2}$ |
| 2% | 0.19 | 0.19 | 4.3 | 13. | 0.13 | 0.20 |
| 5% | 0.057 | 0.057 | 4.3 | 13. | 0.067 | 0.060 |
| 10% | 0.020 | 0.020 | 4.3 | 13. | 0.023 | 0.020 |

dependent on both $Er^{3+}$ concentration and temperature.  At very low
temperatures (6K) and low concentrations (0.5% to 2%) a lifetime
of 0.60 msec is observed[9] for $^4S_{3/2}$ and is believed to be nearly
the radiative lifetime of $^4S_{3/2}$ since other decay processes should
be much slower at this low temperature.  Table 1 shows that the
room temperature lifetime is shorter and decreases with increasing
concentration becoming only 0.020 msec for an $Er^{3+}$ concentration
of 10%.

The quenching of the $^4S_{3/2}$ manifold with increasing temperature and $Er^{3+}$ concentration is believed to occur via the resonant
cross relaxation process indicated in Fig. 2.  This process has
also been observed in $Er^{3+}:LaF_3$[10].  The level positions are
taken from low temperature absorption measurements[11] of Er:YLF
together with values calculated by Wortman et al.[12]  The
$^2H_{11/2}$ manifold is in thermal equilibrium with $^4S_{3/2}$ and the
lifetimes of both manifolds were measured to be equal  as shown
in Table 1.  This cross relaxation can occur in two ways as shown
in Fig. 2.  An $Er^{3+}$ ion initially in the $^2H_{11/2}$ manifold can relax
to $^4I_{13/2}$ with the simultaneous excitation of a second ion from
the ground state to $^4I_{9/2}$.  The $^4I_{9/2}$ excitation relaxes to
$^4I_{11/2}$ by emitting photons to the lattice.  Alternatively an $Er^{3+}$

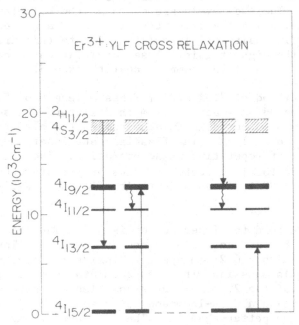

Figure 2.  Cross relaxation processes for quenching of $^2H_{11/2}$ and
$^4S_{3/2}$ in Er:YLF.

ion initially in $^2H_{11/2}$ can relax to $^4I_{9/2}$ with simultaneous excitation of another ion from the ground state to $^4I_{13/2}$. The ion in $^4I_{9/2}$ quickly relaxes to $^4I_{11/2}$. In either case one ion is left in $^4I_{11/2}$ and another in $^4I_{13/2}$ since the storage time of $^4I_{11/2}$ and $^4I_{13/2}$ is much longer than the $^4I_{9/2}$ lifetime. The build-up time for $^4I_{13/2}$ is listed in Table 1 as a fucntion of $Er^{3+}$ concentration and is equal to the fluorescent lifetime of $^2H_{11/2}$ for each concentration. These build-up times are consistent with the above cross relaxation scheme. The build-up time for $^4I_{11/2}$ is also listed in Table 1 and is equal to the fluorescent lifetime of $^4S_{3/2}$ for each concentration. This data indicates that cross relaxation is the dominant process in populating $^4I_{11/2}$.

This cross relaxation process could not achieve inversion between singlet levels since it would populate them equally. Inversion is achieved here because of the favorable occupation factors between the Stark levels of the upper and lower laser manifolds.

The Stark component energy levels of the $^4I_{11/2}$ and $^4I_{13/2}$ multiplets were identified from polarized high resolution spectroscopic analysis. The laser transition at 3559.5 $cm^{-1}$ in air was measured to an accuracy of $\pm 0.5$ $cm^{-1}$. On the basis of our energy level determinations and laser wavelength measurements, we have identified the Stark levels participating in laser oscillation. The two levels are shown by the bold-face arrow in Fig. 1 indicating the laser transition. Measurements of the fluorescence intensity and linewidth indicate this transition should have the highest gain. Laser oscillations occur over about a 1 $cm^{-1}$ linewidth, pumping near threshold.

The laser rod of 7.5% $Er^{3+}$:YLF has a length of 5.5 cm and polished uncoated end faces 0.5 cm in diameter. Pulsed laser performance was measured in an aluminum-plated elliptical cavity with a length of 5.1 cm. The flashlamp and laser rod were water-cooled for repetitive pulse operation. The optical cavity consisted of a total reflector (radius of curvature C=10 m, reflectivity R>99%) and an output mirror (C=4 m, R=94%) spaced 50 cm apart.

The threshold for laser action is 6 J. The laser output is greater than 85 mJ when the pump energy is 40 J. The divergence of the laser with a 0.25 cm aperture inside the cavity is 1 mrad and the beam is Gaussian within experimental error. The polarization of the 2.8 μm laser transition was determined to be π-polarized using a Glan-Thompson polarizer that allowed selection to be better than 1%. The best performance reported previously for a laser based on the $^4I_{11/2} \rightarrow {}^4I_{13/2}$ transition was

achieved using 30% Er:YAG.  The threshold for 30% Er:YAG using
total reflectors was 25 J which is more than 4 times the 6 J
threshold achieved here for 7.5% Er:YLF using a 95% output coupler.

REFERENCES

1.  M. Robinson and D.P. Devor, Appl. Phys. Lett. 10 : 167
(1967).
2.  S. Kh. Batygor, L.A. Kulevskii, A.M. Prokhorov, V.V.
Osiko, A.D. Sovel'ev and V.V. Smirnov, P.N. Lebeder, Sov. J.
Quant. Electr. 4(12) : 1469 (1975).
3.  G.V. Gomelauri, L.A. Kulevskii, V.V. Osiko, A.D.
Savel'ev and V.V. Smirnov, Sov. J. Quant. Electron. 6(3) : 341
(1976).
4.  P.A. Arsenev and K.E. Bienert, Phys. Stat. Sol. (a),
10 : K85 (1972).
5.  E.V. Zharikov et al., Sov. J. Quant. Elec. 4 : 1039 (1975).
6.  A.M. Prokhorov et al., Phys. Stat. Sol. A40 : K 69 (1977).
7.  T.A. Arsenev, A.V. Potemkin, V.V. Fenin and I. Senff,
Phys. Stat. Sol. 43 : K15 (1977).
8.  Kh. S. Bagdasarov, V.P. Danilov, V.I. Zhekov, T.M. Murina,
A.A. Manenkov, M.I. Timoschechkin and A.M. Prokhorov, Sov. J.
Quant. Elect. 8(1) : 83 (1978).
9.  H.P. Jensson, "Phonon Assisted Laser Transitions and
Energy Transfer in Rare Earth Laser Crystals", Crystal Physics
Laboratory Technical Report No. 16, Center for Material Sciences
and Engineering, Mass. Inst. of Technology (Sep 1971).
10.  E. Okamoto, M. Sekita and H. Masui, Phys. B11(12) : 5103
(1975).
11.  M.R. Brown, K.G. Roots and J.W. Shand, J. Phys. Chem.
Ser 2, 2 : 2 (1969).
12.  D.E. Wortman, N. Karayianis and C.A. Morrison, Report
No. HDL-TR-1770, Harry Diamond Lab. 1976.

# DUAL WAVELENGTH LASER OPERATION IN Ho,Er:YLF

R.C. Eckardt, L. Esterowitz, and R.E. Allen

Naval Research Laboratory

Washington, D.C. 20375

Laser oscillation was obtained on the 0.75-$\mu$m $Ho^{3+}$ line and 0.85-$\mu$m $Er^{3+}$ line using a doubly doped crystal of $Ho,Er:LiYF_4$ at room temperature which was optically pumped with the second harmonic of tunable Nd:Glass laser radiation. Tuning the pump wavelength across overlapping Ho and Er absorptions near 0.54 $\mu$m allowed separate operation on either of the two laser lines or both simultaneously depending on the exact pump wavelength. Laser action has previously been obtained in Er:YLF at 0.85 $\mu$m [1] and Ho:YLF at 0.75 $\mu$m [2], and laser action has been obtained in both simultaneously in a flashlamp pumped system with a doubly doped crystal at 77°K [3]. The unique features of the work reported here are simultaneous operation at room temperature in a laser pumped system.

The concentration of both Ho and Er were nominally 2% in the $LiYF_4$ (YLF) crystal used in this experiment. At these relatively low concentrations the time for resonant cross transfer between the two ions is much greater than the buildup time of laser oscillations. Consequently in order to get simultaneous laser action on both the 0.75-$\mu$m, $^5S_2 \rightarrow {}^5I_7$ Ho transition and the 0.85-$\mu$m, $^4S_{3/2} \rightarrow {}^4I_{13/2}$ Er transition, it is necessary to simultaneously pump the two species of ions. The shortest wavelength absorption peak of the Er $^4S_{3/2}$ manifold lies at 0.5407 $\mu$m. The Ho $^5S_2$ and $^5F_4$ absorption overlaps the Er absorption and extends to shorter wavelengths. Pumping at wavelengths between 0.5320 and 0.5400 $\mu$m generated the Ho line at 0.75 $\mu$m. Pumping at 0.5407 $\mu$m generated the Er line at 0.85 $\mu$m. Both Er and Ho ions absorb 0.5404-$\mu$m radiation and both the 0.75- and 0.85-$\mu$m laser lines were generated when the crystal was pumped at this wavelength.

The pump laser wavelength was tuned with dispersing prisms internal to the cavity of that laser. Second harmonic generation was performed with an intracavity nonlinear crystal. The Ho,Er:YLF laser was pumped longitudinally with the second harmonic beam. The pump beam was focused to a 0.4-mm diameter spot to match the mode size of the Ho,Er:YLF laser. The resonant cavity mirrors for that laser were 95% and 99% reflecting between 0.75 and 0.85 μm with one flat and the other 1 meter concave. Threshold for Ho and Er dual laser action was reached at 2 mJ pump energy.

## REFERENCES

1.  E.P. Chicklis, C.S. Naiman, and A. Linz, "Stimulated Emission at 0.85 μm in $Er^{3+}$:YLF," IEEE J. Quantum Electron., QE-8, p 535 (1972). E.P. Chicklis, R.C. Folweiler, J.D. Kuppenheimer, C.S. Naiman, D.R. Gabbe, and A. Linz, "Er:YLF Laser Development Part II," AFAL-TR-75-64 (November 1975).

2.  E.P. Chicklis, C.S. Naiman, L. Esterowitz, and R. Allen, "Deep Red Laser Emission in Ho:YLF," IEEE J. Quantum Electron., QE-13, p 893 (1977).

3.  E.P. Chicklis, C.S. Naiman, and H.P. Jenssen, "Two Color Laser Operation in Er-Ho:YLF," Electro-Optics/Laser 78 Conference, Boston, Mass. (September 1978), Proc. of the Tech. Program p 531. E.P. Chicklis and C.S. Naiman, "Energy Transfer in a Rare Earth Doped Activator-Activator System," Technical Report, Contract Number F44620-76-C-0111 (October 1977).

# AUTHOR INDEX

## A

Achard, J. C., 445, 585, 599
Achiwa, N., 279
Aleonard, R., 341
Allen, R. E., 625, 631
Alperin, H. A., 313
Amado, M. M., 273, 341
Anantharaman, V., 475
Anderson, D., 539
Ausloos, M., 273
Auzel, F., 503, 619

Bourne, D. R., 291
Braga, M. E., 341
Bragal, M. E., 273
Bratland, D., 31
Brauer, G. K., 187
Breslin, J. T., 355
Brossard, L., 221
Brun, T., 569
Bucher, E., 239
Buckmaster, H. A., 469
Bünzli, J.-C. G., 99, 133

## B

Bates, J. L., 553
Beaudry, B. J., 33, 53, 423
Beaury, L., 215
Becker, P., 117
Beda, S., 51
Beeken, R. B., 415
Bénazeth, S., 223
Bergner, R. L., 355
Berrada, A., 307
Berringer, B. W., 457
Beyens, Y., 87
Bilal, B. A., 83, 117
Blanchard, M., 77
Bocquillon, G., 209
Boehm, L., 441
Bornstein, A., 463
Boucherle, J. X., 261
Bourges, J. Y., 159

## C

Cadieu, F. J., 247
Caird, J. A., 111
Carcaly, C., 545
Carnall, W. T., 111
Caro, P., 215
Carré, D., 223
Carter, F. L., 299, 599
Cater, E. D., 415
Chang, W. J., 469
Chateau, C., 209
Chess, C. A., 451
Chevalier, B., 403
Chirico, R. D., 381, 387
Choppin, G. R., 69
Chopra, D., 489
Chrysochoos, J., 475, 481
Clarke, D. R., 547

## SUBJECT INDEX

### A-B

$A_2B_{17}$ intermetallic compounds, 355
Acetates, hydridic, 605
Acetonitrite, 133
Acids, action on metals, 605
Actinide, complexes, 69
Actinide hydrides, 577
Actinide-noblemetal-borides, 173
Alloys-amorphous, 305
Aluminum chloride complexes, 111
Aluminum-lanthanide sulfide
  glasses, 463
Americium(III) humate, 69
Analytical chemistry of the rare
  earths, 517
Angular overlap parameters, 77
Anion substituted lanthanide
  hydroxides, 181
Anisotropy fields in $A_2B_{17}$, 355
Antiferromagnetic coupling be-
  tween Yb(III) and Gd(III), 607
Antiferromagnetism, 415
Anti-$Th_3P_4$ structure, 239, 247
Apatite structure, 547, 613
Appearance potential spectra, 489
Arsenates, $Na_3Ln(XO_4)_2$, 195
Atomic arrangement in crystalline
  and amorphous alloys, 305
Atomic ordering, 315, 333
Barycenter of electron configura-
  tion, 425
Bastnaesite, 511
Bio-Rex-70 resin binding, 69
Borides of the rare earths, 173
Bound magnetic polaron, 445
Bulk diffusion, 571

### C

$CaF_2$-$YbF_3$, 167
$CaF_2$-$YF_3$, 167
$CaIn_2$-type phases, 39
Callen-Callen theory, 321
$CaNi_5$, density of states, 563
Carbides,
  4f electrons in, 291
  pendulum hardness of, 291
  vickers hardness of, 291
  of yttrium, 247
Catalysis, 71, 239, 533
$CdF_2$:Ln, 503
$CeCd_2$-type phases, 39
$CeCl_3$, 387
$CeCu_4Al$, 347
$CeCu_2$-type phases, 39
$CeO_2$, formation of, 59
$CePO_4$, solid-state reactions, 59
Cerium oxidation, 487
Cerium 3+, 4+, oxides, 189
Cerium (III) sulfite-sulfate-
  hydrate, 65
Cerium, zone refining, 33
$Ce_{3-x}S_4$
  heat capacity, 423
  magnetic susceptibility, 423
$Ce_2(SO_3)_2SO_4 \cdot 4H_2O$, 65
Charge-transfer bands, 105
Chemical bonding in rare earth
  compounds, 353
Chevral phases, 225, 245
Chloride-iodide, 51
Cold-boat refining, 25
Complex formation, 83
Complex formation, temperature
  effect, 117

639